WITHDRAWN

CONTAMINANT PROBLEMS AND MANAGEMENT OF LIVING CHESAPEAKE BAY RESOURCES

The Pennsylvania Academy of Science Publications

Books and Proceedings

Book Editor: Shyamal K. Majumdar
Professor of Biology
Lafayette College
Easton, Pennsylvania 18042

1. *Energy, Environment, and the Economy,* 1981. ISBN: 0-9606670-0-8. Editor: Shyamal K. Majumdar.

2. *Pennsylvania Coal: Resources, Technology and Utilization,* 1983. ISBN: 0-9606670-1-6. Editors: Shyamal K. Majumdar and E. Willard Miller.

3. *Hazardous and Toxic Wastes: Technology, Management and Health Effects,* 1984. ISBN: 0-9606670-2-4. Editors: Shyamal K. Majumdar and E. Willard Miller.

4. *Solid and Liquid Wastes: Management, Methods and Socioeconomic Considerations,* 1984. ISBN: 0-9606670-3-2. Editors: Shyamal K. Majumdar and E. Willard Miller.

5. *Management of Radioactive Materials and Wastes: Issues and Progress,* 1985. ISBN: 0-9606670-4-0. Editors: Shyamal K. Majumdar and E. Willard Miller.

6. *Endangered and Threatened Species Programs in Pennsylvania and Other States: Causes, Issues and Management,* 1986. ISBN: 0-9606670-5-9. Editors: Shyamal K. Majumdar, Fred J. Brenner, and Ann F. Rhoads.

7 *Environmental Consequences of Energy Production: Problems and Prospects.* 1987. ISBN:0-9606670-6-7. Editors: Shyamal K. Majumdar, Fred J. Brenner and E. Willard Miller.

8. *Contaminant Problems and Management of Living Chesapeake Bay Resources.* 1987. ISBN: 0-9606670-7-5. Editors: Shyamal K. Majumdar, Lenwood W. Hall, Jr. and Herbert M. Austin.

9. *Proceedings* of the Pennsylvania Academy of Science. Two issues per year; current volume (1987) is 61. ISSN: 0096-9222. Editor: Daniel Klem, Jr.

CONTAMINANT PROBLEMS AND MANAGEMENT OF LIVING CHESAPEAKE BAY RESOURCES

EDITED BY
SHYAMAL K. MAJUMDAR, Ph.D.
Professor of Biology
Lafayette College
Easton, Pennsylvania 18042

LENWOOD W. HALL, Jr., M.S.
Senior Staff Biologist
The Johns Hopkins University
Applied Physics Laboratory
Aquatic Ecology Section
Shady Side, Maryland 20764

HERBERT M. AUSTIN, Ph.D.
Professor of Marine Science
The College of William and Mary
Virginia Institute of Marine Science
Gloucester Point, Virginia 23062

Founded on April 18, 1924

A Publication of
The Pennsylvania Academy of Science

Library of Congress Cataloging in Publication Data

Contaminant Problems and Management of Living Chesapeake Bay Resources

   Bibliography
   Index

Library of Congress Catalog Card No.: 87-62940

ISBN-0-9606670-7-5
   Copyright © 1987 By The Pennsylvania Academy of Science

   All rights reserved. No part of this book may be reproduced in any form without written consent from the publisher, The Pennsylvania Academy of Science. For information write to The Pennsylvania Academy of Science, Attention: Dr. S.K. Majumdar, Editor, Department of Biology, Lafayette College, Easton, Pennsylvania 18042. Any opinions, findings, conclusions, or recommendations expressed are those of the authors and do not necessarily reflect the views of The Pennsylvania Academy of Science.

   Printed in the United States of America by

   Typehouse of Easton
   Phillipsburg, New Jersey 08865

PREFACE

This book was developed from the proceedings of a session titled, "Chesapeake Bay Fisheries and Contaminant Problems," organized at the 152nd National Meeting of the American Association for the Advancement of Science (AAAS) in Philadelphia, Pennsylvania. The session was sponsored by the American Fisheries Society (Potomac Chapter), National Association of Academies of Science, AAAS Section G and co-sponsored by the AAAS Office of Science and Technology Education.

The Chesapeake Bay is the nation's largest and most productive estuary. The unique physical, chemical and biological characteristics of the Bay create a favorable environment for numerous species of aquatic life. In recent years various species have been declining in this estuarine system. One hypothesis for the decline of these biota is the presence of contaminants in the Chesapeake Bay. Contaminants such as anthropogenic trace metals, synthetic organic compounds (including pesticides and herbicides), biocides (chlorine), petroleum hydrocarbons and acidic conditions have been reported at potentially toxic levels in the Chesapeake Bay and its tributaries. Attempting to relate the presence of these contaminants with declining populations of aquatic species is a difficult and often impossible task. Strong circumstantial evidence is often the only method available to establish whether a contaminant is responsible for an observed biological effect in the environment.

The objectives of this book are to provide a general description of the Chesapeake Bay ecosystem and its resident biota, present current information on the various types of contaminants likely affecting Chesapeake Bay biota and discuss actions and management strategies needed to improve contaminant problems impairing Chesapeake Bay resources.

The book is divided into three parts. The opening part is devoted to the Chesapeake Bay and its ecosystems. The chapters in this section include the physical description of the Bay, discuss the socio-economic perspectives of the Bay Fisheries, and cover areas of Chesapeake Bay used or inhabited by early life stages of fishes, mollusk culture, and the factors affecting the distribution of the blue crab. In addition, this section analyzes the Chesapeake Bay Fisheries, waterfowl population, aquatic vegetation, phytoplankton, the American oyster, meso- and microzooplankton and the benthic resources.

The second part is concerned with the contaminant problems and impacts. It includes the chapters on contaminants' source and types, nutrients and anoxia in Chesapeake Bay and their effects on the Bay's primary producers, zooplankton, finfish and shellfish.

The third part is comprised of seven chapters. It covers Maryland, Pennsylvania and Virginia's response to Chesapeake Bay ecological problems, considers public involvement in the Bay program, discusses the restoration efforts of the Bay resources, offers ideas to reduce contamination problems and describes an aquatic biological testing program to reduce toxic pollution of the Chesapeake Bay.

The editors express their appreciation for the excellent cooperation of the contributors. Heartfelt thanks are extended to Drs. George C. Shoffstall and Dean Rosebery, Past-Presidents of the National Association of Academies of Science and the Board Members of the Pennsylvania Academy of Science for their encouragement and help in the preparation of this book. The editors also extend their gratitudes to Lafayette College, Johns Hopkins University, Applied Physics Laboratory, and The College of William and Mary, Virginia Institute of Marine Science for providing facilities to them for the editorial work.

S.K. Majumdar
L.W. Hall, Jr.
H.M. Austin

Editors
November, 1987

FOREWORD

On December 8, 1983, history was made when the Governors of Pennsylvania, Maryland and Virginia, and the Mayor of the District of Columbia joined the Administrator of the United States Environmental Protection Agency in signing the Chesapeake Bay Agreement. The Chesapeake Executive Council, instituted by the Agreement, was a pioneering experiment in intergovernmental relations. Its mission is to guide a multi-state effort to restore and protect the Chesapeake Bay—the nation's, and perhaps the world's, most productive estuary.

This effort has already expanded the pool of scientific information available on the Bay, its ecological mechanisms, and the sources and effects of those pollutants that threaten it. The elusive links between habitat and water quality and the life histories and population patterns of the organisms we seek to replenish are becoming better understood. Practiced and novel technologies for lessening the burden that the by-products of human habitation imposed on the Bay are being evaluated and implemented.

The public participation program undertaken by the Council with the cooperation of the Citizens Program for the Chesapeake Bay, Inc. has raised the awareness of the people in the Chesapeake Bay area to the threat to this unparalleled resource and the need to protect it. Farmers in central Pennsylvania, many miles north of the Bay itself, are now aware that better manure management along the Susquehanna River can lead to better oyster harvest in Virginia.

The chapters that follow will further extend that base of knowledge, as this book, *Contaminant Problems and Management of Living Chesapeake Bay Resources* will serve to expand the public awareness of the problems we face.

The twenty million people who live in the 64,000 square mile Chesapeake Bay basin—farmers, homeowners, factory workers, and sportsmen—all have a stake in the restoration and protection of the Bay. The United States Environmental Protection Agency is proud of the role that it plays, and I hope that everyone who reads this book will become a player, too.

James M. Seif
Regional Administrator
United States Environmental
Protection Agency
Region III
Philadelphia, Pennsylvania
May, 1987

SYMPOSIUM

CHESAPEAKE BAY FISHERIES AND CONTAMINANT PROBLEMS

Organized by Lenwood W. Hall, Jr. (*Senior Staff Biologist, Johns Hopkins Univ., Applied Physics Laboratory, Shady Side, MD*)

152nd National Meeting of the American Association for the Advancement of Science, May 26, 1986, Philadelphia, Pennsylvania
  Presiding: Lenwood W. Hall, Jr.

Chesapeake Bay Fisheries: An Overview
  Herbert M. Austin (*Department Head, Fisheries Science, College of William and Mary, Virginia Institute of Marine Science, Gloucester Point, VA*)

Utilization of Chesapeake Bay by Early Life Stages of Fishes
  John E. Olney (*Assistant Professor of Estuarine and Coastal Ecology, College of William and Mary, Virginia Institute of Marine Science, Gloucester Point, VA*)

Contaminants in Chesapeake Bay
  George R. Helz (*Professor of Chemistry, Univ. of Maryland, College Park, MD*)

Contaminant Effects on Chesapeake Bay Finfishes: A Review of Recent Studies
  Ronald J. Klauda (*Senior Staff Biologist, Johns Hopkins Univ., Applied Physics Laboratory, Shady Side, MD*) and Michael E. Bender (*Professor of Marine Science, College of William and Mary, Virginia Institute of Marine Science, Gloucester Point, VA*)

Actions Needed to Improve Chesapeake Bay Fisheries and Contaminant Problems
  L. Eugene Cronin (*retired, Annapolis, MD*)

Sponsors: American Fisheries Society (Potomac Chapter); National Association of Academies of Science; AAAS Section G. **Cosponsor:** AAAS Office of Science and Technology Education.

Contaminant Problems and Management of Living Chesapeake Bay Resources

Table of Contents

Preface V
Foreword—James M. Seif, US EPA, Region III VII
Symposium—Chesapeake Bay Fisheries and Contaminant Problems VIII
Introduction—Charles McC. Mathias, Jr., Former U.S. Senator of Maryland ... XI

Part One: Chesapeake Bay and its Ecosystems

Chapter 1: A BRIEF PHYSICAL DESCRIPTION OF CHESAPEAKE BAY
J.R. Schubel and D.W. Pritchard 1

Chapter 2: CHESAPEAKE BAY FISHERIES—AN OVERVIEW
Herbert M. Austin 33

Chapter 3: A SOCIO-ECONOMIC OVERVIEW OF THE CHESAPEAKE BAY FISHERIES
James E. Kirkley 54

Chapter 4: UTILIZATION OF CHESAPEAKE BAY BY EARLY LIFE HISTORY STAGES OF FISHES
Eileen M. Setzler-Hamilton 63

Chapter 5: WATERFOWL OF CHESAPEAKE BAY
Matthew C. Perry 94

Chapter 6: SUBMERGED AND EMERGENT AQUATIC VEGETATION OF THE CHESAPEAKE BAY
Carl Hershner and Richard L. Wetzel 116

Chapter 7: PHYTOPLANKTON IN THE CHESAPEAKE BAY: ROLE IN CARBON, OXYGEN AND NUTRIENT DYNAMICS
Kevin G. Sellner 134

Chapter 8: BENTHIC RESOURCES OF THE CHESAPEAKE BAY ESTUARINE SYSTEM
Robert J. Diaz 158

Chapter 9: THE AMERICAN OYSTER (*CRASSOSTREA VIRGINICA*) IN CHESAPEAKE BAY
Dexter Haven 165

Chapter 10: FACTORS AFFECTING THE DISTRIBUTION AND ABUNDANCE OF THE BLUE CRAB IN CHESAPEAKE BAY
W.A. Van Engel 177

Chapter 11: MOLLUSK CULTURE FOR THE CHESAPEAKE BAY
Michael Castagna 210

Chapter 12: MESOZOOPLANKTON AND MICROZOOPLANKTON IN THE CHESAPEAKE BAY
David C. Brownlee and Fred Jacobs 217

Contaminant Problems and Management of Living Chesapeake Bay Resources

Part Two: Problems and Impacts

Chapter 13: CONTAMINANTS IN CHESAPEAKE BAY: THE REGIONAL PERSPECTIVE
George R. Helz and Robert J. Huggett 270

Chapter 14: NUTRIENTS IN CHESAPEAKE BAY
David L. Correll 298

Chapter 15 CONTAMINANT EFFECTS ON CHESAPEAKE BAY FINFISHES
Ronald J. Klauda and Michael E. Bender 321

Chapter 16: CONTAMINANT EFFECTS ON CHESAPEAKE BAY SHELLFISH
Michael E. Bender and Robert J. Huggett 373

Chapter 17: CONTAMINANT EFFECTS ON PRIMARY PRODUCERS IN CHESAPEAKE BAY
James G. Sanders 394

Chapter 18: EFFECTS OF CONTAMINANTS ON ESTUARINE ZOOPLANKTON
Brian P. Bradley and Morris H. Roberts Jr. 417

Chapter 19: ORIGIN, DEVELOPMENT AND SIGNIFICANCE OF CHESAPEAKE BAY ANOXIA
Jon H. Tuttle, Robert B. Jonas, and Thomas C. Malone 442

Part Three: Responses and Management

Chapter 20: PENNSYLVANIA'S RESPONSE TO CHESAPEAKE BAY ECOLOGICAL PROBLEMS AND RESTORATION EFFORTS
Paul O. Swartz, C. Victor Funk and Louis W. Bercheni 473

Chapter 21: VIRGINIA GOVERNMENT RESPONSE TO CHESAPEAKE BAY PROBLEMS
Keith J. Buttleman and Janice Carter-Lovejoy 485

Chapter 22: THE STATE OF MARYLAND'S RESPONSE TO CHESAPEAKE BAY ECOLOGICAL PROBLEMS
Mary Jo Garreis 498

Chapter 23: PUBLIC INVOLVEMENT IN THE CHESAPEAKE BAY PROGRAM
Frances H. Flanigan 512

Chapter 24: USE OF AQUATIC BIOLOGICAL TESTING UNDER THE NPDES PERMIT SYSTEM TO REDUCE TOXIC POLLUTION OF THE CHESAPEAKE BAY
Richard L. Williamson, Jr. and Dennis T. Burton 518

Chapter 25: THE RESTORATION OF LIVING CHESAPEAKE BAY RESOURCES
Charles S. Spooner 541

Chapter 26: ACTIONS NEEDED TO REDUCE CONTAMINATION PROBLEMS IMPAIRING CHESAPEAKE BAY FISHERIES
L. Eugene Cronin 555

Subject Index 567

Academy Officers 573

INTRODUCTION

Charles McC. Mathias, Jr.
Former U.S. Senator of Maryland
Partner in Law Firm of Jones, Day, Reavis and Pogue
655 Fifteenth Street, N.W.
Washington, D.C. 20005

The Algonquin Indian tribe called it "Chesepiooc" — great shellfish bay. The first Europeans to visit America also thought it was a great bay. Today we recognize it as the world's greatest estuary and one of our most valuable national assets.

The history of the Chesapeake since the early English settlements in Virginia and Maryland has been both exciting and romantic. The tides of war have swept up the Bay many times since the sound of gun fire was first heard off Bloody Point. The advance of civilization has been mirrored in the water's surface as travelers have progressed from canoes and polling boats to sail to steam and finally to combustion engines. Habitation along the shore line has grown from teepee to concrete and finally to great steel girders high above the water.

The Bay is such a substantial feature of life to all that know and live near it that it long seemed immutable and inviolate. All sorts of violence could be visited on the lands around its shores and to its tributaries, but no consequences to the Bay itself were ever reckoned. Such majesty and immensity could not be threatened by the construction of a house or a road cr a factory. The humble plow of the farmer could not be a danger. The growth of great cities in the areas that drain into the Chesapeake was viewed as a boon and not a problem.

But all of these developments were having a cumulative and accelerating effect. The endemic species of fish like sturgeon disappeared early, but much of the natural environment survived through the 19th century with only minimal visible deterioration. But after 1900 the signs of trouble were more numerous and easier to see. The economic impact of the changes became measurable; by mid-century the oyster catch was down to a fifth of what it had been at the beginning. The ominous signs began to be evident to many and the dangers were finally perceived.

In all of this unfolding tragedy the political composition of the Bay proved to be a serious handicap. The crabs and rockfish may not have been conscious of a state boundary on the Bay bottom, but Virginians and Marylanders were all too well aware of it. Pride, prejudice and pecuniary interest all stood in the way of a cooperative approach to conservation of the Bay itself and its rich resources.

It was not until the unity of the Bay and its tributaries was recognized that things began to change. The relationship of the quality of water in the rivers and the Bay should have been evident as should the impact of land use on silting and pollution. The comprehensive character of the problems not only reconciled the states to joint action, but also summoned an array of Federal agencies with both jurisdiction and technical skill. A voluntary effort by citizens and private organizations became part of the effort to "Save the Bay."

The mobilization of forces to improve conditions in the Chesapeake Bay is a tribute to our people and to the institutions they have created and supported. But we cannot forget that the problem was allowed to grow and fester because we did not read the signs that were clear and unambiguous. It is a mistake that we cannot afford to repeat.

Contaminant Problems and Management of Living Chesapeake Bay Resources. Edited by S. K. Majumdar, L. W. Hall, Jr. and H. M. Austin. © 1987, The Pennsylvania Academy of Science.

Chapter One

A BRIEF PHYSICAL DESCRIPTION OF THE CHESAPEAKE BAY

J.R. SCHUBEL and D.W. PRITCHARD

Marine Sciences Research Center
State University of New York
Stony Brook, NY 11794

ABSTRACT

The Chesapeake Bay is the largest estuary in the United States and one of the largest in the world. Like all estuaries, it was formed during the most recent rise in sea level and is less than 10,000 years old. The Susquehanna River, the only river that discharges directly into the main body of the Bay, accounts for nearly half the 2280 m^3/s average annual total input of freshwater to the entire estuarine system. Highest fresh water inflows generally occur in March and April, and lowest inflows in August, September and October. There are, however, large year to year departures in the annual average inflow, and in the seasonal distribution of inflow.

The motion of the Bay's waters encompasses a broad range of temporal and spatial scales. The most obvious motions are the tidal currents. When motions of tidal frequencies are removed, residual motions remain. These include the classical estuarine circulation patterns—seaward flow in the upper layer and landward flow in the lower layer—typical of partially mixed estuaries; a Stokes' transport associated with the progressive nature of the tide wave in the Bay; topographically induced horizontal eddy currents; and meteorologically induced motions, both near field and far field.

The residual motions may have amplitudes of the same order as the tidal currents. Since most contaminants are bound to particles, the scales of motion which control the transportation, distribution, and accumulation of fine particles also control the patterns of dispersal, distribution and accumulation of most contaminants. These are the long-term average estuarine circulation and the tidal currents, with topographically induced horizontal eddy currents being of local importance, and meteorologically induced residual flows being important at time scales of from 2 to 10 days.

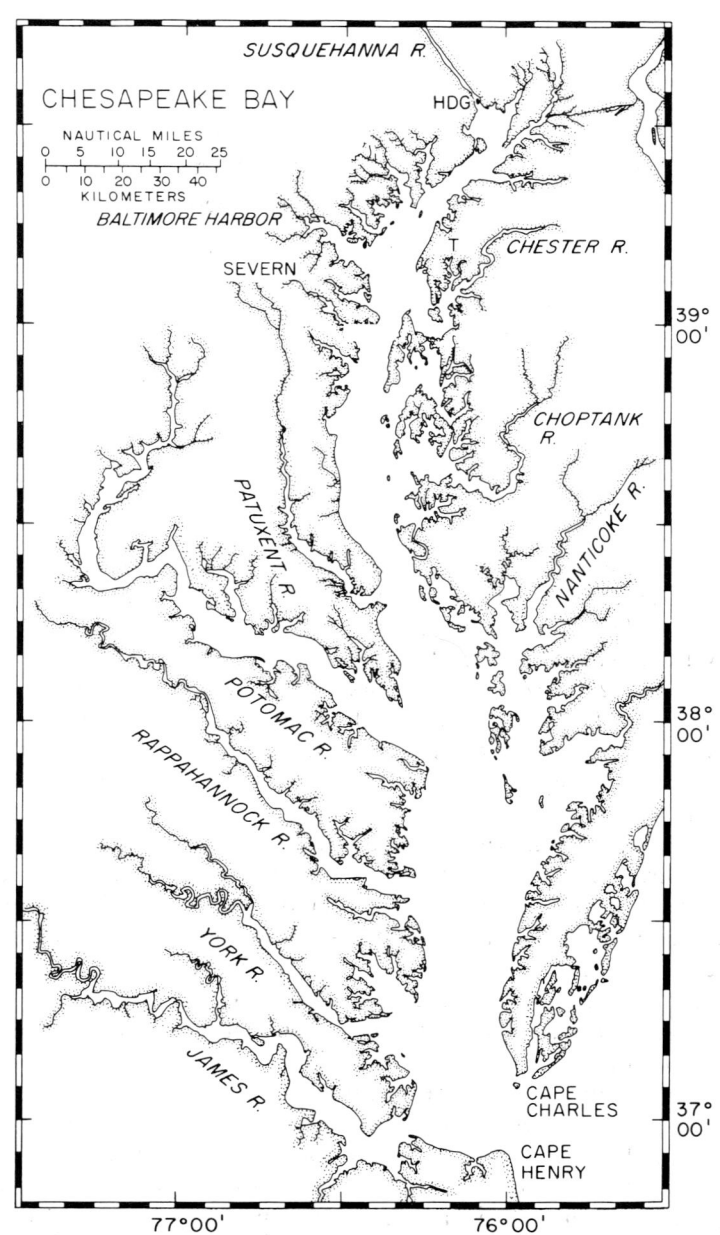

FIGURE 1. Map of the Chesapeake Bay showing the locations of the various tributary estuaries, as well as that of Cape Charles and Cape Henry, which mark the mouth of the Bay, and that of Havre de Grace (HDG), which marks the head of the Bay.

INTRODUCTION

The Chesapeake Bay, the largest estuary in the United States and one of the largest estuaries in the world, stretches for about 320 km from its seaward end at Cape Charles and Cape Henry to the mouth of the Susquehanna River at Havre de Grace, Maryland (Fig. 1). The Chesapeake Bay estuarine system is made up of the Bay proper and its tributary estuaries. The Bay proper has an area of 6.5×10^3 km^2, a mean low water volume of 50 km^3, and a mean depth of 8.42 m. The entire Chesapeake Bay system has a surface area of 11.5×10^3 x km^2, a mean low water volume of 74 km^3, and a mean depth of 6.46 m.

The modern Chesapeake Bay estuary fills a dendritic river valley system carved into the soft coastal plain sediments during the most recent lowstand of sea level. It was converted from a river valley system into an estuarine system during the most recent post glacial period when the rising sea penetrated into and drowned the ancestral river valley system. Like all other estuaries of the world, the Chesapeake Bay estuarine is less than 10,000 years old.

In this chapter we present a description of the physical characteristics of the Chesapeake Bay estuarine system. This description includes information on (a) the sources of freshwater inflow and their seasonal and year-to-year variations; (b) the temporal and spatial distributions of temperature and salinity; and (c) the motion and mixing of waters in the Bay and its tributaries at various scales of time and space.

FRESHWATER INFLOW

Of the 50 major rivers which discharge into the Bay, only one, the Susquehanna River, discharges directly into the main body of the Bay; all the other rivers discharge into their own estuaries well upstream from their junctions with the Bay proper. The total freshwater input to the system averages about 2,280 m^3/s, 71.9 km^3/yr. Nearly half of the total (48.2%) is supplied by the Susquehanna River which has a long-term average discharge of 1,099 m^3/s. The Potomac River with a long-term average discharge of about 310 m^3/s is the second largest river, contributing 13.6% of the total average freshwater input to the entire system.

Other freshwater sources which enter along the western shore of the Bay, and which together with the Potomac River account for 33.4% of the total average freshwater inflow to the Bay, are the James (average flow of 284 m^3/s, or 12.5% of the total), the Rappahannock (average flow of 70 m^3/s, or 3.1% of the total), the York (average flow of 68 m^3/s, or 3.0% of the total), and the Patuxent (average flow of 27 m^3/s, or 1.2% of the total). The two largest freshwater sources which enter along the eastern shore of the Bay are the Choptank (average flow of 27 m^3/s, or 1.2% of the total), and the Nanticoke (average flow of 26 m^3/s, or 1.1% of the total). The combined long-term average flow from all other rivers which

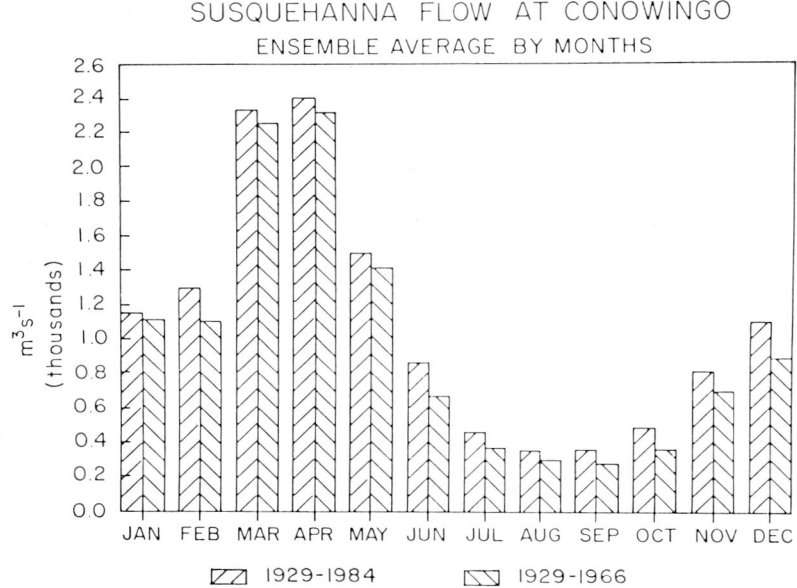

FIGURE 2. The ensemble averaged monthly flow of the Susquehanna River at Conowingo, for the years 1929-1966 and 1929-1984.

enter the Bay along the eastern shore is only 135 m^3/s, or 5.9% of the total freshwater inflow to the Bay. The remaining 234 m^3/s (10.3% of the total average freshwater inflow to the Bay) enters from the 30 or so small streams along the western shore.

All of the major freshwater sources to the Chesapeake Bay exhibit considerable day to day, week to week, month to month, and year to year variability. For our purposes here, we will consider only the month to month and year to year fluctuations in the freshwater inflow. Also, since the statistical characteristics of the inflows from the major rivers are similar, we use the data for the Susquehanna as illustrative of these statistics.

The Susquehanna River has a flow pattern typical of mid-latitude rivers: high discharge in spring produced by snow melt and spring rains followed by low to moderate flow throughout most of the remainder of the year. This is revealed clearly by an ensemble average taken by month of the discharge at Conowingo, the last of three reservoirs along its lower reaches, which is about 15 km above the river's mouth at the northern limit of the Bay. Figure 2 shows the ensemble average monthly mean flows for the period 1929—1984, and also for comparison, for the period 1929 - 1966. This latter data set is the same as that presented by Boicourt,[1] except for a small correction factor as described by Schubel and Pritchard.[2]

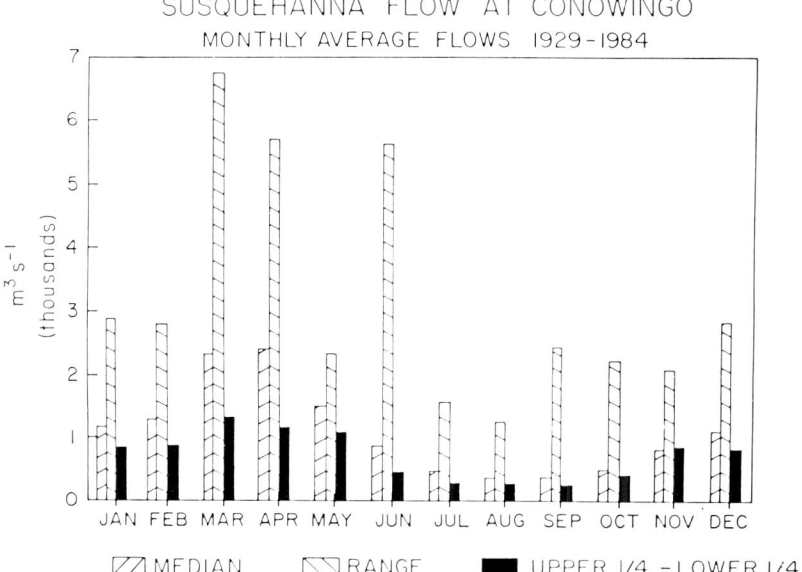

FIGURE 3. Characteristics of the flow of the Susquehanna River at Conowingo. Given are the median, range and the difference, upper quartile minus lower quartile, by month, for the 56-year period 1929-1984.

The similarity of the ensemble averages for the two periods as shown in Fig. 2 supports the statement made above concerning the typical character of the seasonal variation in discharge for long-term average conditions. However, there are relatively large year-to-year variations in monthly average flow values, and this variation is illustrated in Fig. 3, where values of the median, range (maximum value minus minimum value), and the difference between the upper ¼ and lower ¼ values for each month of the year over the 56-year period 1929 - 1984 are shown. This figure shows that the range of monthly average flows for each month over the 56 years of record considerably exceeds the median flow for each month, and that the central 50% of the monthly average flows is contained within a range which for most months is the same order as the median flow.

Both the characteristic seasonal trends and year-to-year variability are illustrated in Tables 1 and 2. For each year of the 56-year record the monthly average flows were ranked, in descending order, from the month having the highest flow to the month having the lowest flow for that year. The frequency of flow rankings, in number of years in the 56-year record that each particular month had the stated ranking, is given in Table 1. Note that the characteristic high flow in late winter and early spring, and low flow in late summer and early

TABLE 1

Susquehanna River at Conowingo 1929 - 1984
Frequency of Monthly Average Flow Ranking
Number of Years in 56 Years of Record

Flow Rank	Jan	Feb	Mar	Apr	May	Jun	Jul	Aug	Sep	Oct	Nov	Dec
1(Max)	2	4	22	22	2	1			1			2
2	2	7	14	14	10	2				2	1	4
3	7	3	10	11	16	1					3	5
4	11	8	6	5	6	3		1		3	5	8
5	10	13	1	1	6	5			2		8	10
6	5	6	2	2	7	11	4		1	3	8	7
7	2	8	1	1	6	9	5	1	3	3	8	9
8	4	5			3	10	12	5		7	7	3
9	5	1				10	13	6	5	4	7	5
10	3	1				3	10	17	8	6	7	1
11	2					1	5	13	18	14	2	1
12(Min)	3						7	13	18	14		1

TABLE 2

Susquehanna River at Conowingo 1929 - 1984
Frequency of Monthly Average Flows Falling Within Indicated Interval

Flow Interval m^3/s	Jan	Feb	Mar	Apr	May	Jun	Jul	Aug	Sep	Oct	Nov	Dec
>3000	1	1	12	13		1						
2500-3000	5	4	9	9	4				1			3
2000-2500	2	3	12	15	12	2				2	2	3
1500-2000	5	11	11	13	10	1	1		1	2	6	7
1000-1500	15	14	10	5	15	10	3	1	1	3	11	16
500-1000	20	18	2	1	14	26	15	12	8	12	16	18
150-500	6	5			1	16	25	20	18	14	15	7
<250	2						12	23	28	23	6	2

fall are revealed clearly by this table, with either March or April having the highest flow of the year in 44 of 56 years, and either August, September or October having the lowest flow of the year in 45 of the 56 years. However, there are some departures from the classical pattern. Maximum mean monthly flow occurred once for June (Tropical Storm Agnes, 1972) and once for September (Hurricane Eloise, 1975). December and January have each been the month with maximum flow of the year twice. Also, January has been the month with minimum flow of the year three times, and December once. In fact, the months of January and December are quite indiscriminate, appearing at least once in each of the 12 ranked positions.

Table 2 gives the number of times a given monthly average flow falls within designated flow ranges. Note that together March and April had flows exceeding

3,000 m³/s in 25 out of the 28 such occurrences, and that August, September and October together had average flows less than 250 m³/s in 74 out of the 96 such occurrences. Note also, however, that January, February, and June each had average flows exceeding 3,000 m³/s once during the 56 years of record. Indicative of the broad spectrum of possible flows for January and December is the fact that monthly average flows greater than 2,500 m³/s occurred 6 times for January and 3 times for December, while each of these months had average flows less than 250 m³/s twice.

On a weekly average basis, the several hydroelectric reservoirs on the Susquehanna River, individually and collectively, have little affect on the discharge of freshwater to the Bay, because they have little storage capacity. Except during high-flow periods, these hydroelectric plants do, however, modify the flow on an hour-to-hour and day-to-day basis. During the summer and fall period of low flow, discharge may be stopped completely on weekends, except for a small ($\sim$ 10 m³/s) leakage.

CHARACTERISTICS OF THE DRAINAGE BASIN

The drainage basin of the Susquehanna River, which accounts for 43% of the Bay's total drainage basin, has an area of about 71,250 km² in New York, Pennsylvania, and Maryland, of which about 70,190 km² are above the tidal reaches of the river. The basin is 402 km long and 275 km wide at its widest point, and lies within four physiographic provinces: the Appalachian, the Ridge and Valley, the Piedmont, and the Blue Ridge. Land use in the drainage basin is predominantly forest and agriculture with no major urban areas.

The drainage basin of the Potomac River above the head of the tide at Little Falls has an area of some 30,000 km². The total drainage area of the Potomac River - Potomac Estuary down to the confluence of this subestuary with the main stem of the Bay has an area of about 36,055 km², or nearly 22% of the Bay's total drainage basin. It is contained within the states of Maryland, Virginia and West Virginia. One major urban area, Washington D.C., lies within this drainage basin, and there is considerable coal mining in the West Virginia portion of the basin. Otherwise, agriculture and forests predominate land usage in this drainage basin.

The drainage basin of the James River and of the James River subestuary above its confluence with the main stem of the Bay has an area of about 26,300 km², or 16% of the Bay's total drainage basin. A major urban area, Richmond, is located at the head of the tide, and considerable industry is found along the tidal river and the upper estuary of the James. Extensive waterfront dependent commerce and industry, including shipyards and U.S. Navy installations, are located within the Newport News - Hampton - Norfolk - Portsmouth complex at the mouth of the James River Estuary. Above Richmond, the predominant land uses are forest and agriculture.

Taken together with their subestuaries, the Susquehanna, Potomac, and James account for 80% of the total drainage area of the Bay.

The major urban area of Baltimore drains into the Bay via the small tributary estuary of the Patapsco River, and the Baltimore business and residential suburbs drain into the Patapsco River and Back River, in close proximity to the Bay. A mixture of residential, farm, and forest usage is found in the several small drainage basins to the north and south of Baltimore along the western shore. Land usage in the remaining drainage areas along the western shore, and in all of the drainage basins of the eastern shore, is predominantly agriculture, residential, and forest.

TEMPORAL AND SPATIAL SCALES OF MOTION IN THE BAY AND TRIBUTARY ESTUARIES

Motions in a natural water body such as the Chesapeake Bay encompass a large range of temporal and spatial scales. The smallest of these scales, that of random molecular motion, is not of concern to us here since turbulent diffusion processes dominate over molecular diffusion in the Bay and its tributaries. Turbulent motions, with space scales ranging from a few centimeters to perhaps ten meters in the vertical and several hundred meters in the horizontal, and with time scales ranging from less than a second to several hundred seconds, provide the mechanism for the horizontal and vertical diffusive spread of pollutants introduced into the Bay, as well as of the phytoplankton and zooplankton, including the eggs and larvae of shellfish and finfish (the ichthyoplankton). While an important mechanism with respect to the fate of pollutants and to the distribution of plankton, we do not have to be concerned with the details of the motion that results in this dispersion, but rather can parameterize the process. We know, for instance, that turbulent dispersion is enhanced by large velocity shear, and that such high shear is normally associated with high velocity, as well as with abrupt changes in bathymetry and topography. We also know that vertical turbulent dispersion is inhibited by a strong vertical density gradient.

The advective transport of pollutants, of plankton, and of fine-grained suspended matter is primarily dependent on the "mean" motion, with the turbulent motions averaged out of the mean. There is, however, a large range of scales of the mean, non-turbulent motions. The most obvious motion is that induced by the astronomical tide. Tidal currents have spatial scales in the Chesapeake Bay of 5 to 10 km (the length of a tidal excursion) and a primary time scale equal to that of the lunar semidiurnal tide, but with smaller contributions from the shorter period M4 and M6 shallow water overtides, and also from longer period tidal constituents such as the diurnal and fortnightly tides.

When motions with tidal frequencies are removed from the record, there remain residual motions which often have amplitudes of the same order as that of the tidal currents. These residual motions can be considered to result from the following processes: (a) there must be a mean flow directed toward the ocean through any cross section equal to, on the long-term average, the volume rate of inflow of freshwater to the estuary landward from the cross section; (b) the mixing of freshwater and sea water in the estuary produces a longitudinal density gradient which drives a depth varying and, to a greater or lesser degree dependent on the intensity of the vertical and lateral mixing, a laterally varying residual circulation pattern; (c) the nonlinear interaction of the longitudinal and vertical components of the oscillatory tidal current produces a Lagrangian "Stokes" transport which must be compensated for by a Eulerian velocity field directed seaward; (d) nonlinear interactions of the oscillatory tidal currents with topographic and bathymetric variations can produce residual horizontal eddy currents which may locally dominate the spatial distribution of nontidal currents in a cross section; and (e) meteorological induced motions, both far field and near field, constitute an important part of the residual velocity field within the Chesapeake Bay.

Each of these types of residual motions includes a range of temporal and spatial scales. It is most convenient to discuss tidal currents separately from the non-tidal residual currents, and to discuss the various types of residual motion in terms of the predominant time scales associated with them.

TIDES AND TIDAL CURRENTS

The tides in the Chesapeake Bay have a predominant lunar semidiurnal component, although there is a sufficient solar semidiurnal component so that the spring tide range is about 50% greater than the mean tide range, and the neap tide range is about 50% less than the mean tide range.

According to Hicks[3] the observed tidal wave progresses from the Bay's entrance at the Capes to the mouth of the Susquehanna River at Havre de Grace in approximately 14 lunar hours. Since the tidal wave entering the Bay from the open ocean has a predominant period of 12 lunar hours, a crest does not quite traverse the length of the Bay before the following crest enters at the Capes. In this respect, Chesapeake Bay is unusual in that it is able to contain a complete semidiurnal tidal wave at all times.

The Bay is wide enough so that rotational effects are important. As a result, in the Bay below the Severn River, the tide has the characteristics of a Kelvin wave, with a slightly larger range on the eastern side than on the western side, and with maximum flood and maximum ebb current speeds occurring at nearly the same time as high water and low water, respectively. North of the Severn, friction and reflection result in characteristics which are intermediate between those of a pure progressive wave and those of a standing wave, but becoming asymmetric to the characteristics of a standing wave as one approaches the head of the Bay. Between the Capes and the mouth of the Severn River, high water and maximum flood are nearly in phase, i.e., in this reach the wave is close to a pure progressive wave, but north of the Severn River, the time between maximum flood and the next high water increases. At Pooles Island, approximately halfway between the mouth of the Severn River and Havre de Grace, maximum flood precedes high water by about one hour, and at Fishing Battery Light, 3 miles below Havre de Grace, high water lags maximum flood by about 2.6 hours, or close to the 3.1 hour time difference characteristic of a standing wave. As a result of these characteristics, it is possible to sample the waters between the Capes and Pooles Island, some 18 miles south of the head of the Bay, at the same phase of the tide if the sampling vessel proceeds northward at a speed of between 10 and 12 knots (~ 480 km/day). In the reach of the Bay North of Pooles Island, slack water occurs at substantially the same time at all locations.

The mean tidal range decreases from about 0.9m (3.0 feet) at the Bay's entrance to a minimum of about 0.3m (1.0 feet) at Annapolis, then rises to 0.7m (2.3 feet) at the head. The maximum range in the system is 1.2 m (3.9 feet) at Walkinton, Virginia on the Mattiponi River. As noted above, the range is significantly larger on the eastern shore of the Bay proper because of rotational effects. The lateral difference is as large as 0.15m (0.5 feet) and averages 0.06m (0.2 feet). Elliot and Wang[4] have reported that velocities are higher on the eastern side of the Bay as well. Although this may be due in part to the fact that the deep channel tends to be nearer the eastern shore, with a consequent lower frictional effect on that side of the Bay.

The most obvious motion of the Bay's waters are the oscillatory tidal currents. Average tidal current amplitudes in the cross section at the mouth of the Bay vary from 0.64 m/s (1.25 knots) to 1.03 m/s (2.00 knots). The tidal current amplitudes decrease within the lower reaches of the Bay, such that at Wolf Trap Light, located between the mouths of the York and the Rappahannock Rivers, the tidal current amplitude is 0.57 m/s (1.10 knots). Over the major portion of the middle reaches of the Bay, the tidal current amplitudes range from 0.13 m/s (0.25 knots) to 0.41 m/s (0.8 knots), with the lower values occurring in the wider cross sections and the higher values in the narrower cross sections. The average tidal current amplitude at the Chesapeake Bay Bridge at Annapolis is about 0.41 m/s (0.8 knots), increasing to 0.59 m/s (1.15 knots) off Worton Point (about 20 km north of the entrance to Baltimore Harbor), and then decreasing to 0.28 m/s (0.55 knots) at Fishing Battery Light on the Susquehanna Flats.

At spring (neap) tides, tidal current amplitudes are about 30% to 40% higher (lower) than the average values given in the previous paragraph.

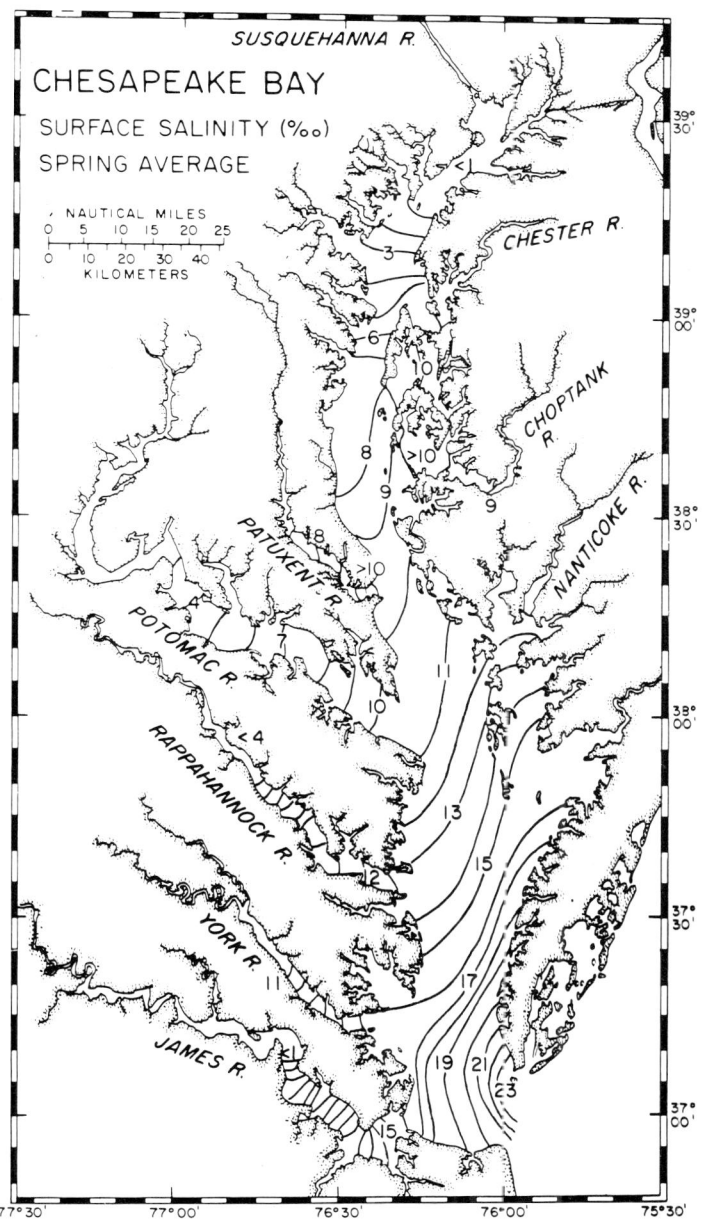

FIGURE 4. Surface salinity distribution in the Chesapeake Bay, averaged for the Spring season, using data collected by the Chesapeake Bay Institute during the years 1949 through 1961.

THE DISTRIBUTIONS OF TEMPERATURE, SALINITY, AND DENSITY

The long-term average residual (non-tidal) circulation in an estuary is best understood in terms of the spatial distribution of density, which is in turn dependent on the distributions of temperature and salinity. For this reason, we describe the distributions of these physical properties before proceeding to a discussion of the residual motions in the Bay.

According to Pritchard,[5] salinity varies more or less regularly along the length of the Bay, from that of nearly full sea water at the mouth to that of the inflowing Susquehanna River water at the head of the Bay. The vertical distribution of salinity is characterized by an upper layer of very slow increase with depth, an intermediate layer of more rapid increase (the halocline) and a deep layer in which the salinity increase with depth is again small. The salinity also varies laterally across the Bay with lower salinities on the western side of the Bay. Although the greater runoff of freshwater from the western shore contributes to this difference, the major cause is the rotation of the earth.

Figures 4 and 5, taken from Pritchard,[5] show the characteristic features of the surface distribution of salinity for spring, a season characterized by high freshwater inflow, and for autumn, a season characterized by low freshwater inflow. Note that minimum salinities occur in spring, with essentially freshwater extending on the average to Pooles Island; and maximum salinities in autumn, when low but measurable ocean-derived salt concentrations extend onto the Susquehanna Flats.

The longitudinal-vertical distribution of salinity along the thalwag of the Bay at a time of high river flow and at a time of low river flow are shown in Figure 6. These figures, taken from Seitz,[6] show that the reach of the mid-Bay between the Severn River and the York River is more stratified than the regions riverward and oceanward. This has important implications for the concentrations of dissolved oxygen concentrations in the lower layer of the mid-Bay in the summer.

There are also marked temporal variations in salinity of monthly and interannual periods which are greatest in the upper reaches of the Bay and its tributary estuaries.[7]

To date, humans have had little affect on the salinity distribution in the Bay or its tributaries. Changes in Bay salinity could result from flow regulation of the Susquehanna or enlargement of the C&D Canal which connects the Chesapeake Bay with Delaware Bay. Pritchard[8] analyzed the effect of widening and deepening the Canal during the early 1970s and concluded that the effect would be the greatest during periods of low river flow when salinities are a maximum and that the average maximum salinity would increase from 17.23 to 17.62‰ at the Bay Bridge, near Annapolis, from 9.00 to 11.58‰ at Pooles Island, and from 2.14‰ to 2.94‰ at Turkey Point at the head of the Bay.

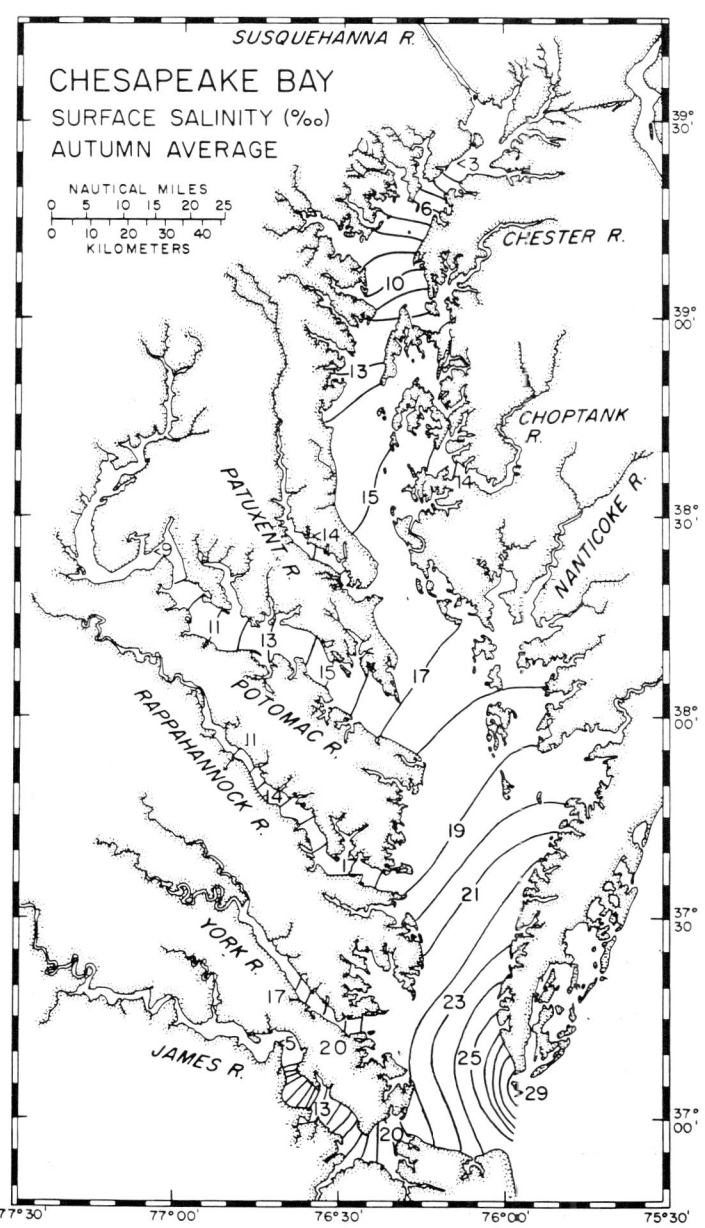

FIGURE 5. Surface salinity distribution in the Chesapeake Bay, averaged for the Autumn season, using data collected by the Chesapeake Bay Institute during the years 1949 through 1961.

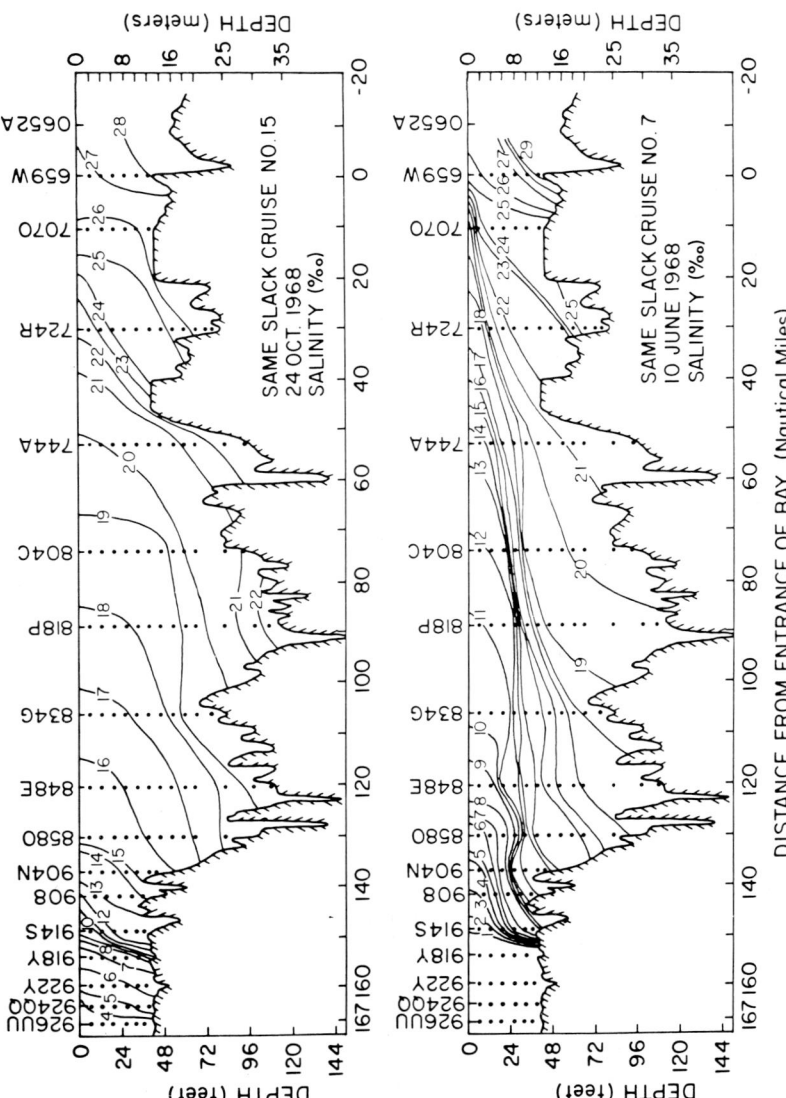

FIGURE 6. Distribution of the salinity in a longitudinal-vertical section taken along the thalweg of the Chesapeake Bay, as observed on a same-slack cruise on 24 October, 1968 (upper diagram), a distribution characteristic of low fresh water inflow, and on 10 June 1968 (lower diagram), a distribution characteristic of high fresh water inflow.

A Brief Physical Description of the Chesapeake Bay 15

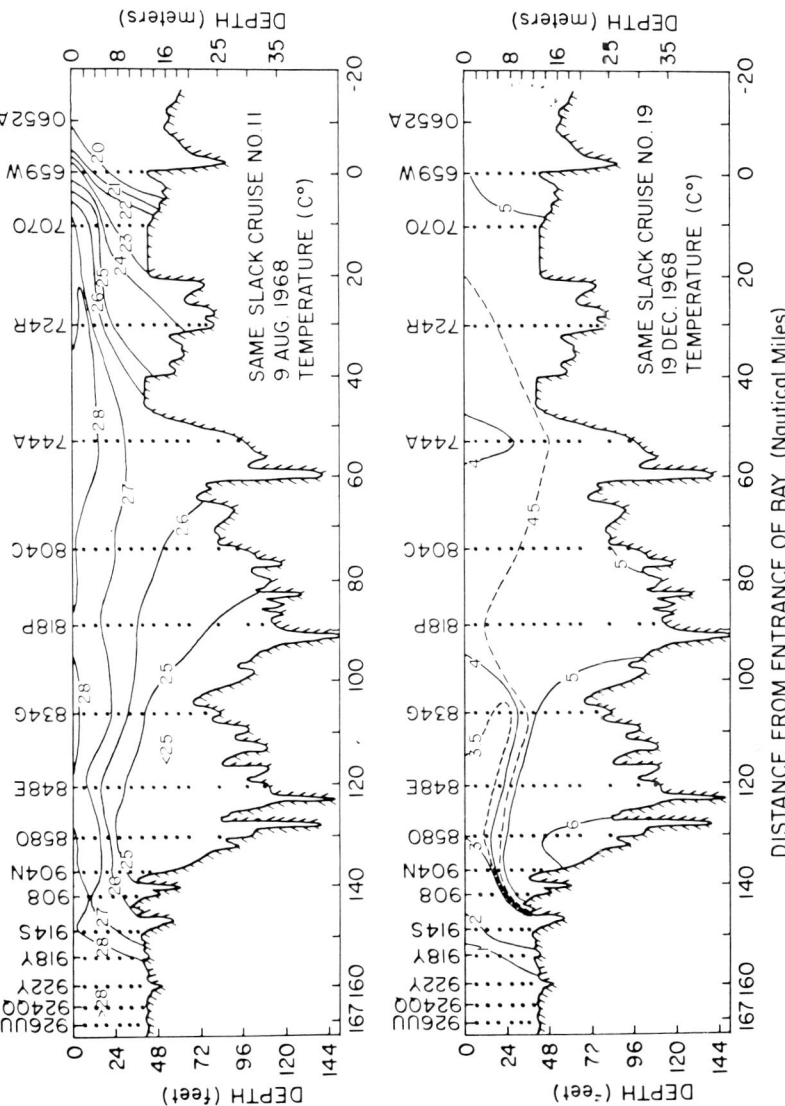

FIGURE 7. Distribution of temperature in a longitudinal-vertical section taken along the thalwag of the Chesapeake Bay, as observed on a same slack cruise on 9 August, 1968 (upper diagram), a distribution characteristic of conditions of maximum temperatures at the surface and decreasing temperatures with depth, and on 19 December, 1968 (lower diagram), a distribution characteristic of low temperatures at the surface, and increasing temperature with depth.

Temperature is an important oceanographic parameter in Chesapeake Bay because of its effect on density, on oxygen solubility, on a number of other physico-chemical properties of sea water, and on biological activity. There are marked natural temporal and spatial variations of water temperature in the Chesapeake Bay system.

According to Seitz,[6] waters in Maryland are somewhat warmer than those in Virginia (0.5-3°C) during the summer months, but cooler during the balance of the year (October-May). On an annual basis the average water temperatures in Virginia are about 0.5°C warmer than in Maryland.[9] Spatially, local horizontal gradients as high as 1°C/km are observed. Vertically, maximum top-to-bottom density differences attributable to temperature occur at mid-Bay stations during the summer months. During June, 1986, for example, temperature accounted for approximately 19% of the total vertical range in density at station 818P (mid-Bay off the mouth of the Patuxent). These features are illustrated in Figure 7, which show the characteristic longitudinal-vertical distribution of temperature along the thalwag of the Bay at a time in late summer and in late winter.

Diurnally, variations as high as 3°C have been observed.[7] The annual range of temperature in the open Bay is from about 1°C to about 28°C. There are also relatively large variations with periods in excess of 1 year. Daily measurements of surface temperature were taken for more than 50 years by the U.S. Coast and Geodetic Survey at selected tidal observation stations in some of the tributary estuaries.[10] Similar data are not available for the Bay proper, but comparison of the data, for say, Solomons, Maryland (Patuxent River) with Fort McHenry (Baltimore Harbor), suggests that these data are quite representative of the Bay system. An analysis of the Fort McHenry data set for the period 1914 - 1962 is given in Fig. 8, taken from Schubel.[9] This figure gives the departure of the annual mean surface temperature at Fort McHenry from the long-term, 49-year mean. The figure shows that the annual mean had a range of 3.5°C; the maximum difference of the annual mean between consecutive years was greater than 1.5°C. It is also apparent from the figure that there are longer term oscillations present in the record.

Superimposed on these natural fluctuations are the thermal effects of human activities, primarily the generation of electricity. The distribution of temperature is affected by man where large fractions of the available "dilution" water is passed through the condensers of a generating station. Significant temperature effects caused by this activity have not been demonstrated for the open Bay as yet, but several other estuaries, including Bay tributaries, have had their temperatures measurably altered.

The density of Bay waters increases with increasing salinity and with decreasing temperature, except at low temperatures and low salinities, where density decreases slowly with decreasing temperature. The temperature of maximum density decreases with increasing salinity faster than does the freezing point. It is 4°C for freshwater, 1.9°C at a salinity of 10‰, - 1.33°C (and equal to the freezing point) at a salinity of 24.7‰. Water temperatures are below the temperature of maximum density for only a short time during the coldest part of the year, and then in only a small fraction of the volume of the Chesapeake Bay estuarine system.

The spatial variations in density result primarily from spatial variations in salinity, and there has been a tendency among investigators to ignore the effects of temperature when modeling the dynamic behavior of the Bay. As has already been noted, vertical variations in temperature do account for as much as 20% of the vertical variation in density in the middle and upper reaches of the Bay during late spring and early summer. More importantly, the onset of the fall overturn of Bay waters is triggered by a change from a condition in which temperature decreases with depth, and hence adds to the effect of increasing salinity with depth to produce a stable water column, to a condition in which the temperature increases with depth, and hence acts in opposition to the effects of the vertical salinity gradient on the vertical stability.[11]

In response to the characteristic longitudinal variation in salinity, the density of the Bay waters increases from just under 1000 kg/m³ at the head of the Bay, and at the upper limits of the estuary proper of each of the tributary estuaries, to values of between 1020 and 1025 kg/m³ at the mouth of the Bay. The vertical gradients of density, as measured by the difference in the densities at the bottom and at the surface at various locations along the thalwag of the Bay, vary seasonally during a given year and also vary from year to year for a given season, depending primarily upon the amount and timing of the

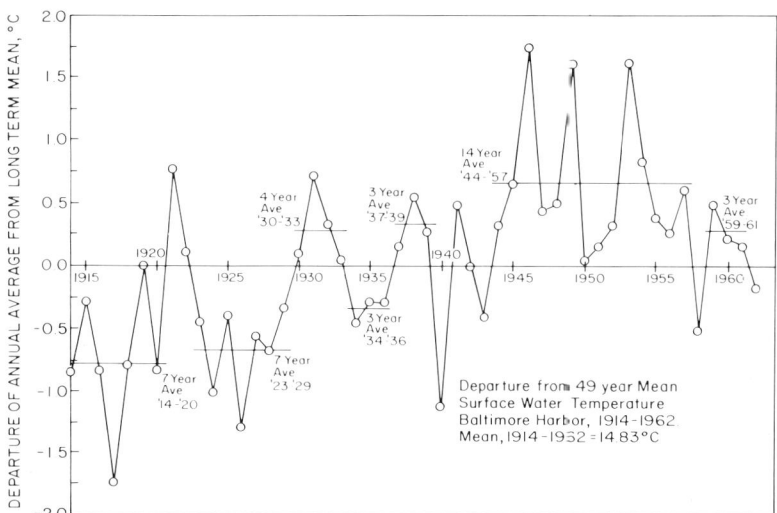

FIGURE 8. Departures of the mean annual surface temperatures (°C) from the 49-year (1914-1962) average surface temperature off fort McHenry, Baltimore Harbor. Mean surface temperatures averaged over periods of several years are also shown.

freshwater inflows to the Bay. Vertical density differences are greater in the middle and upper reaches of the Bay than in the lower Bay. Maximum vertical gradients usually occur in the summer (June through August), with typical top-to-bottom density differences of 7 to 9 kg/m^3. Minimum vertical differences in density occur primarily in October, November and December, with typical values of the surface to bottom density difference ranging from 1 to 4 kg/m^3. Goodrich[11] gave evidence that in the several month period following the autumnal equinox, there are short intervals during and just after large wind events in which the waters in the wide section of the lower Bay below the mouth of the Potomac may become completely mixed, top to bottom. During August and early September, the vertical variation in temperature contributes about 15% to the positive vertical variation in density, reinforcing the effects of the vertical variation in salinity on the vertical density gradient. During a relatively short interval of about two weeks centered on the autumnal equinox, the vertical temperature gradient is reversed, so that the vertical temperature variation acts opposite to the effects of the vertical salinity gradient. Although this counter effect is only about 10% of the effect of the salinity gradient on the vertical stability, this period of the year is characterized by a decreasing vertical gradient in salinity, resulting from the long interval since the high spring inflows of freshwater. As a result, the first strong winds following the shift in the sign of the vertical temperature gradient result in vertical mixing of the waters of the lower Bay. Vertical homogeneity lasts for only a few days, although several mixing episodes may occur over a one-to-two month period. The onset of increased rainfall and consequently increased runoff of freshwater to the Bay during late fall and early winter results in an increase in the vertical salinity and density gradient, and a cessation in the occurrences of vertical overturn of the Bay waters.

LONG-TERM AVERAGE RESIDUAL CURRENTS DENSITY-DRIVEN FLOW PATTERN

The non-tidal (residual) motion in the Bay is driven by the density distribution and by wind forcing, both near-field and far-field. When averaged over periods of 10 days and longer, the predominant motions are density driven, and it is these motions we describe in this section.

The Main Stem of the Bay

The Chesapeake Bay is a partially mixed estuary, that is, an estuary in which (a) the salinity increases with depth; (b) the salinity increases in a seaward direction along the length of the estuary; and (c) over much of the length of the estuary, the longitudinal salinity gradient is only weakly dependent on depth. In such an estuary, the residual current field, when averaged over a sufficient time interval, say 10 days, to filter out most of the meteorologically driven motions, *usually* takes on a characteristic pattern which has long been recognized as the classical estuarine circulation pattern. The word "usually" is included because evidence for some departures to this classical pattern developed with the rapid increase in the data base of direct current measurements in estuaries![2]

Figure 9 is a schematic depiction, in a longitudinal-vertical section and in a horizontal plan view, of the salinity distribution and long-term average residual flow pattern in a strong partially mixed estuary, such as the Maryland portion of the Chesapeake Bay. Partially mixed estuaries exhibit a wide range of vertical stratification conditions. At one extreme of high vertical stratification, these estuaries approach conditions found in a salt wedge estuary. At the other extreme, as the vertical stratification weakens, these waterways approach the conditions found in a vertically homogeneous estuary.

Neither the Chesapeake Bay, nor any of its tributary estuaries, exhibit characteristics close to those of a salt wedge estuary. However, certain parts of the Bay and of some tributaries do, at times, approach conditions of a vertically homogeneous estuary. Figures 10 and 11 demonstrate how this transition can occur. The upper diagram in Fig. 10 shows the salinity distribution in a characteristic cross section in a strong partially mixed estuary. Also shown in this figure is the boundary between the upper, seaward flowing layer, and the lower, upestuary flowing layer. Note that this boundary is sloped slightly,

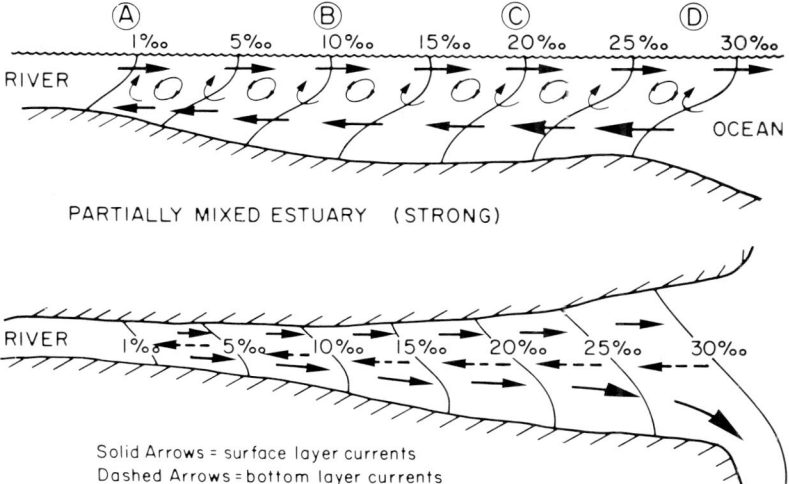

FIGURE 9. Schematic depiction of the distribution of salinity and of the long term (greater than 10 tidal cycle averaged) residual circulation in a partially mixed estuary (strong), characteristic of the northern half of the Chesapeake Bay under nearly all conditions for all seasons, and of the lower Chesapeake Bay and of the major tributaries under most conditions during late winter, spring and summer. The upper diagram is a longitudinal-vertical section taken along the thalwag of such an estuary, and the lower diagram is a plan view.

so that the seaward flowing layer is thicker on the right side of the estuary (looking seaward) than on the left side. The lower diagram in Fig. 10 is a schematic depiction of the salinity distribution in a cross section of a vertically homogeneous estuary. Also shown is the interface between the portion of the cross section in which the long-term average residual flow is directed down the estuary and that part in which this flow is directed up the estuary. Note that the downestuary directed flow is concentrated on the right side of the estuary and the upestuary directed flow is concentrated on the left side of the cross section (looking seaward). The boundary between the two counterflows through the cross section can be thought of as rotating clockwise about a point at mid

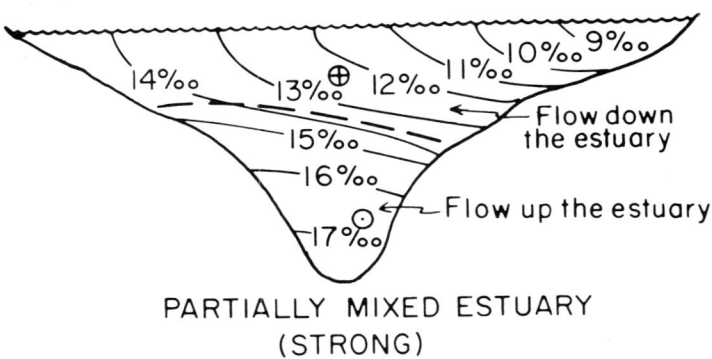

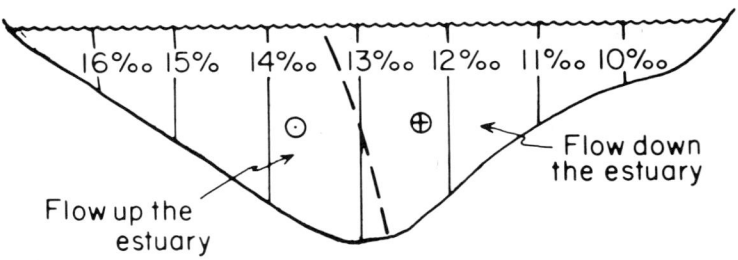

FIGURE 10. Schematic depiction of the salinity distribution in a cross section in the mid-reaches of a partially mixed estuary (strong), upper diagram, and in a vertically homogeneous estuary, lower diagram. Note that the boundary between the seaward flowing waters and the landward flowing waters (the thick broken line) can be considered as rotating in a clockwise direction about its midpoint in the transition from a vertically stratified condition to a vertically homogeneous condition.

depth in the center of the estuary as the estuary shifts from a partially mixed condition to a well mixed condition.

As the vertical stratification decreases from the condition depicted by the upper diagram in Fig. 10 towards the condition depicted by the lower diagram in that figure, conditions intermediate between these two conditions may prevail. Such a situation is shown in the central diagram in Fig. 11. Note that while in the center of the cross section the long term average residual flow pattern is similar to that of a strong partially mixed estuary, with seaward directed flow in the upper layers and upestuary directed flows in the lower layers, the boundary between these counterflows is more sharply sloped than in the case of a strong partially mixed estuary. Consequently, this boundary intersects the surface, and the upestuary directed residual flow extends all the way to the surface over the left hand one-quarter to one-third of the width of the cross section. An estuary or a segment of an estuary, which exhibits this type of salinity and residual flow conditions, might be called a weak partially mixed estuary. The

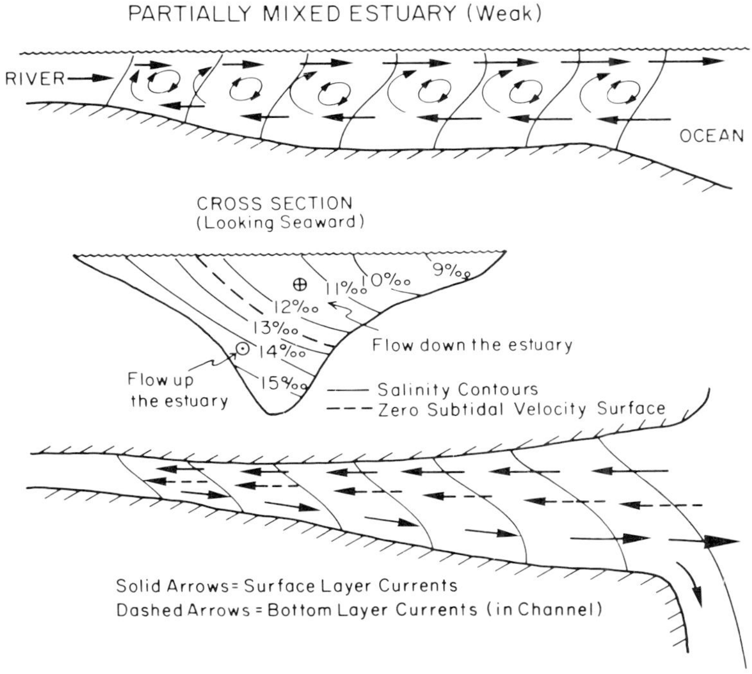

FIGURE 11. The salinity distribution and long term average circulation for a partially mixed estuary (weak) as shown schematically in a longitudinal-vertical section taken down the thalwag of the estuary (upper diagram), in a cross section in the mid-reaches of the estuary (middle diagram), and in a plan view of the estuary (lower diagram). The lower Chesapeake Bay and the James River Estuary, and possibly other major tributary estuaries to the Bay exhibit the characteristics of this type of estuary, particularly during late summer, autumn, and early winter.

upper diagram in Fig. 11, which schematically depicts the salinity distribution and residual flow pattern in a longitudinal-vertical section of such an estuary looks very much like the similar diagram for a strong partially mixed estuary shown at the top of Fig. 9. The lower diagram in Fig. 11, which depicts a horizontal plan view of such an estuary, however, differs from the similar diagram for a strong partially mixed estuary shown at the bottom of Figure 9. Note that the surface layers have a counterflowing pattern with the waters on the right side of the estuary flowing seaward and those on the left side flowing up the estuary. The dashed vectors in this diagram indicate the upestuary directed residual flow in the lower layers of the estuary.

The southern most reaches of the Chesapeake Bay, seaward of about 30° 30' N, exhibit the characteristics of a weak partially mixed estuary, with the upestuary residual flow outcropping in the surface layers along the eastern shore of the Bay. The James River also exhibits these characteristics, at least during low flow periods of the year.

Tributary Estuaries

Tributary estuaries to the Chesapeake Bay exhibit long-term average residual flow patterns which differ from that found in the Chesapeake Bay to a greater or lesser degree depending on size, freshwater inflow, and local topography. The two major tributaries, the Potomac and the James, for the most part, show circulation patterns similar to those described for the Chesapeake Bay, with some local and temporal differences. Boicourt[13] showed that in a stretch of the lower Potomac River estuary the classical estuarine flow pattern was replaced by topographically-induced eddies which provided the mechanism for advective transport of water and salt through the cross sections he studied. At least during a part of the year, the lower James River may exhibit a neap-spring variation in circulation described later in this section.

A number of small tributary estuaries to the upper Chesapeake Bay have relatively small drainage areas. The circulation patterns in these waterways are controlled by physical conditions in the adjacent Bay which are driven primarily by fluctuations in the discharge of the Susquehanna River. Most of these tributaries are relatively shallow and are in direct communication with only the upper layer of the adjacent Bay. For the most part, these tributaries are filled with water from the upper layers of the adjacent Bay. Only in the narrow reaches near the heads of these tributaries are the salinity patterns significantly affected by the local freshwater discharges. The major factor controlling the long-term average residual circulation pattern in this kind of tributary estuary is the temporal variation of the salinity of the adjacent Bay associated with fluctuations in discharge of the Susquehanna.

The salinity of the upper Bay varies seasonally with maximum values in autumn and minimum values in spring. The seasonal salinity variations in these tributary estuaries lag behind the salinity changes in the adjacent Bay. During winter and early spring, when the salinity in the Bay is decreasing with time, the salinity in the tributary is, at any given time, higher than that in the Bay. Consequently, surface waters of the Bay flow into the tributary, and the bottom waters of the tributary flow out to the Bay. During summer and autumn, when the salinity in the Bay is increasing, the salinity in the tributary is less than in the adjacent Bay. Consequently, the surface waters of the tributary flow out to the Bay, while Bay waters flow into the tributary along the bottom. These patterns are shown schematically in Figures 12a and 12b.

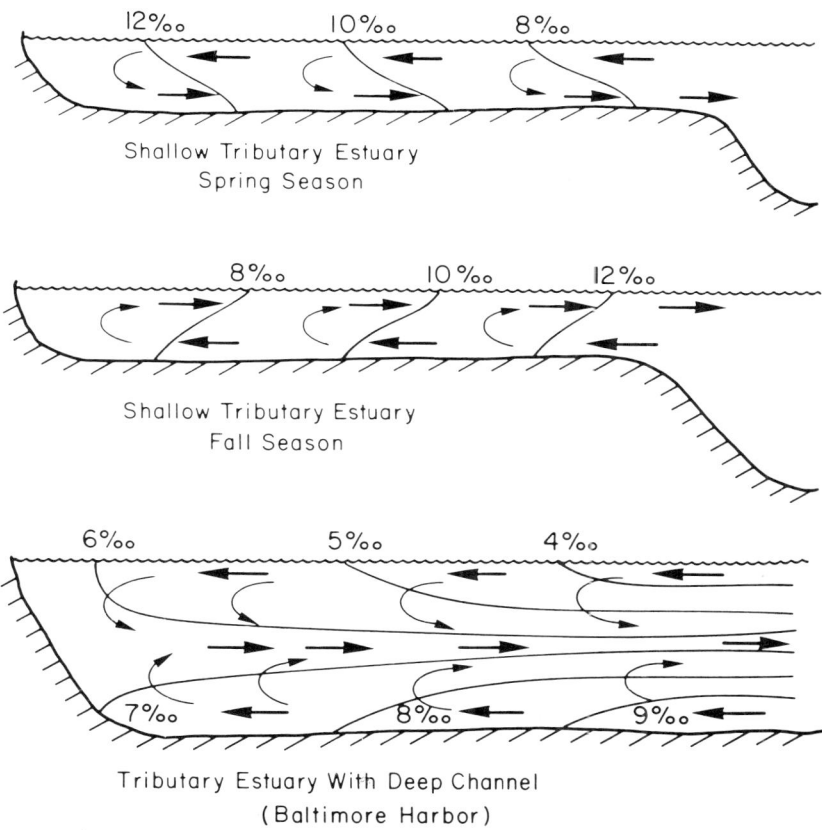

FIGURE 12. Longitudinal-vertical section schematically showing the salinity distribution and long term average residual circulation pattern in: (a) a shallow tributary estuary with a small drainage basin during the spring season of high fresh water inflow to the parent estuary (upper diagram); (b) such a tributary estuary during the autumn season of low fresh water inflow to the parent estuary (middle diagram); and (c) a deep tributary estuary with a small drainage basin. The upper and middle diagrams are characteristic of several small, shallow tributary estuaries along the western shore of the upper Chesapeake Bay. The lower diagram characterizes conditions in Baltimore Harbor.

Baltimore Harbor is the archetype of a class of tributary estuaries having relatively small drainage areas with channel depths about equal to the channel depths in the adjacent parent estuary. Pritchard and Carpenter,[14] were the first to describe the three-layered flow pattern which characterizes Baltimore Harbor, inflow from the adjacent Bay in the surface and bottom layers, and outflow from the Harbor in a mid-depth layer. This pattern is shown schematically in Figure 12c. This is a density driven flow pattern even though the vertically averaged density does not vary significantly along the length of this tributary waterway. A moderately strong stratification is maintained in the waters of the Chesapeake Bay adjacent to the mouth of Baltimore Harbor as a result of the counterflow of the low salinity upper layer and higher salinity deeper layer, in balance with only moderately intense vertical mixing. Within the Harbor, the counterflowing layers, which tend to maintain stratification against the effects of vertical mixing, are lacking and vertical stratification is weakened. In fact, vertical homogeneity usually exists at the inner end of the Harbor. Consequently, the surface waters of the Harbor are more saline than those in the adjacent Bay, while the bottom waters of the Harbor are less saline than the waters at the same depth in the adjacent Bay. The longitudinal density gradients at the surface and at the bottom, therefore, drive the waters in both of these layers into the Harbor; continuity requires that there be an outflow at mid-depth.

Complex interactions between tributary estuaries and the adjacent Chesapeake Bay are not limited to the small tributaries along the Western shore of the upper Bay. Figure 13 is a schematic sketch of the circulation pattern Owen[15] found for significant periods in the Patuxent River estuary. Over a part of the year, this mid-sized tributary estuary shows a two-layered, long-term average residual flow pattern over its entire length. During other times, the estuary appears to be two layered in its upper and middle reaches, and to have a three layered, Baltimore Harbor type circulation pattern near its mouth. Owen attributed this phenomenon to the existence of a constriction in the cross section of the Patuxent estuary at Point Patience, which appears to result in strong tidal currents and a consequent strong vertical mixing. It may be, however, that this

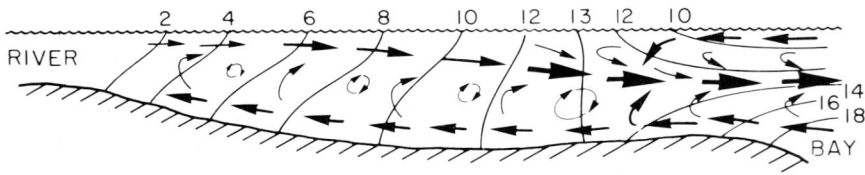

PATUXENT TYPE TRIBUTARY ESTUARY
Longitudinal - Vertical Salinity Distribution and Flow Pattern

FIGURE 13. Schematic depiction of the longitudinal-vertical distribution of salinity and long term average circulation pattern characteristic of the Patuxent River Estuary, and possible of several subestuaries of the Bay in Virginia.

circulation pattern in the lower Patuxent has as its cause the neap-spring phenomenon reported for Virginian subestuaries. In the case of the Patuxent, however, the reversal of the longitudinal salinity gradient in the surface layers of the lower reaches of the tributary occurs over a prolonged period, and is not restricted to a few day period near the time of spring tide.

In the lower reaches of several Virginia subestuaries, notably the Rappahannock and York subestuaries, and possibly the James estuary during periods of low freshwater inflow, Haas,[16] Hayward et. al.,[17] Ruzecki and Evans[18] and Hayward et al.,[19] have shown that a neap-spring cycling of stratification-destratification occurs. It is to be expected that the higher tidal velocities associated with spring tide conditions will result in more intense vertical mixing. However, the destratification phenomenon in the Virginian subestuaries appears to be driven primarily by a mechanism other than simply higher current speeds. The apparent mechanism involves a complex interaction between the tributary and the adjacent Bay. The larger tidal excursion in the Bay during the spring tide period brings lower salinity water from up the Bay to the vicinity of the mouth of the subestuary. Because of the phasing of the ebb and flood flows in the Bay and in the lower subestuary, water with salinities as low, or lower, than that found within the subestuary enters the subestuary from the Bay. This flattens, or reverses, the longitudinal density gradient, thus weakening or even destroying the density driven, two layered estuarine flow, so that upestuary flow of the more saline water in the lower layers of the tributary weakens or ceases. The advective mechanism that favors stratification thus fails, and vertical mixing occurs.

INTERMEDIATE SCALES OF MOTION METEOROLOGICALLY-FORCED TEMPORAL AND SPATIAL DEPARTURES FROM THE LONG-TERM AVERAGE RESIDUAL FLOW PATTERNS

Beginning about 15 years ago, there began a dramatic increase in the number of time series measurements of currents obtained from vertical moorings of *in situ* recording current meters. Simultaneous records of the temporal variations in current velocity over periods of a month, or more, at several depths became available. In several estuaries, but particularly in the Chesapeake Bay, these vertical moorings were often deployed in arrays across a cross section, and in a few cases such arrays were deployed across two cross sections separated by 10 km, or so. The current meters used in these field studies could record observations taken at various, selectable sampling intervals, from a sample every four minutes to a sample every half hour, for record lengths of from one to three months depending on sampling interval. Sampling intervals in this range are quite adequate for time series analysis which permit separation of the various scales of motion, from tidal period to velocity fluctuations having an inverse

TABLE 3

Circulation Patterns According to Direction of Flow* at 3 Levels
(10m, 25m, 40m).[21] Probability of Occurrence and Duration are reported.

Type	Direction of Flows	Frequency of Occurrence	Mean Duration Days
Classical Estuary	($U_{10} > 0$; $U_{40} < 0$; $U_{25} \approx 0$)	43%	2.5
Reverse Estuary	($U_{10} < 0$; $U_{40} > 0$; $U_{25} \approx 0$)	21%	1.6
Three-Layered	($U_{25} > 0$; $U_{10} < 0$; $U_{40} < 0$)	1%	1.0
Reverse Three-Layered	($U_{25} < 0$; $U_{10} > 0$; $U_{40} > 0$)	7%	1.5
Discharge	(All 3 Positive)	6%	1.3
Storage	(All 3 Negative)	22%	1.6
		100%	

*The positive direction is down the estuary.

frequency of 7-to-10 days. By rapid recovery, servicing, and redeployment of the current meters, time series records over a year, or more, could be constructed. This provided greater resolution for periods of 7-10 days, and made it possible to extend the analysis to longer period fluctuations.

Analysis of these records showed that the velocity fluctuations in the meteorologically-forced motions have their largest amplitudes in the period range of 2 to 7 days. In parts of the Bay, some low amplitude fluctuations occurring at periods as long as 20 days, in response to time variations at these periods in the tidal mean sea level in the coastal ocean off the mouth of the Chesapeake were important.

One of the first studies of a long term time series record from current meters in the Chesapeake Bay estuarine system was carried out by Elliott.[20,21] This investigator described an analysis of the records from a mooring in the lower Potomac River estuary consisting of three current meters at nominal depths of 10, 25, and 40 m which were maintained for a full year (1974-75). The records were filtered to remove the major tidal components and the residuals averaged over each calendar day to produce daily estimates of the mean residual flows. An Empirical Orthogonal Function (EOF) analysis in the time domain showed that: (a) 79% of the total velocity variance could be explained by the first two modes; (b) the first mode (classical estuarine or reverse estuarine) was associated with local wind forcing and contained 48% of the total velocity fluctuations. The classical pattern was associated with downstream wind, falling local sea level, and a downstream surface slope. The mode was reversible; and (c) the second mode (storage or discharge) contained 31% of the total velocity variance and was associated with rising (storage) or falling (discharge) sea level. This mode was considered to be caused by non-local meteorological effects which were propagated into the Potomac from the Bay.

The frequencies of occurrence of the six patterns that emerged are summarized in Table 3. Although the classical two-layered flow pattern occurred for a surprisingly small 43% of the time, it always emerged for averaging periods of ten days or longer.

Analysis of a three-month record from a two current meter mooring established in the upper Chesapeake Bay near Howell Point[22] during the fall of 1975 also showed similar residual flow variability. That paper also showed that the variability was related both to local wind forcing which was driven by the component of the wind along the axis of the Bay, and to far-field wind forcing which appeared to be a result of Ekman transport produced by the cross-Bay component of the winds in the middle and lower reaches of the Bay.

Wang and Elliott,[23] and later Elliott and Wang,[4] examined the frequency response and the coupling between the Bay and the Potomac River with spectral and Empirical Orthogonal Function (EOF) analysis techniques. They used data collected during the first two months of Elliott's year-long study[20,21] consisting of wind, surface elevation, and bottom currents in the Potomac. Sea level data from four locations in the Bay and the results of a 3-day intensive current measurement experiment in the Potomac[24] were also considered.

From the records of wind and sea level, they found the Bay wind and sea level spectra to be markedly similar with peaks at 2.5, 5, and 20 days. The dominant sea level fluctuations in the Bay had a period of 20 days and were considered to be the result of Ekman fluxes in the adjacent coastal ocean, i.e., water was driven out of (into) the Bay by northward (southward) winds. These fluctuations were then damped as they propagated up the Bay.

The 5-day fluctuations in Bay sea level had up-Bay phase propagation, but with a slight increase in amplitude. In addition, Annapolis sea level fluctuations were coherent with Patuxent east-west winds, all of which suggests response at this period to both coastal sea level and to local lateral winds (an Ekman response).

The 2.5-day oscillation in Bay sea level had down-Bay phase propagation with a marked decrease in amplitude between Annapolis and Kiptopeake Beach. Surface slopes were coherent with north-south (longitudinal) winds at both Annapolis and Kiptopeake Beach, and are considered, therefore, to be generated within the Bay by north-south winds. On the other hand, if the effective mean depth of the Bay is in fact the depth computed from the observed time of travel of the tidal wave from the mouth to the head of the Bay (14 hours), then the longitudinal seiche period of the Bay is apparently close to 2.5 days. The fact that the 2.5-day oscillation shows an amplitude increase in the up-Bay direction is suggestive of the presence of longitudinal seiching.

In the Potomac River, surface slopes were not coherent with local longitudinal winds (northwest-southwest), but were coherent with surface slope in the Bay, suggesting a coupled Bay-River response to north-south winds. The near-bottom currents in the River had spectral peaks at 20, 5, and 2.5 days.

For periods greater than 10 days, bottom currents were coherent with northeast-northwest (lateral) winds, suggesting a bottom flow compensation for a surface Ekman flux. Bottom currents were not coherent with local sea level at these periods and thus were apparently unaffected by any 20 day fluctuations which propagated into the Bay from the coastal ocean. For periods less than 10 days (5 and 2.5 days), bottom currents were coherent with surface slope and less so with longitudinal wind stress in the River, which is suggestive of local forcing. However, as noted above, since Bay and River surface slopes were coherent, these currents were apparently, at least in part, non-locally forced by longitudinal winds over the Bay. The bottom currents, however, were not coherent with sea level for time scales longer than 3 days. This suggests that the volume flux, i.e., rate of change of sea level, was largely confined to the upper part of the water column (second mode) and, of course, lagged sea level by 90° (15 hours). Since the bottom currents were driven by surface slope, they also led the volume flux, i.e., surface currents, by 90° (15 hours), indicative of an upward phase propagation at periods less than 3 days.

In summary, it has been shown that non-local forcing is an important mechanism for distributing energy between the coastal ocean, the Bay, and its tributaries.[22] It is clear that observational programs greater than 20 days are required if one is to recover all of the components of the residual circulation from the time series record.

Another extensive set of current meter, water level, and wind records were collected just north of the mouth of the Patuxent River in the fall of 1977 by investigators from the Chesapeake Bay Institute.[25] In that study, 22 current meters were deployed in two cross-bay sections, one extending easterly from Kenwood Beach on the western shore; the second located approximately 15 km south of the first section extending easterly from Cove Point. Another eight meters were deployed on three separate moorings located on a diagonal between the two east-west sections. The meters were deployed between 12 October and 2 November 1977. Four tide gauges were located at approximately the four corners of the study area.

A preliminary analysis of the data set was included in Pritchard and Rives.[25] More detailed analyses were made by Vieira,[26] Pritchard and Vieira,[27] and Vieira.[28] Vieira[28] calculated and plotted the laterally integrated residual flux per meter of depth below MLW at the Kenwood Beach section between 16 and 30 October. Vieira's findings were similar to those of Elliot[20,21,22,23,24] as considerable variability in the residual circulation patterns were reported. In 14 days, the circulation pattern changed 8 times. Although the statistics are sparse, both occasions of discharge were followed by a period of classical estuarine flow and two of the three occasions of classical estuarine flow were followed by periods of storage, agreeing with Elliott's comments regarding flow sequences.[20,21] Averaged over the period of the experiment, however, the classical estuarine circulation pattern emerged.

Using the same data set (laterally integrated residual flux per meter of depth)

Vieira[28] removed the mean and trend to reduce the presence of gravitational effects and to highlight the shorter period, meteorologically driven, flows. These data, when plotted and compared to the wind stress calculated from winds from the nearby Patuxent Naval Air Station, showed the surface layers down to about 8 m responding directly to the upbay and downbay components of the winds, and that this direct response propagated down the water column. Conversely, the bottom layers respond somewhat later (8 hours) in the opposite direction, presumably in response to a downwind setup of the sea surface. This counter flow starts at the bottom and propagates upward reaching depths just below the pycnocline in about 20 hours. Calculation of the rotary coherence squared between slope of the sea surface and the wind stress showed significant coherence at all periods greater than about 3 days with maximums at 4 and 7 days.

CLOSING OBSERVATIONS

Most contaminants are relatively insoluble in water and have a high affinity for fine-particulate matter. Because of this, the modes of motion which control the transportation, distribution, and accumulation of fine particles also control the transportation, distribution, and accumulation of most contaminants. The most important of these modes of motion in the Bay proper and its major tributaries is probably the long-term average estuarine circulation, as modified locally by topographically induced eddies. Departures from this mode are driven primarily by near-field and far-field winds, at time scales of from 2 to 10 days, and can affect the transportation, distribution, and accumulation of contaminants at these same time scales. However, following return to average conditions there is a redistribution of sediments and associated contaminants to reflect the effects of the long-term estuarine circulation. The significance of the variability of estuarine circulations matters to the transportation, distribution, and accumulation of sediments and sediment-bound contaminants has not been established. The effects of these fluctuations will depend strongly upon when they occur, how long they persist, and upon the size and geometry of the subestuary. Within the Bay proper, these meteorologically forced fluctuations in the residual currents probably act to increase the effective horizontal dispersion, but have little effect on the average advective transport of pollutants and other waterborne materials.

Tidal currents play an important role in contaminants dispersal and accumulation. Their effects are primarily through the alternate resuspension and redeposition of bottom sediments associated with the waxing and waning of tidal currents.[29] On the average, resuspension of sediments into the bottom layer leads to a net transport of sediment upestuary because of the long-term net upstream flow of the lower layer of partially mixed estuaries. Resuspension coupled with the net upstream motion in the lower layer leads to a redistribution of sediments that may have been carried downstream during conditions of net seaward flow at all depths. The alternate resuspension and redeposition of bottom sediments plays another role in contaminant dispersal. As particles are resuspended, they are transferred into an environment with different physico-chemico characteristics. These charges in environmental conditions may contribute either to a release of surface-bound contaminants into the water column or to a scavenging of contaminants from the water column by the resuspended particles. The mixing of interstitial waters into the overlying water column which accompanies sediment reworking by tidal currents may also lead to a transfer of contaminants from the bottom to the overlying waters.

Extreme events such as Tropical Storm Agnes in June 1972 can result in major inputs of contaminants, both in solution and bound to fine particles. It is unlikely that the largest and most persistent impacts of such events is associated with the input of contaminants. Their residence time in the water column is relatively short before removal by particles and deposition on the Bay floor.

ACKNOWLEDGEMENTS

This paper is Contribution #544 of the Marine Sciences Research Center, State University of New York at Stony Brook.

REFERENCES

1. Boicourt, W. 1969. A numerical model of the salinity distribution in upper Chesapeake Bay. Chesapeake Bay Institute, The Johns Hopkins University, Technical Report No. 54, Ref. 69-7, pp. 55 plus appendices. Baltimore, MD.
2. Schubel, J.R. and D.W. Pritchard. 1986. Responses of Upper Chesapeake Bay to variations in discharge of the Susquehanna River. *Estuaries.* 9(4A): 236-249, December, 1986.
3. Hicks, S.D. 1964. Tidal wave characteristics of Chesapeake Bay. *Chesapeake Science.* 5:103-113.
4. Elliott, A.J., and D.P. Wang. 1978. The effect of meteorological forcing on the Chesapeake Bay: the coupling between an estuarine system and its adjacent coastal waters, pp. 127-143. *In*: J.C.J. Nihoul (Ed.). *Hydrodynamics of Estuaries and Fjords.* Elsevier, Amsterdam, The Netherlands.
5. Pritchard, D.W. 1968. Chemical and Physical Oceanography of the Bay, pp. 49-74. *Proceedings of the Governor's Conference on Chesapeake Bay.* State of Maryland, September 12-13, 1968.
6. Seitz, R.C. 1971. Temperature and salinity distributions in vertical sections along the longitudinal axis and across the entrance of the Chesapeake Bay

(April, 1968 to March, 1969). Chesapeake Bay Institute, The Johns Hopkins University, Graphical Summary, Rep. No. 5, Ref. 71-7. Baltimore, MD.
7. Beaven, G.F. 1960. Temperature and salinity of surface waters at Solomons, Maryland, *Chesapeake Sci.* 1:2-11.
8. Pritchard, D.W. 1971. Chesapeake and Delaware Canal affects environment. Amer. Soc. Civil Eng. Nat. Water Res. Eng. Meeting, Phoenix, Arizona, 14 January, 1971.
9. Schubel, J.R. 1972. The physical and chemical conditions of the Chesapeake Bay, *J. Wash. Acad. Sci.,* 2:56.
10. U.S. Coast and Geodetic Survey. 1965. Surface water temperature and salinity, Atlantic Coast. Coast Geodetic Surv. Publ. 31-1.
11. Goodrich, D. 1985. On stratification and wind-induced mixing in the Chesapeake Bay. Ph.D. Dissertation, Marine Sciences Research Center, State University of New York, Stony Brook, N.Y.
12. Pritchard, D.W. In Press. Estuarine classification—A help or a hindrance. *In:* Bruce Neilson, Albert Kuo, John Brubacker (Eds.) *Estuarine Circulation.* The Humana Press, Clifton, N.J.
13. Boicourt, W.C. 1983. The detection and analysis of the lateral circulation in the Potomac River Estuary. Maryland Power Plant Siting Program and the Chesapeake Bay Institute, The Johns Hopkins University, Report PPRP-66, Baltimore, MD.
14. Pritchard, D.W., and J.H. Carpenter. 1960. Measurements of turbulent diffusion in estuarine and inshore waters, *Bull. Int. Assoc. Sci. Hydrol.* 20:37-50.
15. Owen, W. 1969. A study of the hydrography of the Patuxent River and its estuary. Chesapeake Bay Institute, The Johns Hopkins University, Tech. Rept. No. 53, Ref. 69-6, Baltimore, MD.
16. Haas, L.W. 1977. The effect of the spring-neap tidal cycle on the vertical salinity structure of the James, York, and Rappahannock Rivers, Virginia, U.S.A. *Est. and Coastal Mar. Sci.*, 5:485-496.
17. Hayward, D., C.S. Welch, and L.W. Haas. 1982. York River destratification: An estuary-subestuary interaction. *Science,* 216:1413-1414.
18. Ruzecki, E.P. and D.A. Evans. 1985. Temporal and spatial sequencing of destratification in a coastal plain estuary, pp. 368-389. *In:* Malcolm Bowman, Wm. Peterson and C.M. Yentch (Eds.). *Tidal mixing and plankton dynamics. Lecture notes on coastal and Estuarine studies. Vol. 17.* Springer-Verlag, New York, NY.
19. Hayward, D., L.W. Haas, J.D. Boon III, K.L. Webb, and K.D. Friedland. 1985. Emperical models of stratification variations in the York River Estuary, Virginia, USA, pp. 346-367. *In:* Malcolm Bowman, Wm. Peterson and C.M. Yentch (Eds.) *Tidal mixing and plankton dynamics. Lecture notes on coastal and Estuarine studies, Vol. 17.* Springer-Verlag, New York, NY.

20. Elliott, A.J. 1976. A study of the effect of meteorological forcing on the circulation in the Potomac estuary. Chesapeake Bay Institute, The Johns Hopkins University, Special Report No. 56, Ref. 76-9, Baltimore, MD.
21. Elliott, A.J. 1978. Observations of the meteorologically induced circulation in the Potomac estuary, *Est. and Coastal Mar. Sci.,* 6:285-299.
22. Elliott, A.J., D.P. Wang, and D.W. Pritchard. 1978. The circulation near the head of the Chesapeake Bay, *J. Mar. Res.,* 4:643-655.
23. Wang, D.P. and A.J. Elliott. 1978. Non-tidal variability in the Chesapeake Bay and Potomac River: Evidence for non-local forcing. *J. Phys. Oceanogr.,* 2:225-232.
24. Elliott, A.J., T.E. Hendrix. 1976. Intensive observations of the circulation in the Potomac estuary. Chesapeake Bay Institute, The Johns Hopkins University, Special Report No. 55, Ref. 76-8, Baltimore, MD.
25. Pritchard, D.W., and S.R. Rives. 1979. Physical hydrography and dispersion in a segment of the Chesapeake Bay adjacent to the Calvert Cliffs Nuclear Power Plant. Chesapeake Bay Institute, The Johns Hopkins University, Special Report No. 74, Baltimore, MD.
26. Vieira, M.E.C. 1983. A study of the non-tidal circulation in a segment in the middle reaches of the Chesapeake Bay, Ph.D. Thesis, The Johns Hopkins University, Baltimore, MD.
27. Pritchard, D.W., and M.E.C. Vieira. 1984. Vertical variations in residual current response to meteorological forcing in the mid-Chesapeake Bay, pp. 27-65. *In:* V.S. Kennedy (Ed.) *Estuary As A Filter*, Academic Press, New York, NY.
28. Vieira, M.E.C. 1985. Estimates of subtidal volume flux in mid Chesapeake Bay. *Estuarine, Coastal and Shelf Sci.,* 21:411-427.
29. Schubel, J.R. 1968. The turbidity maximum of the Chesapeake Bay. *Science* 161:1013-1015.

Contaminant Problems and Management of Living Chesapeake Bay Resources. Edited by S. K. Majumdar, L. W. Hall, Jr. and H. M. Austin. © 1987, The Pennsylvania Academy of Science.

Chapter Two
CHESAPEAKE BAY FISHERIES: AN OVERVIEW
HERBERT M. AUSTIN
Virginia Institute of Marine Sciences
College of William and Mary
Gloucester Point, VA 23062

ABSTRACT

The value of the marine resources of the Chesapeake Bay is second only to its value as a transportation corridor. The oyster, blue crab and striped bass or rockfish, along with the sailboat, epitomize our vision of the Bay. Nowhere else do such important renewable natural resources co-exist so closely to man's residential and industrial activities.

Over time, all natural resource distribution and abundance fluctuates in response to a normally fluctuating environment. Man's harvest adds an additional pressure, and in some cases recruitment levels cannot keep pace with consumer demand. In the Bay, pollutants, both intentional point source discharges, and unintentional non-point source run-off degrade the estuarine habitat and further reduce reproductive capabilities. Physical modification to the shoreline including bulkheading and filling, and daming of main tributaries such as the Susquehanna or James changes land run-off patterns thereby reducing the detrital energy source, and block spawning runs.

INTRODUCTION

Long term trends of declining recruitment for several Bay species, including the oyster and rockfish, suggest recruitment failure due to overharvesting and/or pollution. Research efforts by Bay scientists during this decade have focused on the causes of these declines. Stringent management measures have been dictated, often resulting in significant short-term loss of income to the harvesters. Further, regardless of the cause of the decline, either climate, overharvest, or pollution, management has been effected by controlling the harvest. The fishermen have paid the penalty, regardless of the offender.

A major result of the 1979-1983 EPA Bay Program was the development of cooperative programs between citizen groups, politicians, Federal and State agencies and academia. These cooperative efforts have produced an environment of change, one of awareness. This "Bay awareness", with its concern for the marine resources, and resulting state legislation and agency program funding, offers hope for the future of the fisheries.

The Resources

In spite of earlier state, Federal, and Congressional efforts it was not until the EPA Chesapeake Bay Study Report was released in 1983 that there was a focused, knowledgeable, public interest in the Bay and its problems. These efforts, sponsored by the Citizens for the Chesapeake Bay, culminated in December, 1983 with the tri-states' Governors' Conference! The 1984 General Assemblies of the Bay states had a clear mandate of the needed political-legislative reforms and the resources necessary to fulfill them. Both significant reforms and the necessary resources, the Bay Initiatives, were the result of the 1984 General Assemblies.

We have made more progress in Fisheries Management during the past two years (1984-1986) than in the past two decades. Inspite of mandates, reforms, initiatives, and policy statements the ecological cycles in the Bay occur slowly, more so than the political cycles. The public is impatient. Agencies are charged with rectifying more than 350 years of neglect. Activities in the area of resource management are under scrutiny, and unresolved issues still await action. Inevitably, when we speak of the Chesapeake Bay the conversation turns to the resources, their issues, status, and management. More than we, they are the inhabitants of the Bay, and we, as the stewards, must consider their needs over our own.

What are the resources? Several shall be addressed in this chapter, each as a fishery. A fishery is both the resource and the harvester; we cannot separate them. The recreational component ranges from the child on a pier to those running sophisticated high-speed boats along the main stem of the Bay and into the ocean in search of large gamefish. The commercial fin-fishery ranges from the single man in a small boat on a tributary creek fishing a gill-net, to the pound net on major tributaries of the bay, to the multimillion dollar menhaden fishery. The recreational shellfishery for oyster, clam, and in particular the bluecrab is a large, but poorly documented catch.

In many respects the waterman with his gill-net, shaft tongs and crab pots, or the Maryland skipjack is the epitomy of the Chesapeake Bay. The oyster and blue crab, captured so many times in print, on canvass, and on the plate are to many, the Bay.

DISCUSSION

The Recreational Fisheries

A description of the Bay's recreational fisheries can be approached from

Chesapeake Bay Fisheries: An Overview 35

FIGURE 1. A headboat, capable of taking up to 25 recreational fishermen.

FIGURE 2. Small (13-25') boats are popular in the Chesapeake Bay for bluefish (shown here), weakfish, croaker, spot and flounder.

several perspectives, a species by species discussion, age and economic strata, gear type, location along the Bay, or access. The latter provides the best integration of the above perspectives.

Unfortunately, compared to the total available shoreline, there is little public shore access where fishing is permitted. An individual must generally find a public fishing pier, ranging from a small wooden structure to the large sturdy pier on the first island of the Bay Bridge Tunnel. The James River Fishing Pier, the old Highway 17 bridge, is billed as the world's longest fishing pier. Private fishing piers generally charge by the day, and provide bait and tackle for an additional cost. Piers are ideal for families with children, and those who want convenient access to fishing at a minimal expense. However, there is also the "hardcore" pier fisherman that even when given the opportunity to fish from a boat would not give up his or her pier. The method generally employed is that of bottom fishing with dead or live bait. The catch is usually spot, croaker, weakfish, fluke, and bluefish. Skillful anglers, using live bait, can take large bluefish or cobia.

The headboat fishery (Fig. 1) provides half or full day trips for bottom fishing, and bluefish or weakfish.[2] Headboats, like a bus, are generally first-come-first-serve with no reservations. Rods, reels, tackle, and bait are available. The fishing is generally better than from the fixed pier, and the skilled boat captain can put his fishermen on the day's hot spot. Generally the bottom fishing trips take spot, croaker, flounder, and small bluefish and weakfish. The trips that target bluefish or weakfish naturally take larger fish. Mid-Bay (mouth of the Rappahannock and Potomac Rivers) headboat fishermen took roughly 50% as many bluefish in the late 1970's as the commercial fishermen[2] demonstrating the potential impact on the stock by the occasional fisherman. Headboat customers are often out-of-state, and many are from the middle income bracket.

The Chesapeake Bay supports one of the world's largest fleets of small boats ranging from 10' aluminum johnboats with a 2 horse power outboard engine to 20-25' center console fiberglass "fishing machines" equipped with the latest fish finding and navigational electronics and powered by up to 250 horse power inboard/outboard engines (Fig. 2). In some respects these fishermen constitute the "phantom fleet" as their actual number and catch have been undocumented. Maryland and Virginia have, since 1985, participated in the National Marine Fisheries Service Marine Recreational Fishing Survey which will help document the magnitude of this fishery. Maryland has also, in 1979 and 1980, documented the level of participation and catch of the recreational fisheries in Maryland's waters. Maryland, in 1985, instituted a Chesapeake Bay saltwater anglers' license. It is anticipated that this will further help the Maryland fisheries manager keep track of the magnitude of the fleet and catch.

Fishing, when accomplished by anchoring or drifting for bottom fish along the western Bay and tributaries, results in catches of spot, croaker, flounder, weakfish, smaller bluefish and occasionally cobia. Many troll throughout the

Bay for large bluefish during spring and early summer, or cast around the shore and fixed structures for weakfish and rockfish during spring and fall. Deep fishing from small boats around the Bay Bridge-Tunnel at the mouth of the Bay or the Kents Narrows Bridge at Annapolis involves heavy gear, wire lines, and skilled boatsmanship during both day and night fishing trips. The large weakfish, bluefish, flounder, and tautog that are taken make these structures popular; and the Bridge-Tunnel in particular is a favorite late fall site for taking large in-migrating rockfish. Cobia, generally taken drifting a live bait, are also a popular summer fish.

Few people live in homes that have shoreline with a dock. Fortunately, for the boat owners, most boats in this size range are trailerable. They are maintained in backyards, garages, and at marinas in stacked storage. Ownership of a 13-17' outboard powered fiberglass boat is identified with no particular economic class. The very large well equipped boats are more likely to come from those with large spendable surplus incomes, but a 13' outboard may be owned by one on unemployment, a school teacher, or an attorney. Virginians pay a personal property tax on their vessels, regardless of size. The large number of boats in the state must generate indeed, a sizeable revenue for the state.

The larger boats, either a 4-6 person charter boat in the 30-50' range; or the privately owned (Fig. 3) fishing boats, 27-34', constitute a sizable fleet. Boats

FIGURE 3. Boats in the 25-50' ranges are fished in the Bay and because of their speed and size can be used to run offshore also.

FIGURE 4. Menhaden purse seiner. This particular vessel is an older, pre-hydraulic block version.

FIGURE 5. Typical Chesapeake Bay "dead rise". This boat is outfitted for claming with a patent tong.

in this size range are generally not trailerable, and are moored at marinas or private docks. Bottom fishing from these boats is generally only attempted after all efforts towards trolling or casting have failed to yield large bluefish, weakfish, or rockfish. Many spend most time fishing the coastal or offshore bluewater staying in the Bay when weather or time precludes an offshore trip.

An annual Saltwater Fishing Tournament in Virginia provides citations for large fish taken, or in the case of some species (marlin and rockfish), for releases.

Artificial reefs, popular in the ocean as a means of attracting and holding fish over flat sandy bottoms, are attracting an increasingly larger following of anglers in the lower higher salinity region of the Bay.

The Commercial Fisheries

Commercial fishing effort in the Bay can be divided into three major categories by vessel size and man-power. These include the large 150-200' mechanized menhaden purse seiners (Fig. 4) with power blocks, a crew of 10-15, and spotter aircraft; the medium-sized 25-45', and the small (18-43') 1-2 man operation generally in the traditional Chesapeake Bay deadrise (Fig. 5). It is this 1-2 man operation that conjures the image of the "waterman". While many operate a deadrise, the center console outboard driven fiberglass boat in the 17-19' range is becoming increasing popular due to the low maintenance costs. Inspite of the more efficient fiberglass vessels' operation, several Maryland oystermen still operate the classical Chesapeake Bay skipjack (Fig. 6) to harvest oysters,

FIGURE 6. Maryland skipjack.

FIGURE 7. Illustration of hand tonging for oyster. Note the "cull board" on the deck, piled with oysters.

a design in operation for over 100 years.[3] Most skipjacks, in fact, are over 60 years old.

The menhaden vessels in the Bay, fishing only during May through late November, in Virginia's waters, specialize in the taking of menhaden. The watermen on the other hand, in their deadrise fish year-round. During fall and winter they hand tong for market oyster; and then during winter in Virginia, they dredge for the blue crab. The crab fishery switches to potting during spring and summer; and the oystermen to tonging seed oyster after the market oyster season closes. The use of the patent tong in Virginia is limited to hard clams. Gill and pound nets are fished year-round for shad, river herrings, croaker, spot, flounder, bluefish, weakfish, and the rockfish. In recent years the pound net has also taken a substantial catch of menhaden, generally used for crab bait.

THE SHELLFISH

Oyster

The life history and details of the fisheries for the oyster, *Crassostrea virginica* are found in Chapter 9. Generally, the fishery is divided between the "public" and "private" sectors.

Both Maryland and Virginia have public and private oyster beds. Some of Virginia's public beds are "seed beds" where only seed oyster may be harvested for repleation to other public beds, or for sale to lease holders. Public beds are in the common trust, administered by the state, and may be harvested by any citizen. Only hand tongs may be used. Private beds are leased from the state and are repleated and harvested by the leasee.

In Virginia there are about 243,000 acres of public grounds, most of this was

FIGURE 8. Small Maryland deadrise with hand tonger.

surveyed and set aside in 1894 and named the Baylor Grounds after U.S. Navy Lieutenent James Baylor, the surveyor.[5] In 1985, some 111,000 acres were leased from the state. Maryland currently has some 300,000 acres in the public trust, and about 9000 acres under private lease.

Biologically, the oyster fisheries may be separated by size. This separation includes "seed" oyster between ½ and 2½ inches (1-3 years old) that are harvested for transplant from seed beds to either private or public oyster grounds; and market oyster over 3 inches (3-5 years old). In Virginia, seed oyster can only be taken by hand tong (Fig. 7 and 8). Market oyster too are generally taken by hand tong in Virginia, but in some public areas can be harvested by patent tong or dredge in certain management areas (e.g. the Pocomoke-Tangier Sound Management Area). The skipjack may dredge for oyster in Maryland. The vessels must use sail Wednesday through Saturday, and may dredge under power on Monday and Tuesday.

The sizes and terminology of graded oyster are often confusing. The following table identifies average sizes with counts.[4]

TABLE 1

Spat	< ½ inch	1000's/Bushel
Yearling	½-1 inch	1500-2000/Bushel
Small	1-2 inches	1000-1200/Bushel
Seed	up to 2½ inches	500-1000/Bushel
Market	>3 inches	250-280/Bushel
SHUCKED		
Standards		300 and up/Gallon
Selects		210-300/Gallon
Extra selects		160-210/Gallon
Counts		160 or less/Gallon

Clams

Both Maryland and Virginia support a hard clam, *Venus mercinaria* fisheries, and Maryland a viable soft clam, *Mya areneria* fishery. Most hard clams are taken from unassigned ground, those bottom acres not in the Public Oyster Grounds nor leased. The patent tong (Fig. 9-10) is the primary gear employed in both states although rakes and dredges are also used. Maryland permits the use of a hydraulically operated patent tong; Virginia, on the other hand, limits the use of a patent tong to the simple mechanical tong. "Signing" is a rapidly disappearing method, that of feeling the clams with one's bare feet, and picking them up and placing them in a basket with ones toes. The hydraulic escalator dredge (Fig. 11) is used in Maryland for both hard and soft clams, but is illegal in Virginia for the taking of clams. Few soft clams are harvested in Virginia.

Blue crab

The blue crab, *Callinectis sapidus* is one of the most popular crustaceans along the east coast, and certainly the number one species in the Chesapeake Bay. A more detailed discussion of the life history and fisheries, including legislation and regulations is found in Chapter 10. At the turn of the century, and until approximately 1928, when a galvanized wire pot was introduced, the crab trot line was the most widely used gear (Fig. 12). This was replaced during the mid-1940's by the currently used crab pot (Fig. 13-14). A waterman, fishing an average of 300-400 pots a day can take 50 baskets a trip. In Virginia, a winter dredge fishery for crabs has existed since 1900. This fishery, from December through March takes over-wintering female crabs that have migrated toward the mouth of the Bay. The crabs, which bury in the mud when temperatures drop below 47F (8.3C) are dug out with a large dredge.

The peeler crab, a crab about to molt, is sought after both as anglers' fish bait and for the soft crab that it is about to become. Peelers are generally taken in a "peeler pound" (Fig. 15) and held in floats (Fig. 16) until they shed, at which time they are sold as soft shell crabs. During recent years recirculating holding tanks with flow thru troughs for holding the peelers on land have become popular.

THE FINFISH

As with the recreational fisheries, the commercial fisheries are discussed by fishery (i.e. gill net, pound net, purse seine) rather than by species. In the case

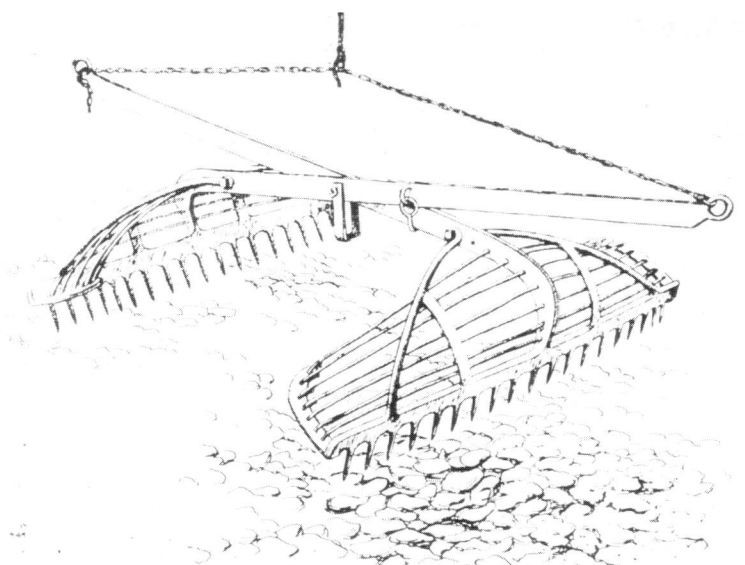

FIGURE 9. Illustration of patent tongs.

of the menhaden, the gear is species specific. The gill net is selective too, but only in that by regulating the mesh size and location of the stand is the catch chosen. The pound net however, is indiscriminate as it takes most species.

Menhaden

The menhaden (Fig. 17) is taken primarily for rendering into oil and fish meal, and to a lesser degree for crab pot bait. As such it is not considered a food fish. Menhaden are taken by purse seine for meal, and by pound net for crab bait. The menhaden purse seine fishery in the Chesapeake Bay goes back to the mid-1800's, prior to that they were taken by beach seine.[6] Until the invention of the hydraulic power block nets were hauled by hand. As many as 25-30 men were required to haul the net (Fig. 18). Prior the mid-1980's several hundred million pounds were taken annually from the Bay, but due to overharvesting, the subsequent sringent management regulations, and competition with the

FIGURE 10. Virginia waterman patent tonging for clams in the York River.

Chesapeake Bay Fisheries: An Overview 45

FIGURE 11. Hydraulic escalator dredge in Maryland. These are illegal in Virginia.

Figure 12. Crab trot-line, pre-1950's.

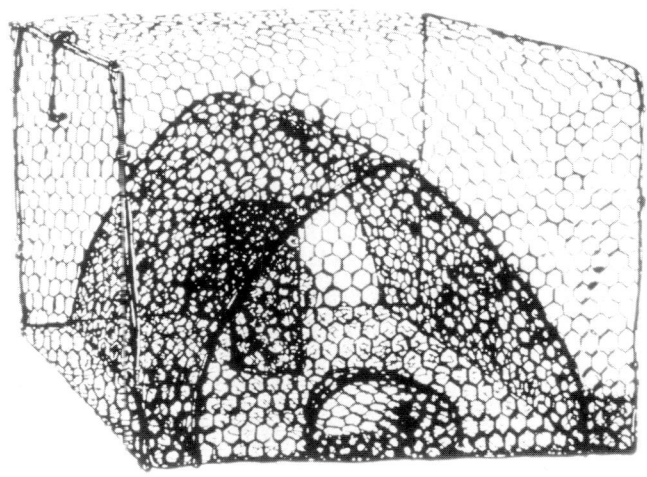

FIGURE 13. Illustration of an early version of the popular chicken wire crab pot.

FIGURE 14. Chesapeake Bay deadrise off Buckroe Beach, Virginia with a load of crab pots.

Chesapeake Bay Fsheries: An Overview 47

FIGURE 15. Peeler crab pound net.

FIGURE 16. Holding pen for peeler crabs, Tangier Island, Virginia.

FIGURE 17. Atlantic menhaden, *Brevoortia tyrannus*.

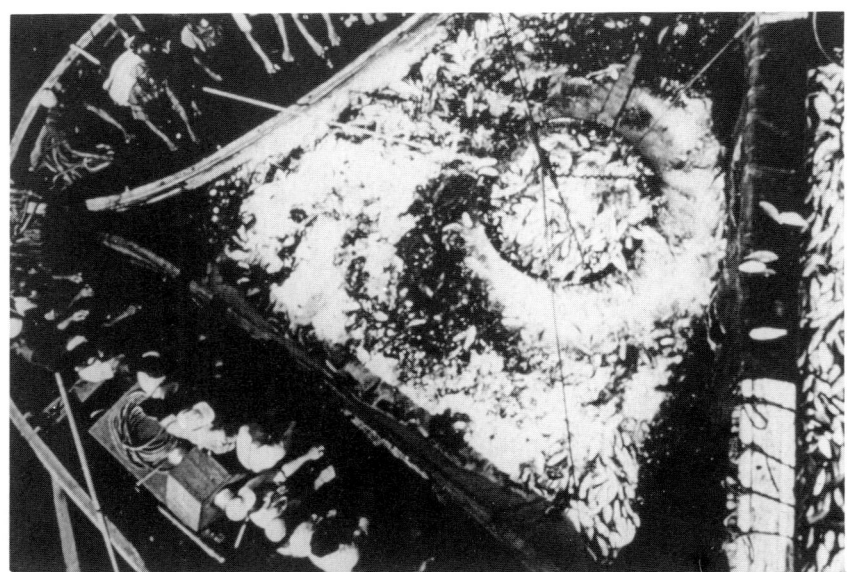

FIGURE 18. Menhaden fishing effort, pre-power block era. Note the labor intensive effort.

Chesapeake Bay Fisheries: An Overview 49

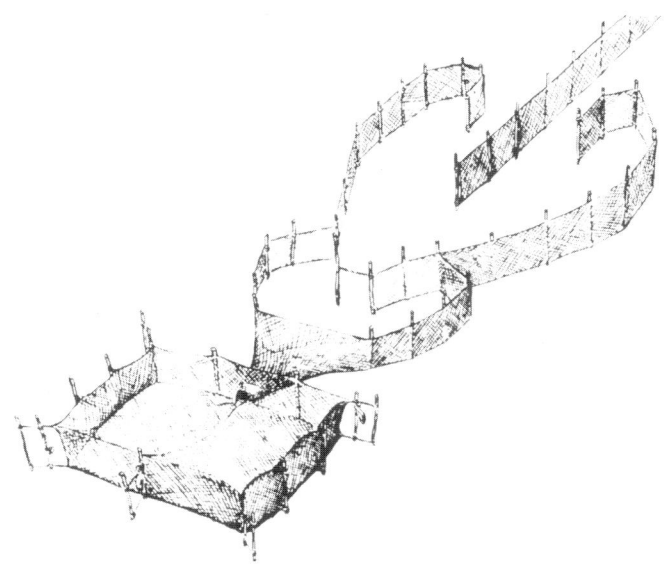

FIGURE 19. Illustration of configuration of pound net.

FIGURE 20. Crew lifting a pound net, York River, Virginia.

soybean industry, one of the Virginia fish meal companies ceased fishing in the Bay during 1986. This company later reopened to provide meal to the Virginian eastern shore poultry industry. The other company never fished in the Bay, taking their fish from the ocean.

Pound Net

The pound net (Fig. 19) is a fixed fishing structure that intercepts fish as they migrate up or down river and Bay. They were first used in the fisheries of New England around 1850, and from there spread southward to Long Island by George Snediker in 1855. It was Snediker that introduced the Bay watermen to the net shortly thereafter, and in 1858 Captain Henry Fitzgerald erected the first pound net in the Bay. By 1880 "...all available sites were taken up."[7] The net consists of the hedge or leader, the long net affixed to poles that runs from near-shore towards the head; the "heart", "bay" or false pound, the heart shaped area seaward of the hedge that concentrates and directs the fish; and the pound or head of the net, the actual impounding area of the net where fish are trapped. Fish migrating along the shore, encountering the hedge, turn and head towards deeper water. They are directed into the pound by the false pound or heart. In the pound they are "corralled" until removed and sorted or culled by the watermen (Fig. 20). A large advantage is that the fish are retained uninjured so that undersized or illegal species can be released.

FIGURE 21. Typical pound net catch of weakfish, bluefish and spot.

Chesapeake Bay Fisheries: An Overview 51

FIGURE 22. Illustration of staked gill net.

FIGURE 23. Illustration of anchor or sink gill net. Striking the water to scare fish into the net is no longer legal.

The catch is seasonal depending upon the migratory habits of the various species. During late winter (February) and through the spring the rockfish or striped bass, shad, and river herrings have been the traditional species that carried the fishery. Menhaden are taken during early spring and are often frozen awaiting the summer crab pot fishery. Bluefish and weakfish also move in during spring followed by spot and croaker. These four species make up the bulk of the food fish catch through the summer (Fig. 21).

Gill net

Gill nets fish by entangling the gills of a fish in the meshes. Old nets were made out of cotton, later nylon twine, but recently monofilament line has been used. Monofilament does not decompose, consequently the State of Maryland has outlawed its use to prevent lost nets from continuing to fish. Monofilament is still legal in Virginia. There are three major types of gill nets. The staked gill net (Fig. 22) fishes the near surface waters, and like the pound net is a fixed structure. The anchor gill net (Fig. 23) is mobile, fished wherever the waterman wants, and is held in place by anchors. It generally fishes the bottom waters. The third type of net is the drift gill net, which is generally fished on the slack water but allowed to drift with the current. It fishes the near surface waters. A fourth net, the anchor-floating net is increasing in popularity. This net is fished anywhere from the surface to the bottom, depending upon water temperature.

CONCLUSION

Mesh size determines the size and species of fish taken. Staked and drift gill nets are most commonly used in early spring to take shad (mesh $4\frac{7}{8}$ - $5\frac{1}{4}$"), river herring (mesh $2\frac{7}{8}$ - $3\frac{1}{4}$"), and rockfish (mesh 4 - 8"), whereas anchored nets are used during late spring through fall to take deeper fish such as spot, croaker, and weakfish (mesh $2\frac{7}{8}$ - 4"). During the late 1970's, and into the early 1980's, there was a move away from using staked gill net stands and towards the use of drift and anchored nets. This was reported to be for economic reasons.[8]

The Bay has, in recent years, received a lot of "bad press". While I am delighted at the increasing level of lay understanding of the Bay system and its problems, and the considerable legislative focus since 1984, the Bay is by no means dead or dying. It is still a strong viable estuarine system that is capable of supporting a substantial *well managed* multi-species fishery for oyster, blue crab, and many species of finfish. Well managed here defined as the ability of a state agency (VMRC or DNR) to react quickly, independently, and in a well informed manner to meet the rapidly changing stocks of the fisheries. The management agencies cannot, and must not, wait for their general assemblies to act. Further, the fishery must be managed; and as indicated early on, the fishery is the stock, the harvester, and even the habitat.

LITERATURE CITED

1. Anonymous. 1983. Choices for the Chesapeake: An Action Agenda, Results and Recommendations of the 1983 Chesapeake Bay Governors' Conference, Fairfax, Virginia, December, 1983.
2. Marshall, A.R. and J.A. Lucy. 1985. Virginia's Charter and Head Boat Fishery, Analysis of catch and socioeconomic impacts, SRAMSOE No. 253, Virginia Institute of Marine Science, Gloucester Point, Va., 90 pages.
3. Tilp, F., 1982. *The Chesapeake Bay of Yore,* Ches. Bay Foundation, Inc., 148 pages.
4. Haven, D.S. and H.M. Austin, 1985. The Virginian Oyster, VMRR-85-4, Virginia Institute of Marine Science, Gloucester Point, Va. 7 pages.
5. Haven, D.S., W.J. Hargis, and P.C. Kendall, 1978. They Oyster Industry of Virginia: Its status, problems and promise. VIMS Spec. Papers in Marine Sci. No. 4, 1024 pages.
6. Frye, J. 1978. *The Men All Singing,* Donning Publ. Norfolk, Va. 242 pages.
7. Reid, G.K., 1955. The Pound-Net fishery in Virginia, History, Gear Description and Catch, Comm. Fish. Review, 17(5):1-15.
8. Austin, H.M. 1986. Status of the 1985 striped bass fisheries in Virginia after implementation of the 1985 Amendment III to the 1981 ASMFC Interstate Management Plan for striped bass. Mar. Res. Rept. 86-3.

Chapter Three

A SOCIO-ECONOMIC OVERVIEW OF THE CHESAPEAKE BAY FISHERIES

JAMES E. KIRKLEY

College of William and Mary
Virginia Institute of Marine Science
Schools of Marine Science and Business Administration
Gloucester Point, Virginia 23062

ABSTRACT

The marine resources of the Chesapeake Bay are believed to provide substantial benefits to residents of the States of Maryland and Virginia. However, the possibility of overfishing and degradation of the marine environment seriously jeopardize the possible benefits. In this section, a brief overview of the economic importance and characteristics of the Chesapeake Bay marine resources is presented. The potential for economic losses are discussed relative to observed economic values.

The resource base of the Chesapeake Bay is important to the economic well being and community structure of many Maryland and Virginia coastal communities. The bay, once important for providing subsistence and a basis for a barter economy, now provides a wealth of commercial and recreational opportunities. Unfortunately, a detailed description and analysis of the importance of the bay is not possible because of inadequate data.

In this section, a brief overview of the socio-economic characteristics of the Chesapeake Bay fisheries is presented. Included in the summary is information on landings, value, employment, and general economic conditions. However, inadequate data preclude a comprehensive overview and limit the period of time over which the characteristics may be summarized.

ECONOMIC IMPORTANCE OF COMMERCIAL FISHERIES

The Commercial Chesapeake Bay fisheries are primarily comprised of many small scale operators.[a] Vessel size is between 11 and 200 feet; however, most of

---

[a]The description of the Chesapeake Bay fisheries is based on a 1985 survey of licensed Virginia Fishermen, Maryland and Virginia State reports, and selected issues of Fishery Statistics of the United States. Also, inadequate data prevent a detailed discussion of the menhaden fishery. Thus, the menhaden fleet is only briefly discussed.

the vessels are between 20 and 40 feet in length (table 1). In comparison, the minimum size of offshore vessels is approximately fifty feet. The traditional fisheries are blue crabs, oysters, hard and soft clams, and various finfish (fluke, sea trout, scup, sea bass, menhaden, spot, and striped bass).

In 1985, approximately sixty percent of the fishermen held crab-pot licenses, and forty-four percent were licensed to fish with a gill net. In terms of species, approximately 55 percent of the fishermen are licensed to harvest oysters, and 68 percent are licensed to harvest crabs. In addition, approximately 80 percent of the fishermen were licensed to fish with more than one gear type. The types of licenses most frequently held by fishermen were various crab gear, oyster gear, crab pot and gill net, and oyster and gill net gear

In 1982, 222.3 million pounds of fish with an ex-vessel or first sale value of $94.6 million were landed in Maryland and Virginia.[b] Landings of fish from the Chesapeake Bay accounted for approximately 64 percent of the total landed value. In comparison to the United States, Chesapeake Bay landings accounted for 2 percent of the total landed value in 1982.

Since 1975, landings have declined while the ex-vessel value has increased (Table 2). The value in 1982 represents a 50 percent increase over the nominal (unadjusted for inflation) value of 1975. The major commercial species, in terms of value, have traditionally been menhaden, blue crabs hard and soft clams, and oysters.

In 1985, Chesapeake Bay fishermen derived approximately 71 percent of their total income from fishing. An estimated fifty-six percent of Bay fishermen earned more than fifty percent of their income from fishing. Thus, slightly less than half of the fishermen are part-time while more than half are full-time fishermen.[c] The U.S. Department of Commerce estimates that annual average earnings were between $3,000 and $4,000 in 1983. A 1985 survey by the Virginia

TABLE 1

*Selected characteristics of the Chesapeake Bay Commercial Fishing Fleet**

Characteristics	Minimum	Maximum	Mean
Vessel length (feet)	11	64	30
Vessel age (years)	new (0)	83	17
Vessel cost ($)	150	30000	6682
Engine cost ($)	150	21800	3234
Age of fishermen (years)	16	79	46
Years of fishing (years)	1	60	19
Percentage of income from fishing	1	100	71

*Summary statistics are based on a 1985 survey of Virginia fishermen. The survey was conducted by Virginia Polytechnic Institute and Virginia Institute of Marine Science. The results are preliminary. The data may not be indicative of the Maryland commercial fishing fleet. Menhaden vessels are excluded from the summary.

[b]Chesapeake Bay landings are defined in this report to be all landings other than ocean landings. Chesapeake Bay landings data are not available for all years.

Polytechnic Institute and the Virginia Institute of Marine Science found annual earnings to be approximately $4,500 in 1985.

Typically, the economic importance of an industry is measured by total income generated. However, the necessary information for assessing the economic importance of the Chesapeake Bay fisheries is not available.[d] Lipton[1] in a 1986 report on the status of Maryland's seafood industry estimated the contribution of fisheries to Maryland's economy to be 3.5 times the value of processed products. Briggs et al.[2] estimated a harvesting sector multiplier of 1.39 for Maine. Norton et al.[3] estimated a multiplier of 2.03 based on commercial and recreational expenditures. Grigalunas[4] estimated the multiplier for the Southern New England harvesting sector to be between 3.30 and 3.56. Cox[5] in a 1986 study estimated the total multiplier for forestry and fishery products in Virginia to be 1.9489. If the multiplier of Cox is applied to the 1982 ex-vessel value of Maryland and Virginia Chesapeake Bay landings, the total economic impact of the harvesting sector was approximately $118.6 million.

The economic impact or contribution of the Bay fisheries, based on Cox's analysis, appears to be highest for the fisheries sector, repair and maintenance business, food products, wholesale and retail trade sectors, and households in the forms of wages and salaries. That is, production of seafood primarily benefits these sectors of the economies of Maryland and Virginia. Alternatively, an increase in expenditures by the fishing industry yields increased earnings primarily for the above listed industries or sectors.

The Chesapeake Bay commercial fisheries are also quite important in terms of employment generated. The exact number of people currently employed in harvesting, processing, and related activities is not available. The most recent official published data on number of fishermen for the Chesapeake region indicate there were approximately 28,000 fishermen in Maryland and Virginia in 1977 (National Marine Fisheries Service).[6] More than seventy percent were Maryland fishermen. In contrast, the Bureau of Census[7] reported 7,500 individuals as full time employees in agricultural services, forestry, and fishing in Maryland in 1984. Fisheries of the United States for 1985 reports 7,704 employees in processing and wholesaling during 1984 for Maryland and Virginia.

Despite the limitations of available data for determining the employment generated by commercial fisheries, the fishery resources are believed to be important in terms of employment opportunities. For example, Norton et al. estimated that one new job was created for every $20,000 of expenditures on only striped bass commercial and recreational opportunities. Alternatively, one job was created for every $40,000 of total output.

---

[c] The National Marine Fisheries Service defines a full-time fisherman as one who earns more than fifty percent of their income from fishing.

[d] Typically, an output multiplier is obtained from an input-output analysis and multiplied by an industry's direct contribution to the economy to determine the total impact.

MAJOR COMMERCIAL SPECIES OF CHESAPEAKE BAY

The Chesapeake Bay is characterized by a wealth of living marine resources. More than fifty species of fish are commercially harvested. However, four species account for most of the income or gross revenues of the Chesapeake Bay fisheries (Table 3). Menhaden accounts for 75 percent of the landings and approximately 24 percent of the ex-vessel value. Excluding menhaden, more than 80 percent of the annual value is from the harvesting of blue crabs, clams, and oysters; all of which are shellfish.

There are differences in the Bay fisheries of the two states. First, there is no major menhaden fishery in Maryland; Virginia fishermen harvest more than 95 percent of the reported harvest of menhaden in the two states. Maryland

TABLE 2

Maryland and Virginia commercial landings and value[a]

Year	Maryland		Virginia	
	Bay	Total	Bay[b]	Total[b]
	Thousands of pounds			
1975	53662.9	64317	73460.9	172875
1976	45669.0	58094	59081.5	134228
1977	47767.5	62033	143060.0	207803
1978	45312.0	59726	61048.2	138676
1979	49649.9	66283	62496.3	139347
1980	55484.2	79571	96908.5	135566
1981	91194.6	115115	89942.6	113977
1982	68752.6	100478	86145.1	121827
1983[c]		90359		130839
1984[c]		89301		117882
1985[c]		91931		122658
	Thousands of dollars			
1975	20254.8	22898	13742.7	26455
1976	25057.5	30808	15442.1	31739
1977	24404.0	31429	16092.7	36368
1978	27037.5	33557	17373.5	53441
1979	30134.3	36945	20558.7	62531
1980	33625.9	44658	22570.0	59292
1981	45591.3	56640	21575.7	52680
1982	39328.2	51438	21531.4	43170
1983[c]		45497		56909
1984[c]		54979		58957
1985[c]		47418		76535

[a]National Marine Fisheries Service data summarized by Clooney Stagg, University of Maryland, Chesapeake Bay Research Laboratory.

[b]Menhaden landings and values are excluded from data summaries.

[c]Maryland and Virginia Chesapeake Bay landings and value data are not available or are inconsistent with previous data.

TABLE 3

Ratio of species ex-vessel value to total Chesapeake Bay landed value

Year	Species									
	alewives	eel	sea trout	shad	spot	striped bass	white perch	blue crab	clams	oysters
					Percent of ex-vessel value					
1970	1.7	1.1	1.0	1.6	2.1	3.9	1.1	20.3	10.4	54.4
1971	1.0	1.3	1.0	1.0	.1	3.7	1.0	24.1	12.7	52.1
1972	1.1	.7	1.0	1.3	1.1	4.8	.1	26.3	6.5	53.7
1973	1.1	.5	1.8	1.6	1.0	7.1	.7	25.4	4.2	53.0
1974	1.4	1.4	1.1	.8	1.0	3.8	.4	27.0	8.1	51.3
1975	1.3	1.5	1.1	1.0	.6	4.8	.4	28.4	5.5	51.2
1976	.4	.6	1.2	.7	.5	3.6	.4	27.0	8.8	52.8
1977	.1	.4	1.5	1.3	.8	3.7	.4	32.7	7.9	47.4

Source of data: National Marine Fisheries Service, Fisheries Statistics of the United States, Selected issues (1970-1977).

fishermen have traditionally harvested the majority of striped bass and white perch; the harvesting of striped bass has been severely restricted in the past few years. Virginia fishermen harvest considerably more sea trout. In the blue crab fishery, a larger volume of hard blue crabs in landed by Virginia fishermen whereas Maryland fishermen have higher reported landings of soft and peeler blue crabs. The fishery for clams is also quite different; soft clams are primarily landed in Maryland and mostly hard clams are landed in Virginia. Annual landings of oysters in Maryland are more than double the landings of Virginia.

The economic importance of each Bay fishery has not been examined. However, Norton et al. did examine the economic importance of striped bass commercial and recreational activity. They noted that in 1980, $24 million was spent by commercial and recreational fishermen on harvesting striped bass in the Chesapeake Bay. The associated value of total output was $48.7 million. Corresponding employment was 1,214 jobs. In comparison, the Virginia Marine Resources Commission reports more than 8,000 fishermen in 1980; this excludes the number of fishermen in the Menhaden fishery.

Fishing occurs in several counties of Maryland and Virginia. However, Dorchester County in Maryland and Northumberland County in Virginia are two of the major counties in terms of Chesapeake Bay fishing activity. Northumberland county, though, is home to the menhaden fleet which accounts for its importance. More than 15 percent of the total landed value of all Maryland and Virginia fisheries are menhaden landings in Northumberland County.

In 1983, personal income per capital in Northumberland was $9,959 (Overman[8]). Average income per capita for individuals employed in agriculture, forestry, and fishing was approximately $12,000 (Virginia Employment Commission[9]). In comparison, reported per capita income for Virginia was $12,122 in 1983. The Bureau of Census' County Business Patterns provide statistics for Maryland which are comparable to Virginia. The per capita income of individuals employed in agriculture, forestry, and fisheries was $12,600 in 1983; it was $16,700 per capita for all individuals employed in Maryland.

THE RECREATIONAL FISHERIES

Although the commercial fisheries of the Chesapeake Bay are important to the coastal economies of Maryland and Virginia, the recreational fisheries are also quite important. The recreational fisheries are perhaps more important in terms of generating income and benefits. However, inadequate information prohibits documentation of the importance of recreational fisheries.

In the Chesapeake Bay, several species of fish are recreationally harvested. Bluefish, white perch, sea trout, croaker, spot, flounder, and striped bass are the more common species sought by recreational anglers. In past years, striped bass was the major recreational species. There also is likely to be substantial harvesting of blue crabs by recreational anglers although the number of anglers and the catch has not been documented.

Despite the problem that necessary data are not available to determine the value of recreational fishing, fishing for pleasure is likely to be very important to coastal economies. Murray and Lucy[10] estimated the number of registered power boats in Virginia to be 139,734 in 1980. More than 90 percent were less than 26 feet in length which is a typical size range for recreational fishing boats. Murray and Lucy also indicated that the average boater uses a boat for recreational fishing about 44 percent of the time. However, recreational fishing also occurs from beaches, piers, jetties, charter boats, and party boats. Thus, it would appear that recreational fishing may be quite important to coastal economies.

Results presented in the study by Norton et al. illustrate the potential economic importance of recreational fisheries. They estimated that Chesapeake Bay recreational anglers spent $22,861,000 in 1979 trying to catch striped bass. In comparison, the total value of commercial landings of all species in the Chesapeake Bay was $72.8 million in 1979. Norton et al. also estimated that the economic value, total benefits less total costs, of the recreational fishery for striped bass in the Chesapeake Bay was approximately double the economic value of the commercial fishery.

DECLINING RESOURCES AND ENVIRONMENTAL DEGRADATION

Declining resources and overfishing are inherent results of open-access common property resources such as fisheries. However, industrial expansion, manufacturing, and increased standards of living in the last twenty years have brought a new source of pressure on fishery resources. Pollution or man's

TABLE 4

Annual landings and maximum sustainable yield of selected Chesapeake Bay species

Species	Year								Maximum sustainable yield[a]
	1970	1971	1972	1973	1974	1975	1976	1977	
	-- Thousand of pounds --								
Shad	5134	2460	3006	3018	1741	1279	803	1487	4000
Alewives	21110	13079	12141	11300	14730	12078	4363	1491	25000
Striped bass	4702	3683	5181	7323	5331	3934	2752	2620	60000
Menhaden	392712	303103	513309	435119	354144	278049	400045	489981	300000
Spot	5900	468	2833	2260	2054	1449	1029	1590	3000
Croaker	121	210	384	1007	971	3015	3927	7095	n/a
Sea trout	2025	2058	2140	4394	2439	3031	3234	3632	2000
Fluke	154	140	162	453	425	426	306	289	3000
Bluefish	641	660	1125	2839	3315	3123	4437	3357	n/a
White perch	1925	1965	1388	1009	677	700	607	831	3000
Yellow perch	110	88	107	37	42	38	24	17	250
Oysters	23742	24690	23455	24799	24424	21882	20772	17519	30000
Blue crabs	67564	74570	71362	55738	63039	56809	46152	57231	65000
Soft clams	6221	5986	1950	668	2099	1246	1751	1654	60000
Hard clams	468	938	868	747	948	723	838	399	n/a

[a]Estimated maximum sustainable yields are taken from U.S. Army, Corps of Engineers, Future Chesapeake Bay Conditions Report (1978).

Source of Data: Selected issues of Fishery Statistics of the United States (1970-1977).

degradation of the natural marine environment may very well reduce the reproductive capabilities of some species of fish and inflict high mortality on other marine species. In some instances, neither the reproductive capability nor stock size changes as a result of pollution, but severe economic hardships and losses still occur because harvesting in prohibited. The economic effects can also extend well beyond the local fishery directly affected as a result of consumer resistance to a species which is contaminated.

OVERFISHING

Overfishing is a term with many definitions. It may be defined in biological or economic terms. It may imply recruitment failure, an inadequate age class distribution, or simply more fish are being removed from a fish stock than are being added to the stock. In economic terms, overfishing usually implies that more resources are used to harvest a given level of fish than are necessary. Consequently, the cost of harvesting is higher than it need be and profit approaches the value of zero.

For the Chesapeake Bay fisheries, many species are believed to be biologically and economically overfished. Irrefutable scientific evidence of either case, however, does not appear to be available. Reductions in landings are most frequently cited as indicators of overfishing in the absence of survey abundance indices. However, commercial landings data also reflect economic responses, and thus, may be inadequate indicators of the fish stocks.

In a 1978 study, "Chesapeake Bay Future Conditions Report," by the U.S. Army Corps of Engineers,[1] nine species were identified as being biologically overfished based on catch statistics, survey measures of abundance, and estimates of maximum sustainable yield (MSY). The species identified as being overfished were shad, alewives, striped bass, menhaden, sea trout, croaker, oysters, soft-shell clams, and hard-shell clams (Table 4).

As indicated in table 4, landings exceeded MSY in several years for some species. It should be noted that landings less than MSY does not imply no overfishing; low landings could be the result of overfishing in previous years, economic conditions, pollution, regulations, natural factors, or other conditions.

ENVIRONMENTAL DEGRADATION

Mortality from pollution or environmental degradation has not been rigorously examined. Little is known about the relationship between the population dynamics and environmental conditions. A possible reason why such information is not widely available is the difficulty of determining the individual effects of overfishing, environmental degradation, climatic conditions, disease, and predation. Species which are believed to have been affected by environmental degradation include shad and striped bass. However, many sedentary species such as clams are believed to be very susceptible to environmental contamination (Moore[12]).

Mortality, however, is not the only effect of environmental pollution. Frequently, fishing areas are closed and the removal of a species is prohibited. This results in lost earnings or income. On occasion, the mere announcement of a contaminated product from one area is accompanied by lower U.S. sales of that product as well as similar products. Thus, the impacts of environmental degradation can extend well beyond mortality.

The combined effect of overfishing and environmental degradation is a reduction in current and future economic opportunities. Simply, man will have fewer resources to exploit and incomes and economies dependent on the marine environment will decline.

However, early recognition of the problems by society offers some hope for preventing economic losses. Efforts to improve the environmental conditions of the Bay are well underway. Maryland and Virginia are both exploring more sound and rational approaches to fisheries management. In the final analysis, though, only the passage of time will determine the success of these efforts.

CITED REFERENCES

1. Lipton, D. 1986. "The status and health of the Maryland seafood industry." Maryland Department of Natural Resources. Annapolis, Maryland.
2. Briggs, H., R. Townsend, and J. Wilson. 1983. "An input-output analysis of Maine's Fisheries." Marine Fisheries Review 1:1-8.
3. Norton, V., T. Smith, and I. Strand. 1983. "The economic value of the Atlantic coast commercial and recreational striped bass fisheries." (UM-SG-TS-83-12, University of Maryland, Sea Grant.
4. Grigalunas, T.A. 1978. "Input-output study of marine related activity in southern New England." Unpub. manuscript, University of Rhode Island, Kingston, Rhode Island.
5. Cox, R.W. 1986. "An input-output table for Virginia." Tayloe Murphy Institute, The Colgate Darden Graduate School of Business Administration. Univ. of Virginia.
6. U.S. Department of Commerce, National Marine Fisheries Service. Selected issues of Fishery Statistics of the United States (1970-1977). U.S. Government Printing Office, Washington, D.C.
7. U.S. Bureau of Census. 1986. County Business Patterns, 1984, Maryland. U.S. Government Printing Office, Washington, D.C.
8. Overman, W.C. and Associates. (1986) "Virginia counties and cities databook," 1986. Cleveland Place, Virginia Beach, Virginia.
9. Virginia Employment Commission, Bureau of Economic Analysis. 1985. "Virginia personal income by city and county, 1978-1983." Reproduced by Tayloe Murphy Institute.
10. Murray, T., and J. Lucy. 1981. "Recreational boating in Virginia: a preliminary analysis." Special Report in Applied Marine Science and Ocean Engineering No. 251, Virginia Sea Grant Program, Virginia Institute of Marine Science. College of William and Mary, Gloucester Point, Virginia.
11. U.S. Army Corps of Engineers. 1978. "Chesapeake Bay Future Conditions Report." Volumes 3 and 9. Dept. of the Army, Baltimore District.
12. Moore, J. 1979. "A preliminary oil spill impact analysis for Eastport, Maine." Unpublished manuscript.

Contaminant Problems and Management of Living Chesapeake Bay Resources. Edited by S. K. Majumdar, L. W. Hall, Jr. and H. M. Austin. © 1987, The Pennsylvania Academy of Science.

Chapter Four
UTILIZATION OF CHESAPEAKE BAY BY EARLY LIFE HISTORY STAGES OF FISHES[1]
EILEEN M. SETZLER-HAMILTON
Chesapeake Biological Laboratory
Center for Environmental and Estuarine Studies
University of Maryland
Solomons, Maryland 20688

ABSTRACT

The objective of this chapter is to discuss fishes which either spawn within Chesapeake Bay or its tributaries or spawn in Atlantic continental shelf waters and utilize Chesapeake Bay as nursery grounds for larval and/or juvenile stages. Discussion is limited to species important in commercial and recreational fisheries, prominent species in estuarine food webs, and characteristic species of major habitats found within the Bay. Fishes are grouped by spawning habitat: 1) marine spawners; a) offshore spawners, b) lower Bay and coastal water spawners; 2) estuarine spawners; a) species which spawn throughout much of the Bay, b) oyster-reef spawners, c) aquatic vegetation spawners; 3) tidal freshwater and oligohaline spawners; a) anadromous species, b) freshwater species tolerant of brackish waters and c) oligohaline (salinities between 0.5-5.0 ppt.) and tidal freshwater spawners.

Forty-eight species of fish eggs and/or larvae were collected in an eight-year study of upper Chesapeake Bay. Bay anchovy, *Anchoa mitchilli,* and hogchoker, *Trinectes maculatus* eggs were most abundant. Most larvae and juveniles were caught in summer months; naked goby, *Gobiosoma bosc*, bay anchovy, and white perch, *Morone americana* larvae dominated catches. Bay anchovy, *Anchoa mitchilli*, dominated lower Bay ichthyoplankton collections. Eggs and larvae of several sciaenid species, especially the weakfish, *Cynoscion regalis*, ranked second in abundance.

Eighteen marine and estuarine species (excluding anadromous and catadromous species) were reported from the tidal fresh and oligohaline waters of five Virginia rivers on the Western Shore.

---
[1]Contribution No. 1771, Center for Environmental and Estuarine Studies of The University of Maryland.

INTRODUCTION

This chapter discusses fishes which either spawn within Chesapeake Bay or its tributaries or spawn in Atlantic continental shelf waters and utilize Chesapeake Bay as nursery grounds for larval and/or juvenile stages. Discussion is limited to species important in commercial and recreational fisheries, prominent species in estuarine food webs, and characteristic species of major habitats found within the Bay. Fishes are grouped by spawning habitat: 1) marine spawners; a) offshore spawners, b) lower Bay and coastal water spawners; 2) estuarine spawners; a) species which spawn throughout much of the Bay, b) oyster-reef spawners, c) aquatic vegetation spawners; 3) tidal freshwater and

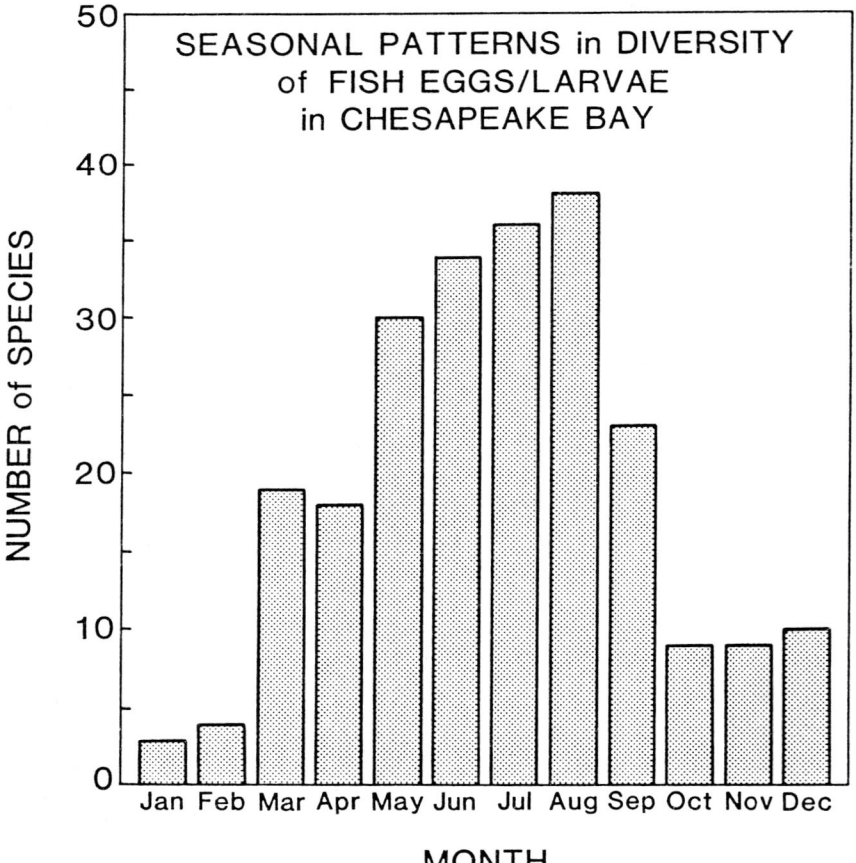

FIGURE 1. Seasonal patterns in diversity of fish eggs/larvae in Chesapeake Bay.
(Data for figure compiled, summarized, and drawn by John E. Olney, Virginia Institute of Marine Science of the College of William and Mary, Gloucester Point, VA).

oligohaline spawners; a) anadromous species, b) freshwater species tolerant of brackish waters and c) oligohaline (salinities between 0.5-5.0 ppt.) and tidal freshwater spawners. Readers interested in descriptions of early life history stages of fishes found in Chesapeake Bay are referred to the comprehensive series, *The Development of Fishes of the Mid-Atlantic Bight*,[1,2,3,4,5,6] which compiles descriptions of the egg, larval, and juvenile stages of over 300 species. Other useful guides to early life history stages are available for Delaware Estuaries,[7] and the Potomac River Estuary.[8] Environmental atlases of the upper Chesapeake Bay[9] and the Potomac River Estuary[10] contain extensive information on the spawning habits and use of estuarine nursery areas by Chesapeake Bay fishes. Alice Jane Lippson, senior author of these environmental atlases, and her husband, Robert, have also written an excellent, popular account of invertebrate and fish species found within the Bay.[11]

This chapter is organized around extensive, tabular summaries of fishes which spawn in lower Bay and coastal estuaries, and tidal freshwater and oligohaline regions of the Bay. Key species within these habitats are discussed in further detail in the text.

SEASONAL AND HABITAT OVERVIEW

Fish spawn in Chesapeake Bay throughout the year (Figure 1). Greatest spawning activity occurs throughout the summer, peaking in August, when eggs and/or larvae of 38 species can be found in the Bay. Few species occur in January; only spot, *Leiostomus xanthurus,* Atlantic croaker, *Micropogonias undulatus* and Atlantic menhaden, *Brevoortia tyrannus* larvae can be found in the Bay. Figure 2 shows seasonal distributions of fish eggs and larvae by estuarine habitat. Olney[12] studied the polyhaline (salinity greater than 20 ppt.) lower Bay. Dovel's work[13] dealt with the mesohaline (5.1-20 ppt.) and oligohaline waters of the upper Bay. Massmann[14] studied the tidal freshwater and oligohaline waters of Virginia's Western Shore tributaries of the Bay. Greatest diversity of fish eggs and larvae is found in the polyhaline lower Bay during July and August. Greatest diversity in upper Bay waters occurs in June, whereas diversity in Virginia's oligohaline tributaries peaks in July. The number of species spawning in the polyhaline lower Bay exceeded those spawning in other parts of the Bay by nearly a factor of two during July and August (Figure 2).

Bay anchovy, *Anchoa mitchilli,* dominated lower Bay ichthyoplankton collections, comprising 96% of all eggs and 88% of all larvae collected. Eggs and larvae of several sciaenid (drum) species, especially the weakfish, *Cynoscion regalis,* ranked second in abundance.[12] Other important species in lower Chesapeake Bay are the feather blenny, *Hypsoblennius hentzi;* seaboard goby, *Gobiosoma ginsburgi;* hogchoker, *Trinectes maculatus;* blackcheek tonguefish, *Symphurus plaguisa;* and the summer flounder, *Paralichthys dentatus.*[12]

Dovel[13] found 48 species of fish eggs and/or larvae in an eight-year study of upper Chesapeake Bay (Patuxent River to the Susquehanna River and the Delaware State Line in the C&D Canal). Half of the fish eggs collected were from mesohaline waters of 12-16 ppt.; 99% of these were bay anchovy and hogchoker eggs. Most of the larvae and juveniles (83%) were caught in summer months at temperatures from 22-29°C. Naked goby, *Gobiosoma bosc,* bay anchovy, and white perch, *Morone americana* larvae dominated catches in the upper Bay. Yellow perch, *Perca flavescens;* winter flounder, *Pseudopleuronectes americanus;* and striped bass, *Morone saxatilis* ranked next in abundance. It is important to note that these studies were conducted from 1960-1968, prior to drastic declines of all anadromous fishes (especially striped bass) in upper

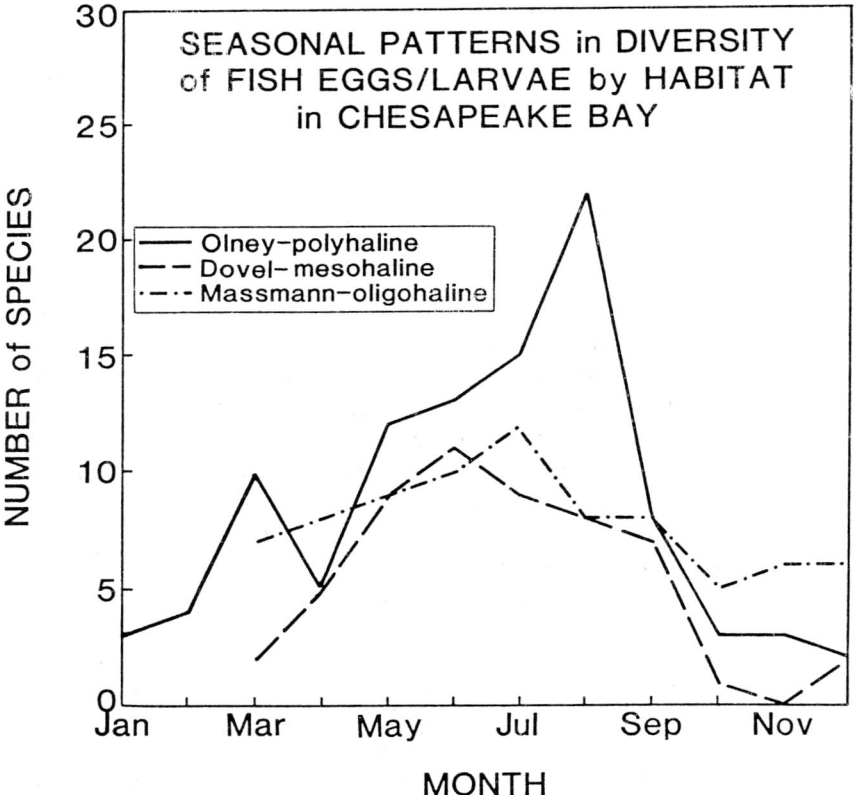

FIGURE 2. Seasonal patterns in diversity of fish eggs/larvae in oligohaline, mesohaline, and polyhaline waters of Chesapeake Bay. (Olney[12]—polyhaline; lower Bay; Dovel[13]—mesohaline and oligohaline; upper Bay; and Massmann[14]—oligohaline reaches of Virginia tributaries).
(Data for figure compiled, summarized, and drawn by John E. Olney, Virginia Institute of Marine Science of the College of William and Mary, Gloucester Point, VA).

Chesapeake Bay. Massmann[14] found 18 species of marine and estuarine fishes (excluding anadromous and catadromous species) in the tidal fresh and oligohaline waters of five Virginia rivers. Mummichogs, *Fundulus heteroclitus,* and tidewater silversides, *Menidia peninsulae* (*M. beryllina* of many studies) were collected farthest upriver.

Generally, mortality rates of fish eggs and larvae are very high. For example, Leggett[15] reported that on the average, only 0.00083% of American shad, *Alosa sapidissima,* produce sexually mature adults. The majority of this mortality (of American shad and most species) occurs from the time of deposition of the egg to the time of the juvenile stage.

MARINE SPAWNERS

Marine fishes found in estuaries can be placed into two groups: 1) estuarine-dependent species which require estuaries for successful completion of their life cycle (menhaden, bluefish, sciaenids, and flounders) and 2) summer transients that are often found in estuaries during summer months, although they survive equally well in coastal or oceanic waters. Of the marine spawners listed in Table 1, most are estuarine-dependent; however, sea robins, puffers, harvestfish, tautogs, striped anchovies, and lookdowns can be considered summer transients.

Atlantic menhaden (Table 1), a schooling, pelagic clupeid, ranks first in total tonnage caught in U.S. commercial fisheries. It is processed into oil and fish meal. Menhaden rank second in total biomass of Chesapeake Bay pelagic fishes.[19] Known locally in southern Maryland as "alewives", in Virginia as "little stinkers", "peanuts", or "bunkers", and by numerous other local names, menhaden spawn up to 64 km offshore from the mouth of Chesapeake Bay. Spawning primarily occurs during the northward spring and southward fall migrations in the Chesapeake Bay Region.[1] The pelagic larvae spend from one to two months at sea and move into estuaries before metamorphosing into juveniles. Movement from offshore spawning areas into inshore nursery grounds such as Chesapeake Bay, may depend in part upon zonal (eastward or westward) wind-driven (Eckman) transport. Eckman transport in the mid-Atlantic Bight generally has a stronger, westward component from November through February, the time of peak immigration of menhaden into mid-Atlantic estuaries.[20] Although there are no documented physiological requirements for estuarine dependence, metamorphosing larvae are rarely caught in the ocean; an indication that ecological requirements (food, shelter, etc.) provided by estuaries are essential in the life cycle of menhaden.

Menhaden larvae occur in upper Chesapeake Bay tributaries from late May to late June and in November. Peak abundances are found in oligohaline waters from <1 to about 3 ppt. Smallest juveniles occur farthest upriver, as far as 56

TABLE 1

Marine Spawners Whose Larval and Juvenile Stages Utilize Chesapeake Bay as a Nursery Area.
O = Oligohaline (0.5 - 5.0 ppt.)　M = Mesohaline (5.1—20.0 ppt.)　P = Polyhaline (>20.00 ppt.)

Species	Spawning Season and Location	Habitat Eggs	Habitat Larvae	Habitat Juveniles	Occurrence of Larvae/Juveniles	References
OFFSHORE SPAWNERS						
Clupeidae — herrings. Atlantic menhaden, Brevoortia tyrannus	During northward spring and southward fall migrations	Pelagic, mostly offshore; P (July-Aug.)	Pelagic, enter estuary ≈ 10 mm P, M, few O	O, tidal freshwater	Larvae lower Bay, Feb, April-May, Aug.; upper Bay May-June, Nov. Juveniles, spring-summer; migrate south in fall, some overwinter	1,12,14
Gadidae —codfishes. Spotted hake, Urophycis regius	Sept.-Feb. or March; peak Oct.	Offshore	Offshore; few P	Inshore tributaries, lower Bay 6-7 months. P, M	Juveniles mainly March-June; peak April and May, York River	2,12
Carrangidae — jacks. Florida pompano, Trachinotus carolinus	Peak April - June	Offshore	Pelagic, offshore	M, P (most)	Juvenile lower Bay, late summer-fall	4
Pomatomidae — bluefishes. Bluefish, Pomatomus saltatrix	June-Aug. (possibly May)	Pelagic, offshore; few mouth of Bay	Pelagic, offshore; few mouth of Bay	Coastal waters. P, M, O, tidal freshwater	Juveniles, summer and fall	3
Sciaenidae — drums. Spot, Leiostomus xanthurus	Late fall, winter	Pelagic, well offshore	Most frequently at bottom, coastal waters	Enter Bay ≈ 4 mo. P, M, O (smaller juv)		4
Atlantic croaker, Micropogonias undulatus	Offshore over wide area; extends some distance offshore, Aug-Dec; peak Aug-Sept; VA may occur in all months. Spring-fall peaks		O, M	Conc. at ≈ 18 ppt; bottom waters of relatively deep channels. M, P	Larvae taken during winter; Juveniles, spring and summer. Croaker larvae and juveniles upper Bay 0-21 ppt; 0-24°C	4,13
Mugilidae — mullets. Striped mullet, Mugil cephalus		Offshore	Offshore	Enter Bay as pre-juveniles	Pre-juveniles enter March-June; peak April and May; Juveniles, summer	6

TABLE 1 (Continued)
Marine Spawners Whose Larval and Juvenile Stages Utilize Chesapeake Bay as a Nursery Area.
O = Oligohaline (0.5 - 5.0 ppt.) M = Mesohaline (5.1–20.0 ppt.) P = Polyhaline (>20.00 ppt.)

Species	Spawning Season and Location	Eggs	Habitat - Larvae	Habitat - Juveniles	Occurrence of Larvae/Juveniles	References
Stomateidae — butterfishes. Butterfish, *Peprilus tricanthus*	Late spring, early summer	Pelagic	Move inshore to bays and other protected waters.	Frequently associated with jellyfish to ≈ 30 mm. M, P	Larvae July, lower Bay; Juveniles, summer; leave by Nov.	6,12
INSHORE, COASTAL WATER SPAWNERS (may include lower portion of Bay)						
Clupeidae — herrings. Atlantic threadfin herring, *Opisthonema oglinum*	May-June (NC)	Pelagic	Coastal waters	M, P (most)	Juveniles, summer months	1
Engraulidae — anchovies. Striped anchovy, *Anchoa hepsetus*	June-Aug. mouth of Bay and coastal waters	Pelagic, buoyant	Recorded along edge of continental shelf off DE and VA. P			1,8
Exocoetidae — flyingfishes. Halfbeak, *Hyporhamphus unifasciatus*	Summer, lower Bay, coastal waters	Semi-buoyant	O, M	Found in schools. M, P	Summer, move offshore and possibly south in fall (adults)	2
Serranidae — sea basses. Black sea bass, *Centropristis striata*	Late May, coastal waters	Pelagic, buoyant	Migrate inshore; > 13 mm demersal or estuarine	Inshore near inlets, in eelgrass, over oyster beds. P	Juveniles, summer; leave estuarine nursery areas in fall; migrate offshore and southward	3
Rachycentridae — cobias. Cobia, *Rachycentron canadum*	June-Aug. vicinity of Virginia Capes (possibly lower Bay)	Pelagic		Along bays, beaches, over shoals; found 64 km up Hudson River. P		3
Sciaenidae — drums. Silver perch, *Bairdiella chrysoura*	May-July, May-July Chesapeake Bay. Delaware Bay, early June (peak)—July;	Pelagic	O, M, P	Shoal areas; move to deeper waters (channels) by fall. O, M, P	Juvenile, June-Nov; York River, May	4

Species	Spawning	Egg type	Larval location	Juvenile habitat	Notes	Ref
Spotted seatrout, *Cynoscion nebulosus*	Spring and summer, peak April and May lower Bay and coastal waters	Initially buoyant; demersal after 12 hr	P, coastal waters; found in bottom vegetation	Mostly within 50m of shoreline in vegetation. P, M	Juveniles, June-Nov. move offshore to deeper waters in winter	4
Weakfish (trout), *Cynoscion regalis*	Late spring and summer; two peaks, June and July, Delaware Bay	Initially buoyant. M, P	Sink to bottom by 8-10 mm. M, P	Soft, muddy bottoms, low salinity areas. O,M,P.	Larvae, May-Aug., Chesapeake Bay. Juveniles, Mar-Oct; may remain Nov-Dec; most seek warmer offshore waters; upper York R. most abundant July; migrating downstream Sept.-Nov.	4
Southern kingfish, *Menticirrhus americanus*	Largely or entirely offshore; spring and early summer		P. Lower Bay	Most common habitat <50 mm open surf and sandy beaches, 50-150 mm estuaries or inshore; P,M,O	Larvae, June-Sept; may be transported far up tidal rivers by high salinity bottom currents; actively move to higher salinity areas as fish grows	4
Northern kingfish, *Menticirrhus saxatilis*	Offshore June-Aug., NJ; April-May (or longer) NC and SC	Buoyant	Beaches; Long Island, June-July	Delaware R., M, P common >16 ppt; York R. only in upper reaches < 5 ppt	Juveniles Delaware River estuary, late summer, late Oct.; Chesapeake Bay, summer and fall; off VA July-Sept.	4
Black drum, *Pogonias cromis*	March-May Chesapeake, Delaware Bays; peak Chesapeake Bay, April-May; Delaware, May, possibly 2nd spawn Sept.	Pelagic	Upper reaches of bays and tidal creeks. O.	Tidal creeks and ditches preferred habitat first 3 mos; remain in shallow bay & shore areas. O,M,P	Juveniles—June, Delaware River. July-Oct. Delaware and Chesapeake Bays.	4
Red drum, *Sciaenops ocellata*	Atlantic coast from VA southward to St. Lucie Inlet, FL, July-Dec.; peak Sept.-Oct.	Pelagic	Upper Chesa. Bay	Shoal, estuarine waters. M,P.	Juveniles, Sept.-Nov.	4

TABLE 1 (Continued)
Marine Spawners Whose Larval and Juvenile Stages Utilize Chesapeake Bay as a Nursery Area.
O = Oligohaline (0.5 - 5.0 ppt.) M = Mesohaline (5.1–20.0 ppt.) P = Polyhaline (>20.00 ppt.)

Family/Species	Adult Habitat	Egg Type	Larval Habitat	Distribution/Comments	Ref.
Labridae — wrasses. Tautog, *Tautoga onitis*	Coastal waters; move into estuarine embayments for spawning, mid May-mid August; eggs, May, lower Bay	Buoyant	Surface; P.	Among eelgrass & algae in coves and channels	5,12
Ammodytidae—sand lances. American sand lance, *Ammodytes sp.*	Inner half of continental shelf, Nov.-March; peak Dec. off Chesapeake Bay. Temp. <9°C	Demersal	Most abundant off so. New England & off Delaware; move seaward with size increase. O,M,P (most)	Larvae—Jan.-March, lower Bay; seasonally present when water column is homogenous	5,12,16
Triglidae — sea robins. Northern searobin, *Prionotus carolinus*	May-Aug, coastal waters. Limited spawning in lower Bay.	Buoyant	*Prionotus* spp. P	M, P (based on adult distributions	*Prionotus* spp. larvae, lower Bay, Aug.	5,12
Stomateidae—butterfishes. Harvestfish, *Peprilus paru*	Generally coastal waters, lower Chesapeake Bay, spring and early summer	Pelagic	Frequently associated with jellyfish P	Frequently associated with jellyfish. M, P	Larvae and Juveniles, summer; frequently encountered in estuaries. Larvae, lower Chesapeake Bay, Aug.	6,12,17
Tetraodontidae — puffers. Northern puffer, *Sphoeroides maculatus*	Shoal waters near shore, May-Aug.	Demersal, attached	Phototrophic, float on surface. M, P	Semidemersal; commonly at shoreline. M, P	Larvae and Juveniles, summer	6
Cynoglossidae — tonguefishes. Blackcheek tonguefish, *Symphurus plagiusa*	Coastal waters, lower Bay and Delaware Bay, June-Aug.	Pelagic	Demersal, M, P (most)	Demersal, M, P	Larvae—lower Bay, July-Aug.; present only at stations with average salinity ≥ 16.5 ppt.	6,12,18

TABLE 1 (Continued)

Marine Spawners Whose Larval and Juvenile Stages Utilize Chesapeake Bay as a Nursery Area.
O = Oligohaline (0.5 - 5.0 ppt.) M = Mesohaline (5.1–20.0 ppt.) P = Polyhaline (>20.00 ppt.)

Bothidae—lefteye flounders. Summer flounder, *Paralichthys dentatus*	Occurs during offshore migration; north of Chesapeake Bay, Sept.-Dec.; south of Ches. Bay, Nov.-Feb.	Near surface	Mostly 22-83 km offshore. P	Estuarine. O, M, P	Larvae—lower Chesapeake Bay, March	6,12
Gobiidae — gobies. Darter goby, *Gobionellus boleosoma*	Mid March-July or Aug., lower Chesapeake Bay, coastal waters. P.	Demersal, attached				5
Diodontidae—porcupine fishes. Striped burfish, *Chilomycterus schoepfi*	Presumed offshore, early spring to July; adults most often in grass beds; 7-47 ppt.	Demersal, non-adhesive		Shallow, sandy areas, often associated with grass beds. M, P		6

km upstream from brackish water in the Rappahannock River.[14] During June and July, 40-50 mm juveniles are generally found from freshwater to 15 ppt. salinity. Most emigrate southward to warmer oceanic waters in the fall, though some menhaden juveniles over winter in lower Chesapeake Bay.[1]

Bluefish, another offshore spawner (Table 1), comprise two principal populations along the Atlantic coast. Spring spawners are present during April and May in the Gulf Stream from Florida to North Carolina; summer spawners occur during June, July (peak), and August along the continental shelf from Cape Hatteras to Cape Cod.[21] In the Chesapeake Bight, bluefish primarily spawn over the outer half of the continental shelf between 55 and 148 km offshore where water temperatures are at least 22°C.[22] Juvenile bluefish from the spring spawning are carried northward and enter Chesapeake Bay and other estuaries of the mid-Atlantic Bight during early summer. Juveniles from the summer spawning population generally do not enter coastal estuaries and move southward out of the middle-Atlantic Bight in the fall as the water cools. During the following spring, these juveniles enter Chesapeake Bay and other mid-Atlantic estuaries as immature "snappers".[10,21]

With the exception of spot and Atlantic croaker, the 13 sciaenid (drum) species which occur in the Chesapeake Bay region are nearshore spawners. Black drum spawn in the spring; spotted sea trout, southern kingfish, weakfish, and silver perch are late spring/summer spawners; northern kingfish primarily summer spawners; and red drum, summer/fall spawners (Table 1). Spot spawn offshore during the late fall and winter[4] or during winter and spring,[7] and Atlantic croaker (hardhead) spawn from late summer through December (Table 1).

Larval croakers are transported up-Bay in the higher salinity bottom currents and overwinter from the tidal freshwater to mesohaline regions of the Bay.[13] Croaker larvae and juveniles must survive winter in estuarine nursery areas where they are susceptible to rapid and unfavorable temperature changes. Colder-than-normal winters result in high mortalities and poor recruitments. Climatic factors apparently play a major role in the fluctuations of Atlantic croaker stocks in the Middle-Atlantic states (northern boundary for croaker spawning).[23,24,25,26]

Other sciaenid larvae and/or juveniles are found in low salinity Bay waters. Larval spot utilize upper Bay waters as nursery areas in the spring and are often taken simultaneously in plankton tows with Atlantic croaker.[8] Weakfish (often called trout by fishermen) larvae become demersal by 8-10 mm and possibly as early as 1.5-5.0 mm. They are found from May through September,[4,13] primarily in the lower and mid-Bay. Juvenile weakfish prefer soft, muddy bottoms in low salinity areas of brackish coves and creeks. Weakfish, spot, and croakers are the most abundant sciaenid larvae and juveniles found in upper Chesapeake Bay.

Juvenile sciaenids are among the most abundant migratory finfishes in the York River.[27] Weakfish, silver perch, Atlantic croaker, and spot dominate sciaenid catches. Juvenile spot, silver perch, weakfish, and croaker enter the York River consecutively from April on; all but croaker leave by December. Juvenile croaker are found throughout the year in the York River estuary. Occasionally, juvenile southern kingfish, northern kingfish, spotted seatrout, black drum, and banded drum utilize York River nursery grounds. Juvenile southern kingfish are found in upper Chesapeake Bay[13] and juvenile black drum are common residents of oligohaline tributaries of Delaware Bay in summer months.[28]

ESTUARINE SPAWNERS

Spawning habits of estuarine fishes common in mesohaline and polyhaline waters of Chesapeake Bay are summarized in Table 2. Fishes adapted for spawning primarily in oyster beds (i.e. blennies, gobies, and clingfish) and in shallow, grassy areas (i.e. pipefish, seahorses, and sticklebacks) are enumerated separately.

Bay anchovy, a pelagic, schooling species, is the most abundant fish in estuarine waters of Chesapeake and Delaware Bays[7,12,13,38,39,40] and dominates total biomass of pelagic fishes in the Bay.[9] Bay anchovies are an euryhaline species, found primarily in estuaries and coastal waters, and present in Chesapeake Bay throughout the year. As with most year-round residents of the Bay, they concentrate in deeper waters during colder months. Bay anchovies are major forage species for white perch, striped bass, bluefish, and other piscivorous fishes of the Bay. They typically spawn in early evening throughout mid and lower Bay waters, at salinities greater than 9 ppt. Peak spawning probably occurs between salinities of 13-15 ppt.[13] Mean abundance of anchovy eggs during periods of peak spawning in the lower Bay at higher salinities are comparable to abundances reported in the upper Bay.[12] Anchovy larvae are transported upstream from spawning areas in Chesapeake Bay. They occur predominantly in surface waters from early May through mid-October with greatest abundances found in salinities of 3-7 ppt. Juvenile anchovies have been found in Virginia tributaries 64 km above brackish water.[14] They are widely distributed from shallow, grassy areas to deeper waters and are most abundant near the salt-freshwater interface from June to September and in deeper waters from October through March.[1]

Other important pelagic estuarine fishes are the three species of silversides: tidewater, *Menidia peninsulae;* Atlantic, *M. menidia;* and rough, *Membras martinica.* Silversides are surface, schooling fishes strongly oriented toward the shore. Seasonal shoreward movements begin as early as March. Silversides remain inshore at least until November, and are found in deeper waters downstream only during the coldest periods.[10] There is considerable overlap in the microgeographical distribution of the three species. Tidewater silversides are particularly abundant in areas of submerged vegetation and are more common in brackish than in salt water, whereas Atlantic silversides are common along

TABLE 2
Estuarine Spawners Whose Larval and Juvenile Utilize Chesapeake Bay as a Nursery Area
O = Oligohaline (0.5 - 5.0 ppt.) M = Mesohaline (5.1—20.0 ppt.) P = Polyhaline (>20.00 ppt.)

Species	Spawning Season and Location	Eggs	Habitat Larvae	Habitat Juveniles	Occurrence of Larvae/Juveniles	References
Engraulidae—anchovies. Bay anchovy, *Anchoa mitchilli* (also spawns in coastal marine waters)	Late April-late Sept. (peak, July) throughout mid and lower Ches. Bay. O, M, P; most >9 ppt.; peak 13-15 ppt. (though eggs most abundant in most saline portion of lower Bay)	Pelagic, con. at surface. M,P	O, M, P surface waters	O, M, P, euryhaline, ascend rivers to freshwater	Larvae early May-mid Oct; greatest abundance 3-7 ppt. Juveniles most abundant brackish water near salt-fresh interface June-Sept.; deeper waters Oct.-March	1,12,13
Belonidae—needle-fishes. Atlantic needlefish, *Strongylura marina*	May-Aug.; O, M; max recorded salinity 18ppt.; may spawn in freshwater	Demersal, attached	O, M	O, M	Yolk-sac larvae reported from freshwater feeders of Delaware-Raritan Canal; larvae, upper Bay June - July	2,13
Cyprinodontidae—killifishes. Sheepshead minnow, *Cyprinodon variegatus*. Euryhaline prefer salinities <20 ppt.	Throughout summer Chesapeake Bay; May (poss. April)-Aug. Delaware Bay; males occupy territories; may or may not construct nests O, M, P	Demersal, adhesive or semi-adhesive	Demersal, O, M, P	Schooling; O, M, P	Larvae 4.0 mm at outlets of springs at exact edge of water. Juveniles large schools often left in shallow depressions in sand by receding tide	2,8
Mummichog, *Fundulus heteroclitus*; euryhaline prefer salinities 0-25 ppt.	Chesapeake Bay, April-Aug; Delaware Bay, May-mid Aug. Several peaks at or near new moon high tide; shallow areas with sparse to dense vegetation. Upper tidal marsh among *Spartina* roots; freshwater O, M, P	Demersal; filamented eggs attached; eggs with reduced filaments, inside vertically-oriented ribbed mussel shells	Remain off bottom; attracted to light; O	Among eelgrass, shallow pools and ditches; O, M, P		2, 8, 29, 30
Spotfin killifish, *Fundulus luciae*. Euryhaline, 0-28 ppt; weedy, shallow tidal creeks.	Mid April-mid Aug. (VA); fibrous substrate in shallow water. O, M, P	Demersal	O, M	Tidal creeks and pools; O, M, P		2,8

Species	Spawning	Eggs	Larvae	Juveniles/Adults	Ref.
Striped killifish, *Fundulus majalis*. Euryhaline, prefer 22-37 ppt.	April-Sept; still, shallow water near shore (reaches peak 2 weeks earlier than *F. heteroclitus*); female, several batches of eggs/season. O, M, P	Deposited close inshore during low tide. May be buried	Shallow water among vegetation; O, M, P		2,8
Rainwater killifish, *Fundulus parva*. Euryhaline, 0-48 ppt.; freshwater habitats, medium to high conductivity.	Early April-July; may be anadromous migrating to fresher water to breed; O, M, P	Demersal, deposited on fibrous substrate.	Demersal, O, M, P		2,8
Marsh killifish, *Fundulus confluentus*. Euryhaline prefer 10-30 ppt.	April-May; among vegetation, both fresh and brackish water; O, M, P	Demersal, possibly attached to vegetation.	O, M, P	O, M, P	2,8
Atherinidae — silversides. Atlantic silverside, *Menidia menidia*	March-July, Chesapeake region; May-Nov, Chesapeake Bay; intertidal zone or shallow waters; estuarine areas prob. seaward of *M. peninsulae*; O, M, P (mostly M and P).	Demersal, attached	Shallow water near shore; school at 8-10 mm TL; O, M, P	Chesapeake Bay, O, M; 1-14 ppt; mode = 7 ppt. Also P	Larvae and Juveniles schools at surface, follow incoming and outgoing tides	6,13
Rough silverside, *Membras martinica*	Along sandy beaches, mid and lower Bay; coastal waters just outside breaker zone; May-early Aug. M, P	Demersal, adhesive in clumps	Taken at all depths, C&D Canal. O, M, P	Along exposed shorelines & beaches. O (smallest fish), M, P	Larvae March-Aug, lower Bay. Juveniles large schools near surface; common in lower Ches. Bay; Mar-Aug.; habitat not different from that of adults	6,12
Pleuronectidae — righteye flounders. Winter flounder, *Pseudopleuronectes americanus*	Mid Dec-May; peak March; estuaries, sandy bottoms, 1.8-3.6 m; 11.4-33 ppt. M, P. 1-10°C; peak 2-5°C		Pelagic, strongly bottom oriented before metamorphosis; O, M, P	Benthic—remain in estuaries 2+ yrs; inshore, except for temperature extremes	Larvae 3.5-27.7 ppt; peak abundance 6-15 ppt. Juveniles 4-30 ppt; normal growth at 20, but not 30 ppt; 0-25°C; normal growth 12-16°C.	6
Soleidae — soles. Hogchoker, *Trinectes maculatus*	May-Sept. primarily in estuaries; 0-24 ppt. peak 10-16 ppt. O, M, P	Near surface, higher salinities; near bottom, lower salinities	Move into low salinity waters. O	Shore zone, move upstream to 240 km inland; over-winter in bays. O, M	Larvae concentrate near salt-freshwater interface remain overwinter; 0-8 ppt; Juveniles move downstream in spring but don't reach spawning area	6,31

TABLE 2
Estuarine Spawners Whose Larval and Juvenile Utilize Chesapeake Bay as a Nursery Area
O = Oligohaline (0.5 - 5.0 ppt.) M = Mesohaline (5.1–20.0 ppt.) P = Polyhaline (>20.00 ppt.)

Species	Spawning Season and Location	Eggs	Habitat Larvae	Juveniles	Occurrence of Larvae/Juveniles	References
OYSTER-REEF SPAWNERS						
Gobiidae — gobies. Naked goby, *Gobiosoma bosc*; shallow, less saline waters, 0-27 ppt.	May-mid Nov; upper Chesapeake Bay, clam and oyster shells; M, P; 10-30 ppt.	Demersal, attached, guarded by male	Day, mid-depth and near bottom; nearer surface at night. O, M	Benthic, O, M	Larvae, upriver or in waters <18.5 ppt; appear in Patuxent first week of May, and in Potomac mid-May; York River, May-Oct.	5,7,12,13, 32,33,34
Seabord goby, *Gobiosoma ginsburgi*; deeper more saline waters than *G. bosc*, 15-31 ppt.	June-Dec. (based on presence of larvae) oyster reefs of lower Bay, also coastal waters M, P (most)	Demersal, attached	M, P (most)	M, P (most)	Larvae, present June-Dec. Larvae and juvenile 17-22 ppt. Upper Bay; Larvae 12-28 ppt; 90% ≥ 16 ppt.	5,12
Green goby, *Microgobius thalassinus*	April-Nov, deeper oyster reefs and large, dense growths of sponge, *Microciona prolifera*; M, P	Demersal, attached	M, P	M, P	Larvae and juveniles 9-17 ppt., upper Bay. Larvae ≥ 16.5 ppt. Most >18.5 ppt.	5,13,35
Blenniidae — combtooth blennies. Striped blenny, *Chasmodes bosquianus*	April-Aug.; June-late July, seaside bays, MD; empty oyster shells, sometimes clams and scallop shells. M, P	Demersal, attached; guarded by male	Pelagic. M, P	Demersal. M, P	Larvae, 11-21 ppt.	5,8
Feather blenny, *Hypsoblennius hentz*	May-Aug; empty oyster and possibly clam and scallop shells. M, P and coastal waters	Demersal, attached; guarded by male	Surface in open waters. M, P	Pelagic at 13 mm; on bottom by 24 mm; M, P	Larvae, 11-21 ppt.	5
Gobiesocidae—Clingfishes. Skilletfish, *Gobiesox strumosus*	April-Aug.; peak late April-early May; empty oyster shells M, P and coastal waters	Demersal, attached; guarded by male	In water column only when swimming; drop to bottom between swimming bouts. M, P	Demersal, M, P	Larvae, May-June. Juveniles, May-Sept. or Oct.	6,36

78 Contaminant Problems and Management of Living Chesapeake Bay Resources

Species	Spawning	Eggs	Larvae/Location	Notes	Ref
Batrachoididae—toadfishes. Oyster toadfish, *Opsanus tau*	April-July or Aug.; cavities among shells or rocks, tin cans, broken bottles, etc. in nests. M, P and coastal waters	Demersal, attached; guarded by male	Yolk-sac larvae remain attached to substrate of nest site until yolk absorbed; cared for by male. M, P	Demersal; become free-swimming between 16-18 mm	Juveniles, summer	6,37

SPECIES WHICH SPAWN PRIMARILY AMONG AQUATIC VEGETATION

Species	Spawning	Eggs	Larvae	Notes	Ref
Gasterosteidae—sticklebacks. Four-spine stickleback, *Apeltes quadracus*	April-mid May; in conical nests constructed of plant material; brackish water of weedy, shallow swamps and ditches. O, M, P	Demersal, deposited in nests in clump, guarded by male	Relatively shallow water. O, M, P	Probably restricted to place of origin. O, M, P	Larvae found near mouth of Patuxent River	2
Syngnathidae—pipefishes & seahorses. Dusky pipefish, *Syngnathus floridae*; most abundant 17-22 ppt.; salinity range 12-38.8 ppt.	Males with eggs from May-Oct; spawning as early as April, lower Chesapeake Bay; peak July and Aug. M, P	Deposited in brood pouch of male	Carried in brood pouch of male until ≈11.5 mm. M, P	Harbors and bays. M, P		2
Northern pipefish, *Syngnathus fuscus*; most abundant 13-20 ppt.; salinity range 0-31.3 ppt.	Lower Chesapeake Bay, Mar-Aug.; peak May and early June; also reported as late as Oct. eelgrass beds. O, M, P	Deposited in brood pouch of male	Retained in brood pouch of male until 8-12 mm; larvae pelagic for several weeks after parturition. O, M, P	Pelagic or semi-pelagic; O, M, P	Recently born larvae (10-12 mm) mid Bay; upper Bay Juveniles found near oligohaline waters (collected by author in Potomac) to a considerable distance offshore, May-June; inshore in York Riv. in June	2,13
Lined seahorse, *Hippocampus erectus* usually associated with or clinging to aquatic vegetation	June-mid Sept.; M, P	Deposited in brood pouch of male. M, P	Hatch in brood pouch of male; remain for indefinite period. M, P	Among masses of floating seaweed, lower Bay. M, P	Juvenile recently born recorded from a branch of the Potomac and from upper Chesapeake Bay; July-Oct., 16-22 ppt.	2,8,13

open beaches. Rough silversides are also an open, shallow water species usually found along exposed shorelines and beaches with little vegetation, but more often at higher salinities than *M. menidia*. All three species spawn in salinities less than 15 ppt. (Table 2). Tidewater silversides spawn primarily in tidal freshwater or brackish water in shallow areas with plenty of vegetation. Atlantic silversides spawn in the intertidal zone or shallow, estuarine waters seaward of tidal silversides. Rough silversides spawn along sandy beaches from mesohaline waters to the polyhaline lower Bay and in coastal waters just outside the breaker zone.[6] In the tidal Potomac River, larval tidewater silversides are centered around the tidal freshwater-oligohaline region around Douglas Point, whereas abundances of larval Atlantic and rough silversides are greatest just below Morgantown in mesohaline waters.[10] Rough silverside larvae have been caught in the lower Bay from March through August.[12]

Important euryhaline, shallow water fishes associated with tidal marshes, creeks, ditches, and more open shores include killifishes and sheepshead minnows (Table 2). These abundant fishes are major prey items of larger fishes and shore birds, and control mosquitoes and other insects. Although the banded killifish is more common in tidal fresh and oligohaline waters and is listed with such spawners in Table 4, it is discussed here for comparative purposes.

Mummichogs, sheepshead minnows, and striped, rainwater, marsh, and spotfin killifishes are all widely distributed from fresh to salt water. Mummichogs, an ubiquitous species, are frequently associated with vegetation in muddy waters or over muddy bottoms, whereas striped killifish are shallow water, near-shore species found along sandy or pebble beaches. Rainwater killifish frequent open shores (as does the oligohaline banded killifish), or habitats with aquatic vegetation; spotfin killifish prefer weedy, shallow tidal creeks and shallow backwaters of mesohaline and polyhaline tidal pools. Marsh killifish, a more southern species occurring from Virginia southward, frequent salt marshes and salt pools with soft, muddy bottoms. Sheepshead minnows are ubiquitous in shallow water habitats, preferably at salinities above 20 ppt. Mummichogs and spotfin killifish can burrow in the mud if left stranded by receding water in a tidal pool or marsh; striped killifish and sheepshead minnows over winter burrowed in the mud.[2] Mummichogs may maintain home ranges throughout the summer. The majority of mummichogs in a Delaware tidal creek maintained a home range of 36 m near a creek bank from July through September.[41] Salinity ranges of these species are summarized in Table 3.[2,10] Spawning is accompanied by elaborate courtship behavior among these killifishes. Both sheepshead minnows and banded killifish exhibit territorial behavior during courtship and mating.

Hogchokers and winter flounder are the only flatfishes which spawn in the mesohaline portions of Chesapeake Bay (Table 2). Hogchokers, one of the most abundant fishes in the mid to upper reaches of the Bay, spawn from May to September; peak egg abundances occur in July. Although hogchoker eggs have been recorded from freshwater to 24 ppt., greatest abundances are found from

TABLE 3

Salinity ranges of abundant killifishes in Chesapeake Bay and its tidal tributaries.[8,10]

	DENSITIES		
	High	Moderate	Low
Banded killifish	fresh-5 ppt.	5-10 ppt.	10-15 ppt.
Mummichogs	0-25	tidal fresh	fresh
Striped killifish	5-37	0-5	tidal fresh
Rainwater killifish	0-48	—	tidal fresh
Sheepshead minnow	5-35	0-5	tidal fresh
Marsh killifish	10-30	—	0-10
Spotfin killifish	10-28	—	0-10

salinities of 10 to 16 ppt. After hatching, hogchoker larvae move upstream and concentrate in oligohaline nursery areas where they remain during winter. As spring approaches, juveniles move farther downstream. In the fall, they move upstream. These spring/fall movements occur throughout the life cycle, but as they mature, hogchokers progressively increase their range of travel away from the nursery area.[31]

Winter flounder spawn during winter and spring months, peaking in March, at water temperatures from 2-5°C from Chesapeake Bay to Cape Cod. Larvae are benthic and usually remain in estuaries at least two years. They are found inshore except during periods of extreme temperature.[6]

Oyster Reef Spawners

Fish, such as gobies, blennies, and skilletfish, typically spawn on the inner surfaces of oyster shells. Toadfish spawn both among oyster shells and around piers, docks, etc. These fish utilize both natural substrates and man's discarded trash for egg attachment. Male parents typically guard eggs of oyster reef spawners and may provide care for the young (toadfish).[5,6] Reef spawners are typically rather sedentary, benthic fishes, cryptically camouflaged to blend into the reef habitat. Skilletfish (so named because of their skillet-like shape) posses a ventral, adhesive disc used for attachment to relatively stable objects (i.e. oyster shells, pilings, rocks, and loggerhead sponges).[42] Gobies also have pelvic fins united to form a sucking disc.[5] Blennies, gobies, and skilletfish can rapidly change colors patterns. When kept in aquaria, they often turn black, especially when attempting to guard a newly-found oyster shell from an intruder.

Four species of gobies are typically found in the Chesapeake Bay region. Naked, seaboard, and green gobies are estuarine species found in waters of varying salinities. Darter gobies are considered a marine species, though they are also found in brackish waters.[5,8] Naked gobies (named because adults lack scales) were the most abundant fish species collected in upper Chesapeake Bay and several tributaries from 1960-1966. Adult naked gobies are found around both oyster bars and shallow grass flats. They move to deeper, higher salinity waters

TABLE 4

Tidal Freshwater and/or Oligohaline (0.5 - 5.0 ppt) Spawners Whose Larval and Juvenile Stages Utilize Chesapeake Bay as a Nursery Area.

Species	Spawning Season and Location	Eggs	Habitat Larvae	Habitat Juveniles	Occurrence of Larvae/Juveniles	References
ANADROMOUS SPAWNERS						
Acipenseridae—sturgeons. Atlantic sturgeon, *Acipenser oxyrhynchus*	Brackish or fresh water; spawning migrations, April, Chesapeake Bay; late April, Delaware River. O	Demersal, adhesive	O	May remain in fresh or brackish water until 3-5 yrs. of age. O		1,7
Clupeidae—herrings. Blueback herring, *Alosa aestivalis*	Fresh and brackish waters not far above tidewater; April-June	Essentially pelagic	Fresh. O	Leave nursery grounds ≈ 50 mm mid Sept.-Oct.; O, M, P	Larvae April-June. Juveniles May-Nov.; James R., downstream migration almost complete by Nov. Some over-wintering in Del. and Chesapeake Bay	1,10
Hickory shad, *Alosa mediocris*	Tidal freshwater late April-early June	Demersal or pelagic	Fresh. O	Leave nursery areas early summer. O, M, P	Juveniles of age group I found sporadically throughout most of year, Ches. Bay and tributaries	1
Alewife, *Alosa pseudoharengus* (Spawning generally precedes blueback herring by 3-4 wks.; spawning peaks separated by 2-3 weeks)	Ascend freshwater streams further than blueback herring; late March-mid May	Essentially pelagic	Freshwater; form schools at <10 mm	Pass slowly down Ches. drainage system. O, M, P	Juveniles main seaward migration, fall	1,10
American shad, *Alosa sapidissima*	Tidal freshwater April-July	Demersal or pelagic; absent at <5 ppm. DO	Fresh. O. Most abundant at surface	Form schools 20-20 mm. O, M, P	Juveniles gradually move downstream; some remain in Chesapeake Bay for first year	1
Gizzard shad, *Dorosoma cepedianum*	Freshwater, near surface; April-June	Demersal, adhesive	Smallest larvae most abundant at surface; freshwater. Larger larvae surface day; midwater night	Greatest abundance well upstream from brackish water	Juveniles < 70 mm only in freshwater	1

Species	Spawning habitat/time	Egg type	Larvae	Juveniles/Adults	Ref.
Threadfin shad, *Dorosoma petenense*	Freshwater or 0 April-July	Demersal, attached	Freshwater; diel migration inshore	Prefer <15 ppt.; most common <5 ppt. O, M	1
Percichthyidae—temperate basses. White perch, *Morone americana*	Freshwater or O; late March-early June	Demersal, attached	Freshwater O; downstream movement with development	Estuarine populations move toward brackish water, Aug.-Nov. O (most), M	Larvae freshwater to at least 8 ppt.; greatest abundances at mid depths in water column (day). Juveniles ≈ 20-25 mm, move inshore shoal areas	3,46,47,48
Striped bass, *Morone saxatilis*	Tidal freshwater April-early June; peak Chesapeake Bay last half April-first week May	Pelagic	Freshwater O, move inshore to shoal areas ≈17 mm	Schools, more abundant areas with pronounced current. O, M	Larvae April-early June; concentrate at bottom; Juveniles general downstream movement late May	3,46,49,50,51

FRESHWATER/TIDAL FRESHWATER SPAWNERS

Species	Spawning habitat/time	Egg type	Larvae	Juveniles/Adults	Ref.
Esocidae – pikes. Chain pickerel, *Esox niger*. Adults maximum salinity 22 ppt.	Tidal freshwater; weedy, shallow areas; early spring, Feb. and March. DE	Initially demersal and attached; become free and semi-buoyant	Tidal freshwater; mostly tidal pools	Shallow waters	Larvae hide among vegetation. Juveniles lie motionless near shore or burrow in mud. NOTE: Juvenile redfin pickerel, *Esox americanus* are frequently found in impounded and tidal creeks of Delaware, though they are not common in the Potomac.	1,7,8
Cyprinidae — minnows and carps. Carp, *Cyprinus carpio*	Shallow marshes and flats, tidal freshwater, May-June	Demersal, attached	Freshwater O; in vegetation	In vegetation	Tidal tributaries of Chesapeake Bay	1,8
Silvery minnow, *Hybognathus nuchalis*	Tidal freshwater over soft bottoms; April-May	Essentially nonadhesive	Small schools among emergent vegetation		Tidal tributaries of Chesapeake Bay	1,8
Golden shiner, *Notemigonus crysoleucas*	Tidal freshwater; often in tidal pools; late March-Aug. possibly Oct.	Demersal, adhesive	Freshwater, O		Tidal tributaries and upper Chesapeake Bay	1,8

TABLE 4 (continued)
Tidal Freshwater and/or Oligohaline (0.5 - 5.0 ppt) Spawners Whose Larval and Juvenile Stages Utilize Chesapeake Bay as a Nursery Area.

Species	Spawning Season and Location	Eggs	Habitat — Larvae	Habitat — Juveniles	Occurrence of Larvae/Juveniles	References
Spottail shiner, *Notropis hudsonius*	Tidal freshwater, possibly low, brackish water; April-late Aug. or early Sept.	Demersal, clumps or scattered		Shallow water abundant vegetation	Tidal tributaries of Chesapeake Bay	1,8
Ictaluridae — freshwater catfishes. Channel catfish, *Ictalurus punctatus*	May-June; tidal freshwater, in nests, depressions or other protected areas	Demersal, adhesive	Guarded by male 2-5 days after hatching Freshwater. O	Freshwater, O, M	Larvae upper salinity tolerance ≈ 8 ppt.; Juveniles grow at salinities ≤ 11 ppt. at 5-6 mo.	1,8
White catfish, *Ictalurus catus*	May-June; tidal freshwater; large nests near sand or gravel banks	Demersal, adhesive, aerated by male parent		Initially guarded by parents. O, M	Juveniles remain in schools until end of first summer; capable of growth ≤11 ppt.	1,8
Brown bullhead, *Ictalurus nebulosus*	May-June; tidal freshwater in nests	Demersal, adhesive, aerated by parents		School throughout summer. O	Early juveniles herded about in schools by one or both parents	1,8
Percidae — perches. Tessellated darter, *Etheostoma olsmuedi*	May (perhaps April)-June; tidal freshwater	Demersal, guarded by male		Freshwater		3,8
Yellow perch, *Perca flavescens*	March-April; tidal freshwater; O (max salinity 2.5 ppt.)	Attached; long flat demersal, semibuoyant or rarely floating bands or ribbons	Pelagic, phototrophic, O	Large schools initially pelagic, then demersal; O, M	Larvae, end of March-mid May; Potomac and Patuxent estuaries Juvenile salinity range to 9.5 ppt.	3,46,48

Species	Spawning	Egg type	Parental care	Juvenile habitat	Ref.
Centrarchidae — sunfishes. Pumpkinseed, *Lepomis gibbosus*	Nests, slow-moving streams, mid May-mid Aug. (peak mid May-June); DE; O, M (max salinity 10 ppt.)	Demersal, attached	Remain in nests 2 days. O, M	O, M	3
Smallmouth bass, *Micropterus dolomieui*	April-July; nests over bottoms of gravel, rocks, or rubble near shore	Demersal, guarded by male	Remain in nests 4-9 days; guarded by male	Guarded by male parent to ≈ 25 mm	Smallmouth bass not recorded from tidewater of Potomac	3,8
Largemouth bass, *Micropterus salmoides*	Spring and early summer; warm, shallow bays or coves, usually in nests, tidal freshwater, O. Potomac May-July, peak early June	Demersal, attached	Leave nests for short intervals, 4th day; rise from nests 5-10 days; O	O. Max salinity ≈ 7 ppt.	Juveniles, shallow areas associated with aquatic vegetation; tight schools to ≈ 90 days	3,8
White crappie, *Pomoxis annularis*	Late May-Aug; peak June in Potomac; freshwater nests 2-20 ft.	Demersal, attached	Guarded by male parent	Freshwater. O	Juveniles recorded along beaches in Delaware River ≥ 10.5 mm; salinity < 2 ppt.	3,8
TIDAL FRESHWATER AND OLIGOHALINE WATER SPAWNERS					
Banded killifish, *Fundulus diaphanus*	April-Sept. fresh and brackish water. O. associated with aquatic vegetation	Demersal, attached to vegetation	Demersal, O	Among low weeds. O		2,8,10
Poeciliidae—livebearers. Mosquitofish, *Gambusia affinis*	May-Sept.; tidal and non-tidal freshwater ponds, marshes, and ditches	Livebearer; 2-8 broods per season	Surface schooling	In schools, high thermal tolerance. O		2,7
Atherinidae—silversides. Tidewater silverside, *Menidia peninsulae*. Typically estuarine or tidewater; now separated from *M. beryllina*, inland silverside	Mid April-mid Sept; tidal freshwater or breakish water; O; rooted vegetation preferred	Demersal, attached	Large schools, vegetated shallows; O, M surface waters	Abundant in shallow waters. O, M	Larvae and Juveniles 0-15 ppt;; peak conc. 2-8 ppt.	6

during winter months. Naked goby larvae are transported upstream from spawning areas into low salinity nursery areas.[13,33] Densities of naked goby larvae ranged from 12,700 to 71,500 per 1000m^3 of water filtered from late June through early August at two stations in oligohaline waters of the Patuxent River Estuary.[32] During late June, 1977, densities of naked goby larvae averaged over 18,000 per 1000m^3 of water filtered over a 28 km stretch of the Patuxent River.[33]

Feather and striped blennies are widely distributed around shallow flats and oyster reefs of the mid and lower Bay (12-25 ppt.) during spring and summer months. These species inhabit deeper, more saline waters during winter. Both species are abundant in oyster trays suspended off the Chesapeake Biological Laboratory pier.

Grass Bed Spawners

Northern and dusky pipefishes, lined seahorses, and fourspine sticklebacks typically reside among submerged aquatic vegetation (SAV). Northern pipefish ranked second in abundance (14% of total numbers) in a 14-month study of fishes associated with eelgrass beds of lower Chesapeake Bay.[43] Abundances of fishes associated with grass beds have drastically declined since the reduction and almost total disappearance of SAV from mid and upper Chesapeake Bay and its tributaries during the late 1960's and early 1970's.[43,44,45] Today their areas of abundance are restricted to surviving grassbeds (mostly eelgrass and widgeon grass) in the lower Bay, primarily on the Eastern Shore.

Seahorses and pipefishes are unique among fishes in that males posses brood pouches in which eggs are incubated and hatched. Young fish reside in these pouches during most or all of their larval stage.

FRESHWATER/OLIGOHALINE SPAWNERS

Fishes which spawn and utilize tidal freshwaters and/or oligohaline waters as nursery areas include anadromous species (species which spend most of their lives in marine waters and migrate to freshwaters to spawn), freshwater species (which tolerate low to mid-salinities), and resident (primarily oligohaline) fishes (Table 4). Larvae and juveniles of these fishes are found in the upper reaches of Chesapeake Bay and its tidal tributaries. These areas are most subject to environmental degradation and pollution. The early life history stages of fishes are the most sensitive to and adversely affected by environmental pollution.

Anadromous Species

True anadromous species which undertake spawning migrations from oceanic to freshwaters include alewives and blueback herrings, American shad, hickory shad, striped bass, and Atlantic sturgeon. Semi-anadromous species which migrate from the lower estuary to freshwaters to spawn include white perch, gizzard shad, estuarine populations of threadfin shad, and yellow perch. (Yellow perch are included with the freshwater species in Table 4). The deleterious effects of degrading water quality and destruction of spawning habitat in the Bay's freshwater spawning reaches, tidal freshwater, and oligohaline areas, are most evident in the Bay's anadromous species. All of these species are currently experiencing reduced populations.[52,53,54,55]

Alewives and blueback herring are often called river herring because of their spawning migrations. Alewives begin their anadromous migrations up Chesapeake Bay in late February or early March. They spawn primarily in large rivers and freshwater streams much farther upriver than blueback herring. Spawning in Maryland waters occurs from late March through April; peak spawn occurs the last two weeks of April (Table 4;[1,10]). Blueback herring spawn in fresh and brackish waters of the Bay and its tributaries, above tidewater from about mid-April to mid-May. Adults of both species gradually move downstream after spawning. By mid-summer these species have left for the ocean.[1,10]

Atlantic sturgeon and the endangered semi-anadromous shortnose sturgeon, *Acipenser brevirostrum* move into oligohaline and tidal fresh-waters to spawn. Spawning migrations of Atlantic sturgeon begin in early April in Chesapeake Bay and late April in the Delaware River. Atlantic sturgeon populations have been greatly reduced as a result of overfishing, habitat alteration, and pollution. This species is now regarded as threatened.[52,56]

American and hickory shad have also suffered great reductions in population size as a result of overfishing, environmental degradation, and habitat destruction.[52,53,55] Chesapeake Bay populations of both American and hickory shad are so low that in 1980 the fishery was closed in Maryland.[55] Striped bass have also declined so drastically that beginning 1 January, 1985 there has been a total mortorium on the catch, sale, or possession of striped bass in Maryland. Virginia has reduced commercial catches of striped bass by enacting a six-month closure of Chesapeake Bay fisheries from 2 December through 31 May, and raising the legal size limit to 24 inches total length (TL).

Most striped bass spawning within Chesapeake Bay and its tidal tributaries occurs in Maryland waters. Historically, in Maryland about half of striped bass spawning activity occurred in upper Chesapeake Bay, approximately 25% in the Potomac River, and the remainder primarily in the Choptank and Nanticoke Rivers of the Eastern Shore. Striped bass spawn in tidal freshwaters beginning when water temperatures reach about 11°C. Peak spawning occurs between 14 and 19°C (Table 4). In Maryland waters, spawning usually begins in early to mid-April, peaks from the third week of April to the first week of May, and is essentially completed by the end of May.[46,49,51] Spawning in Virginia Rivers may precede that in Maryland waters by one to two weeks. During 1982 and 1983, estimates of striped bass egg production in the Rappahannock River were an order of magnitude above estimates of striped bass egg production in the

James River and two orders of magnitude above those from the Pamunkey River.[57] The pelagic striped bass eggs hatch in about two days. During the next five to seven days, larvae depend upon the yolk-sac with its large oil globule for nourishment. During this yolk-sac stage, the gut completes development and the eyes become pigmented. Striped bass larvae begin feeding at about 5 mm, initially on rotifers, cladocerans, and copepodite stages of copepods. By 7 to 8 mm, copepodite and adult stages of copepods, primarily *Eurytemora affinis* and cyclopoid copepods; and cladocerans, primarily *Bosmina* and *Daphnia*, are the mainstay of the diet.[58] Striped bass larvae are concentrated near the bottom in deeper waters,[50] when they reach about 17 mm they move into shallow shoal areas.[59] Juveniles spend their first summer in shallow, near-shore waters. They gradually move downstream into oligohaline and mesohaline waters as the season progresses. Juvenile striped bass are opportunistic, generalist feeders which prey on copepods, amphipods, mysid shrimp, insect larvae, larval fish and polychaete worms.[51,60]

White perch spawn in both large rivers and smaller tributary streams from the end of March through June.[47] Eggs are not all released at once, and ovulation may continue for 10-12 days.[3] White perch larvae begin feeding at a smaller size (about 3 mm) than striped bass larvae. They initially feed on copepod mauplii and copepodites, and rotifers, but switch primarily to adult and copepodite stages of copepods and cladocerans by about 8 mm.[58] Juvenile white perch move toward brackish water from August through November and are generally found along the shoreline in shallow, sluggish water.[3]

Freshwater Species

Yellow perch, a freshwater species, have adapted exceptionally well to estuarine conditions. Estuarine populations apparently grow faster and live longer than those found in fresh water (maximum 10 to 12 years vs 5 to 7 years). Females also mature later in estuarine water.[61] Adult yellow perch are usually found in oligohaline and mesohaline waters of the Bay and its tributaries (up to 11 or 12 ppt. salinity) except during spawning season. There is evidence that adults may be indigenous to their own river system. If they are transported into another area, they leave the new stream almost immediately.[62] Currently, stocks of yellow perch in Maryland tidal waters are very low.[54]

Many freshwater fishes regularly descend into the tidal fresh and oligohaline waters of the Potomac River and other Bay tributaries, especially during winter months. Spawning usually occurs in fresh or tidal freshwaters, though some species may spawn in oligohaline waters (Table 4). Freshwater fishes common in oligohaline waters of major and smaller tributaries of the Bay include yellow perch, silvery minnow, spottail shiner, golden shiner, carp, chain pickerel, white catfish, channel catfish, and brown bullhead. Freshwater fishes common in tidal freshwater and oligohaline portions of smaller tributaries include threadfin shad, longnose gar, bluegill, pumpkinseed, goldfish, satinfin shiner, mosquitofish, and tessellated darter.[10] Those species whose larval and/or juvenile stages are commonly found in the oligohaline and tidal freshwater portions of Chesapeake Bay and its major tributaries are listed in Table 4.

REFERENCES

1. Jones, P.W., F.D. Martin, and J.D. Hardy, Jr. 1978. *Development of Fishes of the Mid-Atlantic Bight: An atlas of egg, larval, and juvenile stages.* Vol. 1. Acipenseridae through Ictaluridae. Fish Wildl. Serv., USDI, Biological Services Program, FWS/OBS-78/12, Washington, D.C., 366 pp.
2. Hardy, J.D., Jr. 1978. *Development of Fishes of the Mid-Atlantic Bight: An atlas of egg, larval, and juvenile stages.* Vol. II. Anguillidae through Syngnathidae. Fish Wildl. Serv., USDI, Biological Services Program, FWS/OBS-78/12, Washington, D.C., 458 pp.
3. Hardy, J.D., Jr. 1978. *Development of Fishes of the Mid-Atlantic Bight: An atlas of egg, larval, and juvenile stages.* Vol. III. Aphredoderidae through Rachycentridae. Fish Wildl. Serv., USDI, Biological Services Program, FSW/OBS-78/12, Washington, D.C., 394 pp.
4. Johnson, G.D. 1978. *Development of Fishes of the Mid-Atlantic Bight: An atlas of egg, larval, and juvenile stages.* Vol. IV. Carangidae through Ephippidae. Fish Wildl. Serv., USDI, Biological Services Program, FWS/OBS-78/12, Washington, D.C., 314 pp.
5. Fritzsche, R.A. 1978. *Development of Fishes of the Mid-Atlantic Bight: An atlas of egg, larval, and juvenile stages.* Vol. V. Chaetodontidae through Ophidiidae. Fish Wildl. Serv., USDI, Biological Services Program, FWS/OBS-78/12, Washington, D.C., 340 pp.
6. Martin, F.D. and G.E. Drewy. 1978. *Development of Fishes of the Mid-Atlantic Bight: An atlas of egg, larval, and juvenile stages.* Vol. VI. Stromateidae through Ogcocephalidae. Fish Wildl. Serv., USDI, Biological Services Programs, FWS/OBS-78/12, Washington, D.C., 416 pp.
7. Wang, J.C.S. and R.J. Kernehan. 1979. Fishes of the Delaware estuaries: A guide to the early life histories. EA Communications, Ecological Analysts, Inc., Sparks, MD, 410 pp.
8. Lippson, A.J. and R.L. Moran. 1974. Manual for the identification of early developmental stages of fishes of the Potomac River Estuary. Power Plant Siting Program, Maryland Dept. Nat. Res. PPSP-MP-13, Annapolis, MD., 282 pp.
9. Lippson, A.J. (Ed. and Ill.). 1973. Chesapeake Bay in Maryland: An atlas of natural resources: for Nat. Res. Inst., Univ. of Maryland, The Johns Hopkins Univ. Press, Baltimore, MD, 55 pp.
10. Lippson, A.J., M.S. Haire, A.F. Holland, F. Jacobs, J. Jensen, R.L. Moran-

Johnson, T.T. Polgar, and W.A. Richkus. 1979. Environmental atlas of the Potomac Estuary. Environmental Center, Martin Marietta Corp. for Power Plant Siting Program, Maryland Dept. Nat. Res., Annapolis, MD, 279 pp.
11. Lippson, A.J. and R.L. Lippson. 1984. Life in the Chesapeake Bay. Johns Hopkins University Press, Baltimore, MD, 229 pp.
12. Olney, J.E. 1983. Eggs and early larvae of the bay anchovy, *Anchoa mitchilli*, and the weakfish, *Cynoscion regalis*, in lower Chesapeake Bay with notes on associated ichthyoplankton. *Estuaries.* 6:20-35.
13. Dovel, W.L. 1971. Fish eggs and larvae of the upper Chesapeake Bay. Nat. Res. Inst., Univ. Maryland Spec. Rept. No. 4, College Park, MD, 71 pp.
14. Massmann, W.H. 1954. Marine fishes in fresh and brackish waters of Virginia Rivers. *Ecology.* 35:75-78.
15. Leggett, W.C. 1977. Density dependence, density independence and recruitment in the American shad (*Alosa sapidissima*) population of the Connecticut River, pp. 3-17. *In:* W. Van Windle (Ed.). *Proc. Conf. on Assessing the Effects of Power Plant-Induced Mortality and Fish Populations.* Pergamon Press, New York.
16. Norcross, J.J., W.H. Massmann, and E.B. Joseph. 1961. Investigations of inner continental shelf waters off lower Chesapeake Bay. Part II. Sandlance larvae, *Ammodytes americanus. Chesapeake Sci.* 2:49-60.
17. Mansueti, R.J. 1963. Symbiotic behavior between small fishes and jellyfishes, with new data on that between the stromateoid, *Peprilus alepidotus*, and the Scyphomedusa, *Chrysaora quinquecirrha. Copeia.* 1963:40-80.
18. Olney, J.E. and G.C. Grant. 1976. Early planktonic larvae of the blackcheek tonguefish, *Symphurus plagiusa* (Pisces: Cynoglossidae) in the lower Chesapeake Bay. *Chesapeake Sci.* 17:229-237.
19. Richkus, W.A., D. Mulryan, K. Zankel, and B. Kobler. 1979. Calvert Cliffs acoustic finfish surveys—1977. Calvert Cliffs Monitoring Program, Report Series, Ref. No. CC-79-8. Maryland Power Plant Siting Program, Annapolis, MD, 55 pp.
20. Nelson, W.R., M.C. Ingham, and W.E. Schaaf. 1977. Larval transport and year-class strength of Atlantic menhaden, *Brevoortia tyrannus. Fish. Bull.* 75:23-41.
21. Kendall, A.W., Jr. and L.A. Walford. 1979. Sources and distribution of bluefish, *Pomatomus saltatrix*, larvae and juveniles off the East Coast of the United States. *Fish. Bull.* 77:213-227.
22. Norcross, J.J., S.L. Richardson, W.H. Massmann, and E.B. Joseph. 1974. Development of young bluefish (*Pomatomus saltatrix*) and distribution of eggs and young in Virginia coastal waters. *Trans. Am. Fish. Soc.* 103:477-497.
23. Joseph, E.B. 1972. The status of the sciaenid stocks of the Middle Atlantic coast. *Chesapeake Sci.* 12:87-100.
24. Norcross, B.L. 1983. Climate scale environment factors affecting year-class

fluctuations of Atlantic croaker (*Micropogonias undulatus*) in the Chesapeake Bay. Ph.D. Dissertation, College of William and Mary, Gloucester Point, VA, 388 pp.
25. Norcross, B.L. and H.M. Austin. 1981. Climate scale environmental factors affecting year-class fluctuations of Chesapeake Bay croaker *Micropogonias undulatus*. Virginia Institute of Marine Science. Tech. Rept. No. 110, Gloucester Point, VA.
26. Norcross, B.L. and G.H. Shaw. 1983. The VIMS trawl survey: Juvenile Atlantic croaker. Virginia Institute of Marine Science. Data Rept. No. 22. Gloucester Point, VA.
27. Chao, L.N. and J.A. Musick. 1977. Life history, feeding habits, and functional morphology of juvenile sciaenid fishes in the York River Estuary, Virginia. *Fish Bull.* 75:657-702.
28. Thomas, D.L. and B.A. Smith. 1973. Studies of young of the black drum, *Pogonias cromis*, in low salinity waters of the Delaware Estuary. *Chesapeake Sci.* 14:124-130.
29. Able, K.W. and M. Castagna. 1975. Aspects of undescribed reproductive behavior in *Fundulus heteroclitus* (Pisces: Cyprinodontidae) from Virginia. *Chesapeake Sci.* 16:282-284.
30. Taylor, M.H., G.J. Leach, L. DiMichele, W.M. Levitan, and W.F. Jacob. 1979. Lunar spawning cycle in the mummichog, *Fundulus heteroclitus* (Pisces: Cyprinodontidae). *Copeia.* 1979:291-297.
31. Dovel, W.J., J.A. Mihursky, and A.J. McErlean. 1969. Life history aspects of the hogchoker, *Trinectes maculatus,* in the Patuxent River Estuary, Maryland. *Chesapeake Sci.* 10:104-119.
32. Setzler, E.M., K.V. Wood, D. Shelton, and G. Drewry. Chalk Point Stream Electric Station Studies: Patuxent Estuary Studies: Ichthyoplankton Population Studies. 1978 Data Rept. Univ. of Maryland Center for Environ. and Estuarine Studies, Ches. Biol. Lab., CEES Ref. No. 79-20-CBL. Solomons, MD, 111 pp.
33. Shenker, J.M., D.J. Hepner, P.E. Frere, L.E. Currence, and W.W. Wakefield. 1983. Upriver migration and abundance of naked goby, *(Gobiosoma bosci)* larvae in the Patuxent River Estuary, Maryland. *Estuaries.* 6:36-42.
34. Massmann, W.H., J.J. Norcross, and E.B. Joseph. 1963. Distribution of larvae of the naked goby, *Gobiosoma bosci* in the York River. *Chesapeake Sci.* 4:120-125.
35. Richardson, S.L. and E.B. Joseph. 1975. Occurrence of larvae of the green goby, *Microgobius thalassinus,* in the York River, Virginia. *Chesapeake Sci.* 16:215-218.
36. Dovel, W.L. 1963. Larval development of the clingfish, *Gobiesox strumosus,* 4.0 to 12.0 mm TL. *Chesapeake Sci.* 4:161-166.
37. Dovel, W.L. 1960. Larval development of the oyster toadfish, *Opsanus tau. Chesapeake Sci.* 1:187-195.

38. Dovel, W.L. 1967. Fish eggs and larvae of the Magothy River, Maryland. *Chesapeake Sci.* 8:125-129.
39. Schauss, R.P., Jr. 1977. Seasonal occurrence of some larval and juvenile fishes in Lynnhaven Bay, Virginia. *Am. Midl. Nat.* 98:275-282.
40. Pearson, J.C. 1941. The young of some marine fishes taken in lower Chesapeake Bay, Virginia, with special reference to the grey sea trout, *Cynoscion regalis* (Black). *U.S. Fish Wildl. Serv. Fish Bull.* 50:79-102.
41. Lotrich, V.A. 1975. Summer home range and movements of *Fundulus heteroclitus* (Pisces: Cyprinodontidae) in a tidal creek. *Ecology.* 56:191-198.
42. Runyan, S. 1961. Early development of the clingfish, *Gobiesox strumosus. Chesapeake Sci.* 2:113-141.
43. Orth, R.J. and K.L. Heck. 1980. Structural components of eelgrass (*Zostera marina*) meadows in the lower Chesapeake Bay—Fishes. *Estuaries.* 3:278-288.
44. Boynton, W.R. and K.L. Heck. 1983. Ecological role and value of submerged macrophyte communities: A scientific summary, pp. 431-501. *In:* E.G. Macalaster, D.A. Barker, and M. Kasper (Eds.), *EPA Chesapeake Bay Program. Technical Studies: A Synthesis.* U.S. Environmental Protection Agency, Annapolis, MD.
45. Orth, R.J. and K.A. Moore. 1983. Distribution and abundance of submerged aquatic vegetation in Chesapeake Bay: A scientific summary, pp. 384-430. *In:* E.G. Macalaster, D.A. Barker, and M. Kasper (Eds.), *EPA Chesapeake Bay Program. Technical Studies: A Synthesis.* U.S. Environmental Protection Agency, Annapolis, MD.
46. Setzler-Hamilton, E.M., J.A. Mihursky, K.V. Wood, W.R. Boynton, D. Shelton, M. Homer, S. Kerig, and W. Caplins. 1980. Potomac Estuaries Fisheries Program. Ichthyoplankton and juvenile investigations. 1977 Final Rept., Univ. of Maryland CEES Ref. No. 79-202 CBL. Solomons, MD, 175 pp.
47. Mansueti, R.J. 1964. Eggs, larvae, and young of the white perch, *Roccus americanus,* with comments on its ecology in the estuary. *Chesapeake Sci.* 5:3-45.
48. Mihursky, J.A., K.V. Wood, S. Kerig, and E.M. Setzler-Hamilton. 1980. Chalk Point Steam Electric Station Studies: Patuxent Estuaries Studies; Ichthyoplankton Population Studies, 1979 Final Rept. Submitted to Maryland Dept. Nat. Res., Power Plant Siting Program, Univ. of Maryland Center for Environ. and Estuarine Studies, Ches. Biol. Lab., Ref. No. UMCEES 80-39-CBL, Solomons, MD, 101 pp.
49. Setzler-Hamilton, E.M., W.R. Boynton, J.A. Mihursky, T.T. Polgar, and K.V. Wood. 1981. Spatial and temporal distribution of striped bass eggs and larvae in the Potomac Estuary. *Trans. Am. Fish. Soc.* 110:121-136.
50. Mihursky, J.A., E.M. Setzler-Hamilton, and F.D. Martin. 1981. Spatial and temporal patterns of utilization of the Potomac Estuary by fish larvae, pp.

20-27. *In:* C.F. Bryan, J.V. Conner, and F.M. Truesdale (Eds.), *Proc. 5th Ann. Larval Fish Conf* (March 2-3, 1981). Louisiana State Univ., Baton Rouge, LA.
51. Sertzler, E.M., W.R. Boynton, K.V. Wood, H.H. Zion, L. Lubbers, N.K. Mountford, P. Frere, L. Tucker, and J.A. Mihursky. 1980. Synopsis of biological data on striped bass, *Morone saxatilis (*Walbaum). USDC, NOAA, Tech. Rept., NMFS Cir. 433, 69 pp.
52. Wiley, M.L. 1984. Endangered and threatened marine and estuarine fishes in Maryland, pp. 281-286. *In:* A.W. Norden, D.C. Forester, and G.H. Fenwick (Eds.), *Threatened and endangered plants and animals of Maryland.* Maryland Natural Heritage Program, Spec. Publ. 84-1, Maryland Dept. Nat. Res. Annapolis, MD.
53. Stagg, C. 1985. An evaluation of the information available for managing Chesapeake Bay Fisheries: Preliminary stock assessments, Vol. I. Chesapeake Bay Commission. Univ. of Maryland Center for Environ. and Estuarine Studies, Ches. Biol. Lab., Solomons, MD. UMCEES[CBL] 85-29, Solomons, MD.
54. Stagg, C. In press. An evaluation of the information available for managing Chesapeake Bay Fisheries: Preliminary stock assessment, Vol. III. Chesapeake Bay Commission. Univ. of Maryland Center for Environ. and Estuarine Studies, Ches. Biol. Lab., Solomons, MD.
55. Richkus, W.A. and G. DiNardo. 1984. Current status and biological characteristics of the anadromous alosid stocks of the eastern United States: American shad, hickory shad, alewife, and blueback herring. Atlantic States Marine Fisheries Commission, Fisheries Management Report 4, Wash. D.C. XIX + 225 pp.
56. Deacon, J.E., G. Kobetich, J.D. Williams, S. Contreras, and other members of the Endangered Species Committee of the North American Fisheries Society. 1979. Fishes of North America, endangered, threatened, or of special concern: 1979. *Fisheries.* 4:29-44.
57. Olney, J.E., G.C. Grant, and G. Hill. 1985. Relative contribution of three Virginia Rivers to spawning activity of striped bass, *Morone saxatilis.* Project AFC-14-1, Completion report to USDC, NOAA, Natl. Mar. Fish. Serv., P.L. 89-304, Virginia Institute of Marine Science and School of Marine Science, College of William and Mary, Gloucester Point, VA. 53 pp.
58. Setzler-Hamilton, E.M., P.W. Jones, F.D. Martin, K. Ripple, and J.A. Mihursky. 1981. Comparative feeding habits of white perch and striped bass larvae in the Potomac Estuary. pp. 139-157. *In:* W.A. Richkus (Ed.), *Proc. 5th Ann. Meeting Potomac Chapter, American Fisheries Soc.,* Harpers Ferry, WVA, May 28-29, 1981.
59. Martin, F.D. and E.M. Setzler-Hamilton. 1983. Assessment of larval striped bass stocks in the Potomac Estuary. A final report for the years 1980, 1981, and 1982. Submitted to the U.S. Natl. Mar. Fish. Serv. UMCEES Ref. No.

83-55-CBL. Solomons, MD.
60. Boynton, W.R., T.T. Polgar, and H.H. Zion. 1981. Importance of juvenile striped bass food habits in the Potomac Estuary. *Trans. Am. Fish. Soc.* 110:56-63.
61. Muncy, R.J. 1962. Life history of the yellow perch, *Perca flavescens,* in estuarine waters of Severn River, a tributary of Chesapeake Bay, Maryland. *Chesapeake Sci.* 3:143-159.
62. Mansueti, R.J. 1951. Comparison of the movements of stocked and resident yellow perch, *Perca flavescens,* in tributaries of Chesapeake Bay. *Chesapeake Sci.* 1:21-35.

Contaminant Problems and Management of Living Chesapeake Bay Resources. Edited by S. K. Majumdar, L. W. Hall, Jr. and H. M. Austin. © 1987, The Pennsylvania Academy of Science.

Chapter Five

WATERFOWL OF CHESAPEAKE BAY
MATTHEW C. PERRY

U.S. Fish and Wildlife Service
Patuxent Wildlife Research Center
Laurel, MD 20708

ABSTRACT

The long-term average (1948-86) of Chesapeake Bay waterfowl populations during January has been approximately one million birds. During the 1980's, this average was still one million, although major changes in species composition had occurred. For example, Canada goose and snow goose populations during the 1980's were 75% and 250% higher, respectively, than their average populations before 1980. This is directly related to their ability to utilize the abundant field crop resources (mainly corn) on the eastern shore.

Among the ducks, only the mallard, gadwall, and bufflehead populations during the 1980's were higher than their average populations during the 32-year period of 1948-79. All other species of ducks have shown significant declines, which probably is a direct result of the degradation of waterfowl habitat in Chesapeake Bay. Duck populations in Chesapeake Bay can be expected to remain at low levels until submerged aquatic vegetation beds recover in the Bay, and production improves in the breeding areas.

INTRODUCTION

Few areas in the world are as well-known for wintering waterfowl as Chesapeake Bay. This 180-mile long bay, with 4,000 miles of shoreline and extensive shoal water areas, provides optimal foraging habitat for millions of waterfowl, and numerous rest areas in protected coves and rivers. Chesapeake Bay waterfowl are of longstanding interest to sportsmen and nature lovers who have been acutely aware of changes in the distribution and abundance of their favorite species.

Many factors are responsible for yearly changes in migratory waterfowl populations, and the causes of these fluctuations are complex and often difficult to investigate. Natural phenomena (e.g., water conditions on the breeding grounds) result in annual fluctuation of waterfowl production. Mortality due to disease, weather, and other natural causes can further alter population densities. Hunting regulations are evaluated yearly and can be changed in response to natural production.

From approximately 1870 to 1910, waterfowl wintering on Chesapeake Bay sustained the largest market hunting business known to man.[1,2] Accounts by sportsmen and naturalists relate how the water areas were covered with ducks.[3,4,5] Waterfowl were killed by the thousands, stuffed into barrels for transport by train to the major cities in the East, and subsequently sold.

A decreasing number of waterfowl in Chesapeake Bay early in the 20th century aroused concern among Americans. In 1918, market hunting was outlawed with the historically important treaty between the United States of America (USA) and Great Britain (for Canada). Waterfowl populations began to slowly increase in North America until the drought of the 1930's. Coupled with excessive drainage of northern breeding areas, the drought resulted in population declines that again aroused the public to the plight of waterfowl. New hunting regulations in 1935 outlawed the use of live decoys and bait while hunting. The now well-known "duck stamp" program was initiated in 1935 in order to provide funds for the establishment of additional waterfowl refuges in the USA.

During the 1960-80's the public became increasingly concerned about effects of chemical contaminants and other pollutants on waterfowl habitats. Chesapeake Bay, with huge metropolitan areas on the western shore, received the brunt of this abuse, resulting in continued habitat degradation. During this period biologists became aware that submerged aquatic vegetation—prime waterfowl food—had disappeared in many areas of the Bay.[6,7,8,9] Parts of some rivers, especially in the Upper Bay region, became totally devoid of plants. Turbidity and herbicides are among the many factors implicated as the cause of vegetation declines.[10]

Although there have been long-term declines in native submerged aquatic vegetation, some exotic species of aquatic plants had phenomenal increases. Eurasian watermilfoil *(Myriophyllum spicatum)* was reported in 1881 in the Potomac River[11] and water chestnut (*Trapa natans*) was observed in the Potomac in 1923.[12] Both plant species increased their distribution dramatically and restricted growth of wildcelery (*Vallisneria americana*) and other preferred submerged plants. The eventual declines of watermilfoil and water chestnut enabled preferred aquatics to reappear. During the 1980's the exotic plant *Hydrilla verticillata* became dominant in the Potomac River,[13] but unlike other exotics it appears to be a beneficial food item for ducks.

Some species of invertebrates have increased their distribution and abundance in recent years. The brackish-water clam (*Rangia cuneata*), first reported in the Bay by Pfitzenmeyer and Drobeck,[14] is now found throughout the Bay in areas of low salinities (< 15 ppt). The exotic Asiatic freshwater clam (*Corbicula manilensis*) was first reported in the James River by Diaz.[15] *Corbicula* has since been found in the Potomac River[16] and in the Susquehanna Flats.[17] Five species of puddle ducks from the Chesapeake Bay area have used *Corbicula manilensis* as a food item.[18]

The objective of this report is to discuss the present status of the major waterfowl species in the Chesapeake Bay on the basis of 39 years (1948-86) of winter survey data.

METHODS

Adequate censusing of the numbers and distribution of waterfowl using Chesapeake Bay from year to year has been important to waterfowl managers. These surveys were developed by the U.S. Fish and Wildlife Service (USFWS) in cooperation with the states, in order to assess the relative numbers of wintering waterfowl in various regions of the United States.[19]

All survey data used to determine waterfowl trends were obtained from published[19] and unpublished data in files of the Office of Migratory Bird Management, USFWS, Laurel, Maryland. Chesapeake Bay data represent estimates from Maryland and Virginia combined, and were collected by personnel from both states. Aerial surveys were flown at low levels (25-100 m) with single engine aircraft (twin engine for a few years in 1950's) of the USFWS and various state wildlife agencies. Surveys in Chesapeake Bay have been conducted

TABLE 1

High, low, and average populations of 15 waterfowl species in Chesapeake Bay, 1948-86, as determined by aerial winter surveys.

Species	High Count	(Year)	Low Count	(Year)	39-year mean	1980's mean
Tundra swan	75,854	(1955)	18,216	(1948)	36,710	35,065
Canada goose	701,470	(1981)	62,130	(1948)	382,760	590,335
Snow goose	126,000	(1985)	17	(1955)	26,484	63,941
Black duck	281,485	(1955)	28,820	(1979)	84,197	51,365
Mallard	182,195	(1956)	8,235	(1949)	51,212	57,553
Gadwall	21,183	(1980)	100	(a)	2,325	4,408
Wigeon	144,350	(1955)	900	(1984)	29,246	5,226
Pintail	78,211	(1956)	400	(1970)	16,282	3,935
Canvasback	399,320	(1954)	34,300	(1986)	104,012	52,931
Redhead	118,800	(1956)	800	(1983)	35,288	3,506
Scaup	403,658	(1954)	10,700	(1982)	65,909	28,973
Goldeneye	40,518	(1956)	2,445	(1976)	19,659	17,513
Bufflehead	36,023	(1977)	2,502	(1959)	14,813	16.840
Ruddy duck	124,740	(1953)	4,703	(1976)	33,642	15,729
Scoter	130,900	(1971)	1,551	(1981)	16,760	6,565

(a) Less than 100 recorded in 1948, 1951, 1953, 1956, 1959, 1969, and 1970.

since January 1948. Surveys are conducted in early January because waterfowl populations are more stable and concentrated than at other times during the winter.

Each major waterfowl species is compared to the status of populations in the Atlantic Flyway and North America in order to determine if changes detected in Chesapeake Bay during the 39-year period (1948-86) were related to conditions in the Bay, or to Flyway or Continental population levels. The average number of waterfowl recorded between 1980-86 was compared with the average number during the 32 years before 1980 (1948-79) to determine present status of waterfowl. Survey data in graphs are presented as five-year averages (except for the four-year period, 1983-86); these averages tend to minimize annual fluctuations and to emphasize long-term trends. Further discussion on survey techniques and data analysis are given in Perry et al.[20]

SPECIES ACCOUNTS

Tundra Swan—*Cygnus columbianus*

Tundra swan populations in Chesapeake Bay were variable during the 39-year

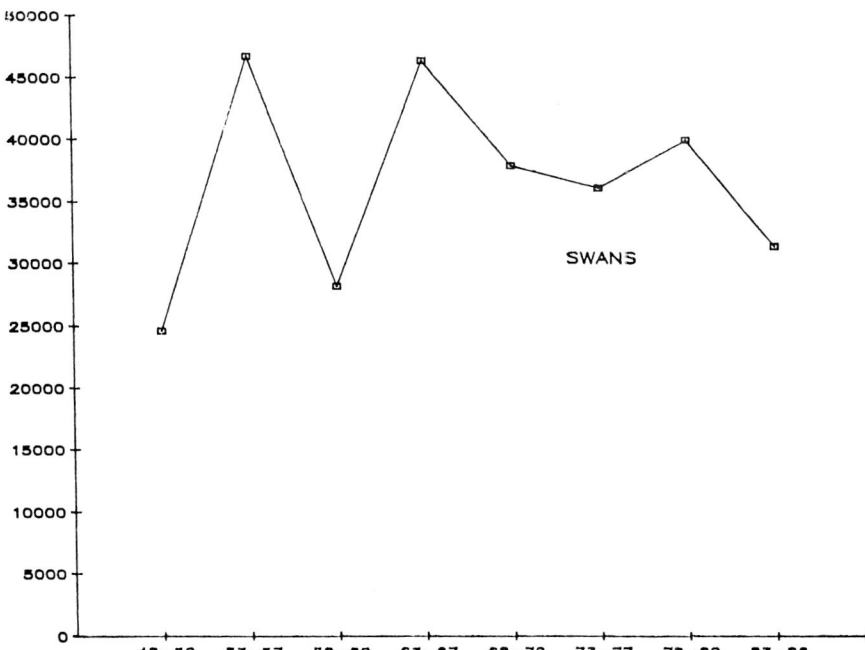

FIGURE 1. Long-term population trends (1948-86) of tundra swans in Chesapeake Bay during 8 periods.

period of aerial surveys (Table 1; Fig. 1). Lowest numbers occurred at the beginning of the surveys in 1948 (18,216) and then peaked in 1955 (75,854). Populations were also high in the mid-1960's. The long-term average population was 36,710. The average recorded during the 1980's was 35,065 which was only 5% less than the pre-1980 average of 37,070.

In the early years of the surveys, swans in Chesapeake Bay were found mostly in the lush aquatic vegetation beds in the central portions of the eastern shore. In the late 1960's and early 1970's, however, tundra swans began to feed extensively in agricultural fields on waste corn and winter cover crops. Although most of this feeding occurred on the Eastern Shore one large flock of approximately 1,000 was seen regularly in farm fields near Benedict, Maryland, not far from the Patuxent River. The use of fields for feeding areas occurred in the Bay area[21] at the same time that submerged aquatic vegetation beds in the Bay were disappearing. Stewart[22] did not mention field feeding by swans. The swans have apparently adapted to an alternate feeding pattern which appears to be to their advantage.

Chesapeake Bay historically has been the most important wintering area for tundra swans in North America, and in the early years of the survey, population trends in Chesapeake Bay, Atlantic Flyway, and North America were similar. During the 1979's and 1980's however, an increasing number of tundra swans were recorded in North Carolina. During this period more than half of the Atlantic Flyway population was recorded in North Carolina. This change in distribution was most likely due to increased number of agricultural areas in North Carolina and to less human disturbance in these feeding areas. Agricultural fields in North Carolina tend to be larger than in the Chesapeake Bay area, which also may favor the large swans during take off. The increased population and purported damaged to agricultural areas (eating and trampling winter wheat) by the swans were two reasons given for establishing special hunting regulations for tundra swans in North Carolina during the 1984-85 and 1985-86 hunting seasons.

Mute Swan—*Cygnus olor*

The European mute swan was introduced into the New York-New Jersey area prior to 1920.[23] mated pairs escaped from impoundments along tributaries of east Chesapeake Bay in 1962 and the population rapidly increased to 18 in 1968, 151 in 1974, and over 300 in 1979.[24,25,26]

Mute swans were first recorded during aerial winter surveys of Chesapeake Bay in 1975, but the maximum number recorded has been only 100. Mute swans may be mistaken for the more common tundra swans that also occur in the same habitat. Surveys conducted during summer months give a more accurate assessment of mute swan populations, which appear to be stabilizing due to high juvenile mortality from turtles.

Although there have been no known food habits analyses conducted with mute swans, observations have been made of their foraging behavior. Mute swans appear to be completely vegetarian, and this is similar to observations recorded in Rhode Island populations.[27] Because they are large and remain in the Bay throughout the summer, they have been blamed for causing declines in amounts of submerged aquatic vegetation. Based on limited observations, mute swans appear capable of denuding aquatics in some small coves but because of their current low numbers, are an insignificant factor throughout the Bay.

Canada Goose—*Branta canadensis*

Canada goose populations in the Chesapeake Bay area have undergone phenomenal changes during the 39 years of winter surveys (Table 1; Fig. 2). As was the case with tundra swans, lowest numbers occurred at the beginning of the survey in 1948.[62,130] The population steadily increased and peaked in 1981 (701,470). The long-term average was 382,760. The average during the 1980's was 590,335 geese, which was 75% higher than the pre-1980's average (337,352). Overall, population trends in Chesapeake Bay during the 39-year period were not similar to trends in the Atlantic Flyway and North America. Populations south of Chesapeake Bay (mainly North Carolina and Florida) declined during this period, whereas continental trends have been variable.

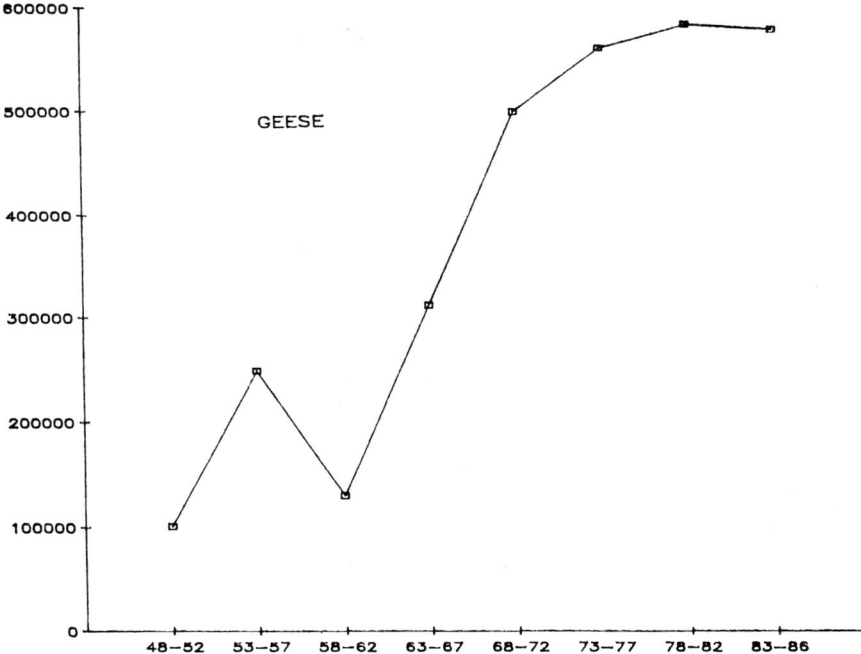

FIGURE 2. Long-term population trends (1948-86) of Canada geese in Chesapeake Bay during 8 periods.

This dramatic increase in Canada goose populations appears to be directly related to their ability to capitalize on abundant food in the agricultural areas of the eastern shore of the Chesapeake Bay. Waste corn available to geese after harvesting provided the necessary high energy food for geese at the same time that their traditional foods of emergent and submergent plants were declining throughout the Bay. Corn provides 1560 kcal/$_{kg}$ of food which is the most of all major seeds. Goose populations continued to increase during the 1970-80's despite liberal hunting regulations for this species. By feeding on high energy food, geese were able to feed less frequently and were therefore exposed to less hunting pressure than species constantly searching for food. By maintaining excellent body condition throughout the winter, geese were in good breeding condition when they reached nesting areas in northern Canada. Geese usually improve their breeding condition by feeding at James Bay, Canada, before the final flight north.

There is concern among hunters and biologists that increased hunting pressure on the eastern shore of the Bay and abundant food resources in Pennsylvania and New York may result in fewer geese migrating to the Delmarva peninsula. This "shortstopping" phenomenon began during the 1960's and resulted in reductions to the historically large goose populations in the Southeast, especially North Carolina. Canada geese will minimize their southward migration, especially to areas of high human habitation that are heavily hunted, as long as snow-free corn fields and ice-free water areas are abundant in more northern areas.

An increasing number of non-migratory (resident) Canada geese are recorded in the Chesapeake Bay area. Although no surveys are conducted for these geese during summer months, the resident population is estimated to be approximately 2,000. These geese breed, feed, and rest on refuges, golf courses, and large estates in the Bay area. They sometimes cause problems in these areas, and hazing and deterrent programs are conducted to lessen their adverse impacts.

Snow Goose—*Chen caerulescens*

Snow goose populations in the Chesapeake Bay area are composed of two subspecies: the greater snow goose (*Chen caerulescens atlantica*), and the lesser snow goose (*Chen caerulescens caerulescens*). The lesser snow goose has traditionally occurred in small flocks where up to 50% of the flock was the blue color phase. These flocks have been recorded at Blackwater National Wildlife Refuge (NWR) in Maryland and in fields near Presquile NWR on the James River in Virginia. Flock size in each of these two areas has ranged from 300 to 1000 geese. Lesser snow geese are usually observed in harvested corn fields where they most likely are feeding on waste corn and various green herbaceous plants. Stewart[22] reported the food contents of the gullet and gizzard of one blue goose collected in Dorchester County during November was composed entirely of leaf fragments of saltmarsh cordgrass (*Spartina alternifolia*).

The greater snow goose traditionally has wintered along the Atlantic coast from New Jersey to North Carolina where they feed exclusively on saltmarsh cordgrass. During the 1980's, large flocks of greater snow geese have been recorded in harvested corn fields and winter wheat fields in Maryland, especially in the Ruthsburg, Sudlersville, and Barclay areas. These flocks have been observed flying to and from Bombay Hook NWR where they spend the night. Some flocks of snow geese spend the night in Maryland on farm ponds and rivers (e.g., Corsica and Chester) near areas where they feed.

Greater snow geese populations in the Chesapeake Bay area (including coastal bays) have ranged from a minimum of less than 100 in 1951, 1954, and 1955 to a maximum of 126,000 in 1985. The long-term average population was 26,484 geese. During the 1980's the population averaged 63,941, which is 250% greater than the pre-1980's average of 18,290. More snow geese were recorded during the 1980's in Maryland than in any other state of the Atlantic Flyway.

American Black Duck—*Anas rubripes*

The black duck has traditionally frequented the eastern coast, and large numbers have been recorded in Chesapeake Bay (Table 1; Fig. 3). The highest number of black ducks in the Bay was recorded in 1955 (281,485) and the lowest in 1979 (28,820). The long-term average population was 84,197 ducks. During the 1980's the population averaged only 51,365 ducks, which was 44% lower than the pre-1980's average of 91,379.

During the 1950's approximately half the Atlantic Flyway black ducks were recorded in Chesapeake Bay. During the 1960's and 1970's only one third were recorded in the Bay, and during the 1980's less than one fourth of the Atlantic Flyway black ducks wintered in Chesapeake Bay. Although black duck populations have declined most dramatically in Chesapeake Bay, declines have been noted in all wintering areas of the Flyway. Surveys now record black ducks most frequently in coastal areas of New Jersey.

During the 1950's, approximately 85% of the Maryland black ducks were recorded in sections of the eastern shore of the Bay, especially the Chester River.[22] With the demise of the submerged aquatic vegetation, black ducks did not have an alternate food source readily available. Most black ducks in Chesapeake Bay during recent years were found on fresh water areas of the Patuxent River and tributaries of the York and James rivers. In these areas, black ducks feed on seeds of smartweeds (*Polygonum* spp.), rice cutgrass (*Leersia oryzoides*), and other marsh plants.[8] Small flocks of black ducks are also recorded throughout the saltmarsh cordgrass marshes in the brackish areas of the Bay, where the salt marsh snail (*Melampus bidentatus*) is their predominant food. Black ducks have also been observed feeding in corn fields in agricultural areas from the Chester to the Nanticoke Rivers.

The black duck was the main breeding duck in the Chesapeake area, and nesting habitat included woods, fields, marshes, and duck blinds. Stotts and

Davis[28] found that 60% of the black ducks observed throughout Chesapeake Bay (1953-59) nested in wooded areas. Nest density varied from 0.6 to 15.2 nests per acre. In 1986, nesting areas originally surveyed by Scotts[29] were resurveyed. Many of the islands had disappeared or were reduced in size by erosion. Housing and marina developments have greatly altered habitats. Black duck production in general was greatly reduced.

Mallard—*Anas platyrhynchos*

The mallard has traditionally been mainly a Mississippi Flyway duck, but populations tend to spill over to other flyways. Mallard population trends in the Bay are similar to those of the black duck (Table 1; Fig. 3). Populations in Chesapeake Bay were lowest in the late 1940's and early 1950's, with a low count in 1949 (8,235). Excellent breeding conditions in the prairie provinces of Canada in the mid-1950's caused populations to rise, and a peak wintering population in Chesapeake Bay occurred in 1956 (182,195). Drought conditions in the late 1950's and early 1960's cause populations to decrease and to remain relatively low and stable throughout the 1970's.

In the mid-1970's, large numbers of game-farm mallards were released in Chesapeake Bay with releases continuing throughout the 1980's. The release

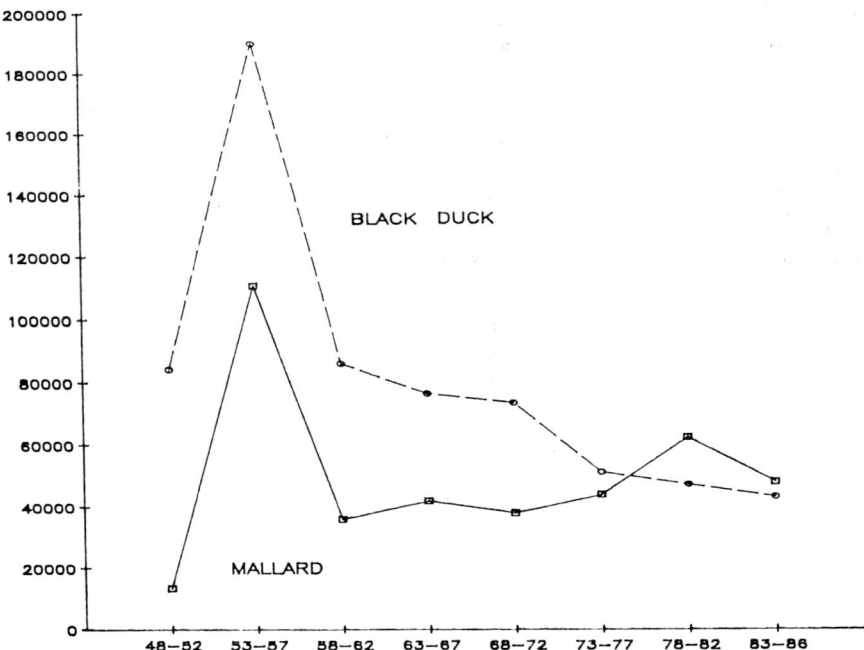

FIGURE 3. Long-term population trends (1948-86) of black ducks and mallards in Chesapeake Bay during 8 periods.

program is probably a major reason that mallard numbers in Chesapeake Bay were 16% higher during the 1980's (57,553) than the pre-1980 average (49,826). Many of these game-farm mallards are found in close association with man, and appear to adapt to changing environmental conditions more readily than the closely related black duck. Mallards were more numerous than black ducks in Chesapeake Bay during eight of the last ten years (1977-1986). The long-term average population of mallards in Chesapeake Bay was 51,212.

Stewart[22] found that seeds of smartweeds, bulrushes *(Scirpus* spp.), and burreed (*Sparganium americana*) predominated in the mallard diet in freshwater areas. In brackish areas, seeds, leaves, and stems of submerged aquatic vegetation were more important as food sources. Rawls[30] found submerged aquatic vegetation as the predominant food during the 1960's, whereas Munro and Perry[31] found only 5% of the food eaten by mallards was submerged aquatic vegetation during the 1970's. Seeds of a variety of marsh plants (over 100 species) were the predominant foods.

Mallards on the Potomac River near Washington, D.C., fed on the aquatic plant *Hydrilla verticillata* during 1984-85. Approximately 85% of the food in the gullet and 72% of the gizzard food of nine mallards was *Hydrilla*. The part of the plant most commonly eaten was the turions (winter buds), but leaves

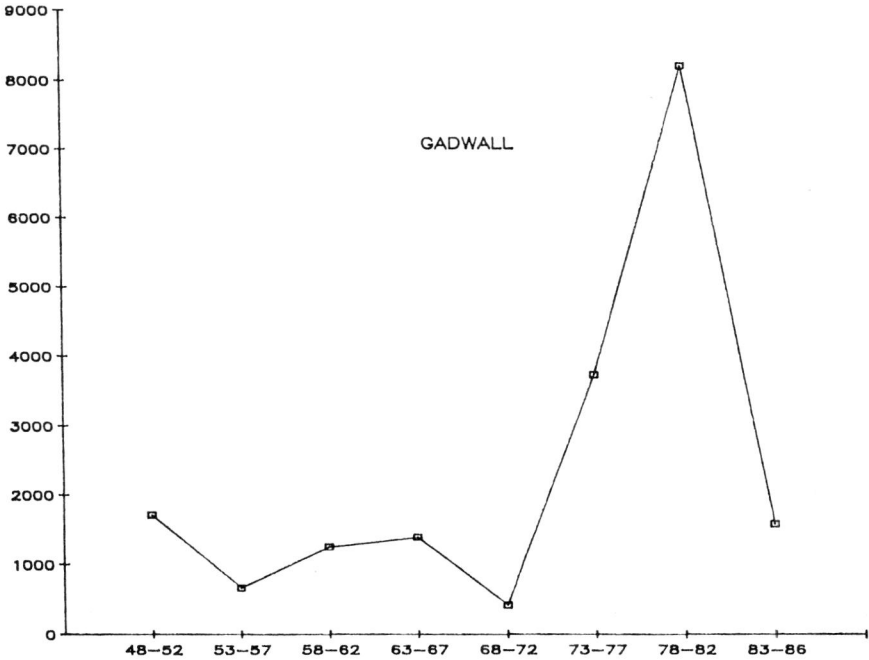

FIGURE 4. Long-term population trends (1948-86) of gadwall in Chesapeake Bay during 8 periods.

and stems were also commonly eaten. *Hydrilla* could therefore become a very important food item for future mallard populations in the Chesapeake Bay area.

Gadwall—*Anas strepera*

Gadwall populations in Chesapeake Bay have always been relatively low when compared to other duck species. Less than 100 gadwall were recorded in the Bay in 1948, 1951, 1953, 1956, 1959, 1969, and 1970 during aerial winter surveys. Peak population occurred in 1980 (21,183) and the long-term average was 2,325. During the 1980's, the winter population averaged 4,408 which is 236% more than the pre-1980's average population of 1,869 (Fig. 4).

Stewart[22] found that wigeongrass (*Ruppia maritima*) was the predominant food of 24 gadwall from brackish areas of Chesapeake Bay, followed by eelgrass (*Zostera marina*) and muskgrass (*Chara* sp.). Animal food in the form of fish was found in only 4% of the gadwall. Although Stewart[22] reported gadwall in areas where eurasian watermilfoil occurred, he did not report it as a food item. Florschutz[32] found that gadwall consumed milfoil more than any other dabbling ducks that were analyzed.

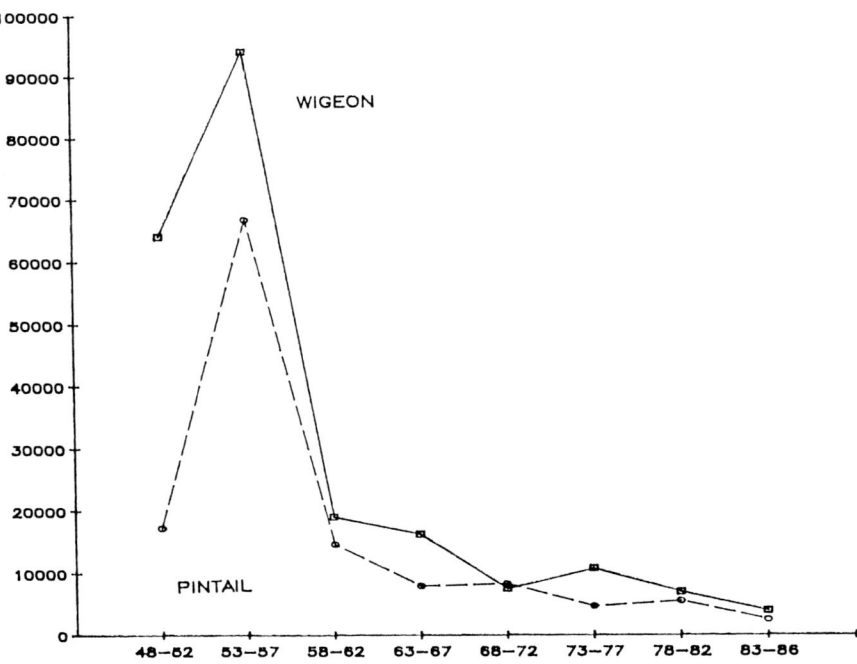

FIGURE 5. Long-term population trends (1948-86) of wigeon and pintail in Chesapeake Bay during 8 periods.

American Wigeon—*Anas americana*

Wigeon populations in Chesapeake Bay declined during the years of aerial surveys (Table 1; Fig. 5). Peak populations occurred in 1955 (144,350), most likely due to excellent production in the breeding provinces of Canada. Wigeon numbers declined to a low of only 900 ducks in 1984. The long-term average winter population was 29,246. During the 1980's the winter population averaged only 5,226 ducks, which was 85% lower than the pre-1980's average of 34,500. Population declines of wigeon in Chesapeake Bay occurred at a faster rate than those in the Atlantic Flyway and in North America.

Wigeon in Chesapeake Bay have traditionally been associated with canvasbacks and tundra swans. This species usually feeds in vegetated areas. During the 1950's, over 80% of the total Bay population was recorded along the eastern shore of the Bay.[22] Wigeon typically ate the upper vegetated parts of plants that were discarded or dislodged by canvasbacks or other waterfowl, although they also fed on winter buds of wildcelery.[22] During the 1960's, wigeon ate eurasian watermilfoil more than any other duck species[30] and during this period there were more wigeon in the mid-Bay area than other ducks. Because the wigeon was apparently unable to change to alternate food sources, as some other species did, wigeon numbers in the Bay have declined as the amount of vegetation decreased.

Northern Pintail—*Anas acuta*

The pintail is mainly a Pacific Flyway duck although large numbers have occurred in Chesapeake Bay (Table 1; Fig. 5). Peak populations occurred in 1956 (78,211) but numbers declined to a low of only 400 in 1970. The long-term average number of pintail in Chesapeake Bay was 16,282. During the 1980's an average of only 3,935 ducks were recorded which was 70% lower than the pre-1980's average of 18,982. The average number of pintail in the Atlantic Flyway during the 1980's was 52,657; most were recorded in the Carolinas. Continental pintail populations reached an all-time low in 1986.

The pintail, like the wigeon, was most common in Chesapeake Bay during periods of good breeding conditions in Canada and excellent winter habitat in Chesapeake Bay. Pintail populations have decreased with the loss of submerged aquatic vegetation in the Bay. It seems that this species was unable to take advantage of alternate food sources, with one notable exception. Perry and Uhler[18] found that pintail from the James River had fed on the Asiatic freshwater clam more than any of the other duck species examined. However, umbrella sedges (*Cyperus* spp.), rice cutgrass, and smartweeds were predominant foods.

Canvasback—*Aythya valisineria*

Large numbers of canvasbacks have wintered in the Chesapeake Bay, especially in the Susquehanna Flats area (Table 1; Fig. 6). During the heyday of market hunting the canvasback continually commanded top price among ducks in the market. It is not known how many canvasbacks once frequented the Bay, but aerial surveys since 1948 showed that peak numbers were recorded in 1954 (399,320). Canvasback populations plummeted shortly afterwards to a low of 48,120 in 1958.

Populations increased in the mid-1960's as a result of better conditions on the breeding grounds and restrictive hunting regulations.[20] Canvasback numbers, however, decreased again in the late 1960's, and in 1972 the hunting season on canvasbacks was closed. The long-term average population of canvasbacks in Chesapeake Bay was 104,012. In the 1980's the population averaged 52,931, which was 54% lower than the pre-1980's average of 115,811. In 1986, canvasbacks in the Bay were at an all time low of 34,400. Canvasback populations in Chesapeake Bay during the 1970's were relatively stable despite increasing number of canvasbacks in the Atlantic Flyway and North America. Perry et al.[20] speculated that habitat degradation in the Bay was adversely affecting numbers of wintering ducks.

When submerged aquatic vegetation beds in the Bay declined, the canvasback

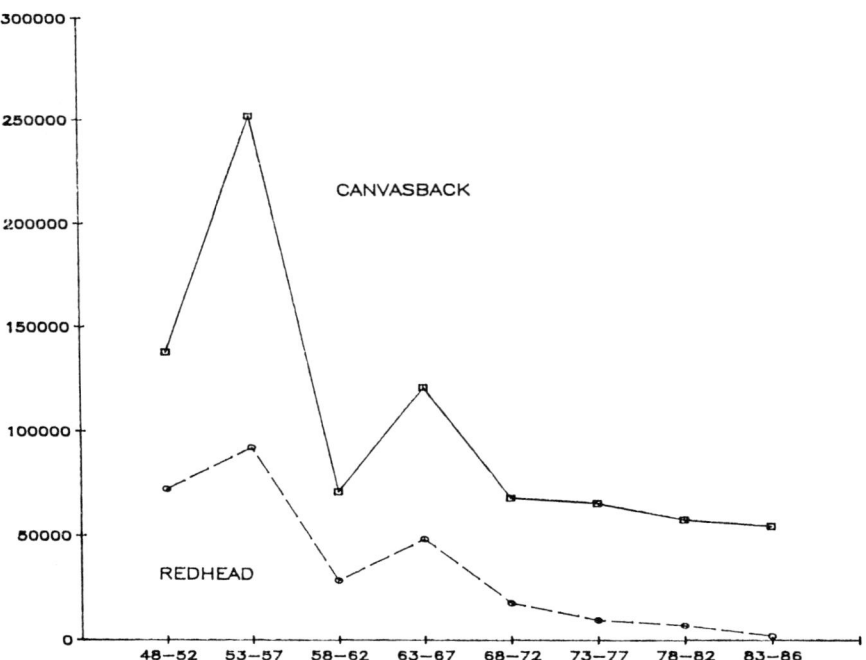

FIGURE 6. Long-term population trends (1948-86) of canvasback and redhead in Chesapeake Bay during 8 periods.

was forced to seek alternate food sources. This species was much more effective at changing diets than other duck species. Presently Chesapeake Bay canvasbacks feed predominantly on mollusks.[20] This food source, however, is not considered to be as nutritionally sound as the high energy plant tubers upon which it formerly fed.[33]

Redhead—*Aythya americana*

Redhead population numbers in Chesapeake Bay are on a long-term downward trend (Table 1; Fig. 6). Although redhead numbers peaked in 1956 (118,800), this population has steadily declined to a low of only 800 ducks in 1983. The long-term average was 35,288 redheads. During the 1980's the average winter population recorded in the Bay was only 3,506 which was 92% less than the pre-1980's average of 42,240. Most of these ducks were in the Tangier Island area. An average of 97,914 redheads were recorded in the Atlantic Flyway during the 1980's, thus indicating that population declines in Chesapeake Bay have been more drastic than in other areas.

Unlike the canvasback, the redhead did not change its food habits as habitat conditions changed. It still feeds on the upper vegetative parts of submerged aquatics.[20] Redhead populations in the Bay declined with the loss of submerged aquatic vegetation in the Bay. Presently redheads are most abundant in North Carolina, Florida, and Texas where submerged aquatic vegetation is abundant.[34] Because the redhead is now wintering in different areas than the canvasback, hunting regulations are no longer the same for these two species. Regulations are now more liberal for the redhead.

Scaup—*Aythya* spp.)

Scaup populations in Chesapeake Bay consist of two species, the greater scaup (*A. marila*) and the lesser scaup (*A. affinis*). Scaup (Table 1; Fig. 7) in Chesapeake Bay peaked in 1954 at 403,658 and then declined in the late 1950's. Populations increased in the 1960's and then declined steadily to a low of 10,700 in 1982. The long-term average population size was 65,909. In the 1980's, the population was 28,973 which is 61% lower than the pre-1980's average of 73,988. Trends of scaup populations in Chesapeake Bay have not been similar to those in North America and the Atlantic Flyway. For unknown reasons, scaup populations in Chesapeake Bay in the early 1960's did not reflect the record 2.6 million recorded throughout North America.

Scaup usually feed on molluscs and crustaceans.[22,31] Current food habits indicate similar food preferences. It is doubtful whether the loss of submerged aquatic vegetation in the Bay has significantly affected the distribution of abundance of scaup. Although, the diversity and numbers of invertebrate food organisms have probably declined due to the loss of submerged aquatic vegetation.[20]

Common Goldeneye—*Bucephala clangula*

The goldeneye is a hole-nesting duck that breeds in the forested wetlands of southern Canada. Wintering populations in Chesapeake Bay peaked in 1956 at 40,518 and reached a low of 2,445 in 1976 at 2,445 (Table 1; Fig. 8). The long-term average population in the Bay is 19,659. In the 1980's the average population in the Bay was 17,513 which was 13% lower than the pre-1980's average of 20,128. Trends of goldeneye populations in North American and Atlantic Flyway are similar during survey years.

Goldeneye feed mainly on invertebrates.[22,31] Changes in the distribution and abundance of submerged aquatic vegetation have probably not affected goldeneye populations. Goldeneye feed in deep water and usually not in shallow areas where submerged aquatic vegetation occurs. The amount of vegetation eaten by goldeneye has declined, however, during the hundred years in which food habits analyses have been conducted.

Bufflehead—*Bucephala albeola*

Although the bufflehead and goldeneye both breed and winter in similar habitat, their wintering population trends are different (Table 1; Fig. 8). Bufflehead numbers have been steadily increasing from a low of 2,502 in 1959 to

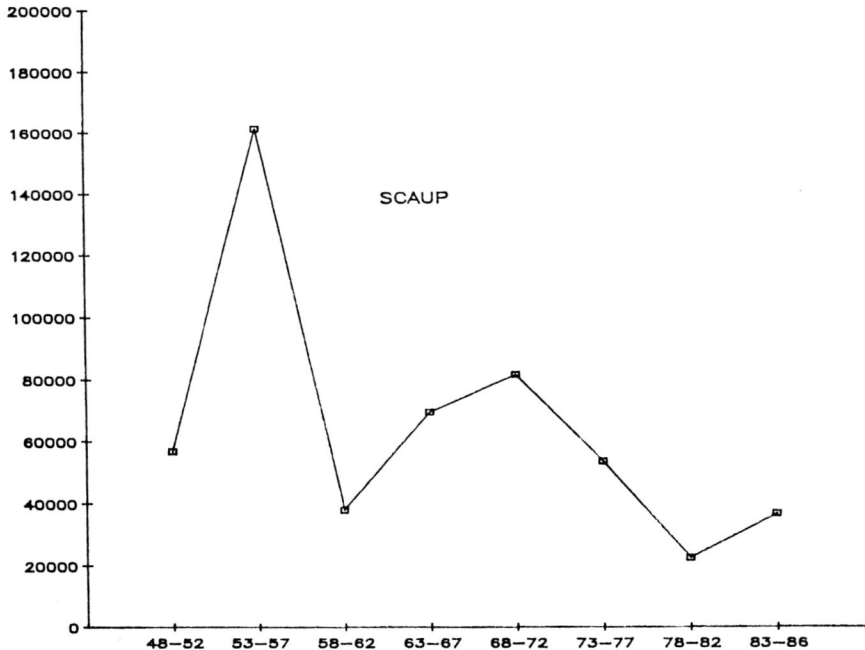

FIGURE 7. Long-term population trends (1948-86) of scaup in Chesapeake Bay during 8 periods.

a peak of 36,023 in 1977. The long-term average population was 14,813. During the 1980's the average population was 16,840 ducks; 17% higher than the pre-1980's average of 14,444. Population trends in Chesapeake Bay buffleheads have been similar to those in the Atlantic Flyway and in North America.

The bufflehead has traditionally been an invertebrate feeder, although vegetation has formed a more important part of its diet in the past than it does now. During the 1970's, the predominant food eaten by buffleheads was the duck clam (*Mulinia lateralis*).[31]

Ruddy Duck—*Oxyura jamaicensis*

The ruddy duck has shown a significant decline in numbers in Chesapeake Bay during years of aerial surveys (Table 1; Fig. 9). Peak numbers occurred in 1953 (124,740) and declined to a low in 1976 (4,703). The long-term population average for Chesapeake Bay was 33,642. In the 1980's the average population was 15,729, which was 58% less than the pre-1080's average of 37,560. Trends of Chesapeake Bay ruddy duck populations have been different from those of the Atlantic Flyway and North America. Highest numbers of ruddy ducks in the Atlantic Flyway are now recorded in North Carolina.

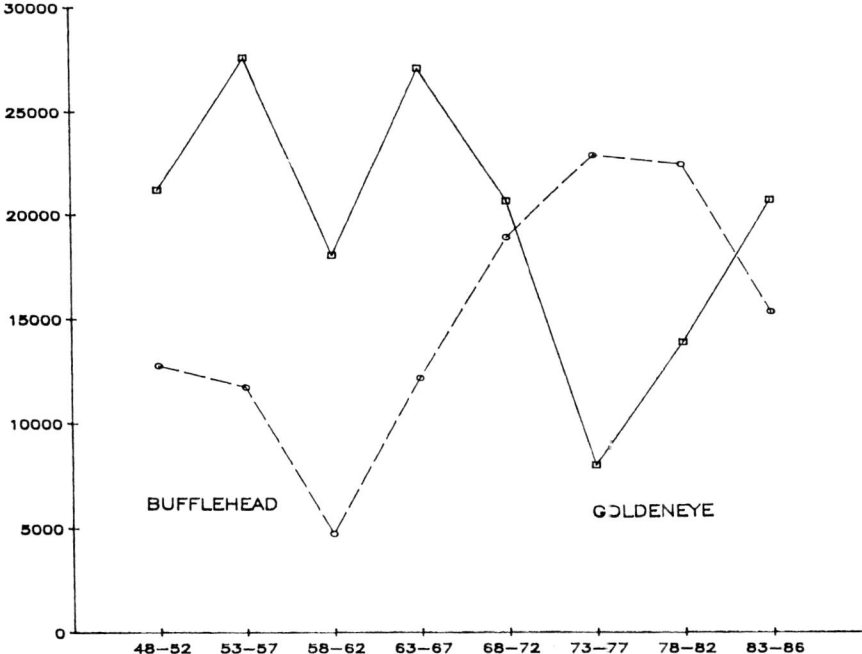

FIGURE 8. Long-term population trends (1948-86) of goldeneye and bufflehead in Chesapeake Bay during 8 periods.

The ruddy duck was traditionally a vegetative feeder.[35] It now feeds on invertebrates to a greater extent.[20] Increasing numbers of ruddy ducks were recorded around cities like Baltimore and Washington, D.C.[36] This species was probably feeding on tubificid worms (*Tubificidae*).[37]

Scoter—*Melanitta* spp.

Scoter populations in Chesapeake Bay consist of three species, the black scoter (*M. nigra*), surf scoter (*M. perspicillata*), and white-winged scoter (*M. fusca*). All three species are often seen in similar habitat and are surveyed together as one species.

Scoter populations in Chesapeake Bay have been variable (Table 1; Fig. 9). Peak population occurred in 1971 (130,900), and then reached a low of 1,551 in 1981. The long-term average was 16,760. The average in the 1980's was 6,565; 65% lower than the pre-1980's average of 18,990. The average Atlantic Flyway scoter population during the 1980's was 57,386.

Scoters are traditionally invertebrate feeders,[35,38] although no record of their food habits was made by Stewart,[22] Rawls,[30] or Munro and Perry[31] for Chesapeake Bay. Changes in scoter distribution within Chesapeake Bay may be due to changing food resources and should be investigated.

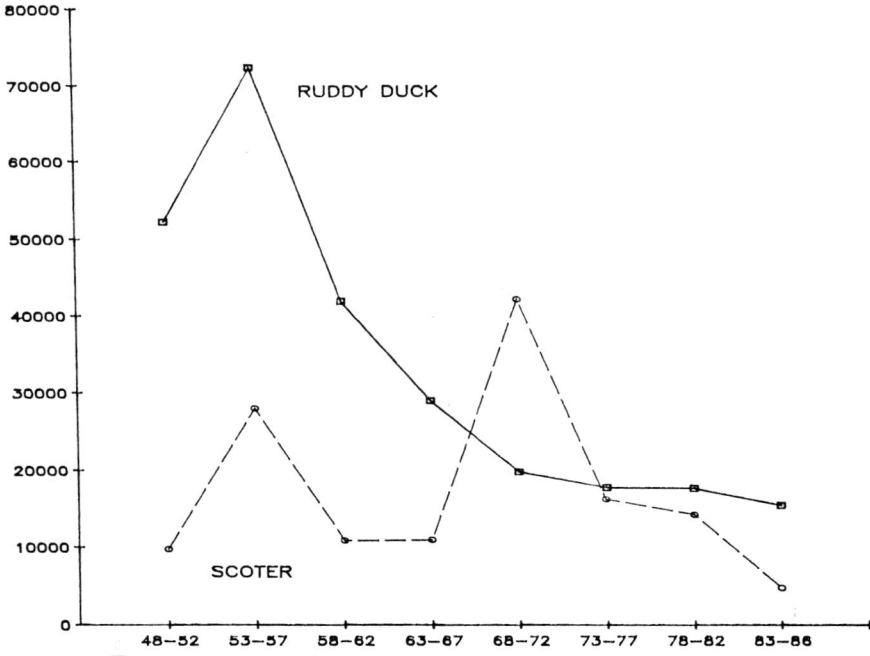

FIGURE 9. Long-term population trends (1948-86) of ruddy duck and scoter in Chesapeake Bay during 8 periods.

POLLUTANTS AND WATERFOWL

Although there have been many studies showing the effects of various pollutants (nutrient enrichment, turbidity, herbicides) on waterfowl habitat,[7,8,10] there also have been studies suggesting possible direct toxic effects of pollutants on waterfowl. Dieter et al.[39] determined the amounts of lead and polychlorinated biphenyls (PCB's) in canvasbacks by direct measurement of residues of these pollutants in the blood and by measuring blood enzyme levels that are affected by these pollutants. They found that 17% of the blood samples from Chesapeake Bay canvasbacks had less than one-half the normal delta-aminolevulinic acid dehydrogenase (ALAD) activity. This enzyme is known to be specifically inhibited by lead. There was a highly significant inverse correlation between ALAD and blood lead concentrations. Lead residues were four times higher than background levels.

Dieter et al.[39] also found that 11% of these same blood samples exhibited abnormal aspartate aminotransferase or lactate dehydrogenase activities. These enzymes are indicative of the presence of organochlorine contaminants like PCB's. A weak but significant inverse correlation between plasma aspartate aminotransferase activity and blood PCB concentrations was found. However, pollutant-induced changes in enzyme levels, even substantial changes, are difficult to correlate with actual harm to birds. Such changes are sometimes best viewed as indicators of exposure to a contaminant, rather than proof that exposure was high enough to harm survival or reproduction.

Studies of canvasbacks from Chesapeake Bay by White et al.[40] showed that organochlorine pesticides and PCB's were at relatively low concentrations in carcasses. These levels generally were lower than those known, through experimental studies with other waterfowl species, to have adverse effects on avian reproduction and survival. A few samples did contain DDE (a metabolite of DDT) or PCB's close to the concentrations where adverse effects may be expected to begin. Elevated lead levels were found in 10% of the wingbones and copper and zinc residues were high in livers.

Residues of cadmium, lead, copper, and zinc were measured in the livers, kidneys, and ulnar bones from 15 species of ducks from the Chesapeake Bay region.[41] Liver and kidney concentrations of cadmium were highest among oldsquaw (*Clangula hyemalis*) and white-winged scoters. These species mainly feed on animal food items. Lead concentrations, however, were generally highest in species like the mallard, black duck, and gadwall which are largely herbivorous. Lead in the tissue of ducks is probably coming from ingestion of spent shot of hunters. No marked trends were detected between food habits of ducks and the concentrations of copper and zinc in their tissues.

Eggs of black ducks were collected from Chesapeake Bay in 1971 and 1978 to assess possible effects of environmental pollutants on reproduction.[42,43] The mean concentration of DDE for Maryland eggs in 1978 was 0.10 ppm and was not detected in eggs from Virginia. Eggs from Maryland and Virginia contained no detectable PCB's. Eggshells, however, in 1964 and in 1971 were 7.5 and 1.2 percent thinner, respectively, than normal. Thinner eggshells in 1964 were probably associated with higher DDE residues that were found in black duck eggs during the 1964 survey.[42] In experimental studies with black ducks and mallards, DDE in the diet was shown to cause shell thinning, cracked eggs, changes in mineral composition of eggshells, and lowered reproductive success.[44]

The major food organisms of canvasbacks were analyzed for heavy metals by White et al.[40] because of the known ability of mollusks to bioconcentrate pollutants. Relatively low metal residues were found in clams from Chesapeake Bay. Cadmium and lead levels were similar in all species, but copper and zinc were found at higher levels in the brackish-water clam than in other species, perhaps because this species occurs in less saline water where major industries are located.

Plant and invertebrate food items of ducks were analyzed in another study[45] to determine concentrations of cadmium, lead, copper, and zinc. Cadmium and lead concentrations were generally greater in whole plants than in soft tissues of clams, although the reverse occurred with zinc. No marked trend in copper concentrations was detected between plants and animals. Higher cadmium concentrations in plants versus clams did not follow patterns of cadmium concentrations in tissues of ducks.[41] Higher concentrations of cadmium would be expected in herbivorous duck species, and the cause of this inconsistency is unknown.

Although some of the studies of pollutants in Chesapeake Bay waterfowl have shown some cause for concern, in general, pollutants in tissue and eggs of waterfowl are below levels normally considered to cause adverse effects. Monitoring of contaminants in waterfowl should continue, especially in the most contaminated areas of the Bay, but the direct impact of these pollutants on birds is probably less important than the indirect effects on waterfowl habitat from pollutants such as nutrients, suspended sediments, and perhaps herbicides. These types of pollutants, which result in the reduction of light and poisoning of plants, are a threat to vegetation in the Bay and the ducks which need vegetation for food. Populations of ducks in Chesapeake Bay will not occur in historic numbers until submerged aquatic vegetation is restored. This will be possible only if there is major improvement in the water quality of the Bay in future years.

ACKNOWLEDGEMENTS

The assistance of Kinard Boone and Edward Burton in the preparation of graphs, and of Jean Higgis, Valerie Lumsden, and Deborah Sumeriski in word processing is appreciated. Robert Munro assisted with analysis and interpretation of data. Drafts of the manuscript were reviewed by ronald Eisler, James

Fleming, Gary Heinz, Larry Hindman, Robert Munro, Jan Reese, Fairfax Settle, Daniel Stotts, and Vernon Stotts. The assistance of these persons and those of the Office of Migratory Bird Management and Maryland and Virginia state agencies who assisted in conducting surveys is appreciated.

LITERATURE CITED

1. Kimball, D., and J. Kimball. 1969. The market hunter. Dillon Press, Inc., Minneapolis, Mn. 132 pp.
2. Walsh, H.M. 1971. The outlaw gunner. Tidewater Publishers, Cambridge, Md. 178 pp.
3. Herbert, H.W. 1873. Frank Forester's field sports of the United States and the British Provinces of North America. George E. Woodward, New York, 383 pp.
4. Grinnell, G.B. 1901. American duck shooting. Forest and Stream Publ., New York. 628 pp.
5. Sanford, L.C., L.B. Bishop, and T.S. Van Dyke. 1903. The water-fowl family. Mac Millan, New York. 597 pp.
6. Kerwin, J.A., R.E. Munro, and W.W.A. Peterson. 1976. Distribution and abundance of aquatic vegetation in the Upper Chesapeake Bay, 1971-1974. Pages 393-400 *in* J. Davis, Coord., The effects of tropical storm Agnes on the Chesapeake Bay estuarine system. Chesapeake Res. Consortium Publ. No. 54. The Johns Hopkins Univ. Press, Baltimore, Md. 639 pp.
7. Bayley, S., V.D. Stotts, P.F. Springer, and J. Steenis. 1978. Changes in submerged aquatic macrophyte populations at the head of Chesapeake Bay, 1958-1975. Estuaries 1:171-182.
8. Orth, R.J., and K.A. Moore. 1981. Submerged aquatic vegetation of the Chesapeake Bay: past, present, and future. Trans. North Am. Wildl. and Natur. Resour. Conf. 46:271-283.
9. Haramis, G.M., and V. Carter. 1983. Distribution of submersed aquatic macrophytes in the tidal Potomac River. Aquatic Bot. 15:65-79.
10. Stevenson, J.C., and N.M. Confer. 1978. Summary of available information on Chesapeake Bay submerged vegetation. U.S. Fish and Wildl. Serv., FWL/OBS-78/66-3. 35 pp.
11. Ward, L.F. 1881. Guide to the flora of Washington and vicinity, U.S. Natl. Mus. Bull. No. 22. 764 pp.
12. Gwathmey, J.H. 1945. Potomac River cleared of "floating islands". Md. Conservationist. 22:21-23.
13. Steward, K.K., T.K. Van, V. Carter, and A.H. Pieterse. 1984. Hydrilla invades Washington, D.C. and the Potomac. Amer. J. Bot. 71:162-163.
14. Pfitzenmeyer, H.T., and K.G. Drobeck. 1964. The occurrence of the brackish water clam, *Rangia cuneata*, in the Potomac River, Maryland. Ches. Sci. 5:209-215.

15. Diaz, R.J. 1974. Asiatic clam, *Corbicula manilensis* (Phillippi), in the tidal James River, Virginia, Ches. Sci. 15:118-120.
16. Dresler, P.V., and R.L. Cory. 1980. The Asiatic clam, *Corbicula fluminea* (Muller) in the tidal Potomac River, Maryland. Estuaries 3:150-151.
17. Stotts, V.D. 1979. Vegetation survey of Susquehanna Flats. Progress Report PR-W-30-R Md. Dept. Nat. Res. 16 pp.
18. Perry, M.C., and F.M. Uhler. 1981. Asiatic clams (*Corbicula manilensis*) and other foods used by waterfowl in the James River, Virginia. Estuaries 4:229-233.
19. Voelzer, J.F., E.Q. Lauxen, S.L. Rhoades, and K.D. Norman, eds. 1982. Waterfowl status report, 1979. Spec. Sci. Rep. Wildl. 246. U.S. Fish and Wildl. Serv., Washington, D.C. 96 pp.
20. Perry, M.C., R.E. Munro, and G.M. Haramis. 1981. Twenty-five year trends in diving duck populations in Chesapeake Bay. Trans. North Am. Wildl. Natur. Resour. Conf. 46:299-310.
21. Munro, R.E. 1980. Field feeding by *Cygnus columbianus columbianus* in Maryland. Pages 261-272 *in* G.V.T. Matthews and M. Smart, eds., Proc. Second Internat. Swan Symp., Internat. Waterfowl Res. Bur., Slimbridge, England. 396 pp.
22. Stewart, R.E. 1962. Waterfowl populations in the Upper Chesapeake Region. Spec. Sci. Rep. Wildl. No. 65. U.S. Fish Wildl. Serv., Washington, D.C. 208 pp.
23. Cooke, M.T., and P. Knappen. 1940. Some birds naturalized in North America. N. Amer. Wildl. Conf. 5:176-183.
24. Reese, J.G. 1969. Mute swan breeding in Talbot County, Maryland. Maryland Birdlife 25:14-16.
25. Reese, J.G. 1975. Productivity and management of feral mute swans in Chesapeake Bay. J. Wildl. Manage. 39:280-286.
26. Reese, J.G. 1980. Demography of European mute swans in Chesapeake Bay. Auk. 97-449-464.
27. Willey, C.H., and B.F. Halla. 1972. Mute swans of Rhode Island. Rhode Island Dept. Nat. Res., Div. Fish Wildl., Pamphlet No. 8. 35 pp.
28. Stotts, V.D., and D.E. Davis. 1962. The black duck in the Chesapeake Bay of Maryland: breeding behavior and biology. Ches. Sci. 1:127-154.
29. Stotts, V.D. 1955. A nesting study of the black duck in the Kent Island area of Maryland. MS Thesis. Univ. of Illinois, Urbana, Il. 95 pp.
30. Rawls, C.K. 1978. Food habits of waterfowl in the Upper Chesapeake Bay, Maryland. Unpub. Rept. Univ. of Maryland, Chesapeake Biol. Lab., Solomons, Md. 140 pp.
31. Monro, R.E., and M.C. Perry. 1981. Distribution and abundance of waterfowl and submerged aquatic vegetation in Chesapeake Bay. U.S. Fish and Wildl. Serv., FWS/OBS 78-D-X0391. 180 pp.
32. Florschutz, O., Jr. 1972. The importance of eurasian milfoil (*Myriophyllum*

spicatum) as a waterfowl food. Proc. Annu. Conf. Southeast. Assoc. Game and Fish Comm. 26:189-193.
33. Perry, M.C., W.J. Kuenzel, B.K. Williams, and J.A. Serafin. 1986. Influence of nutrients on feed intake and condition of captive canvasbacks in winter. J. Wildl. Manage. 50:427-434.
34. Perry, M.C., and F.M. Uhler. 1982. Food habits of diving ducks in the Carolinas. proc. Southeast Assoc. Fish and Wildl. Agencies 36:492-504.
35. Cottam, C. 1939. Food habits of North American diving ducks. Tech. Bull. No. 643, U.S. Dep. Agriculture, Washington, D.C. 139 pp.
36. Wilds, C. 1979. The Washington, D.C. Christmas bird count as an indicator of environmental changes. Pages 10-11 *in* J.F. Lynch, ed., Bird populations—a litmus test of the environment. Proc. Mid-Atlantic Nat. Hist. Symp., Audubon Naturalist Soc., Washington, D.C. 48 pp.
37. Stark, R.T. 1978. Food habits of the ruddy duck (*Oxyura jamaicensis*) at the Tinicum National Environmental Center. M.S. Thesis, Pennsylvania State Univ., University Park, Pa. 68 pp.
38. Martin, A.C., H.S. Zim, and A.L. Nelson. 1951. American wildlife and plants—a guide to wildlife food habits. McGraw-Hill, New York. 500 pp.
39. Dieter, M.P., M.C. Perry, and B.M. Mulhern. 1976. Lead and PCB's in canvasback ducks: relationship between enzyme levels and residues in blood. Arch. Envi. Contam. Toxicol. 5:1-13.
40. White, D.H., R.C. Stendell, and B.M. Mulhern. 1979. Relations of wintering canvasbacks to environmental pollutants—Chesapeake Bay, Maryland. Wilson Bull. 91:279-287.
41. DiGiulio, R.T., and P.F. Scanlon. 1984. Heavy metals in tissues of waterfowl from the Chesapeake Bay, USA. Environ. Pollut. Series A. 35:29-48.
42. Longcore, J.R., and B.M. Mulhern. 1973. Organochlorine pesticides and polychlorinated biphenyls in black duck eggs from the United States and Canada—1971. Pestic. Monit. J. 7:62-66.
43. Haseltine, S.D., B.M. Mulhern, and C. Stafford. 1980. Organochlorine and heavy metal residues in black duck eggs from the Atlantic Flyway, 1978. Pestic. Monit. J. 14:53-57.
44. Ohlendorf, H.M. 1981. The Chesapeake Bay's birds and organochlorine pollutants. Trans. North Am. Wildl. and Natur. Resour. Conf. 46:259-270.
45. DiGiulio, R.T., and P.F. Scanlon. 1985. Heavy metals in aquatic plants, clams, and sediments from the Chesapeake Bay, USA. Implications for waterfowl. The Science of the Total Environment. 41:259-274.

Contaminant Problems and Management of Living Chesapeake Bay Resources. Edited by S. K. Majumdar, L. W. Hall, Jr. and H. M. Austin. © 1987, The Pennsylvania Academy of Science.

Chapter Six

SUBMERGED AND EMERGENT AQUATIC VEGETATION OF THE CHESAPEAKE BAY

CARL HERSHNER and RICHARD L. WETZEL

Virginia Institute of Marine Science
School of Marine Science
College of William and Mary
Gloucester Point, Virginia 23062

ABSTRACT

Chesapeake Bay supports a diverse assemblage of submerged and emergent aquatic vegetation. The distribution of species of each kind of vegetation is governed largely by salinity. The functions of both submerged and emergent vegetation in the Bay ecosystem include contributing to total net primary production, service as habitat and performance in both water quality and sedimentation processes. Research on submerged aquatic vegetation is focused on its role in the estuarine system and determinants of its distribution and abundance. Research on emergent vegetation still concerns basic questions of structure and function, but has also branched into methodologies for utilization by man to meet water quality, erosion control and habitat objectives.

INTRODUCTION

Aquatic vegetation in the Chesapeake Bay has been the focus of numerous management and research efforts over the last two decades. In the case of submerged aquatic vegetation (SAV), interest was spurred by dramatic declines in both the distribution and abundance during the 1970's. This led directly to a wide variety of research projects designed to document the function and importance of submerged grasses in the Chesapeake Bay. While regulatory programs specific to SAV have not been established in the Bay, the value of this resource is now routinely considered by other existing management programs. As discussed in the following sections, research on SAV continues along several lines. The major accomplishment to date may be the heightened awareness of the resource and its potential values.

Emergent wetlands have been widely appreciated for a somewhat longer time. Research in the 1960's led to passage of protective legislation at both state and federal levels of government. As a consequence, losses of emergent tidal wetlands in Chesapeake Bay due to man's activities have apparently been significantly reduced. Current research efforts are aimed at expansion of understanding of certain types of these wetlands (particularly lower salinity types) and also at appropriate utilization of emergent wetlands for specific goals (e.g., water quality improvement, erosion control, habitat).

Understanding and appropriate management of aquatic vegetation within Chesapeake Bay seems increasingly important. Our experience to date indicates that both submerged and emergent vegetation are sensitive to man's activities. Indeed, as a consequence of existence in an aquatic habitat, even activities spatially removed from these resources can produce significant impacts. This circumstance combined with recognition of the importance of both submerged and emergent vegetation to the natural processes of the Bay continues to provide impetus for improved understanding and management.

SUBMERGED AQUATIC VEGETATION

Submerged aquatic vegetation, i.e. vascular plants growing in shallow bottom sediments and carrying out their life cycles submersed, are important components of many riverine, estuarine and coastal marine ecosystems throughout the world. In some aquatic habitats, these generally inconspicuous plant communities produce the majority of organic matter used either directly or indirectly as food by higher organisms. In all systems where submerged aquatics are present, the vascular plants contribute to organic matter production, provide shelter and nursery areas for resident and non-resident species, act to stabilize sediments by dissipating tidal and storm generated wave energy, and, influence both sediment and water chemistry.[1,2] It has become widely accepted among aquatic scientists and resource managers over the past decade that the general environmental health of a water body is reflected by the relative growth and vigor of resident submerged aquatic plant populations.

Submerged plants occur in all aquatic environments. Over broad geographical scales, the distribution of submerged plant species is governed largely by salinity and temperature regimes. Salt tolerant species, seagrasses, generally occur in estuarine and coastal marine areas that have mean annual salinities greater than 10 to 15 parts per thousand (ppt). Brackish and freshwater species are restricted to areas with little or no salinity which include both inland lakes and streams and coastal rivers. Within a particular geographical area, submarine light, temperature, dissolved inorganic nutrients, and, general water quality conditions govern the distribution, relative abundance, and, growth of a given submerged aquatic species.[3]

Chesapeake Bay, because of its size and variety of aquatic environments, has submerged plant communities occurring in habitats that range from freshwater to marine. Specific habitats are generally dominated by a few species. Freshwater and brackish areas have the highest plant species diversity whereas estuarine and marine areas generally have only one or at most two dominant species. Here we intend to present the past and current status of submerged aquatic vegetation within Chesapeake Bay and its tidal tributaries and discuss current thinking regarding environmental factors which control their distribution and abundance.

SUBMERGED AQUATIC SPECIES IN CHESAPEAKE BAY

For the Chesapeake Bay and its tributaries, 20 species of submersed vascular plants in seven families have been reported.[4,5,6] Table 1 provides a list of their common and scientific names, and, a general note on their preferred habitat. Of the twenty species listed only one, *Zostera marina* or eelgrass, is considered a true seagrass and it is the dominant submersed plant in middle and lower portions of the Chesapeake Bay. *Ruppia maritima,* or widgeon grass, is a

TABLE 1

Common and Scientific Names of Submerged Aquatic Plants Occurring in the Tributaries and Main Stem of Chesapeake Bay (after Carter et al., 1985; Hotchkiss, 1972; Orth et al., 1986).

Scientific Name	Common Name	Habitat
Chara braunii	Muskgrass	Freshwater
Chara zeylanica		Freshwater
Nitella flexilis	Nitella	Freshwater
Potamogeton perfoliatus	Redhead grass	Freshwater/Brackish
Potamogeton pectinatus	Sago pondweed	Freshwater/Brackish
Potamogeton crispus	Curly pondweed	Freshwater/Brackish
Potamegeton pusillus	Slender pondweed	Freshwater/Brackish
Najas quadalupensis	Souther naiad	Freshwater/Brackish
Najas gracillima	Naiad	Freshwater/Brackish
Najas minor		Freshwater/Brackish
Zannichellia palustris	Horned pondweed	Freshwater/Brackish
Ruppia maritima	Widgeon grass	Freshwater/Mesohaline
Vallisneria americana	Wild celery	Freshwater/Brackish
Elodea canadenis	Common celery	Freshwater/Brackish
Eqeria densa	Water weed	Freshwater/Brackish
Hydrilla verticillata	Hydrilla	Freshwater
Ceratophyllum demersum	Coontail	Freshwater
Myriophyllum spicatum	Eurasian watermilfoil	Freshwater/Brackish
Heteranthera dubia	Water stargrass	Freshwater
Zostera marina	Eelgrass	Mesohaline/Marine

cosmopolitan species in the Bay due to its wide physiological salt tolerance range and co-dominates with eelgrass many mesohaline areas in the Bay. In lower salinity (brackish) and freshwater areas of the middle and upper Bay and coastal rivers, *R. maritima, Myriophyllum spicatum* (eurasian water milfoil), *Potomageton perfoliatus* (redhead grass), *Vallisneria americana* (wild celery), and, the recently introduced eurasian species *Hydrilla verticillata* (hydrilla), are the more common SAV communities.

Based on a recent report by Orth et al.,[6] 15.6%, 25.7% and 58.7% of the total SAV in Chesapeake Bay occur in the upper, middle and lower portions of the Bay respectively. Based on these aerial mapping and coverage data, the species that presently dominate SAV communities for the Bay as a whole are *Z. marina* and *R. maritima* in the lower and middle portions, *Myriophyllum spicatum* and *Potamegeton perfoliatus* in the upper and brackish to freshwater sections, and, *Hydrilla verticillata* in local areas of the Bay and particularly the upper tidal Potomac River. With the exception of recent research interest on hydrilla, the vast majority of studies on submerged macrophyte population ecology and community dynamics has focused on the other four species.

HISTORICAL CONSIDERATIONS

Prior to the mid 1970's, knowledge of and research on submerged aquatics in Chesapeake Bay was sparse. Stevenson and Confer[7] in summarizing the data available at the time indicated that most available information concerned species abundance and distribution and few reports addressed topics relative to population ecology, production, or, environmental regulation and control of plant growth. What had become evident by this time to not only aquatic scientist and resource managers but increasingly the general public was that submerged aquatic plants were rapidly declining throughout the Bay and available information was not adequate to address solutions to the apparent problem or indicate probable causes. These declines in SAV abundance took on added significance because species throughout the Bay were similarly affected and the problem appeared unique to the Chesapeake.[8] A significant, previously recorded decline in the 1930's affected only eelgrass, *Z. marina,* and occurred throughout the North Atlantic basin in both North America and Western Europe.[9,10,11]

In response to this and as a result of congressional mandate, the U.S. Environmental Protection Agency established the Chesapeake Bay Program which in turn sponsored both basic and applied research on SAV's in Virginia and Maryland waters. The research carried out within this program and which continues in part to the present has focused on a wide range of subjects. They include studies on bay-wide assessment of submerged aquatic plant distribution and abundance, plant population biology and community ecology, and, environmental factors that either do or potentially might control plant growth, productivity and long term survival. The results of these federally sponsored programs are summarized in Macalaster et al.[12]

Except for relatively recent efforts, assessment and documentation of SAV abundance throughout the Bay has been sporatic. Earlier reports summarized historically available data and concluded that not only had SAV's once been a principal component of the Bay's shallow water benthic environments but the present declines (1970's) were the most severe and widespread.[13] It is very likely though impossible to quantify, that SAV's in many sections of the Bay and its tributaries once produced the major fraction of organic matter used to support the trophic structure and their loss could potentially impact higher level consumer resources. The clearest example of this is the loss of the Bay scallop fishing industry on the Eastern Shore following the 1930's demise of eelgrass.

Beginning in the 1970's both Maryland and Virginia initiated monitoring programs to assess trends in SAV distribution and abundance. In the upper and middle portions of Chesapeake Bay, the Maryland Department of Natural Resources (Md.DNR) established a network of regularly sampled stations. Figure 1 illustrates, beginning in 1971, the percentage of stations sampled that had SAV's present.[14] The upper and lower panels give data on selected river systems for the western and eastern shores respectively. The periods of decline and regrowth are apparent and the severe losses of submerged aquatics in the Chester, Choptank, and, Honga Rivers after 1980 on the eastern shore are particularly evident. For all cases, SAV's as judged by the Md.DNR surveys have not returned to 1971 levels. Time series data such as these are not available for the lower Chesapeake Bay. However, Orth et al.[13] compiled available aerial shoreline mapping information dating back to the 1930's and have since 1978 conducted aerial mapping of SAV distribution and abundance for the lower Bay. Figure 2 illustrates the abundance of submerged aquatics (primarily *Z. marina*) at selected study sites in the lower York River (Mumfort Islands and Jenkins Neck) and the Mobjack Bay area (East River) on the western shore and for a single, historically stable eastern shore site (Vaucluse Shores). As for the upper and middle bay, declines in SAV abundance are particularly evident at the York River sites beginning in the early 1970's. One site, Mumfort Island which was historically vegetated prior to this time, lacked all submerged vegetation by 1974 and has not recovered to the present. Additionally, attempts to revegetate the site artificially by transplanting have not been successful (unpublished data). At the lower Bay eastern shore site (Vaucluse), the vegetation has been relatively stable since at least the 1930's. Although an overall trend of reduced abundance is indicated by these data, the cause is thought due to be natural.[14]

The conclusion that emerges from analysis of data such as these as well as

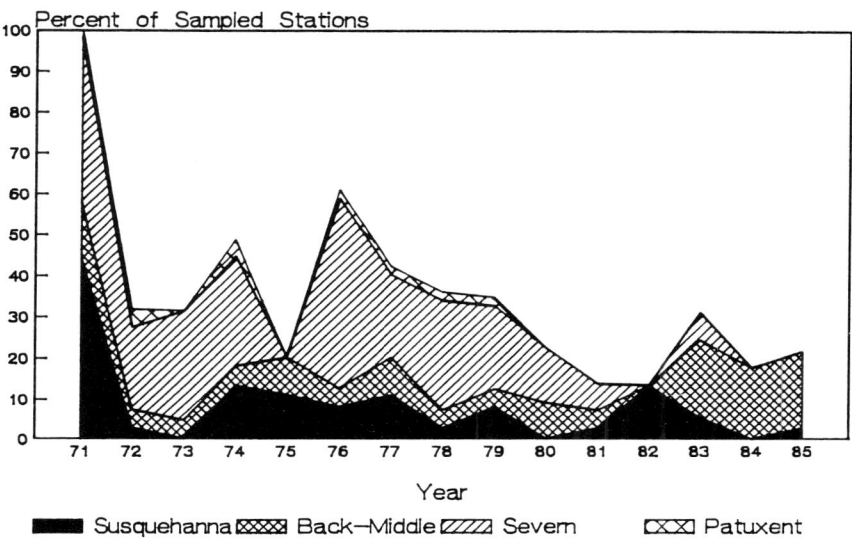

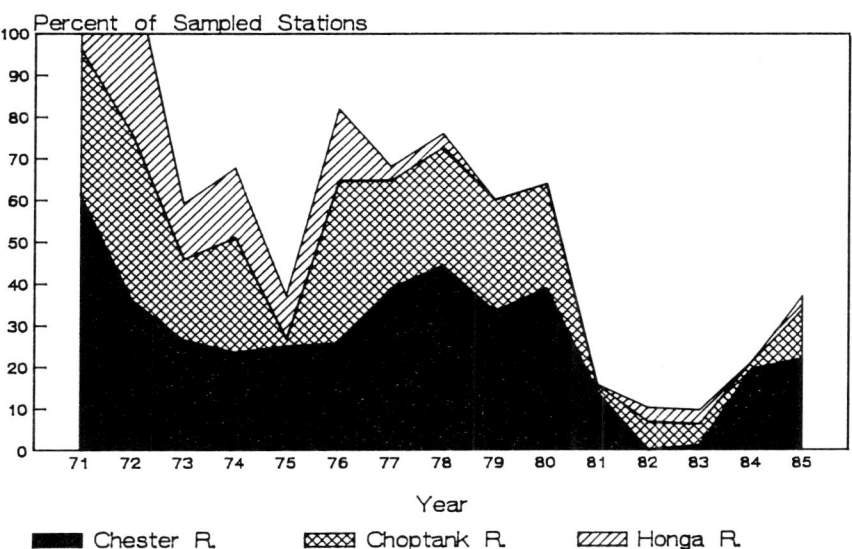

FIGURE 1. Trends in the abundance of SAV for upper bay western shore tributaries (upper panel) and upper bay eastern shore tributaries (lower panel). Source data are reported in Orth et al.[6]

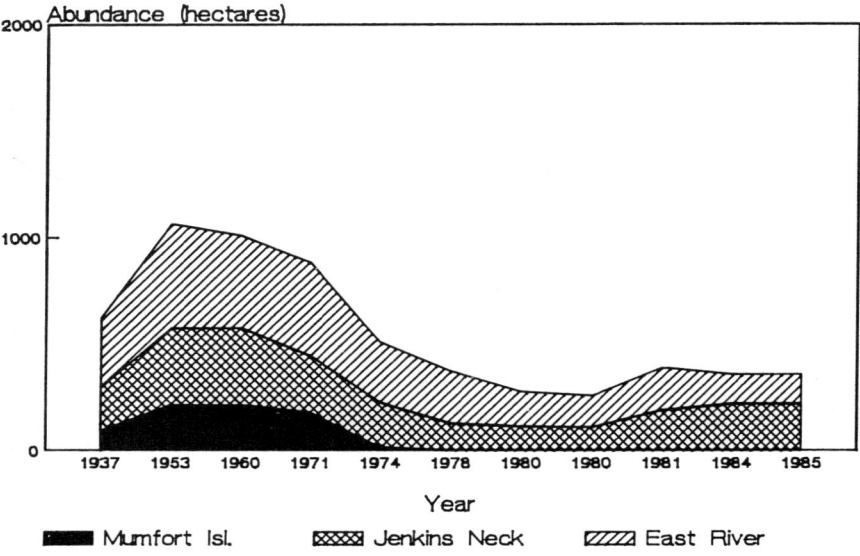
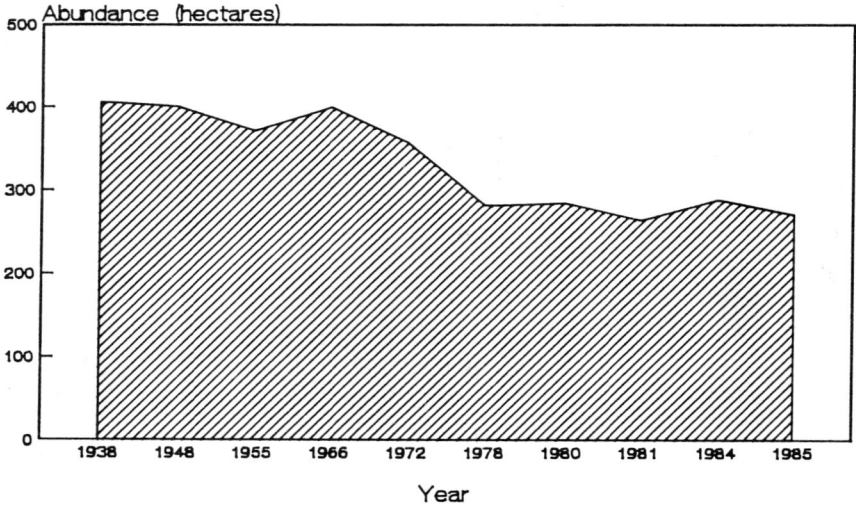

FIGURE 2. Trends in the abundance of SAV for lower bay western shore areas in and around the York River—Mobjack Bay (upper panel) and for a historically, stable eastern shore lower bay area (lower panel). Source data are reported in Orth et al.[6]

other information available through the published literature is that Chesapeake Bay has had over approximately the past decade an unprecedented decline in submerged aquatic plant communities throughout the Bay. The historic distribution and abundance data indicate that the loss of submerged aquatics has occurred in a "down-Bay, down-river" pattern.[13] This strongly suggests that long term changes in riverine water quality have played a key role. There is every indication that there has been a concomitant reduction in the maximum depth to which SAV's are distributed which would further suggest that the principle environmental factor is submarine light. The overall result is that a valuable natural resource has declined to the point that its long term survival can be questioned and the most probable cause is anthropogenic rather than attributable to natural causes. The impact, environmentally or economically, of total loss remains problematic but without question would be significant.

CURRENT STATUS OF SAV IN CHESAPEAKE BAY

Since 1984 both state (Maryland and Virginia) and federal funds have been granted for Bay-wide aerial mapping and determining submerged aquatics distribution and abundance. Orth et al.[6,13,14] have provided these data and summarized the current status of SAV in a series of papers and grant reports.

For 1985, Orth et al.[6] estimated that total SAV abundance was 19,390 hectares (ca. 48,000 acres) which represents an increase of 3990 hectares (9855 acres) over 1984 estimates.[14] The areas of significant increase were in the middle Bay region and were accounted for to large extent by establishment and growth of widgeon grass, *R. maritima,* in areas of the eastern shore. The most probable explanation for this large regional increase between inventories is the appreciable lack of rainfall in 1985 that allowed salinities in the region to remain relatively high and reduced runoff.[15]

The second major area of SAV expansion was in the upper Potomac River region.[16] SAV increased from 3,446 to 13,557 hectares between 1984 and 1985. The remaining regions of the Bay showed either small gains or losses to which little significance can be attached until a longer term record is available. For the lower Bay region, the areas that had but historically supported submerged aquatics remain non-vegetated. With few exceptions, those areas that have been artificially re-vegetated by transplanting have not been successful and the reasons remain the subject of current research.

Overall, the current status of submerged aquatics in the Chesapeake Bay remains at levels below historic abundances; however an accurate estimate of actual loss goes beyond the data available. For at least the past several years it would appear that total abundance within the Bay has varied between 15,000 and 20,000 hectares which may be much less than historically present. Relatively small changes in environmental conditions that favor or not submerged aquatic plant growth can result in large annual variations when compared to assessment of SAV resources over the past few years.[17] Trends in abundance and longer term assessment of the effectiveness of management programs related to SAV's can only be determined by continued Bay-wide annual inventories.

EMERGENT AQUATIC VEGETATION

The Chesapeake Bay drainage basin supports a large inventory of wetlands with emergent vegetation. These wetlands can be classified as either tidal or nontidal. Here we discuss only tidal wetlands, those surrounding the Bay and its tributaries and influenced by astronomical tides. These wetlands have been the focus of a great deal of research and management effort during the last several decades. During that time period emergent tidal wetlands have become a valued, if not thoroughly understood, resource. As discussed in greater detail below, these wetlands are believed to play important roles in the maintenance of an estuary's productivity, water quality and shoreline stability.

TYPES AND EXTENT OF EMERGENT WETLANDS IN CHESAPEAKE BAY

The emergent tidal wetlands of the Chesapeake Bay have been classified a number of different ways. The U.S. Fish and Wildlife Service classification scheme is arguably the most complete and/or complex.[18] It separates wetland types on the basis of both physical and biological characteristics. Simpler classification schemes are currently in use by both Maryland and Virginia as part of their respective resource management programs. Maryland recognizes seven major groups of emergent tidal wetlands with a total of thirty one individual types based on dominant vegetation and salinity regime (Table 2). Virginia recognizes twelve types of emergent tidal wetlands based on the dominant vegetation (Table 3).

Classification of wetlands on the basis of average salinity experience generally results in about four classes of tidal wetlands. Following the terminology of Cowardin et al.,[18] these classes would include: polyhaline (salinities <30 ppt and >18 ppt); mesohaline (<18 ppt and >5 ppt); oligohaline (<5 ppt and >0.5 ppt); and tidal freshwater (<0.5 ppt). Species composition of wetlands changes gradually along this salinity gradient with the most distinct differences occurring between oligohaline and tidal freshwater systems. In the Chesapeake Bay polyhaline and mesohaline tidal wetlands support almost identical plant assemblages. Typically *Spartina alterniflora* is a dominant with *Juncus roemerianus, Spartina patens* and *Distichlis spicata* as co-occurring dominants. In oligohaline systems *Spartina cynosuroides* frequently becomes dominant

TABLE 2

Types of emergent tidal wetlands recognized in Maryland (McCormick and Somes, 1982).

1. Shrub swamps	
—swamp rose	*Rosa palustris*
—smooth alder/black willow	*Alnus servulata/Salix nigra*
—red maple/ash	*Acer rubrum/Fraxinus* spp.
2. Swamp forests	
—bald cypress	*Taxodium distichum*
—red maple/ash	*Acer rubrum/Fraxinus* spp.
—loblolly pine	*Pinus taeda*
3. Fresh marshes	
—smartweed/rice cutgrass	*Polygonum* spp./ *Leersia oryzoides*
—spatterdock	*Nuphar advena*
—pickrel weed/arrow arum	*Pontedaria cordata/Peltandra virginica*
—sweetflag	*Acorus calamus*
—cattail	*Typha* spp.
—rosemallow	*Hibiscus* spp.
—wildrice	*Zizania aquatica*
—bulrush	*Scirpus* spp.
—big cordgrass	*Spartina cynosuroides*
—common reed	*Phragmites communis*
4. Brackish high marshes	
—meadow cordgrass/spikerush	*Spartina patens/Distichlis spicata*
—marsh elder/groundselbush	*Iva frutescens/Baccharis halimifolia*
—needlerush	*Juncus roemerianus*
—cattail	*Typha* spp.
—rosemallow	*Hibiscus* spp.
—switchgrass	*Panicum virgatum*
—threesquare	*Scirpus* spp.
—big cordgrass	*Spartina cynosuroides*
—common reed	*Phragmites communis*
5. Brackish low marshes	
—smooth cordgrass	*Spartina alterniflora*
6. Saline high marshes	
—meadow cordgrass/spikegrass	*Spartina patens/Distichlis spicata*
—marsh elder/groundselbush	*Iva frutescens/Baccharis halimifolia*
—needlerush	*Juncus roemerianus*
7. Saline low marshes	
—smooth cordgrass, tall growth form	*Spartina alterniflora*
—smooth cordgrass, short growth form	*Spartina alterniflora*

in the assemblage. The move to tidal freshwater systems is marked by the appearance of a very diverse plant community composed of grasses, rushes, broad-leafed plants and shrubs, with no typically dominant species.

Emergent wetlands can also be broadly classified by the physical conformation of the vegetated area. As an example Virginia's wetlands inventory program utilizes descriptive terms such as creek marsh, delta marsh, extensive marsh and fringe marsh to indicate the shape and/or location of a particular wetland.

TABLE 3
Types of emergent tidal wetlands recognized in Virginia.

1. Saltmarsh cordgrass	*Spartina alterniflora*	
2. Saltmeadow	*Spartina patens/Distichlis spicata*	
3. Black needlerush	*Juncus roemerianus*	
4. Saltbush (gallbush)	*Iva frutescens/Baccharis halimifolia*	
5. Big cordgrass	*Spartina cynosuroides*	
6. Cattail	*Typha angustifolia/Typha latifolia*	
7. Arrow arum/pickrel weed	*Peltandra virginica/Pontedaria cordata*	
8. Reed grass	*Phragmites communis*	
9. Yellow pond lily	*Nuphar luteum*	
10. Saltwort	*Salicornia* spp.	
11. Freshwater mixed	may include:	*Scirpus* spp.
		Carex spp.
		Polygonum spp.
		Osmunda regalis
		Pontedaria cordata
		Peltandra virginica
		Zizania aquatics
		Bidens spp.
		Leersia oryzoides
12. Brackish water mixed	may include:	*Spartina alterniflora*
		Spartina patens
		Distichlis spicata
		Juncus roemerianus
		Iva frutescens
		Baccharis halimifolia
		Scirpus spp.
		Spartina cynosuroides
		Typha spp.

The shape and location of a wetland in part determine the principal roles it may play in the estuarine system (see following section).

It is interesting to note that much of the research on emergent wetlands, which has resulted in protective legislation for these systems, was conducted in "extensive" wetlands (e.g. the very large marshes behind the barrier island chain on the Atlantic coast). By contrast, a majority of the regulatory effort in the Chesapeake Bay is focused on fringing wetlands, the narrow bands of intertidal vegetation which stretch along much of the vegetated shoreline. Using Virginia as an example, the importance of fringing marshes in the regulatory program is demonstrated by the following calculations. According to Hobbs et al.[9] Virginia has approximately 5123 miles of shoreline. Marshes occur along 4048 miles of that total (79%). Fringing marshes (defined in this case as marshes less than 400 feet wide) occur along 2265 miles of shoreline (44% of the total shoreline, 56% of the marsh shoreline). The average width of "fringe" marshes in Virginia is considerably less than 400 feet, probably something less than 150 feet. Using 150 feet as an average width, the 2265 miles of fringe marsh would comprise 41,185 acres of wetlands or only about one third of Virginia's complement of wetlands. It is estimated[20] that over 90% of the shoreline development projects in Virginia involving vegetated wetlands involve fringe marsh-wetlands. Given the relative effort invested in protecting fringe wetlands, it is somewhat surprising that more effort has not been invested in investigation of the structure and function of fringing wetlands.

Both Virginia and Maryland have undertaken extensive inventory programs of their emergent tidal wetlands. Maryland's inventory is relatively recent and fairly comprehensive.[21] Virginia's inventory has been approached on the basis of political jurisdictions (i.e. counties and cities) and despite being initiated in 1972 remains incomplete as of this date. As a consequence, it is impossible to document total emergent wetland acreage with great certitude. Combining the McCormack and Somes[21] report for Maryland with the available published Virginia inventories (individual reports available from the Virginia Institute of marine Science, Gloucester Point, VA 23062) and best estimates for uninventoried Virginia areas[22] we have derived the estimates presented in Table 4 on a river basin basis.

The reader should be aware, or perhaps aggrieved, that this estimate does not correspond well with other extant measures of the Chesapeake Bay wetland complement (e.g. 23). The problem lies primarily in definition of the wetlands included in any given inventory. For example, Tiner[23] provides estimates of "estuarine wetlands" which do not include tidal freshwater wetlands (which are included in our estimates), but do include aquatic beds and tidal flats (which are not included in our estimates). Neither approach is necessarily correct, but it is important the user understand the working definitions prior to accepting any one estimate.

FUNCTIONS OF EMERGENT WETLANDS

Emergent tidal wetlands have been widely recognized for the past several decades to play a number of important roles in coastal systems. While the roles have been identified in many different ways, they can be generally grouped into five areas: 1) primary production centers; 2) habitat; 3) modifiers of water quality; 4) erosion buffers; and 5) flood buffers. Within Chesapeake Bay both Maryland and Virginia have recognized these roles of emergent tidal wetlands in their respective management programs.

It is noteworthy that continuing research on tidal wetlands has modified the early dogmatic acceptance of wetlands universal value in each of the five general roles. While every marsh might be argued to have potential value in each role, the biological and physical structure as well as the siting of each marsh are now recognized to constrain its potentials. This has led to a variety of methods for estimating the "value" of individual marshes for management purposes.

It is not our purpose here to review evaluation methodologies, but an indication of the range of complexity or sophistication in evaluation methodologies can be appreciated by comparing approaches in Virginia and Maryland. Virginia's management guidelines were developed in the early 1970's and represent a relatively simple, straight forward evaluation. Vegetated wetlands are classified on the basis of the dominant macrophytes into twelve types. The twelve types are then grouped in five categories based on "the estimated total environmental value of an acre of each type".[24] Maryland's wetland evaluation procedure was developed in the early 1980's. It is based on a fairly comprehensive review of the available information on wetland systems and tries to address the variation among individual wetlands with a quantitative scoring system. The Maryland procedure recognizes thirty-one types of wetlands based on dominant vegetation and then evaluates individual wetlands using specific information about factors such as area, production, wildlife food value, and diversity.[21] Both states recognizes that judicious use of their respective evaluation procedures is what determines the final usefulness of each.

The point here is that continued research into wetland system function has not made management of the resource easier. Investigators, while shedding a great deal of light on natural processes in emergent wetlands, have also rendered simplistic generalizations about wetlands less acceptable. Consequently, the following brief discussion of the five general values of emergent wetlands is appropriate only insofar as it represents the range of potential functions of individual wetlands in the Chesapeake Bay system.

Primary Production.—Emergent wetlands exhibit a wide range of macrophytic primary production. Net annual production of aerial plant parts generally ranges from several hundred to two or three thousand grams dry weight per square meter. The maximum values represent some of the highest production rates known for natural systems. In estuarine systems, it has long been assumed that a large proportion of this material becomes available to organisms

TABLE 4

Aerial extent of emergent tidal wetlands in the Chesapeake Bay.

	Fresh		Brackish	
	acres	km^2	acres	km^2
Patuxent River	3086	12.49	3315	13.42
Potomac River	6085	24.63	7592	30.72
Rappahannock River	5785	23.41	7088	28.68
York River	6514	26.36	15276	61.82
James River	13524	54.73	18278	73.97
Bay main stem	45025	182.21	192137	777.55
TOTAL by Type	80019	323.83	243686	986.16
TOTAL for Bay	323,705 acres	1309.99 km^2		

at higher trophic levels through the detrital based food web. The assumption has been that high production of commercially important finfish and shellfish is to some degree dependent on the tremendous primary production occurring in emergent wetlands. Recent research has focussed on quantification of the contribution emergent wetlands make to total estuarine primary production. Findings seem to indicate that the connection between emergent wetlands and commercially important secondary production is not as direct as originally believed.[25] Further, the size of the contribution may not be as large as originally believed.[26] Nixon's analysis of pertinent data suggests that if one assumes an annual export of 100 grams of carbon from each square meter of salt marsh, those areas are contributing approximately 10 percent of the total primary production in Chesapeake Bay, the balance being provided by phytoplankton. The distribution of emergent wetlands would indicate they may play a significantly greater role in Bay tributaries, but their importance on that scale has yet to be specifically investigated.

It is of interest to note that most estimates of production in emergent wetlands are limited to analysis of the above ground plant parts. This is generally due to the difficulty of developing accurate estimates of net production in belowground tissues. Unlike the above ground plant parts which typically senesce each year, many wetland species have extensive perennial belowground tissues which pose intractable sampling and analytical problems. The net production rates of these tissues are assumed to range from a fraction of above ground rates in some annual species to several times above ground rates in certain perennial species (e.g. 27). Belowground production certainly is important in the maintenance of marsh systems and contributes to their value in other roles such as erosion and flood buffering, but its importance in support of higher trophic levels has not been generally established.

Habitat. — The value of emergent wetlands for resident and migrating wildlife is relatively well documented. Wetlands provide not only shelter and nesting sites, but also support many floral and faunal species used as food sources by wildlife. In general, freshwater emergent wetlands are of greater importance as potential food sources than other types of tidal wetlands by virtue of the greater number of edible plants. Within the Bay the value of mesohaline and polyhaline marshes as habitat is probably enhanced by their frequent co-occurrence with productive shoal areas in which waterfowl forage.

Water Quality. — The high levels of primary production in emergent wetlands suggest at least a potential for significant nutrient uptake from waters passing over or through these systems. Investigations of this hypothesis, however, suggest the macrophytic production plays a relatively minor role in a wetland's ability to reduce nutrient loadings in incident water masses. Instead microflora and sediments seem to dominate this aspect of emergent wetland functions. The specific capacity of a wetland to modify nutrient loadings is affected by the hydrologic setting (residence time of water masses is particularly important), the organic content of the soils, sedimentation processes (rates of accumulation affect retention capacity), and microbial activity. In general emergent wetlands have the ability to utilize or immobilize some nutrients or other pollutants. Ongoing investigations suggest however that they are not bottomless sinks.

Emergent wetlands are also valued as sediment traps. Water passing through a marsh or swamp is slowed by its encounter with aerial plant parts and consequently loses some of its ability to transport sediments. This is true whether the water is tidal water or runoff. In either case emergent wetlands can be efficient filters depending on the size and type of the wetland, the velocity of the water and the sediment load. Indeed the ability of emergent wetlands to trap sediments is sometimes crucial to maintenance of an intertidal position in the face of rising sea levels.

Erosion Buffer. — The ability of emergent wetlands to slow incident water masses, thus dissipating some of their energy, serves not only to trap sediments but also to retain them. A second attribute of emergent wetlands which is important to their success as erosion buffers is the extensive root and rhizome mat some produce. This mat physically holds sediments in place. Not all emergent wetlands are equally good in this role. Tidal freshwater marshes are distinctly less valuable than oligohaline or mesohaline wetlands as erosion buffers. This is due to the disappearance of most above ground tissues during winter months and the relatively less well developed belowground rhizome mat.

Investigations of emergent wetlands as erosion buffers in the Chesapeake Bay have demonstrated the efficacy within certain limits. As with most other roles of wetlands, physical and biological factors combine to make wetlands' values vary greatly.[28]

Flood Buffer. — Emergent wetlands value in mitigation of flood damage has been documented in several investigations. One of the first and perhaps most widely recognized of these is the evaluation of wetlands on the Charles River in Massachusetts.[29] In that and subsequent studies in other areas, riverine wetlands have been shown to moderate peak discharges associated with major runoff events.

The importance of wetlands flood buffering abilities in the Chesapeake Bay is probably relatively low. While Bay tributaries do have significant amounts of tidal wetlands, the flooding in these river reaches is primarily caused by coastal storm surges (with their effectively limitless supply of water) as opposed to runoff events.

Within Chesapeake Bay, extensive wetlands may play a role in the "normal" hydrologic cycle of tributaries by virtue of the same capabilities which make them good flood buffers in other settings. The ability to effectively absorb large quantities of water during floods (or perhaps spring tides) and then slowly release it during lower water levels (such as neap tidal cycles) could make wetlands an important hydrologic factor in some tributaries.

TRENDS IN EMERGENT WETLAND DISTRIBUTION AND ABUNDANCE

Information on long term trends in distribution and/or abundance of emergent wetlands in Chesapeake Bay is somewhat difficult to assemble. This results from the lack of a comprehensive, consistent, and continuing inventory program for the resource. The U.S. Fish and Wildlife Service's national Wetland Inventory (NWI) produced preliminary estimates for Maryland and Virginia which indicated both states losing "estuarine wetland" acreage at about the same rate over a period from the mid 1950's to the early 1980's.[23] In total the NWI estimate for losses in the Chesapeake Bay is about 20,000 acres during that 20 to 25 year period.

For the sake of discussion, this estimate might be compared to an estimate generated just for Virginia. Byrne and Anderson[30] estimated that marsh shoreline in Virginia had been changing at a rate of approximately $-7,361$ acres per 100 years. It is not clear to us exactly how these estimates might be directly compared, but they do suggest several possible interpretations. First, if loss estimates over the longer term are not significantly greater than loss estimates over the shorter term, the rate of loss may have been accelerating. Second, if man's activity and impact on the shoreline has been similarly escalating, anthropogenic impacts may be more important than natural changes. The data presented here certainly is not sufficient to draw either conclusion but both have been suggested on other bases as well.

The important points are that we do not have a good estimate of trends in distribution or abundance of emergent tidal wetlands in the Chesapeake Bay and we do not know the balance between natural and man-made impacts. The establishment of protective regulatory programs for emergent wetlands is widely believed to have significantly reduced the rate of loss since the early 1970's but no one can do more than estimate the effects. Compared to SAV in Chesapeake Bay, documentation of trends in the resource is minimal. We believe we continue to lose emergent wetlands in Chesapeake Bay but we are not certain at what rate nor are we certain of the relative importance of the causes.

REFERENCES

1. Kemp, W.M., W.R. Boynton, R.R. Twilley, J.C. Stevenson and L.G. Ward.

1984. Influences of submersed vascular plants on ecological processes in upper Chesapeake Bay. pp. 367-394 *In:* V.S. Kennedy (ed.), The Estuary as a Filter. Academic Press, Inc., N.Y.
2. Ward, L.G., W.M. Kemp and W.R. Boynton. 1984. The influence of waves and seagrass communities on suspended sediment dynamics in an estuarine embayment. Mar. Geol. 59:85-103.
3. Wetzel, R.L. and P.A. penhale. 1983. Production ecology of seagrass communities in the lower Chesapeake Bay. Mar. Tech. Soc. J. 17:22-31.
4. Carter, V., N.B. Rybicki, R.T. Anderson, T.J. Trombley and G.L. Zynjuk. 1985. Data on the distribution and abundance of submersed aquatic vegetation in the tidal Potomac River and transition zone of the Potomac estuary, Maryland, Virginia, and the District of Columbia, 1983 and 1984. U.S. Geological Survey Open-File Report 85-82. 65 pp.
5. Hotchkiss, N. 1972. Common Marsh, Underwater and Floating-leaved Plants of the United States and Canada. Dover Publications, Inc., N.Y.
6. Orth, R.J., J. Simons, V. Carter, L. Hindman, S. Hodges, K. Moore and N. Rybicki. 1986. Distribution of submerged aquatic vegetation in the Chesapeake Bay and tributaries—1985. Final Report. Coop. Agreement Z-003301-01.
7. Stevenson, J.C. and N.M. Confer. 1978. Summary of available information on Chesapeake Bay submerged vegetation. U.S. Dept. of Interior, Fish and Wildlife Service FWS/OBS-78/66.
8. Orth, R.J. and K.A. Moore. 1983. Chesapeake Bay: an unprecedented decline in submerged aquatic vegetation. Science 222:51-53.
9. Cottam, C. 1934. Past periods of eelgrass scarcity. Rhodora 36:261-264.
10. Cottam, C. 1935. Wasting disease of *Zostera marina.* Nature 135:306.
11. Cottam, D., and D.A. Munro. 1954. Eelgrass and environmental relations. J. Wildlife Management 18:449-460.
12. Macalaster, E.G., D.A. Barker and M. Kaspar. 1982. Chesapeake Bay Program Technical Studies: A Synthesis. U.S. EPA, Washington, D.C.
13. Orth, R.J. and K.A. Moore. 1981. Submerged aquatic vegetation of the Chesapeake Bay: past, present and future. pp. 271-283 *In:* Transactions of the 46th North American Wildlife and Natural Resources Conference, Wildlife Management Institute, Washington, D.C.
14. Orth, R.J., J. Simons, R. Allaire, V. Carter, L. Hindman, K. Moore and N. Rybicki. 1985. Distribution of submerged aquatic vegetation in the Chesapeake Bay and tributaries—1984. Final Report. Coop. Agreement Z-003301-01.
15. Stevenson, J.C., K. Staver, L.W. Staver and D. Stotts. 1986. Revegetation of submerged aquatic vegetation in mid-Chesapeake Bay—1985. Univ. Maryland CEES, Tech. Ser. No. TS-50-86.
16. Carter, V., and N. Rybicki. 1986. Resurgence of submersed aquatic macrophytes in the tidal Potomac River, Maryland, Virginia, and the

District of Columbia. Estuaries 9(4B):368-375.
17. Wetzel, R.L. and H.A. Neckles. 1986. A model of *Zostera marina* L. photosynthesis and growth: simulated effects of selected physical-chemical variables and biological interactions. Aquatic Botany 26:307-323
18. Cowardin, L.M., V. Carter, F.C. Golet, and E.T. LaRoe. 1979. Classification of wetlands and deepwater habitats of the United States. U.S. Fish and Wildlife Service, Washington, D.C. FWS/OBS-79-31, 103 pp.
19. Hobbs, C.H., D.W. Owen, L.C. Morgan. 1979. Summary of shoreline situation reports for Virginia's Tidewater localities. SRAMSOE No. 209, Virginia Institute of Marine Science, Gloucester Point, VA.
20. Barnard, T.A. (Personal Communication). Wetlands Advisory Program. Virginia Institute of Marine Science, Gloucester Point, VA.
21. McCormack, J. and H.A. Somes, Jr. 1982. The coastal wetlands of Maryland. Report prepared for Maryland Department of Natural Resources, Annapolis, MD 21401.
22. Silberhorn, G.M. (Personal Communication). Wetlands Inventory Program, Virginia Institute of Marine Science, Gloucester Point, VA.
23. Tiner, R.W., Jr. 1985. Wetlands of the Chesapeake Bay watershed: an overview. pp. 16-24 *In:* D.M. Burke, H.A. Groman, T.R. Henderson, J.A. Kusler, and E.J. Meyers (eds.), Proceedings of the Conference—Wetlands of the Chesapeake. The Environmental Law Institute, Washington, D.C.
24. Virginia Marine Resources Commission. 1982. Wetlands Guidelines. VMRC, Newport News, VA. 57 pp.
25. Haines, E.B. 1977. The origins of detritus in Georgia saltmarsh estuaries. Oikos 29:254-260.
26. Nixon, S.W. 1980. Between coastal marshes and coastal waters—a review of twenty years of speculation and research on the role of salt marshes in estuarine productivity and water chemistry. pp. 437-525 *In:* P. Hamilton and K.B. McDonald (eds.), Estuarine and Wetland Processes. Plenum Publishing Corp., New York, N.Y.
27. Whigham, D.F., J. McCormick, R.E. Good, and R.L. Simpson, Jr. 1978. Biomass and primary production in freshwater wetlands of the Middle Atlantic coast. *In*: R.E. Good, D.F. Whigham and R.L. Simpson (eds.), Freshwater Wetlands: Production Processes and Management Potential. Academic Press, New York, N.Y.
28. Hardaway, C.S., G.R. Thomas and A.W. Zacherle. 1982. Vegetative erosion control project: first annual report, 1981. Virginia Institute of Marine Science, Gloucester Point, VA.
29. U.S. Army Corps of Engineers. 1972. Charles River Watershed, Massachusetts. New England District, Waltham, MA.
30. Byrne, R.J. and G.L. Anderson. 1979. Shoreline erosion in Tidewater, Virginia. Chesapeake Research Consortium Report No. 8. SRAMSOE NO. 111, Virginia Institute of Marine Science, Gloucester Point, VA 102 pp.

Chapter Seven

PHYTOPLANKTON IN CHESAPEAKE BAY: ROLE IN CARBON, OXYGEN AND NUTRIENT DYNAMICS

KEVIN G. SELLNER

Benedict Estuarine Research Laboratory
The Academy of Natural Sciences
Benedict, MD 20612

ABSTRACT

Phytoplankton production dominates carbon dynamics and flux in Chesapeake Bay, ultimately resulting in high carbon flow through microheterotrophic assemblages and development and maintenance of hypoxia (<2 mg O_2 l^{-1}) and anoxia in sub-pycnocline waters of mid-Chesapeake Bay. Annual phytoplankton production ranges from 74-851 gC m^{-2} with much of the material supporting microheterotrophic metabolism in the water column and sediments. Phytoplankton remain largely ungrazed in the Bay leading to high sedimentation rates to the pycnocline and subpycnocline waters and benthos. Phytoplankton carbon supports high bacterioplankton densities ($>10^{10}$ l^{-1}) and oxygen demand in the water column and sediments leading to seasonal oxygen depletion in the deep trough of the main Bay and its tributaries. Oxygen consumption from oxidation of phytoplankton material proceeds aerobically as well as anaerobically, yielding high concentrations of phosphate and several reduced chemical species (such as ammonium and sulfide) in near-bottom waters. The interaction between meteorological events and vertical displacements of nutrient-rich sub-pycnocline waters into shallow littoral zones of the Bay may effectively result in positive feed-back for continued phytoplankton production in the Bay and additional supplies of carbon for microheterotrophic metabolism in the system.

INTRODUCTION

Chesapeake Bay is one of the most productive estuaries in the world with primary production exceeding 400 gC m^{-2} yr^{-1} ^{1} indirectly resulting in the world's largest production of blue crab and greater than one-half and one-quarter, respectively, of the U.S. catch in soft-shelled clams and oysters.[2] In recent years, however, this high phytoplankton production in the Bay has been implicated as one of the primary reasons for general deterioration of the ecosystem; phytoplankton production apparently supports a microheterotrophic food web that is partially responsible for the development of anoxic bottom waters,[3,4,5,6] in turn, an unfavorable habitat for commercially valuable stocks in Chesapeake Bay.[7,8]

Two areas will be addressed in the present paper: 1) a general summary of the distributions of phytoplankton taxa, densities, biomass and production; and 2) a discussion of the role of phytoplankton in carbon, oxygen and nutrient dynamics of Chesapeake Bay and its tributaries.

DISTRIBUTIONS OF PHYTOPLANKTON, CHLOROPHYLL AND PRODUCTIVITY

Phytoplankton in Chesapeake Bay are typical of temperate zone estuarine assemblages. In general, nanoplankton are responsible for most of the primary production, chlorophyll and, in meso- and polyhaline waters of the Bay, total suspended particle concentrations;[9,10,11,12,13,14] however, as will be discussed below, aperiodic blooms of large diatoms and dinoflagellates result in biomass and productivity dominated by these larger cells. The nanoplankton includes a diverse assemblage of phytoplankton taxa, including microflagellates and small diatoms. Most recently, data has been collected suggesting that small, 1-2 μm coccoid cells may be the largest contributor to total cell numbers in this group.[15] These cells reach maximum densities in the spring and late summer with densities of 140×10^6 cells l^{-1} recorded in the upper Bay; densities in the oligohaline and mesohaline region generally approximate 10^6 - 10^7 l^{-1}. Although identification has not been verified, these cells may represent the ubiquitous cyanobacteria *Synechococcus* common to oceanic and neritic waters of the world's oceans[16,17] as well as the cyanobacteria observed in the lower Bay[18] and York River.[19,20]

The distribution of phytoplankton in Chesapeake Bay reflects distributions of two environmental parameters, turbidity and nutrient concentrations (Fig. 1). Phytoplankton densities, chlorophyll concentrations and productivity vary with the longitudinal distributions of these two parameters in the Bay. High turbidities[21] limit phytoplankton production from the Susquehanna Flats to just below the turbidity maximum. As a result of light-limited production[22,23] and increasing salinities, high phytoplankton densities entering the Bay in the Susquehanna decrease downbay; light-limitation also prohibits growth of the marine taxa in saline waters moving upbay. The net effect of light-limited production in the upper Bay is the development of a chlorophyll minimum in the oligohaline region of the upper Bay.[23] Further downbay, north of the upper extent of the deep trough (in the vicinity of the bay Bridge), high nutrients from the Susquehanna, increasing water clarity and water column turbulence associated with the passage of internal waves frequently observed in the region[24,25,26] result in chlorophyll and productivity maxima for the Bay (Fig. 1). Phytoplankton densities, pigment levels and productivity decrease further downbay, presumably due to nutrient limitation of phytoplankton production (see below), to lowest levels in polyhaline waters of the lower Bay.

Phytoplankton biomass, as chlorophyll and cell densities, and production in Chesapeake Bay follow slightly different seasonal distributions. In general, cell densities (Fig. 2) and chlorophyll concentrations have a unimodal distribution with a large spring peak associated with the spring bloom. In contrast, primary productivity has a bimodal distribution with high primary productivity accompanying the spring bloom throughout the Bay and a secondary maximum in summer. Variations on these themes are abundant, due principally to surface aggregations or blooms of phytoplankton taxa, including diatoms and dinoflagellates, throughout the year and high chlorophyll concentrations at depth in the winter.

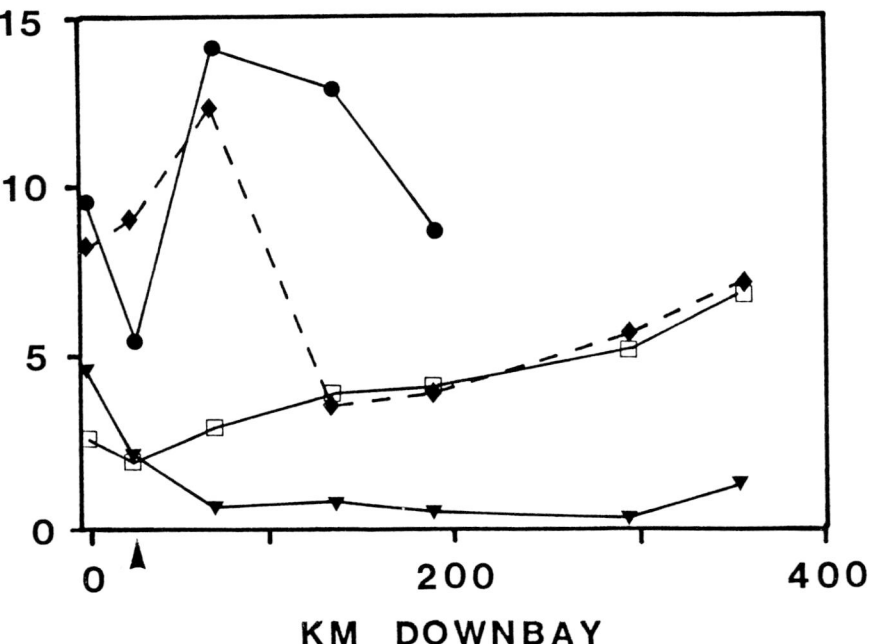

FIGURE 1. Longitudinal distributions of euphotic zone depth (m; □——□), dissolved inorganic nitrogen in the surface mixed layer (μg-at l^{-1} x 10^{-1}; ▼——▼), chlorophyll concentration at 0.5 m (μg l^{-1}; ♦ — — ♦) and carbon fixation (gC m^{-2} d^{-1} x 10; ●——●), Chesapeake Bay, 19-21 August 1985. The arrow (▲) approximates the position of the turbidity maximum.

Highest phytoplankton biomass is observed in the spring, generally due to the typical spring diatom bloom of temperate estuaries. Depending on location, and therefore salinity and bottom topography in the Bay, other phytoplankton groups make substantial contributions to the spring flora. For example, from the head of the Bay into mesohaline waters, *Cyclotella* sp. and *Thalassiosira* sp. dominate total cell numbers in April and early May and diatoms contribute approximately 60% of total eucaryote cell numbers.[15] However, at the head of the Bay, pennate diatoms and chlorophytes, particularly *Scenedesmus* spp., are sub-dominants in the spring bloom. With increasing salinity downbay, chlorophytes and pennate diatoms become less important and dinoflagellates increase. For example, just north of the northern end of the deep trough, the two centric diatoms *Cyclotella* sp. and *Thalassiosira* sp. still dominate but dinoflagellates, principally *Prorocentrum minimum (P. mariae-lebouriae)* and to a lesser extent, *Katodinium rotundatum,* make substantial contributions to total eucaryote numbers, forming 40% of the phytoplankton assemblage in the spring bloom.[15] *Prorocentrum* sp. is delivered to the upper end of the deep trough from upbay transport of the population at depth during the winter[27] and then through mixing into surface, lighted waters, the cell forms dense aggregations in the vicinity of the Bay bridge in April-May.

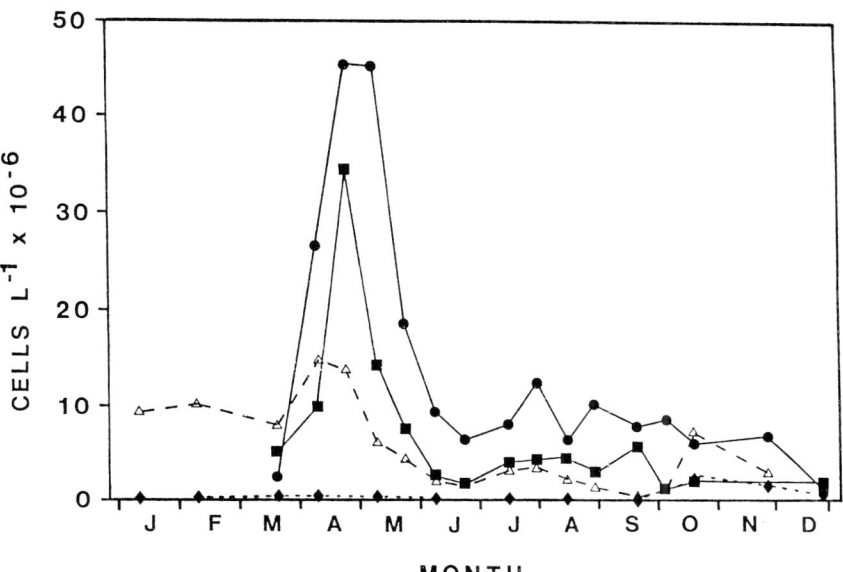

FIGURE 2. Seasonal distributions of phytoplankton in Chesapeake Bay. Total densities are presented for eucaryotes in stations sampled in 1985[15] for the head of the Bay (●——●), in the oligohaline region coincident with the turbidity maximum (■——■) and mesohaline region (△ — — △). Densities for the polyhaline region of the lower Bay (◆——◆) were obtained in 1978.[34]

Further downbay in mesohaline and polyhaline waters, more marine diatom species become dominant, including *Cerataulina pelagica, Rhizosolenia fragilissima, Asterionella japonica, Leptocylindrus* spp., *Nitzschia pungens, Skeletonema costatum, Cyclotella glomerata* and *Thalassiosira nordenskioldii*;[13,15,18,28-35] flagellate taxa, primarily cryptophytes, also become more important.

This general longitudinal distribution of phytoplankton taxa is accompanied by a distinct gradient in biomass. For example, during the spring bloom from late April-early May, chlorophyll concentrations in surface waters ranged from 6.8-12.7 μg l^{-1} at the head of the Bay, 8.8-24.3 μg l^{-1} in the vicinity of the turbidity maximum (oligohaline region) to maximum concentrations of 26.4-36.3 μg l^{-1} at the head of the deep trough.[15] Further downbay in the mesohaline region over the deep trough, chlorophyll concentrations declined to minimum levels (approximately 3 μg l^{-1}) only to increase in shallower polyhaline regions of the lower Bay, 16.4-21 μg l^{-1}, in April 1985 and 1986 (36, T. Malone, pers. comm.).

The longitudinal pattern in surface chlorophyll concentrations is similar for total chlorophyll in the water column. In general, chlorophyll concentrations (mg m^{-2}) are low at the head of the Bay and increase to maximum concentrations in the vicinity of the northern end of the deep trough. For example, in 1985, chlorophyll increased from 74 mg m^{-2} at the head of the Bay to 169 mg m^{-2} in the vicinity of the turbidity maximum to 923 mg m^{-2} in the vicinity of the Bay bridge.[15] Further downbay, chlorophyll concentrations decreased to levels of 174 mg m^{-2}.[36]

Phytoplankton biomass declines following the spring bloom; cell densities decline to < 10 x 10^6 cells l^{-1} (Fig. 2) as diatoms become less important in the total phytoplankton assemblage. By August, dinoflagellates, flagellated taxa and small centric diatoms are dominant in saline, surface waters; a cyanophyte *Merismopedia* and a chlorophyte *S. quadricauda* are major contributors to phytoplankton assemblages at the head of the Bay.[15] From the Susquehanna Flats to the mouth of the Bay, surface chlorophyll concentrations are generally < 10 μg l^{-1} and decline to several micrograms l^{-1} in waters below the pycnocline. In 1985, integrated chlorophyll concentrations from late May to December were < 240 mg m^{-2} throughout the Bay and in the most productive summer months of July and August, 1985 and 1986, chlorophyll concentrations were < 100 mg m^{-2} in mesohaline waters of the Bay.[15]

A small increase in diatom densities in mesohaline and polyhaline regions of the Bay is observed in the fall (late September-October). For example, in 1985, diatom densities increased from 1.1 x 10^6 l^{-1} in late September to 6.6 x 10^6 l^{-1} in early October in the mesohaline region of the Bay.[15] Marshall[35] reported a similar pattern in 1977 in the lower Bay; phytoplankton densities increased from 0.14 x 10^6 cells l^{-1} in September to 2.7 x 10^6 l^{-1} in October, principally due to an increase in the centric *S. costatum*. Since fall destratification occurs at this time,[37] rapid growth of diatoms might be favored through the introduction of nutrients from near bottom waters, higher water column turbulence and comparable depths for the surface mixed layer and euphotic zone.

Phytoplankton densities rapidly decline to near annual minima for the year in November-December, 1-2 x 10^6 cells l^{-1} in the upper Bay and ≤ 0.8 x 10^6 l^{-1}

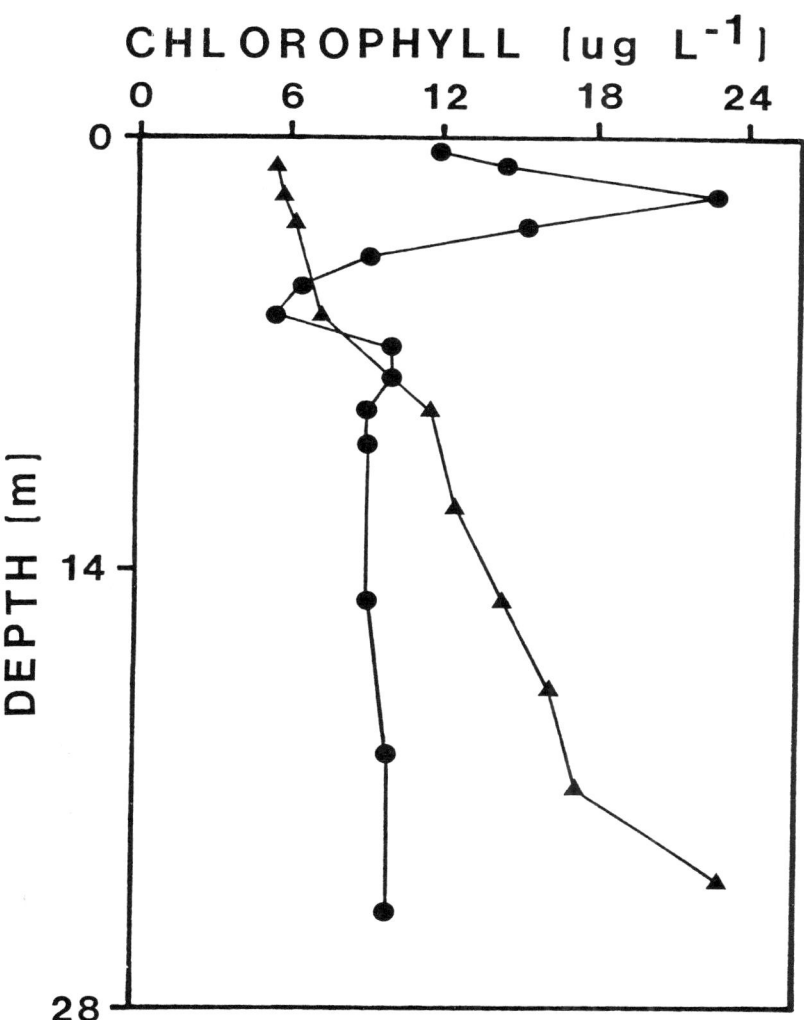

FIGURE 3. Vertical distributions of chlorophyll in August, 1984 (●——●) and March, 1985 (▲——▲) in the mid-channel of mesohaline Chesapeake Bay.[15]

in the lower Bay.[15,35] Chlorophyll, in surface waters and the entire water column, also reach low levels (<5 µg l^{-1}, <40 mg m^{-2}, respectively, in the upper Bay in 1985).

In polyhaline regions of the lower Bay, highest chlorophyll concentrations typify January-early April. In 1985, average monthly surface chlorophyll concentrations ranged from 6.21-13.20 µg l^{-1}.[36] However, due to high chlorophyll concentrations at depth, chlorophyll concentrations over the water column ranged from 109-174 mg m^{-1} for January through early April.

Elevated chlorophyll concentrations at depth in the winter are also observed throughout the mesohaline portion of the Bay due to upbay transport of water and diatoms (*Skeletonema, Cerataulina, Rhizosolenia*) from the shelf and *Prorocentrum* populations reaching the mouth of the Bay in late fall.[27] As a result, chlorophyll concentrations are low in surface waters and high at depth, opposite to vertical distributions noted in the summer (Fig. 3). These subpycnocline populations are transported upbay resulting in winter diatom blooms (discussed below) and early spring surface blooms of *Prorocentrum*[27,38] and diatoms in the vicinity of the Bay bridge. In addition, following the spring freshet and a marked increase in vertical stratification in the Bay, high chlorophyll levels in the bottom waters probably serve as initial substrate for microheterotrophic oxygen demand and development of hypoxia/anoxia in the mesohaline area of the Bay.

In addition to the typical seasonal distribution of phytoplankton for a temperate estuary described above, saline waters of Chesapeake Bay are characterized by aperiodic blooms of other taxa throughout the year. These blooms may remain largely ungrazed by macrozooplankton in the system[39] and could provide large substrate pools for either microzooplanktonic grazers[40,41] or, via sedimentation, heterotrophs in pycnocline or subpycnocline waters. Aperiodic blooms of diatoms and dinoflagellates have been observed in all seasons. In December-February, blooms (e.g. 90 x 10^6 l^{-1}) of *Skeletonema costatum* and *Chaetoceros* spp., primarily *C. atlanticum,* have been observed in a 11 m water column off Calvert Cliffs, MD (mesohaline region[13]). Coldwater diatom blooms have also been observed in other Bay studies[31,32] and several other east coast estuaries.[42,43] Dinoflagellate blooms are also common in winter with *Katodinium rotundatum* and *Heterocapsa triquetra* the principal taxa noted. *K. rotundatum* was observed at $\geq 10^8$ cells l^{-1} in the Patuxent River estuary in February, 1987 (S. Cibik, pers. communication). Comparable densities and chlorophyll concentrations >200 µg l^{-1} have been noted for other dinoflagellate blooms, red or mahogany tides, for the period April through October in the Bay; representative taxa include *Gymnodinium nelsonii, G. stellatum, Gyrodinium uncatenum, G. estuariale, Peridinium* spp., *Cochlidinium catenatum, Prorocentrum* spp., *Oxytoxum* and *Ceratium lineatum.*[13,27,35,38,39,44,45,46,47,48,49]

As noted above, primary production in Chesapeake Bay follows a bimodal distribution over the year with a pronounced peak accompanying the spring bloom and high carbon fixation rates during July and August. In winter, carbon fixation is limited by temperature and light through the Bay. For example, carbon fixation rates of 0.37 ± 0.09 gC m^{-2} d^{-1} were noted in January-March, 1985 and 1986 for the region from the Susquehanna Flats to the mouth of the Potomac River.[15] In the lower Bay in February, Malone (pers. communication) measured similarly low rates, 0.16-0.37 gC m^{-2} d^{-1}, at three stations off the mouth of the Rappahannock River.

With increasing daylength and temperatures, longitudinal differences in carbon fixation become obvious in the Bay. Carbon fixation is light-limited in the upper Bay resulting in lower daily carbon fixation rates at the head of the Bay and in the turbidity maximum region in oligohaline waters; for example, carbon fixation at the head of the Bay and in the turbidity maximum region was 0.96 and 0.49 gC m^{-2} d^{-1} during the May bloom, respectively.[15] Further downbay, water clarity increases and carbon fixation increases to maximum rates in the vicinity of the Bay bridge (5.2 gC m^{-2} d^{-1}), only to decline to rates of 0.45-0.75 gC m^{-2} d^{-1} of the Rappahannock River (April-May, 1986; T. Malone, pers. communic.). High summer rates (>1 gC m^{-2} d^{-1}) are also maintained for mesohaline and polyhaline regions of the Bay. Average rates ($\pm$ s) for June-early September, 1985 in the light-limited upper Bay and mesohaline region and in 1986 for the polyhaline region were 0.92 ± 0.40, 1.43 ± 0.56 and 1.06 ± 0.49 gC m^{-2} d^{-1}, respectively.[15] Rates near the mouth of the Bay are not available.

The fall diatom bloom in mesohaline and polyhaline regions of the Bay is also accompanied by an increase in carbon fixation. For example, in two stations between the Choptank and Potomac Rivers in 1985, the average carbon fixation rate in October was 2.34 gC m^{-2} d^{-1}, a two fold increase from early September rates.[15]

Carbon fixation in aperiodic summer blooms, i.e., red tides, is very high. For example, carbon fixation in a bloom of *Cochlidinium heterolobatum* in the York River was 1546 μgC l^{-1} h^{-1}.[47] Non-bloom rates range from 73-251.8 μgC l^{-1} h^{-1}. (T. Malone, pers. communication). The importance of these blooms in total annual primary production in the Bay cannot be estimated, however, because the spatial and temporal heterogeneity of the blooms cannot be routinely determined. Continuous shipboard underway measurement of chlorophyll currently employed in northern and mid-Chesapeake Bay[15] and remote sensing[19,50,51] might provide finer resolution of bloom frequency. However, both techniques are expensive and only provide *pigment* data dependent on ship schedules and cloud cover, respectively. Therefore, the total contribution of bloom carbon to Bay production is probably underestimated and the effects of bloom carbon on oxygen dynamics remain unresolved.

Over the year, the seasonal and spatial heterogeneity in phytoplankton distributions results in annual primary production rates of 74-86 gC m^{-2} yr^{-1} in the upper Bay[22] and 167-851 gC m^{-2} yr^{-1} in the mid-Bay region.[1,13,22,52,53]

NUTRIENT-LIMITATION

Using several techniques, nutrient limitation in late spring through fall in mesohaline and polyhaline waters of the lower Bay and tributaries has been suggested. Examination of elemental ratios indicates phosphorus and nitrogen limitation in the spring and summer/fall, respectively, in some years[13,54] but not all.[55] Field kinetic studies and the application of published kinetic data to Bay assemblages[13,54,56,57,58,59,60] also support aperiodic nutrient limitation (N, P and silicon) in mesohaline and polyhaline regions of the Bay.

Further support for nutrient-limited phytoplankton production in the Chesapeake Bay system was obtained in a series of microcosm studies conducted in 1983-1986. In summer and fall, the addition of dissolved nitrogen stimulated growth of natural phytoplankton assemblages from mesohaline and polyhaline regions of the Patuxent River and York River estuaries; no effect was seen with the addition of phosphorus. In winter, however, phosphorus additions resulted in increases in phytoplankton while nitrogen additions had little effect.[61,62,63,64]

The importance and frequency of nutrient-limited phytoplankton growth in Chesapeake Bay and its subestuaries is also confounded by estimates of water column regeneration rates for nitrogen and phosphorus in the Bay. Measured and calculated regeneration rates in the water column alone would more than meet phytoplankton demand[13,54,60,65] even without consideration of nutrient flux from the sediments (see later discussion of nutrient-sediment relationships). Because nutrients from sediments and sub-pycnocline waters are not permanently removed from surface waters (see discussion below), their aperiodic introduction (< 7-15 d frequency) would also support relatively nutrient-replete conditions for short periods in the Bay. The duration of nutrient limited conditions for estuarine phytoplankton assemblages in the Bay therefore remains open.

PHYTOPLANKTON RESPONSES TO BAY CIRCULATION

Chesapeake Bay is a partially mixed estuary with Susquehanna River flow controlling vertical stratification in the Bay. Spring runoff in the Susquehanna initially establishes strongest statification in the Bay resulting in limited exchange between surface and bottom water. Officer et al.[6] have estimated that nonadvective vertical exchange decreases by a factor of 10 from early February mixed conditions to post-freshet conditions (i.e. stratified conditions), essentially limiting the exchange of dissolved materials between surface and bottom layers. Their equations also suggest that river flow is directly related to stratification intensity and inversely related to exchange between the two layers.

Stratification intensity has marked effects on chemistry and biology of mid-Bay waters. With high stratification, respiration below the pycnocline (water column and benthos) results in rapid development of low oxygen/anoxic conditions in bottom waters and regeneration of high concentrations of nutrients.[6] Reaeration from diffusive processes is more difficult and anoxia is maintained for longer periods. One mechanism for reaerating subpycnocline waters that results in marked increases in phytoplankton biomass and/or productivity in the Bay is through wind-induced cross-bay displacement of the pycnocline (Fig. 4;[4,5,13,66,67]). Continuous winds predominantly from the west or east force surface waters to the east or west, respectively. Displacement of surface waters to either shore results in anoxic bottom waters rich in nutrients moving upwind into shallow littoral zones on the opposite shore. Once in the shallow littoral zones on either side of the Bay, sub-pycnocline water is reoxygenated and phytoplankton standing crops and productivity increase,[4,68] probably in response to nutrient additions and/or mixing of the water column. Malone et al.[4] suggest that this surface production also apparently accounts for the majority of organic material settling into deeper waters of the Bay since material collected in sediment traps deployed in mid-Bay is more closely correlated with surface production over the flanks than with production in mid-channel.

The introduction of sub-pycnocline waters into shallow waters on either side of the Bay, and therefore presumably enhanced phytoplankton production, appears to be fairly frequent. Sellner and Kachur,[13] Malone et al.[4] and Houde (pers. communication) have measured intrusions at frequencies ≤ 7 d to 15 d. However, it is unlikely that every tilt event will result in a phytoplankton response since 1) sub-pycnocline waters are not always anoxic (compare 1984 and 1985, Tuttle et al., this volume) and nutrient-rich and 2) intruding waters may not always enter the euphotic zone.

Tuttle et al. (this volume) have compared data from 1984 and 1985 in an effort to verify the role of stratification in controlling mid-Bay oxygen and carbon dynamics. They have suggested that strong stratification and the resulting anoxia in 1984 resulted in higher algal standing crops due to low oxygen conditions limiting zooplankton grazing pressure. In 1985, however, lower flow conditions and weaker stratification limited anoxia development, thereby permitting higher zooplankton grazing pressure and lower phytoplankton standing crops. Spring Susquehanna River flow, therefore, may be a fairly accurate predictor for anoxia development and surface production in mid-Chesapeake Bay.

Longitudinal basin morphometry also has a distinct effect on primary production in the Bay. The deep trough region in mid-Chesapeake Bay (30 m) abruptly shallows to water depths approximating 10 m to the north and south. Internal waves are characteristic of the transition region on the northern end of the trough (near the Bay Bridge at Annapolis).[24,25,26] Conceivably, turbulence associated with the passage of or breaking of these internal waves could introduce nutrients and phytoplankton cells from sub-pycnocline water into the overlying euphotic waters, resulting in the elevated phytoplankton densities, blooms, chlorophyll concentrations ≥ 100 mgChl m^{-2} and primary production ≤ 5.3 gC m^{-2} d^{-1} typical of this region of the Bay.[15,53]

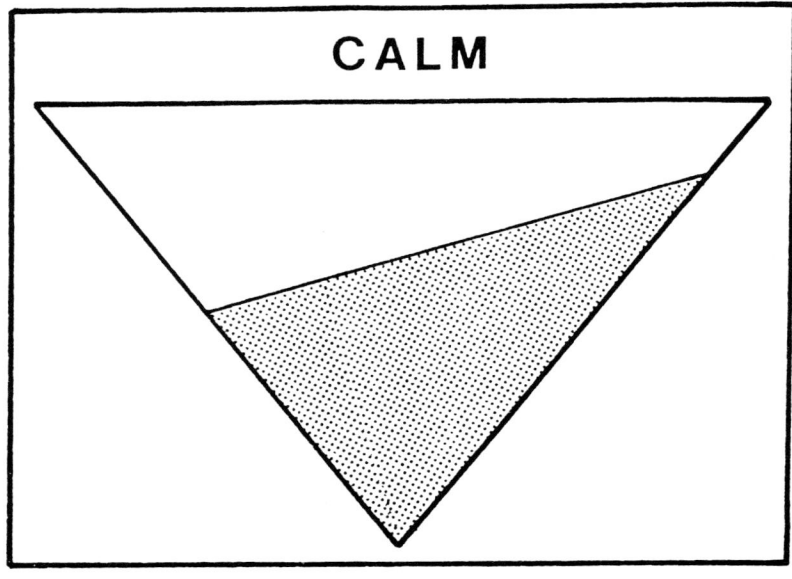
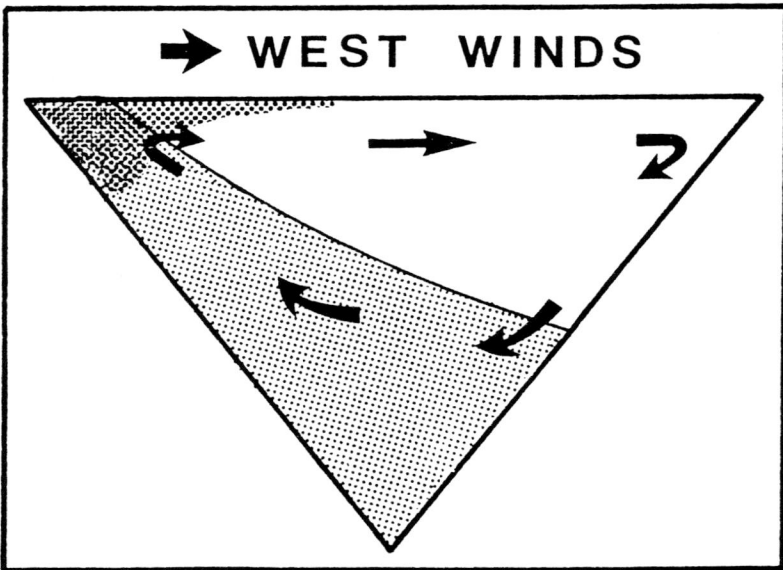

FIGURE 4. Idealized vertical profiles of surface mixed layer, pycnocline and sub-pycnocline waters (shaded area) in mid-Chesapeake Bay under calm conditions and with winds predominantly from the west. Intense shading on the western shore in the lower panel represents surface phytoplankton bloom that may accompany intrusion of sub-pycnocline waters into shallow depths during the summer months (4; see text).

RELATIONSHIPS BETWEEN PHYTOPLANKTON, SECONDARY PRODUCERS, NUTRIENTS AND DISSOLVED OXYGEN

Phytoplankton and Bacteria

Classical food webs almost always include a link between phytoplankton and bacteria. The link usually occurs through dissolved organic carbon excreted or spilled from the primary producers.[69,70,71] In the last three years, strong positive correlations have been established between primary production in the surface mixed layer, bacterial cell densities and specific growth rates in the mesohaline region of Chesapeake Bay.[3,4,5] Malone and co-workers[4] have presented convincing evidence that approximately 60% of summer photosynthetic production in 1984 was metabolized by bacteria in the surface mixed layer and that during stratified periods in the summer, 91% of the variation in bacterial cell densities could be predicted from ambient chlorophyll concentrations. In addition, bacterial densities in the Bay approximated 10^{10} cells l^{-1} in the surface layer in 1984 and 1985 with maximum specific growth rates of 0.9-1.3 d^{-1}. The overall importance of bacteria in the Bay could be seen from the ratio of bacterial biomass to phytoplankton biomass, ranging from 23-58% over the four month study period in 1984. These data suggest that carbon flow in Chesapeake Bay may be proceeding through the microbial loop[70] rather than to higher trophic levels visible as increased fish and shellfish stocks.

The direct effect of high bacterial metabolism in the stratified water column of the Bay is maintenance of high oxygen demand leading to microaerobic/anoxic conditions in sub-pycnocline waters of the Bay. These conditions favor additional bacterial metabolism in denitrification and sulfate reduction, with sulfate reduction ranging from 6-335 m mol m^{-2} d^{-1} in mid-Chesapeake Bay sediments.[72] Tuttle (pers. communication) estimates that subsequent oxidation of sulfide in overlying oxygenated waters may consume sufficient oxygen to equal the concentration of oxygen used in aerobic microbial oxidative processes in the water column. This 50-50% split between aerobic microbial oxygen demand and spatially more isolated sulfide oxidation results in a net oxygen consumption rate of 1 mgO_2 l^{-1} d^{-1} in the vicinity of the Bay Bridge near Annapolis, with lower rates noted at the mouth of the Patuxent River and in the mesohaline portion of the Potomac River (Tuttle et al., this volume; Tuttle, pers. communication). Ultimately, the diversion of up to 60% of photosynthetic production to aerobic bacterial metabolism, i.e. oxygen consumption, noted by Malone et al.[4] only serves to promote microaerobic/anoxic bacterial metabolism and the perpetuation of low oxygen conditions in waters beneath the pycnocline of Chesapeake Bay.

Heterotrophic utilization of phytoplankton production proceeds in the water column and benthos resulting in the oxidation of organic matter and the consumption of oxygen. Planktonic respiration, as the sum of all microaerobic oxidative metabolism in the water column, has been implicated by Taft and co-workers[65] as the process controlling bottom water oxygen concentrations in mid-Chesapeake Bay. The importance of respiration in the water column is supported by several other observations: 1) in the summer and fall of 1984, microheterotrophic respiration in the water column was responsible for most of the oxygen demand in mesohaline waters of Chesapeake Bay;[3,5,73,74] 2) P/R ratios for plankton assemblages from April-October, 1975-1980, in an 11 m water column off Calvert Cliffs were only greater than 1 in 6 of 41 sample dates[75] and 3) P/R ratios were always >1 in a 3 m water column, but in a 10 m water column off Calvert Cliffs, ratios were both >1 and <1 for summer and fall observation periods.[76]

Oxygen demand in the benthos and sediment oxygen demand (SOD) have also been identified as primary factors controlling oxygen concentrations in bottom waters of Chesapeake Bay (e.g. 6). Boynton et al.[77] have summarized data collected from June 1984-June 1985 in mid-channel sediments of the Bay and concluded that upper Bay sediment oxygen demand was higher than observed in the mesohaline region of the system. Highest mean rates were approximately 2 mgO_2 m^{-2} d^{-1} in May and 0.7 mgO_2 m^{-2} d^{-1} in October in the oligohaline region of the Bay; in the mesohaline sediments, the range for the year was 0.2-1.1 mgO_2 m^{-2} d^{-1} with higher rates in the spring and early summer coincident with highest benthic macrofauna biomass (see below). These researchers estimated that mid-Bay sediment oxygen demand would consume approximately 60% of primary production in overlying surface waters.

Phytoplankton and Zooplankton

Microzooplankton-phytoplankton studies have been recently initiated in a long-term monitoring program in the Bay. Preliminary evidence[78,79] suggests that microzooplankton (44-202 μm) grazing pressure may be equal to or greater than macrozooplankton demand at certain times of the year. In addition, seasonal distributions of phytoplankton, micro- and macrozooplankton suggest that carbon flow may include the trophic links [phytoplankton to zooplankton] *and* [microzooplankton to macrozooplankton].

Several investigators have attempted to ascertain the impact of Bay and tributary zooplankton populations (principally copepods) on estuarine phytoplankton production. Slightly different conclusions were reached in each case. Heinle and Flemer[80] concluded that there was insufficient phytoplankton for the herbivorous copepod, *Eurytemora affinis* during early spring in the Patuxent River estuary. In contrast, Sellner and Horwitz[81] estimated that maximum zooplankton grazing pressure would remove, at most, 11% of the available phytoplankton in the vicinity of the turbidity maximum zone in the Patuxent with <3% of the phytoplankton stocks removed in the fresher waters upstream or the mesohaline waters downstream. Similar results were obtained in the lower

Gunpowder River estuary in upper Chesapeake Bay.[12] Using zooplankton densities and published clearance rates, Sellner and Kachur[13] determined that maximum grazing pressure off Calvert Cliffs, Md (mesohaline portion of the Bay) for the period 1975-1980 was limited to the late summer where 20-83% of the water column would be cleared by zooplankton >73 μm. Maximum grazing pressure during other times of the year was always <12%. Similar observations have been made for the period August, 1984-July, 1985 in the same region.[78,79]

Macrozooplankton grazing in the tidal fresh region near the mouth of the Susquehanna was very low for the 1984-1985 period.[78,79] Maximum grazing pressure was observed in November, 1984 with <5% of the available phytoplankton grazed. In the oligohaline/transition zone, highest macrozooplankton grazing was estimated for March and July, 1985 with 52% and 40% of the available chlorophyll removed; 3 to 27% of the ambient chlorophyll would have been grazed by >202 μm zooplankton in the remaining months of the August, 1984-July, 1985 study period.

These data suggest that for most of the year macrozooplankton herbivory removes only a small fraction of the available phytoplankton production in upper and mid Chesapeake Bay. The low spring and high summer grazing pressure noted for mesohaline waters of the Bay (see above) agrees remarkably well with carbon flux estimated from sediment trap collections in the region.[77,82] In the spring, daily sedimentation rates represented 84- >300% of primary production in mid-Bay with C/N values of seston and trap collected material between 6-8. Chlorophyll sedimentation rates were also maximum during spring and summer blooms[77] further suggesting minimal utilization in the water column. These data imply rapid sedimentation of phytoplankton to bottom waters during the spring bloom, typical of estuarine and coastal systems in North America and Europe.[83,84,85,86] Carbon transport to the bottom represented 50% of total surface production for the spring,[82] conditions fueling high sediment oxygen demand and subsequent development of low oxygen/anoxic conditions in the main channel during the stratified summer period.[87]

In the summer, however, more of the surface production is consumed in the upper water column. Using data from Boynton et al.[82] for June, July and August, 1984-1985, daily carbon settling from the surface layer accounted for approximately 32% of the daily production. This suggests remineralization/utilization of nearly 70% of primary production in the upper water column, coincident with the period of highest grazing rates for the year (see above).

Phytoplankton and the Macrobenthic Community

Another possible consumer is the benthic macrofauna community. Kemp and Boynton[88] reported that macroinvertebrate respiration ranged from 1-39% of total community respiration off Calvert Cliffs in 1977-1978 with maximum demand in May and June. Maximum demand was coincident with benthic macroinvertebrate biomass increases due to spring recruitment of benthic juvenile populations.[89] This period coincides with highest benthic demand on phytoplankton in the overlying water column. Heck[90] estimated that macrobenthic assemblages could have cleared 102%, 74% and 72% of a 7 m water column off Calvert Cliffs in June, 1978-1980. On average, only 20% of the water column would have been cleared by the macrofauna in the remaining months of the year.

Relationships between Phytoplankton and Nutrient Cycling

The coupling between phytoplankton production and oxygen dynamics in Chesapeake Bay has direct implications for nutrient dynamics in bottom waters of the region. Heterotrophic utilization of phytoplankton production results in the regeneration of essential nutrients, including nitrogen, phosphorus and silica in the water column and sediments.[91] Nutrient concentrations are sufficient for phytoplankton demand Bay-wide in the winter and in the upper Bay year-round.[54,56] At other times, nutrient regeneration in the water column and benthos should meet phytoplankton demand at least in the upper and mid-Bay regions. Taft[54] has suggested that 73->100% of the required nitrogen and 68->100% of the required phosphorus, respectively, would be regenerated through planktonic respiration in August for Bay phytoplankton north of the Rappahannock River. Similarly, Glibert[60] has estimated comparable ammonium uptake and regeneration rates for plankton assemblages from the mid- and lower Bay in May-June 1980. Using Glibert's remineralization rates, Sellner and Kachur[13] estimated that nitrogen demand off Calvert Cliffs would be met by water column regeneration rates in 23 of 24 monthly sample collections for the period July-October, 1975-1980.

Rapid nutrient regeneration in the water column is also accompanied by high nutrient regeneration rates in bottom sediments of the Bay. Fluxes of silica, nitrogen and phosphorus have been estimated for sediments in mid-Chesapeake Bay and the Patuxent River estuary.[77,82,87,92] Diatoms, with biogenic silicon frustules, are readily remineralized at temperatures >15°C and salinities typical of mid- and lower Chesapeake Bay.[93] Remineralized diatoms could yield 2-3 mol Si m^{-2} yr^{-1},[92] with remineralization independent of oxygen content of the sediments.[93] Boynton et al.[82] reported silicon efflux from the sediments in northern and mid-Chesapeake Bay for the period August, 1984-June, 1985. In the northern Bay, mean silicate efflux was 270 and 329 μg-at Si m^{-2} h^{-1} in October, 1984 and May, 1985, respectively. For three stations in mid-Bay, highest rates were observed in the most saline stations in June; rates for the three stations averaged 779 μg-at Si m^{-2} h^{-1}. During the remaining period of the study, mean rates ranged between 171 and 400 μg-at Si m^{-2} h^{-1}. Substrate for silicon regeneration in the sediments is probably derived from the rapid sedimentation of winter and spring diatom blooms[13,77,82] with sediment efflux producing relatively rapid post-bloom increases in dissolved silicate in mesohaline waters of Chesapeake Bay.

Nitrogen and phosphorus regeneration in Bay sediments also results in the accumulation of these nutrients in bottom waters of the mid- and lower Bay. In the upper Bay, however, nitrate uptake by the sediments was observed in August and October, 1984 and June, 1985 with rates of 8-63 μg-at NO_3-N m^{-2} h^{-1} into the sediments.[82] In mid-Chesapeake Bay, nitrogen flux was dominated by NH_4^+ efflux with highest rates in the spring. In June, 170-435 μg-at NH_4^+-N m^{-2} h^{-1} was released from the sediments at three mid-Bay stations. Lowest rates, in October, 1984, were 19-140 μg-at NH_4^+-N m^{-2} h^{-1}. Phosphate flux was generally low and out of the sediments. Maximum rates were 11 and 9-87 μg-at P m^{-2} h^{-1} for upper (May, 1985) and mid-Bay (June, 1985) sediments, respectively. Rates were $< 6 \mu$g-at P m^{-2} h^{-1} at all stations during August and October, 1984 and for mid-Bay stations in May, 1985.[82]

Introduction of high nutrient concentrations from below the pycnocline into the euphotic zone is dependent on the frequency of mixing events (e.g. reaeration, interval wave breaking, cross-bay displacements of the pycnocline, fall destratification) that displace or disrupt vertical stratification in the Bay. As noted above, the intrusion of sub-surface nutrient rich water into surface waters may be accompanied by an increase in phytoplankton production and biomass in near-shore waters of the Bay[4] as well as surface waters overlying the northern end of the deep trough.[15] Similarly, October diatom peaks in mesohaline and polyhaline regions of the Bay follow fall destratification and water column mixing in the same region.

SUMMARY

Carbon, oxygen and nutrient dynamics in Chesapeake Bay are governed in most part by the interaction between phytoplankton production and circulation patterns in the Bay. Most recent evidence suggests that phytoplankton production primarily supports microheterotrophic metabolism with high standing stocks and production of aerobic bacterial populations. The high oxygen demand by this microbial biomass leads to low oxygen concentrations in stratified waters of the Bay. The low oxygen environment selects for anaerobic microbial metabolism and continued high oxygen demand in near bottom waters and sediments yielding anoxic conditions and elevated concentrations of ammonium and phosphate. Introduction of these nutrient species into overlying waters through physical displacement of bottom waters into shallow littoral zones ultimately supports further phytoplankton production and continuation of the cycle.

Phytoplankton production is seasonally important to other secondary producers, such as micro- and macrozooplankton in the upper and mid-Bay regions, respectively. Highest grazing pressure by macrozooplankton assemblages occurs from July-September. Benthic suspension feeders could remove substantial quantities of phytoplankton in late spring-early summer when macrofauna biomass is highest due to spat and juvenile sets. However, macrozooplanktonic and benthic herbivory appear to be of secondary importance to microheterotrophic utilization of phytoplankton carbon in the water column and sediments.

ACKNOWLEDGEMENTS

Completion of this manuscript was possible through extensive use of unpublished data collected within the last three years. The author would like to thank the following investigators for their recent research efforts in the region: R.W. Alden, W.R. Boynton, D.C. Brownlee, H.W. Ducklow, T.R. Fisher, D.M. Goodrich, L.W. Harding, A.F. Holland, W.M. Kemp, T.C. Malone, R.E. Magnien, H.G. Marshall, M.M. Olson, R.C. Siegfried, J.H. Tuttle and M.A. Tyler. Access to MD-VA Monitoring data through the U.S. EPA Chesapeake Bay Program office is also appreciated. Much of the KGS data presented was obtained with funding from Maryland Office of Environmental Programs; the author gratefully acknowledges the support.

REFERENCES

1. Boynton, W.R., C.A. Hall, P.G. Fallkowski, C.W. Keefe and W.M. Kemp. 1983. Phytoplankton productivity in aquatic systems, pp. 305-327. *In:* O.L. Lange, P.S. Nobel, C.B. Osmond and H. Ziegler (Eds.). *Physiological plant ecology IV.* Encyclopedia of Plant Physiology, New Series, Vol. 12D. Springer-Verlag, Berlin.
2. EPA (U.S. Environmental Protection Agency). 1982. Chesapeake Bay: Introduction to an ecosystem. U.S. Environmental Protection Agency, Washington, D.C. 33 pp.
3. Tuttle, J., T. Malone, R. Jonas, H. Ducklow and D. Cargo. 1985. Nutrient-dissolved oxygen dynamics: Roles of phytoplankton and microhetrotrophs under summer conditions. University of Maryland, Ctr. for Environ. Studies Ref. No. [UMCEES] CBL 85-39. Chesapeake Biological Lab., Solomons, MD.
4. Malone, T.C., W.M. Kemp, H.W. Ducklow, W.R. Boynton, J.H. Tuttle and R.B. Jonas. 1986. Lateral variation in the production and fate of phytoplankton in a partially stratified estuary. *Mar. Ecol. Prog. Ser.* 32: 149-160.
5. Tuttle, J.H., R.B. Jonas and T.C. Malone. 1987. Origin, development and

significance of Chesapeake Bay anoxia. *In:* S.K. Majumdar, L.W. Hall, Jr. and H.M. Austin (Eds.) *Contaminant problems and management of living Chesapeake Bay resources.* Chapter 19. The Pennsylvania Academy of Science, Easton, PA.

6. Officer, C.B., R.B. Biggs, J.L. Taft, L.E. Cronin, M.A. Tyler and W.R. Boynton. 1984. Chesapeake Bay anoxia: Origin, development and significance. *Science* 223: 22-27.
7. Price, K.S., D.A. Flemer, J.L. Taft, G.B. MacKiernan, W. Nehlsen, R.B. Biggs, N.H. Burger and D.A. Blaylock. 1985. Nutrient enrichment of Chesapeake Bay and its impact on the habitat of striped bass: A speculative hypothesis. *Trans. Amer. Fish. Soc.* 114: 97-106.
8. Seliger, H.H., J.A. Boggs and W.H. Biggley. 1985. Catastrophic anoxia in Chesapeake Bay in 1984. *Science* 228: 70-73.
9. McCarthy, J.J., W.R. Taylor and M.E. Loftus. 1974. Significance of nanoplankton in the Chesapeake Bay estuary and problems associated with the measurement of nanoplankton productivity. *Mar. Biol.* 24: 7-16.
10. Van Valkenburg, S.D. and D.A. Flemer. 1974. The distribution and productivity of nanoplankton in a temperate estuarine area. *Est. Coastal Mar. Sci.* 2: 311-322.
11. Van Valkenburg, S.D., J.K. Jones and D.R. Heinle. 1978. A comparison by size class and volume of detritus versus phytoplankton in Chesapeake Bay. *Est. Coastal Mar. Sci.* 6: 569-582.
12. Sellner, K.G. 1983. Plankton productivity and biomass in a tributary of the upper Chesapeake Bay. I. Importance of size-fractionated phytoplankton productivity, biomass and species composition in carbon export. *Est. Coastal Shelf Sci.* 17: 187-206.
13. Sellner, K.G. and M. E. Kachur. In press. Phytoplankton: Distribution, production and integrators of environmental conditions. *In:* K.L. Heck, Jr. (Ed.). *The ecosystem of Chesapeake Bay. Long term studies near Calvert Cliffs Maryland.* Springer-Verlag, NY.
14. Sanders, J.G. 1985. Arsenic geochemistry in Chesapeake Bay: Dependence upon anthropogenic inputs and phytoplankton species composition. *Mar. Chem.* 17: 329-340.
15. Sellner, K.G. and D.C. Brownlee. 1986. Maryland Office of Environmental Programs Chesapeake Bay Water Quality Monitoring Program, Phytoplankton and microzooplankton component. The Acad. Natl. Sci., Benedict Est. Res. Lab., Benedict, MD.
16. Waterbury, J.B., S.W. Watson, R.R.L. Guillard and L.E. Brand. 1979. Widespread occurrence of a unicellular, marine, planktonic, cyanobacterium. *Nature* 277: 293-294.
17. Glover, H.E. 1985. The physiology and ecology of the marine cyanobacterial genus *Synechococcus*. *Adv. Aquatic Microbiol.* 3: 49-107.
18. Marshall, H.G. and R.V. Lacouture. 1986. Seasonal patterns of growth and

composition of phytoplankton in the lower Chesapeake Bay and vicinity. *Est. Coastal Shelf Sci.* 23: 115-130.
19. Exton, R.J., W.M. Houghton, W. Esaias, L.W. Haas and D. Hayward. 1983. Spectral differences and temporal stability of phycoerythrin fluorescence in estuarine and coastal waters due to the domination of labile cryptophytes and stabile cyanobacteria. *Limnol. Oceanogr.* 28: 1225-1231.
20. Ray, R.T. 1986. The role of picoplankton in plytoplankton dynamics. Unpubl. Masters Thesis, Virginia Institute of Marine Science, College of William and Mary, Gloucester Point, VA. 85 pp.
21. Biggs, R.B. 1970. Sources and distribution of suspended sediment in northern Chesapeake Bay. *Mar. Geol.* 9: 187-201.
22. Flemer, D.A. 1970. Primary production in the Chesapeake Bay. *Chesapeake Sci.* 11: 117-129.
23. Harding, L.W., Jr., B.W. Meeson and T.R. Fisher, Jr. 1986. Phytoplankton production in two east coast estuaries: Photosynthesis-light functions and patterns of carbon assimilation in Chesapeake and Delaware Bays. *Est. Coastal Shelf Sci.* 23: 773-806.
24. Brandt, A., C.C. Sarabun, D.C. Dubbel and C.J. Vogt. 1985. Estuarine internal wave stability and breaking. *EOS* 66: 1269.
25. Dubbel, D.C., A. Brandt and C.C. Sarabun. 1985. Internal wave activity in Chesapeake Bay. *EOS* 66: 1269.
26. Sarabun, C.C., C.J. Vogt and A. Brandt. 1985. Internal waves on an estuarine front. *EOS* 66: 1269.
27. Tyler, M.A. and H.H. Seliger. 1978. Annual subsurface transport of a red tide dinoflagellate to its bloom area: Water circulation patterns and organism distributions in the Chesapeake Bay. *Limnol. Oceanogr.* 23: 227-246.
28. Wolfe, J.J., B. Cunningham, N.F. Wilkerson and J.T. Barnes. 1926. An investigation of the microplankton of Chesapeake Bay. *J. Elisha Mitchell Sci. Soc* 42: 25-54.
29. Cowles, R. 1930. A biological study of the offshore waters of Chesapeake Bay. *Bull. Bur. Fish.* 46: 277-381.
30. Mulford, R. 1963. The net phytoplankton taken in Virginia tidal waters. Spec. Rept. 43. Virginia Inst. Mar. Sci., Gloucester Point., VA.
31. Mulford, R.A. 1972. An annual plankton cycle on the Chesapeake Bay in the vicinity of Calvert Cliffs, Maryland, June 1969-May 1970. *Proc. Acad. Natl. Sci. Phila.* 124: 17-40.
32. Patten, R., R. Mulford and J. Warinner. 1963. An annual phytoplankton cycle in the lower Chesapeake Bay. *Chesapeake Sci.* 4: 1-20.
33. Marshall, H.G. 1966. The distribution of phytoplankton along a 140 mile transect in the Chesapeake Bay. *Virginia J. Sci.* 17: 105-119.
34. Marshall, H.G. 1967. Plankton in James River estuary, Virginia. I. Phytoplankton in Willoughby Bay and Hampton Roads. *Chesapeake Sci.*

8: 90-101.
35. Marshall, H.G. 1980. Seasonal phytoplankton composition in the lower Chesapeake Bay and Old Plantation Creek, Cape Charles, Virginia. *Estuaries* 3: 207-216.
36. Siegfried, R.C., R.W. Alden, III and B. Nielsen. Unpublished data from the Virginia Water Quality Monitoring Program, 1984-1985.
37. Goodrich, D.M., W.C. Boicourt, P. Hamilton and D.W. Pritchard. Submitted. Wind-induced destratification in the Chesapeake Bay. *J. Physical Oceanogr.*
38. Tyler, M.A. and H.H. Seliger. 1981. Selection for a red tide organism: Physiological responses to the physical environment. *Limnol. Oceanogr.* 26:310-324.
39. Sellner, K.G. and M.M. Olson. 1985. Copepod grazing in red tides of Chesapeake Bay, pp. 245-250. *In:* D.M. Anderson, A.W. White and D.G. Baden (Eds.) *Toxic dinoflagellates.* Elsevier, NY, pp. 561.
40. Stoecher, D., R.R.L. Guillard and R.M. Kavee. 1981. Selective predation by *Favella ehrenbergii* (Tintinnina) on and among dinoflagellates. *Biol. Bull.* 160:136-145.
41. Sellner, K.G., D.C. Brownlee and L.W. Harding, Jr. 1987. Implications of microzooplankton grazing on carbon flux and anoxia in Chesapeake Bay. Abstract, NOAA Seminar on Hypoxia and related processes in Chesapeake Bay. 21-22 January 1987, College Park, Md. 3 pp.
42. Pratt, D.M. 1963. The winter-spring flowering in Narragansett Bay. *Limnol. Oceanogr.* 10:173-184.
43. Smayda, T.J. 1983. The phytoplankton of estuaries. *In:* B.H. Ketchum (Ed.), *Estuaries and enclosed seas.* Elsevier, Amsterdam, pp. 65-102.
44. Flemer, D.A. 1969. Continuous measurement of *in vivo* chlorophyll of a dinoflagellate bloom in Chesapeake Bay. *Chesapeake Sci.* 10: 99-103.
45. Loftus, M.E., D.V. Subba Rao and H.H. Seliger. 1972. Growth and dissipation of phytoplankton in Chesapeake Bay. I. Response to a large pulse of rainfall. *Chesapeake Sci.* 13:282-299.
46. Seliger, H.H. and M.E. Loftus. 1974. Growth and dissipation of phytoplankton in Chesapeake Bay. II. A statistical analysis of phytoplankton standing crops in the Rhode and West Rivers and an adjacent section of the Chesapeake Bay. *Chesapeake Sci.* 15:185-204.
47. Zubkoff, P.L., J.C. Munday, Jr., R.G. Rhodes and J.E. Warinner III. 1979. Mesoscale features of summer (1975-1977) dinoflagellate blooms in the York River, Virginia (Chesapeake Bay estuary), pp. 279-286. *In:* D.L. Taylor and H.H. Seliger (Eds.) *Toxic dinoflagellate blooms.* Elsevier, NY.
48. Seliger, H.H., K.R. McKinley, W.H. Biggley, R.B. Rivkin and K.R.H. Aspden. 1981. Phytoplankton patchiness and frontal regions. *Mar. Biol.* 61:119-131.
49. Tyler, M.A. and J.F. Heinbokel. 1985. Cycles of red water and encystment

of *Gymnodinium pseudopalustre* in the Chesapeake Bay: Effects of hydrography and grazing, pp. 213-218. *In:* D.M. Anderson, A.W. White and D.G. Baden (Eds.) *Toxic dinoflagellates.* Elsevier, NY, pp. 561.
50. Houghton, W.M., R.J. Exton and R.W. Gregory. 1983. Field investigation of techniques for remote laser sensing of oceanographic parameters. *Remote Sensing of Environ.* 13:17-32.
51. Tyler, M.A. and R.P. Stumpf. 1985. Shipboard and satellite detection of phytoplankton blooms in estuaries. *EOS* 66:1268.
52. Boynton, W.R., W.M. Kemp and C.W. Keefe. 1982. A comparative analysis of nutrients and other factors influencing estuarine phytoplankton production, pp. 69-90. *In:* V.S. Kennedy (Ed.) *Estuarine comparisons.* Academic Press, NY.
53. Taylor, W.R. and J.E. Hughes. 1967. Primary productivity in the Chesapeake Bay during the summer of 1964. Tech. Rept. 34, Ref. 67-1, Chesapeake Bay Institute, The Johns Hopkins University, Shady Side, Md.
54. Taft, J. 1982. Nutrient processes in Chesapeake Bay, pp. 103-149. *In: Chesapeake Bay Program technical studies: A synthesis.* U.S. EPA, Washington, D.C., pp. 634.
55. Magnien, R.E., M.C. Curtis and K.G. Sellner. 1986. Relations between nutrients and plankton in Chesapeake Bay. I. Nutrient dynamics. Abstract, 50th annual meeting, Amer. Soc. Limnol. Oceanogr. June, 1986, University of Rhode Island.
56. Fisher, T.R., Jr., L.W. Harding, Jr., D.W. Stanley, and L.G. Ward. In review. Phytoplankton, nutrients and turbidity in the Chesapeake, Delaware and Hudson estuaries. *Est. Coastal Shelf Sci.*
57. Taft, J.L. and W.R. Taylor. 1976. Phosphorus dynamics in some coastal plain estuaries, pp. 79-89. *In:* M. Wiley (Ed.) *Estuarine processes.* Vol. 1. Academic Press, NY., pp. 541.
58. McCarthy, J.J., W.R. Taylor and J.L. Taft. 1977. Nitrogenous nutrition of the plankton in the Chesapeake Bay. I. Nutrient availability and phytoplankton preferences. *Limnol. Oceanogr.* 22: 996-1011.
59. Taft, J.L., M.E. Loftus and W.R. Taylor. 1977. Phosphate uptake from phosphomonoesters by phytoplankton in the Chesapeake Bay. *Limnol. Oceanogr.* 22: 1012-1021.
60. Glibert, P.M. 1982. Regional studies of daily, seasonal and size fraction variability in ammonium remineralization. *Mar. Biol.* 70: 209-222.
61. D'Elia, C.F., J.G. Sanders and W.R. Boynton. 1986. Nutrient enrichment studies in a coastal plain estuary: Phytoplankton growth in large-scale, continuous cultures. *Can. J. Fish. Aquat. Sci.* 43: 397-406.
62. Sanders, J.G., S.J. Cibik, C.F. D'Elia and W.R. Boynton. 1987. Nutrient enrichment studies in a coastal plain estuary: Changes in phytoplankton species composition. *Can. J. Fish. Aquat. Sci.* 44: 83-90.
63. Sanders, J.G. and S.J. Cibik. 1986. The response of estuarine phytoplankton

to nutrient enrichment under varying light regimes. Rept. to Proctor & Gamble Co., Acad. Natl. Sci., Benedict Est. Res. Lab., Benedict, MD. 9 pp.
64. Virginia Institute of Marine Science. 1986. Executive summary of "Is phosphorus removal an efficient/effective Chesapeake Bay management practice?" Rept. to Council on the Environment, VIMS, College of William and Mary, Gloucester, VA. 36 pp.
65. Taft, J.L., W.R. Taylor, E.O. Hartwig and R. Loftus. 1980. Seasonal oxygen depletion in Chesapeake Bay. *Estuaries* 3: 242-247.
66. Carter, H.H, R.J. Regier, E.W. Schiemer and J.A. Michael. 1978. The summertime vertical distribution of dissolved oxygen at the Calvert Cliffs generating station: A physical interpretation. Spec. Rept. 60, Ref. 78-1. Chesapeake Bay Institute, The Johns Hopkins University, Shady Side, Md.
67. Magnien, R.E., M.C. Curtis and N.L. Matthews. 1985. Reaeration event disrupts Chesapeake Bay deep-trough anoxia during summer, 1984. Abstract, 48th Annual Meeting, Amer. Soc. Limnol. Oceanogr., June, 1985.
68. Tyler, M.A. 1984. Dye tracing of a subsurface chlorophyll maximum of a red-tide dinoflagellate to surface frontal regions. *Mar. Biol.* 78:285-300.
69. Pomeroy, L. 1974. The ocean's food web, a changing paradigm. *Bioscience* 24: 499-504.
70. Williams, P.J. leB. 1981. Incorporation of microheterotrophic processes into the classical paradigm of the plankton. *Kieler Meeresforsch* 5:1-28.
71. Azam., F., T. Fenchel, J.G. Field, J.S. Gray, L.A. Meyer-Reil and F. Thingstad. 1983. The ecological role of water-column microbes in the sea. *Mar. Ecol. Prog. Ser.* 10:257-263.
72. Tuttle, J.H. and C.C. Gilmour. 1985. Microbial sulfate reduction in Chesapeake Bay. *EOS* 66:1319.
73. Ducklow, H.W. 1985. Nutrient-dissolved oxygen dynamics: Roles of phytoplankton and microheterotrophs under summer conditions. Bacterioplankton biomass and production. Horn Point Environ. Labs., University of Maryland, Cambridge, Md.
74. Hill, S.M., H.W. Ducklow, T.C. Malone, W.M. Kemp, B. Wendler, J.H. Tuttle and R.B. Jonas. 1985. Bacterioplankton-phytoplankton couplings as a factor governing anoxia in Chesapeake Bay. *EOS* 66:1319.
75. Sellner, K.G. Unpublished data, Calvert Cliffs Nuclear Power Plant Program, The Academy of Natural Sciences, Benedict, Md.
76. Kemp, W.M. and W.R. Boynton. 1984. Influence of biological and physical processes on dissolved oxygen dynamics in an estuarine system: Implications for measurement of community metabolism. *Est. Coastal Mar. Sci.* 11: 407-431.
77. Boynton, W., M. Kemp, J. Garber and J. Barnes. 1986. Nutrient flux. Supplement to Maryland Office of Environmental Programs Chesapeake Bay Water Quality Monitoring Program, Ecosystems processes component. Chesapeake Biological Laboratory, University of Maryland, Solomons, MD.

78. Brownlee, D.C., F. Jacobs, S.G. Brownlee and K.G. Sellner. 1986. Relationship between nutrients and plankton in Chesapeake Bay. III. Potential roles of micro- and macrozooplankton. Abstract, 50th annual meeting, Amer. Soc. Limnol. Oceanogr., June, 1986, University of Rhode Island.
79. Brownlee, D.C. and F. Jacobs. 1987. Mesozooplankton and Microzooplankton in the Chesapeake Bay. *In:* S.K. Majumdar, L.W. Hall, Jr. and H.M. Austin (Eds.). *Contaminant problems and management of living Chesapeake Bay resources.* Chapter 12. Penn. Acad. Sci., Easton, PA.
80. Heinle, D.R. and D.A. Flemer. 1975. Carbon requirements of a population of the estuarine copepod *Eurytemora affinis. Mar. Biol.* 31: 235-247.
81. Sellner, K.G. and R.J. Horwitz. 1983. Plankton interactions in the Patuxent River estuary: Field studies of community composition and density, with a deterministic model of the effects of zooplankton grazing on phytoplankton carbon, production of fecal matter, sediment oxygen demand and nutrient regeneration. Rept. No. 82-14F, The Acad. Natl. Sci., Benedict Est. Res. Lab., Benedict, MD.
82. Boynton, W.R., W.M. Kemp and J.M. Barnes. 1985. Maryland Office of Environmental Programs Chesapeake Bay Water Quality Monitoring Program, Ecosystems processes component (EPC). Level I data report no. 2. University of Maryland Center for Environ. Estuarine Studies Ref. No. [UMCEES]CBL 85-121, Chesapeake Biological Laboratory, Solomons, MD.
83. Riley, G.A. 1956. Oceanography of Long Island Sound, 1952-54. IX. Production and utilization of organic matter. *Bull. Bingham Oceanogr. Coll.* 15: 324-343.
84. Smetacek, V., K. von Brockel, B. Zeitzschel and W. Zenk. 1978. Sedimentation of particulate matter during a phytoplankton spring bloom in relation to the hydrographical regions. *Mar. Biol.* 47: 211-226.
85. Hargrave, B.T. 1980. Factors affecting the flux of organic matter to sediments in a marine bay, pp. 243-263. *In:* K.R. Tenore and B.C. Coull (Eds.) *Marine benthic dynamics.* University of South Carolina Press, Columbia, SC.
86. Davies, J.M. and R. Payne. 1984. Supply of organic matter to the sediment in the northern North Sea during a spring phytoplankton bloom. *Mar. Biol.* 78: 315-324.
87. Boynton, W.R. and W.M. Kemp. 1985. Nutrient regeneration and oxygen consumption by sediments along an estuarine salinity gradient. *Mar. Ecol. Prog. Ser.* 23: 45-55.
88. Kemp, W.M. and W.R. Boynton. 1981. External and internal factors regulating metabolic rates of an estuarine benthic community. *Oecologia* 51: 19-27.
89. Holland, A.F., N.K. Mountford, M.H. Hiegel, K.R. Kaumeyer, D. Cargo and J.A. Mihursky. 1978. Results of benthic studies at Calvert Cliffs.

Martin-Marietta Corp. Ref. No. CC-78-2 and Chesapeake Biological Laboratory Ref. No. 78-165-CBL, Baltimore, MD.
90. Heck, K.L., Jr. In press. Benthos. *In:* K.L. Heck, Jr. (Ed.) *The ecosystem of Chesapeake Bay. Long term studies near Calvert Cliffs Maryland.* Springer-Verlag, NY.
91. Kemp, W.M. and W.R. Boynton. 1984. Spatial and temporal coupling of nutrient inputs to estuarine primary production: The role of particulate transport and decomposition. *Bull. Mar. Sci.* 35: 522-535.
92. D'Elia, C.F., D.M. Nelson and W.R. Boynton. 1983. Chesapeake Bay nutrient and plankton dynamics: III. The annual cycle of dissolved silicon. *Geochim. Cosmochim. Acta* 47: 1945-1955.
93. Yamada, S.S. and C.F. D'Elia. 1984. Silicic acid regeneration from estuarine sediment cores. *Mar. Ecol. Prog. Ser.* 18: 113-118.

Chapter Eight

BENTHIC RESOURCES OF THE CHESAPEAKE BAY ESTUARINE SYSTEM

ROBERT J. DIAZ

Virginia Institute of Marine Science
Gloucester Point, Virginia 23062

The term benthic is derived from the Greek root "benth" which means the depths of the sea. Benthic has then come to mean that which is associated with the bottom of any body of water from lakes to oceans, and encompasses a broad range of organisms from algae to fish. In the Chesapeake Bay there is a wide variety and combination of benthic environments ranging from intertidal flats of sand or mud, shallow seagrass meadows, subtidal bottoms, and deeper channels. These environments take on a unique biological and chemical character, depending upon where they occur along the gradient of Bay salinities and sediment type.

Salinity not only plays a key role in determining the biological character of the benthic environment, but is also important in determining the distribution of the various benthic environment types around the Bay. For the most part, all the Bay bottoms south of the Rappahannock River and mouths of the York and James Rivers are sandy. This results from the net upbay movement of the heavier salt water entering the Bay and the fact that there isn't very much silt or clay in this water. North of the Rappahannock River the Bay bottom is predominantly muddy, except for the extensive sandy shoal on which Tangier and Smith Islands sit. The majority of the mud (silts and clays) north of the Rappahannock originated from the Susquehanna River drainage basin. As it entered the Bay at Harve de Grase, the silts and clays were suspended in the water. Moving down the Bay they encountered saltier water which caused the silt and clay particles to aggregate and quickly settle to the bottom. This normally occurs in the area between Baltimore and the mouth of the Patuxent River. During periods of high runoff, this newly settled mud is distributed further south.

Depth also plays an important part in the character of the benthic environment. For any given depth, deep or shallow, the bottom can be either muddy or sandy. There are large differences between the same sediment type found at various depths in terms of the organisms present and stability of the bottom. Shallow areas, are for the most part, more dynamic being affected most by wind induced currents and waves. In these shallow areas the sand grains tend to be coarser. These forces cause the surface sediments to move about producing the characteristic rippling seen on intertidal and shallow bars. Deep bottoms, say over 30 feet, are less prone to disturbance from wind effects and if tidal currents are not very strong, the sediments tend to be muddier since the fine silts and clays will have more time to accumulate.

The benthic communities that inhabit shallow areas are composed primarily of smaller species that are capable of at least limited mobility. As depth increases the number of larger sessile (nonmobile) forms increases. This is due, in part, to increased sediment stability with depth. However, other phenomena such as oxygen stress may alter this pattern and completely control the character of the benthic environment. This is the case in the upper Bay. Sections of the deep muddy channel that runs from about Baltimore to the Rappahannock River turn anoxic (no dissolved oxygen in the water) in the bottom water layers every summer. This event appears to be a natural phenomena, although the effects of man's activities may be causing an extension of the area involved and duration of the event. The benthic communities in these areas are very distinct from other deep mud areas that are not oxygen stressed. The communities are generally characterized by tolerant animals or animals which can rapidly reproduce. There are no benthic environments around the Bay that are completely devoid of life. Even the areas that annually become anoxic in the summer will recolonize with the return of oxygenated waters. There may be areas near outfalls or inner harbors that are severely stressed but even here benthic organisms manage to colonize if but for brief periods.

The presence of organisms in and on the bottom is very important to the character of the benthic environment. In the process of day-to-day life, benthic organisms modify the physical structure of the bottom sediments. This modification is generally called bioturbation from the Green root "Bios" (life) and the Latin root "turba" (disturbance). Organisms can literally dig, push, eat, defecate, compact, resuspend, and/or glue together the sediments in which they live causing major disturbance within the benthic environment. When organisms are active, they can greatly influence or even control the movement of compounds (both toxic and nontoxic) between the overlying water and sediments, and even the stability of the sediment. Depending upon the species and its particular life habits, bioturbation activities modify sediments in a number of ways. Through mixing, newly arrived sediment at the surface can be quickly buried or older sediments may be resurfaced. Through burrowing, water ventilates the sediment bringing dissolved oxygen deeper into the sediment and carrying off other dissolved compounds to the water column. Through tube building and secretions, surface sediments are made more stable and less prone to erosion. Through feeding on fine suspended particles and forming pellets, sedimentation rates are greatly increased. Through fluffing up the sediments by increasing its water content, sediment stability is decreased. No one species can do all of these things to the benthic environment. There are literally hundreds of species that do fit into each of the above categories. Some of the more common species are listed in Table 1.

To summarize, the benthic environments of the Chesapeake Bay are incredibly diverse and complex. They take their character from a combination of sediment type, salinity, depth, oxygen stress (all physical components), and species composition (the biological component). Of all these the most incredible are the species that inhabit the benthos. They represent a miriad of organisms from bacteria and algae to crabs and fish that not only reside on or in the bottom but reshape it to suit their individual needs.

Benthic invertebrates are a very large and diverse group of animals that encompass many different life styles. The term benthic (see the chapter on the Benthic Environment) is used to indicate a reliance of the animals on the bottom or some other substrate, such as pilings. The vast majority of benthic invertebrates are small and rather obscure, blending in very well with their surroundings. Even if animals smaller than 0.5 mm are not counted, there are well over 900 different species of invertebrates in the Chesapeake Bay. The breakdown of these species by major taxonomic group is seen in Table 1. It is also likely that there are at least an equal number of invertebrates smaller than 0.5 mm

TABLE 1

Major taxonomic groups of benthic invertebrates occurring in the tidal waters of the Chesapeake Bay.

Group	Number of Species
Sponges	14
Other Colonial Animals	107
Jelly Animals	18
Flat worms	51
Ribbon worms	25
Polychaete worms	195
Oligochaete worms	37
Leeches	11
Other worms	4
Clams	64
Snails	83
Sea Slugs	8
Crabs	57
Shrimp	73
Amphipods	77
Isopods	47
Other crab-like animals	62
Barnacles	9
Starfish like	11
TOTAL	953

in the Bay. Our most well known benthic invertebrates are blue crabs, hard clams, soft clams, and oysters. Unfortunately they tend to be covered as individual species, but they are still part of the benthic system.

The significance of benthic invertebrates is twofold. They serve as a major link in the food web of the Bay, passing energy from primary producers (phytoplankton and plants) to top carnivores (fish and crabs). Much of the fisheries harvest from the Bay each year would not be possible without the even greater production by benthic invertebrates. On average there are approximately 50 to 500 wet weight grams of benthic invertebrates produced per square meter per year throughout the Bay, not counting crabs, clams, or oysters. This then figures to 3.25×10^8 to 10^9 metric tons of potential food available to our fisheries each year.

In addition to functioning as a food source, benthic invertebrates are very important in cycling nutrients back to the water column from the sediments. Most benthic invertebrates spend all of their adult life within the sediment where they feed, dig, burrow, and build tubes. (See the chapter on the Benthic Environment.) Since all benthic invertebrates require oxygen and very little oxygen penetrates down into the sediments, especially muddy sediments, the animals

---

Legend for Figure 1.

Tidal Fresh Water

   Cm—*Corbicula fluminae* — Asiatic Clam; Ch—Chironomid larvae—Non-biting Midges; El—*Elliptio* spp.—Freshwater Mussel; Li—*Limnodrilus* spp.—Freshwater Earthworms; Ph—*Physa* spp.—Pond Snail; An—*Anodonta* spp.—Freshwater Mussel; Ug—*Urnatella gracilis* — Bryozoan; Na—*Nais* sp.—Freshwater Earthworms; Pl—Planorbid Snail.

Low and Middle Salinity

Po—*Phoronis* spp.—Lophophore Worm; Ma—*Mya arenaria* — Soft Clam; Sd—*Streblospio benedicti* — Spionid Worm; Ed—*Ensis directus*—Razor Clam; Pl—*Polydora* spp.—Spionid Worm; Cg—*Cistena gouldii*—Ice Cream Cone Worm; Ns—*Nereis succinea*—Sand Worm; Am—*Ampelisca* spp.—Ampeliscid Amphipod; Rc—*Rangia cuneata*—Common Rangia Clam; Tu—*Tubificoides* spp.—Saltwater Earthworm; Ml—*Mulinia laterlis*—Coot Clam; Hf—*Heteromastus filiformis*—Capitellid Worm; Ne—Nemertean Worm; Mb—*Macoma balthica*—Baltic Macoma Clam; Od—*Odostomia* spp.—Odostome Snail; Sa—*Sertularia argentea*—Colonial Hydroid; Br—*Brachidontes recurvus* — Curved Mussel; Cl—*Corophium lacustre*— Corophid Amphipod; Mm—*Molgula manhattensis*—Sea Grape; Cr—*Crassostrea virginica* — Oyster.

High Salinity

Pw—*Protohaustorius wigleyi* — Haustorid Amphipod; Ct—*Clymenella torquata* — Bamboo Worm; Ta—*Tellina agilis* — Dwarf Tellin Clam; Lm—*Loimia medusa* — Tentacle Tube Worm; Ed—*Ensis directus* — Razor Clam; Np—*Nephtys* spp.—Nephtyid Worm; Sa—*Sertularia argentea* — Colonial Hydroid; Cp—*Chaetopterus variopedatus* — Parchment Tube Worm; Gd—*Glycera dibranchiata* — Blood Worm; Am—*Ampelisca* spp.—Ampeliscid Amphipod; Cg—*Cistena gouldii* — Ice Cream Cone worm; Ez—*Euclymene zonalis* — Bamboo Worm; Pi—*Pinnixa* spp.—Comensual Crab; Me—*Mytilus edulis* — Edible Mussel; Dc—*Diopatra cuprea* — Onuphid Worm; Cd — *Ceriantheopsis* spp.—Burrowing Sea Anemone; Cb—*Corophium turberculatum* — Corophid Amphipod; Sv—*Sabellaria vulgaris*—Reef Worm; Pl—*Polydora* spp.—Spionid Worm

Drawn by H. Burrell, L. Schaffner, and R. Llanso.

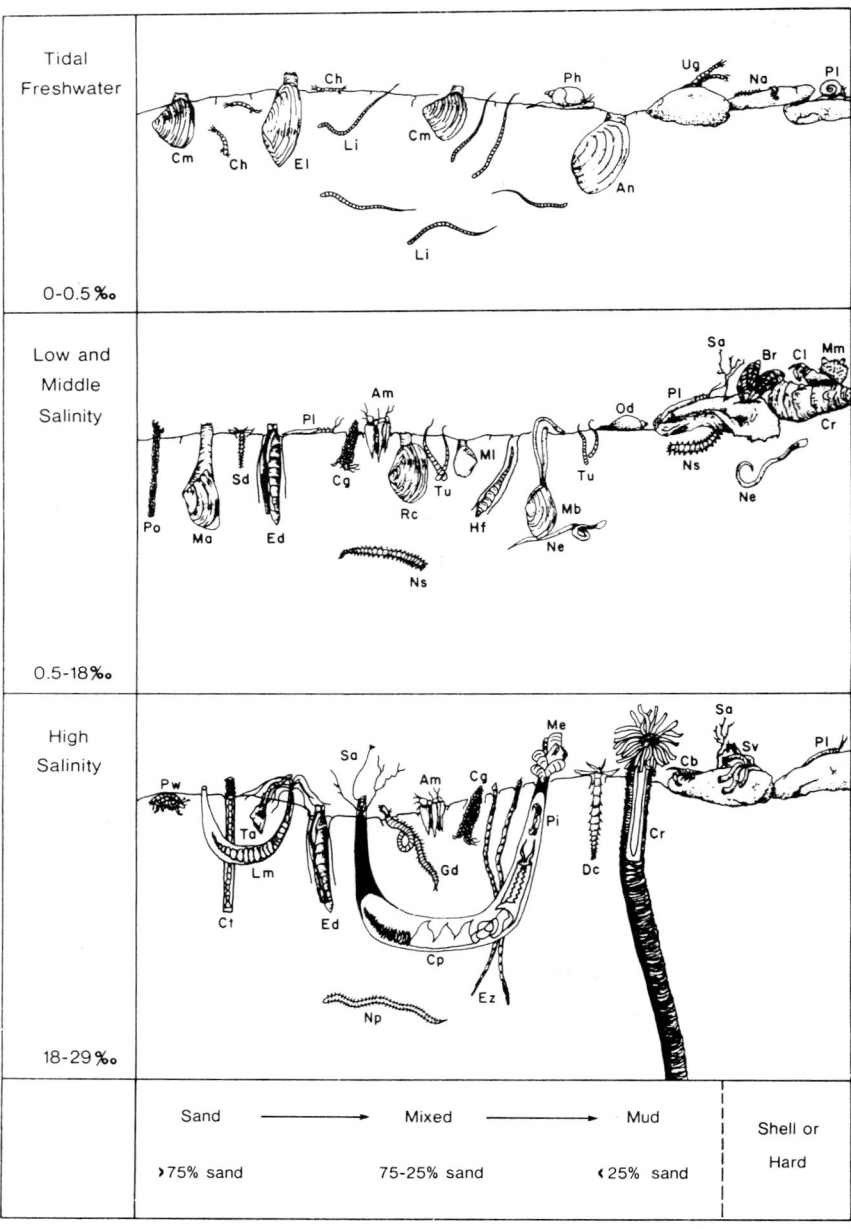

Figure 1. Representation of some common benthic invertebrates from the Chesapeake Bay.

cope with this problem by either living very near the sediment surface or by circulating water through their tubes or burrows to get oxygen. In the process of ventilating their homes, the animals also rid themselves of metabolic wastes (the nutrients plants needs) and any other chemicals that happen to seep into their burrows or tubes. Worms, mainly polychaetes and oligochaetes, are responsible for most of the burrowing and tube building that goes on around the Bay. But there are burrowers among the other taxonomic groups, particularly clams and shrimp.

There is not a uniform blanket of benthic invertebrates from the mouth of the Bay to its head waters some 200 miles away. In fact, within any given small area, the animals tend to be patchy, with high numbers in one spot and lower numbers in the other. Thus, the abundance of benthic invertebrates throughout the Bay can be thought of as a mosaic of different sized patches. If we consider the different kinds or species of benthic invertebrates to be the colors that paint our mosaic, we find that the Bay is not a uniform combination of hues. The species are not uniformly distributed but respond to a combination of physical (salinity, sediment type, depth) and biological (predation, competition) factors.

Salinity is the most important factor to the animals. As the salinity gradually decreases from the mouth to the head of the Bay, there is also a gradual change in the species of benthic invertebrates. This pattern of change is also seen in the major tributaries (James, York, Rappahannock, Potomac, Patuxent, Nanticoke, Choptank, Chester). Thus, the color the species paint for the lower Bay is almost completely different from the upper Bay. There are several extraordinary polychaetes (such as *Nereis, Strebelospio*) that occur over a wide range of salinity and are also widely distributed in the Bay and its tributaries. In addition to the species change with decreasing salinity, the number of species also declines. The majority of the 953 species in the Bay occur above 18°/oo. The fewest species, on the order of 30 to 40, are found in the 1 to 5°/oo range.

Sediment type is second only to salinity in importance to benthic invertebrates since they live in such intimate contact with the bottom. The type of sediment basically controls the life styles of the animals. In sandy sediments the predominant life styles are tube building, filter feeding, and surface feeding. Sand is not conducive to burrowing since unsupported burrow walls would collapse. In mud sediments predominant life styles change to burrowing, subsurface feeding, and surface suspension feeding. Since mud is cohesive (sticky), there is no need for a tube; animals can freely burrow through the sediment. Many of the animals living in mud simply feed by eating the mud. Those animals that feed from the water column (filter feeders) are not favored in mud because the fine clay and silt particles that make up mud clog their filtering apparatus. About the same number of species occur in sand as do in mud, but the highest number of occurrences are in mixed sediments (30 to 70% mud and 70 to 30% sand). Generally, in mixed sediments all life styles can be accommodated. A general idea of what different bottoms look life is depicted in Figure 1. Represented are the more common species from each area of the Bay.

While not that well known by most people who either work or play on the Bay, primarily because of their small size and cryptic habits, benthic invertebrates are the cornerstone of the Bay's high productivity and give it much of its biological character. Without the relentless work and growth of benthic invertebrates, the Bay would be very different from what we know it as today.

Chapter Nine

THE AMERICAN OYSTER *CRASSOSTREA VIRGINICA* IN CHESAPEAKE BAY

DEXTER HAVEN*
Virginia Institute of Marine Science
Gloucester Point, VA 23062

ABSTRACT

The American Oyster (*Crassostrea virginica*) is widely distributed in Chesapeake Bay where it grows in the intertidal zone to depths of about 6.5 m. The salinity range over which it occurs, is from about 5 to 34°/oo. It is most abundant in protected embayments where bottoms are a firm sand-clay mixed with shelly material. This bivalve is a filter feeder, and ingests planktonic material which it strains from the water with its gills. Spawning occurs in Chesapeake Bay from June through September, and the eggs and resulting larvae are widely distributed during their 10-20 day planktonic life. Mortality is high during this stage, and only a very few survive to attach to a firm substrate. After oyster larvae attach and during the 3 to 4 year period required for them to reach market size many more die due to predators, diseases, and other associated factors. During recent years there has been a major decline in commercial landings of oysters in Maryland and Virginia. The cause for this reduction in harvest has been associated with a decline in recruitment, overfishing, poor management practices, and pollution.

The American oyster *Crassostrea virginica* is widely distributed in Chesapeake Bay (Figure 1) where salinities range from about 5.0°/oo to that of seawater, about 34°/oo. Within this range, it occurs from the intertidal zone to depths of about 6.5 m (21 feet). On the seaside of Maryland and Virginia's eastern shore, and between the barrier islands and the shore, it inhabits the margins of channels, salt marshes and elevated intertidal patches of oyster shell in open bays. In this latter environment large beds are often exposed to the air twice daily

*Contribution No. 1382 from the Virginia Institute of Marine Science, School of Marine Science, The College of William and Mary.

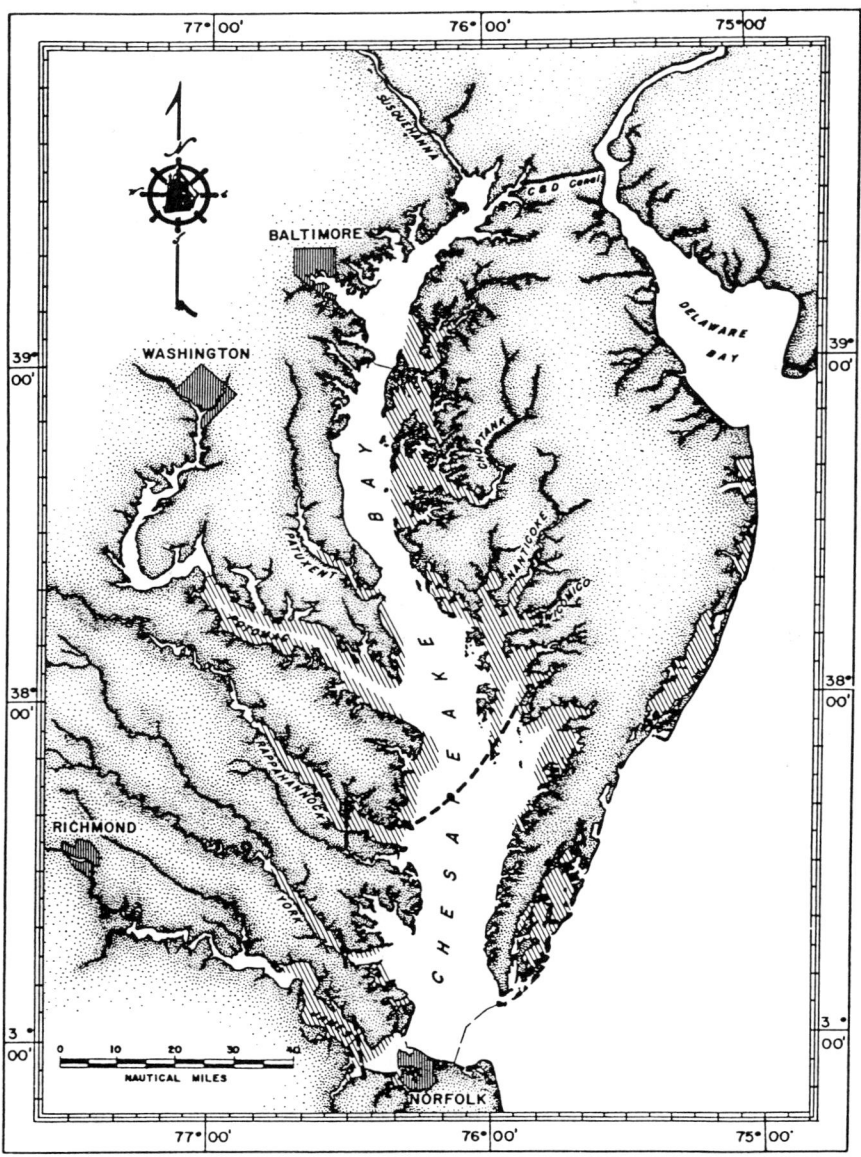

FIGURE 1. Oyster growing areas in Chesapeake Bay. Shaded areas show area where oysters grow naturally or where they are cultivated. The dashed line shows the approximate location in spring of the 15 ppt isohaline.

by tidal action. Oysters occur naturally on a wide range of bottom substrate types, but usually are found where the bottom is firm sand-silt-clay, mixed with oyster shell and oysters.

The hard rough texture of oyster shell is composed of about 94% calcium carbonate; magnesium and other minerals make up an additional 4% while about 2% is organic material. The two shells attached by a leathery hinge at the narrow end enclose the living animal. This mollusc has a mouth, gills for respiration and straining food from the water, and organs of reproduction, digestion and elimination. A strong adductor mussel near the center of the animal closes the shell; when it relaxes the shell gapes allowing entry of seawater (Figure 2).

The oyster obtains its food by filtering minute particles of detritus or living plankton from the water with its gills. These thin, ribbed organs which surround about one-half of the margin of an oyster, also serve as an organ of respiration. They are permeated by many minute pores, and their surfaces are covered

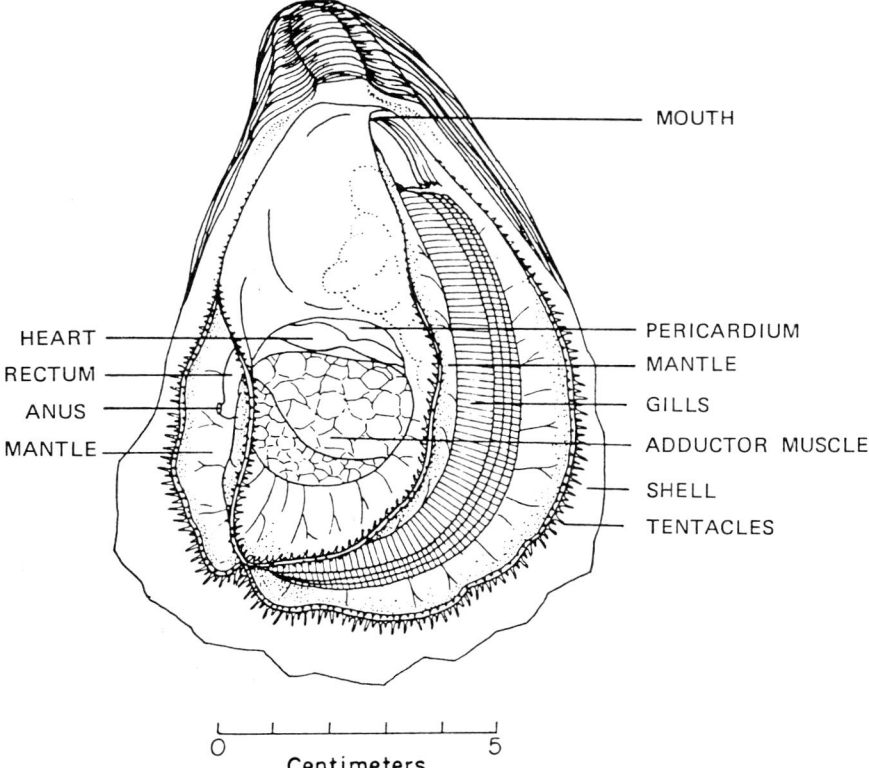

FIGURE 2. Anatomy of the American Oyster showing organs of feeding, respiration, digestion and reproduction (from Galtsoff).

and lined by innumerable hair-like cilia. By beating in unison, the cilia set up water currents through the pores; and by this action respiration is achieved, and food particles are strained from the water. Other groups of specialized cilia transport the food from the gill surface to the mouth and into the oyster's stomach.

The volume of water pumped through the gills of oysters during feeding and respiration depends largely on water temperature. In winter as temperatures approach 5-6°C (41-43°F) water transport and feeding virtually stops. When water temperature exceeds about 10°C (50°F) oysters become fully active. Volumes filtered by large oysters during the warmer summer months may range from 1.1 to 24 liters per hour.[1,2]

Oysters may change their sex from male to female and from female to male. During the first year of growth both male and female gonadal cells occur in the same individual, but at the end of this first growth period the spermatogonia have proliferated more rapidly than the ovogonia, giving them a predominantly male appearance. At the end of the second growth season a further development of the gonadal tissue occurs which results in a population which approaches equality of sexes. Thereafter the sex of individual oysters may or may not change each year.

Spawning typically occurs in Chesapeake Bay over a long period from early June through late September, when water temperature ranges from about 20-32°C (68-90°F).[1,3] Spawning is generally initiated by a sudden sharp increase in water temperature, but other factors such as nutritional levels in the oyster are involved.[4] During spawning a large female oyster 8-12 cm (3.1-4.7 in.) long may on the average, produce up to 2.9 million eggs each year; smaller sizes may release less than 100,000.[5] The mature eggs and sperm are released into the water where fertilization occurs; an unfertilized oyster egg is about 40 microns (0.002 in) in diameter. Fertilized eggs rapidly develop into free swimming larvae termed trochophores about 50 μ in diameter, and then into the veliger stage. Subsequently, they grow into fully mature larvae about 250 μ (0.01 inch) in size termed eyed larvae since they have a pigmented spot in their tissue (Figure 3).[6] These mature larvae sink to the bottom and cement their shell to any hard object which in Chesapeake Bay is usually another oyster or an oyster shell. Only a very few survive to the end of the pelagic stage.

During the 10-20 day period when larvae are developing they are widely distributed by water currents since they are weak swimmers incapable of significant power to move horizontally. They are however, capable of a slow vertical migration (about 1 cm/s) within the water column, by alternately swimming upward or sinking toward the bottom. This behaviorial characteristic may at times modify the direction or speed of transport by water currents.[7]

The time oysters set or strike varies widely and many estuaries having low salinity (5-10 ppt) may receive little or no set over a period of many years. Other areas, where salinities are higher, may or may not receive a good set. Even estuaries where the set is usually satisfactory may experience years of little or no set. The intensity of set in an estuary during a season also varies, but usually for each area there is a period when the set is highest. Typically, the seasonal peak will occur in late June, July or August in the upper Chesapeake Bay above the entrance of the Potomac River, and in July, August or September in the southern half.[3,8,9]

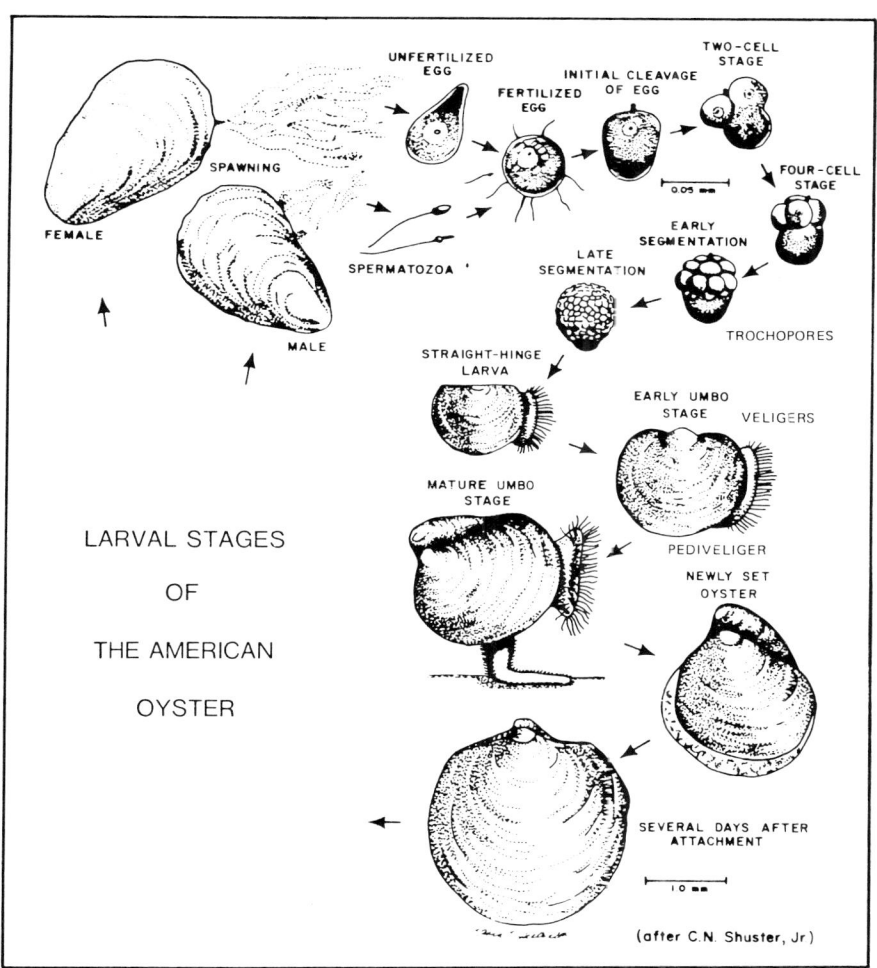

FIGURE 3. Life cycle of the American oyster showing larval stages, and small recently set spat on an oyster shell.[6]

GROWTH AND DEVELOPMENT

After permanent attachment the small oyster is called a spat, and by the end of its first week it averages about 1 mm (.04 in) in length. By the end of this same season, when growth stops in November due to falling water temperature, its length varies in different areas. In the low salinity regions of the James River, Virginia it averages 0.5 to 2.0 cm long (0.2-0.8 inch), and in many other regions of the bay it may average from about 1.0 to 3.5 cm (0.4-1.4 in.) depending largely on salinity, available nutrients, and the time the spat set. The small spat are very susceptible to crowding and predation, and consequently mortality is very high during their first season, sometimes reaching 90% to 100%. After setting, growth will vary with those in low salinity regions of the upper bay and in the upper reaches of many of its estuaries taking 3-5 years after setting to reach 7.5 to 10.0 cm (3-4 inches). In high salinity areas usually in the lower bay, they may grow to a similar length in 3-4 years.

In the market, shucked oysters are graded according to size, and identified as follows. Standards 300 and up/gallon; Selects 210-300/gallon; Extra selects 160-210/gallon; and Counts 160 or less/gallon.

PREDATORS AND DISEASES

During their life oysters are subject to many predators and diseases which may inhibit their growth or kill them and few of the many which originally set on shell survive to maturity. In high salinity waters over about 15°/oo, there are small gastropods (*Eupleura caudata* and *Urosalpinx cinerea*) known as oyster drills which can destroy newly set oysters and even those up to about 5 cm (2 in.) by boring a small hole through the shell and then ingesting the meat. The common blue crab (*Callinectes sapidus*) eats many oysters when they are small. In many areas schools of cow-nosed rays (*Rhinoptera sp.*) may actually crush shells of oysters up to 12.5 cm (5 in.) long and ingest meats. A small soft bodied crab known as the pea crab (*Pinnotheres osterum*) is sometimes found inside oysters. It causes only minor injury to the oyster and in former times was in demand for making soup.[10]

There are three oyster pathogens in the lower bay which may cause extensive mortalities of oysters. These are Dermo (*Perkinsus marinus*), SSO or Seaside organism (*Minchinia costalis*) and MSX (*Haplosporidium nelsoni*). MSX is a haplosporidan systemic disease of oysters which was first observed in 1957 in oysters in Delaware Bay.[11] In that area by mid-summer, it killed nearly half the oysters in high salinity regions and the following year nearly all had died in that area. By late summer 1959 it was found in Chesapeake Bay, and by 1961 nearly all oysters growing in locations where salinities exceed about 15°/oo had died. Where salinities were below this approximate level its impact on mortalities declined sharply. Annual variations in salinity in the Chesapeake Bay and its tributaries above and below 15°/oo, result in a fluctuating zone in which mortalities may or may not occur.[12] Scientists even after many years of study still do not know the full life cycle of MSX and the way it is transmitted from oyster to oyster is still unknown. Today, mortalities from MSX continue but there is some evidence that native populations are developing some resistance.

Perkinsus marinus (Dermocystidium marinum) (Dermo), a protozoan parasite infecting oysters, is found in the Chesapeake Bay and along the mid Atlantic and Gulf of Mexico coats. In these areas it usually causes limited mortalities, but under certain circumstances, its impact may be severe.[13]

Dermo is primarily a disease infecting oysters in high salinity regions and it is active in Chesapeake Bay over the same approximate range as MSX (fall salinities over 15°/oo). Unlike MSX, the life cycle of Dermo is known and the disease is easily transmitted from one oyster to another. Oysters become infected during early summer or fall by ingesting waterborne spores from disintegrating tissues of oysters killed by the disease. It is most destructive during long periods of above average water temperatures and where oysters are closely crowded on the bottom.

While the 15°/oo isohaline in the bay and its estuaries seems to divide the high and low mortality regions for MSX and Dermo, neither disease is associated with significant, or extreme, mortalities in the tidal lagoons of Maryland and Virginia between the barrier island and the mainland of the Eastern Shore. Here salinities usually range from about 24-32°/oo, and the reason for the low incidence of both diseases is not known.

The oyster pathogen SSO or Seaside organism was first observed in the tidal lagoons of the sea side of the Eastern Shore in 1959 and it has probably been in that area for many years. Only part of the life cycle of this organism is known, and like MSX its method of transmittal is unknown. It produces sharply peaked mortalities during May and June and, like Dermo, it seems to impact individual beds of oysters rather than large areas.[14]

THE FISHERY

In Chesapeake Bay management of the oyster fishery is by the two states which border the bay and by a bi-state Maryland-Virginia commission which manages the Potomac River. In Maryland, nearly all the naturally productive oyster bottom (215,000 acres) have been designated as public bottom where oysters may be harvested by the public provided they obtain a license, use the proper gear to obtain the oysters, and follow the laws and regulations related to season, etc. Only about 9,000 acres are presently leased by companies or individuals.[15]

Virginia also has naturally productive bottoms (about 243,000 acres) set aside for public use called Baylor Grounds after Lt. Baylor, USN, who first surveyed the grounds in 1894. They are subject to many of the same type of laws and restrictions as exist in Maryland, but there are also major differences. Unlike Maryland, large areas outside the designated public bottoms may be leased for private use by companies or individuals for a 10 year period for a fee. About 110,000 acres were leased in Virginia in 1985. Leased bottoms in Virginia are usually not naturally productive as are many of the public bottoms. Consequently, to be productive they must be planted with small oysters termed seed oysters which are juvenile oysters ranging from about 0.6 to 2.5 inches long. These seed are transferred from a good setting area, such as the James River, to areas where setting is poor but where growth is good. The volume of seed oysters harvested annually from the James River is large and in the 1984-85 season about 400,000 bushels were tonged and sold to the lease holders; this source, provides from 75 to 80% of all seed planted by lease holders in Virginia.

A bushel of seed oyster sold by tongers working in the James River, Virginia may contain from 500 to 1000 seed 1.5 to 6.5 cm (0.6-2.5 in) long which are about 1 to 4 years old. Counts of current year spat on seed oysters are generally not considered in these counts. Typically, a bushel of seed oysters, planted in Virginia's rivers will yield one bushel of market oysters two to three years later. A bushel of seed, in 1985 cost about $3.50-3.75 to purchase and plant, and a bushel of mature oysters will sell for $12-16 wholesale. Other aspects of the oyster fishery including harvesting are discussed in Chapters 2, 3 and 4.

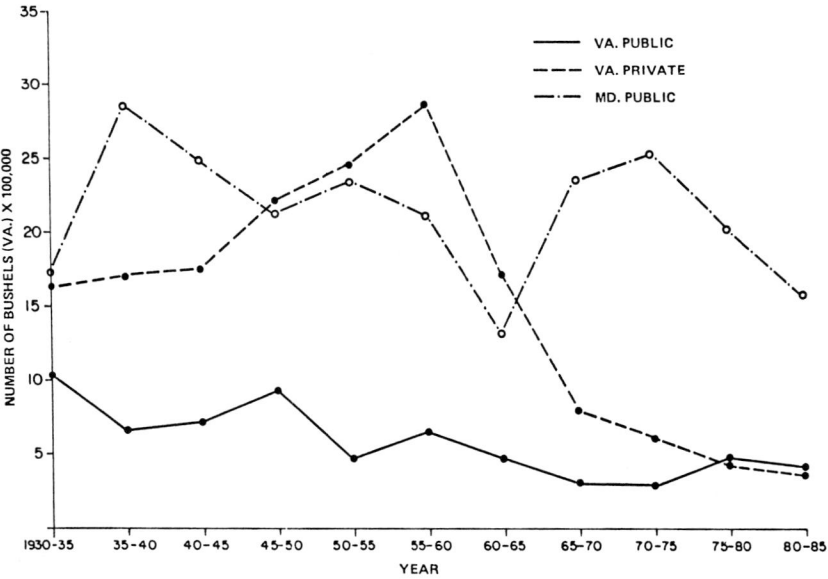

FIGURE 4. Oyster landings in Maryland from public beds, and landings for Virginia for leased areas and public bottoms. All data have been converted to Virginia bushels (3003.9 cu. in.)

A DECLINE IN PRODUCTION AND PRODUCTIVITY

Chesapeake Bay today is the largest oyster producer on the Atlantic and Gulf coasts, but production in recent years has declined sharply both in Maryland and in Virginia. In Maryland, average annual state wide production has fallen since the 1971-75 period from about 2.5 million bushels to about 1.6 million for 1980-85. On Virginia public oyster bottoms there has been a downward trend in landings from about 924,000 bushels during the 1945-50 period to about 416,000 for 1980-85. The major part of the decline in Virginia's statewide production occurred in the private sector. Here landings fell from about 2.8 million bushels during the 1955-60 period to about 500,000 for 1980-85 (Figure 4).

The decline in production from the public bottom in both states has occurred despite extensive repletion programs by both states. In these programs, shells to obtain a set and seed oysters have been planted on suitable bottoms to help increase production; without these extensive and costly efforts, the annual harvest from public bottoms in both states would have been much lower.

Much study has been directed toward determining the basic cause for the decline in landings on the public oyster grounds of Maryland and Virginia, and for the absence of production from Virginia's leased areas, and although there is agreement on several important causes of the decline there is not complete agreement as to their relative importance. One factor however, is clear. There has been a decline in recruitment of small oysters on the public bottoms of both states, as measured by numbers of spat occurring each fall on natural bottom cultch after setting has stopped. In Maryland on representative oyster bars, scientists have shown that, with the exception of the 1961-64 period, the general trend was downward from 1945 to 1976. This decline of course, means fewer oysters when the surviving spat grow to marketable size. Available data suggest however, that other factors in Maryland are involved in the decline in natural productivity, including overfishing.[15,16]

A decline in recruitment has certainly been responsible for at least a portion of Virginia's lowered harvest rates on its public oyster bottoms. Some estuaries have shown little change in recruitment since 1960 but other locations have experienced a major decline. For example, in the James River, Virginia, the source of 75 to 80% of the seed oysters planted on leased bottom, there has been since 1960 a 90% decrease in set on natural bottom cultch on one of the largest oyster reefs[10] (Figure 5). In the upper half of the same estuary there has been a 40-50% reduction. Other Virginia estuaries showing a decline in spatfall include the York River and portions of the Rappahannock River.

The causes of the decline in landings from the leased areas in Virginia which usually must be planted with seed if they are to produce, are complex, and there are several aspects. MSX was the cause for the initial drop in productivity after 1960 since oyster culture was no longer economically profitable in the higher salinity areas where most oysters were cultivated. It is still not possible there today.

After 1960 however, the lease holders did not relocate their growing activities to MSX free areas, or to areas where MSX was only occasionally active or marginal. There were many reasons for this and they are all interrelated. Economic conditions in the post 1960 period were changing and costs of culturing oysters had increased even in good areas, profits were sometimes marginal; costs of seed and planting had increased. Moreover, the private sector had not adopted cost effective culture techniques, such as mechanized planters or harvesters etc. Adding to their problems was competition in the form of imported oysters from Maryland and the Gulf Coasts and the pollution of good growing areas.[10]

The lowered levels of natural recruitment discussed above may be responsible for a major part of the decline in landings from the public bottoms, and factors related to overfishing may also be involved, but the basic cause(s) for the decline is not too apparent. Factors such as the destruction of bottom cultch, increased levels of sedimentation, higher levels of larval and spat mortality, and pollution have been discussed and studied. In respect to pollution, many commonly used petrochemicals, herbicides, pesticides, and heavy metals have been shown in laboratory studies to be toxic or have a sublethal effect on oysters and oyster larvae.[15] Many of these same substances occur in bay waters, where they probably have an adverse impact on survival and growth of oysters and

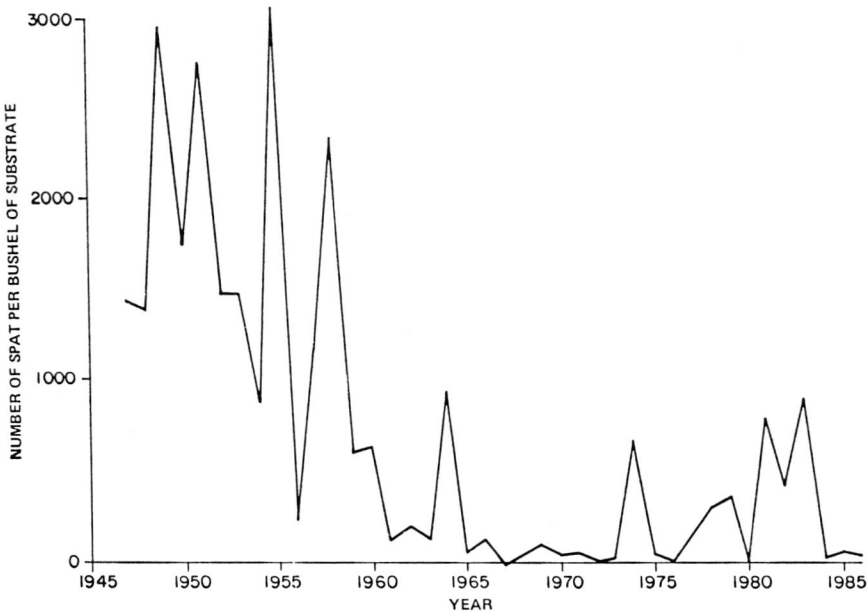

FIGURE 5. Number of spat per bushel on natural bottom cultch during late fall at Wreck Shoals in the Lower James River, Virginia from 1947 to 1986.

oyster larvae. For most of these substances cause and effect relation in the estuary have yet to be fully demonstrated.

In conclusion, oyster production in the Chesapeake Bay has declined in recent years, but large areas are still productive; and these still have an enormous potential for oyster culture. Studies have shown however that unless the present antiquated management policies and practices for the bay are drastically modified, this potential will not be realized. Such changes are possible and practical and have been successfully used in other areas. Moreover, they may be used with little or no damage to the environment.[10] Toward this goal the Virginia Marine Resources Commission in 1986 and 1987 began developing a long range Virginia Fisheries Management Plan to enhance oyster production. If this plan is fully and correctly implemented with sufficient funds and personnel, much improvement will result.

LITERATURE CITED

1. Galtsoff, P.S. 1964. The American oyster *Crassostrea virginica* Gmelin. Fishery Bull. of the Fish and Wildlife Serv. 64. Washington, D.C.; 480 pp.
2. Jorgensen, C.B. 1986. Biology of suspension feeding. Pergamon Press, Oxford; 357 pp.
3. Haven, D.S., and L.W. Fritz. 1985. Setting the American oyster *Crassostrea virginica* in the James River, Virginia, USA: temporal and spatial distribution. Mar. Biol. 86:271-282.
4. Lutz, R.A., H. Hidu and K.G. Drobeck. 1970. Acute temperature increase as a stimulus to setting in the American oyster *Crassostrea virginica* (Gmelin). Proc. Nat. Shellfish Assoc. 60:68-71.
5. Davis, H.C. and P.E. Chanley. 1956. Spawning and egg production of oysters and clams. Proc. Nat. Shellfish Assoc. 46:40-51.
6. Haven, D.S., and V.G. Burrell. 1982. The Oyster...A Shellfish Delicacy. Atlantic States Mar. Fish. Comm. Washington, D.C. Leaflet No. 11:1-5.
7. Wood, J.L. and W.J. Hargis, Jr. 1971. Transport of bivalve larvae in an estuary. pp. 29-44 *In:* D.J. Crisp (ed.), Proceedings of the Fourth European Marine Biology Symposium 1969, Cambridge University Press, Cambridge.
8. Beaven, G.F. 1955. Various aspects of oyster setting in Maryland. Proc. Nat. Shellfish Assoc. 45:29-37.
9. Kennedy, V.S. 1980. Comparison of recent past patterns of oyster settlement and seasonal fouling in Brood Creek and Tred Avon River, Maryland. Proc. Nat. Shellfish. Assoc. 70: 36-46.
10. Haven, D.S., W.J. Hargis, Jr. and P.C. Kendall. 1981. The oyster industry of Virginia: its status, problems and promise. A comprehensive study of

the oyster industry in Virginia, 2nd edition. Spec. Pap. Mar. Sci., Va. Inst. Mar. Sci., No. 4:1-1024.
11. Haskin, H.H., L.A. Stauber and J.A. Makin. 1966. *Minchinia nelsoni;* n.sp. (Haplosporidia, Haplosporidiidae: Causative agent of the Delaware Bay oyster epizootic. Science 153(3742):1411-1416.
12. Andrews, J.D. 1968. Oyster Mortality Studies in Virginia. VII. Review of epizootology and origin of *Minchinia nelsoni.* Proc. Nat. Shellfish Assoc. 58:23-36.
13. Andrews, J.D. and W.G. Hewatt. 1957. Oyster Mortality Studies in Virginia. II. The fungus disease caused by *Dermocystidium marinum* in Chesapeake Bay. Ecol. Monogr. 27:1-26.
14. Wood, J.L. and J.D. Andrews. 1962. *Haplosporidium costalis* (Sporozoa) associated with a disease of Virginia oysters. Science 136(3517):710-711.
15. Kennedy, V.S. and L.L. Breisch. 1981. Maryland's Oysters: Research and Management. Univ. Maryland (Sea Grant) Publ. No. Um-SG-TS-81-04:1-186.
16. Kranze, G.E. and D.W. Meritt. 1977. An analysis of trends in oyster spat set in the Maryland portion of Chesapeake Bay. Proc. Natl. Shellfish Assoc. 67:53-59.

Contaminant Problems and Management of Living Chesapeake Bay Resources. Edited by S. K. Majumdar, L. W. Hall, Jr. and H. M. Austin. © 1987, The Pennsylvania Academy of Science.

Chapter Ten

FACTORS AFFECTING THE DISTRIBUTION AND ABUNDANCE OF THE BLUE CRAB IN CHESAPEAKE BAY

W. A. VAN ENGEL

Virginia Institute of Marine Science
School of Marine Science
College of William and Mary
Gloucester Point, Virginia 23062

ABSTRACT

That environmental conditions in the Chesapeake Bay are optimal for the blue crab population is suggested by the fact that hard crab landings by Virginia and Maryland watermen accounted for almost 48% of the total of East and Gulf coast landings in 1985. Estimates of total mortality from the egg to the adult stage range from 0.999973 to 0.999996. Commercial fishing removes an additional 0.0000031 to 0.0000251, leaving 0.0000024 to 0.000001 as the rates of removal by other sources. Physical and chemical pollutants, predators, and plants and animals symbiotic with the blue crab are part of the environment that must be acknowledged as actual or potential factors affecting the rates of reproduction, growth and survival, and the behavior and distribution of the blue crab population. The impact of parasites and disease, predation, salinity, temperature, dissolved oxygen, heavy metals, polynuclear aromatic hydrocarbons, and halogenated substances on the blue crab are described.

THE FISHERY

Since the beginning of the commercial fishery for blue crabs in the Chesapeake Bay over 100 years ago, large interannual and longer oscillations in landings have occurred (Fig. 1). The perspective provided by the longer record does not indicate any long-term change. The decline in landings from an all-time high in 1966 to the recent low in 1976 has been overemphasized by some correspondents; similar events occurred from 1930-1941 and from 1950-1959.

Contribution No. 1383 from the Virginia Institute of Marine Science, School of Marine Science, The College of William and Mary.

Comparison of 1981-1985 hard crab landings with those from any earlier period is feasible only on a state-by-state basis, since the data acquisition method in Maryland was changed in 1981. Maryland's 5-year mean landings were 45.5 million (M) pounds, double the mean of 21.2 M for 1976-1980 and of 23.2 M for the 21 years of 1960-1980. These numbers do not include estimates of the recreational crabbers catch. No perceptible change in the accounting method occurred in Virginia: the 1981-1985 mean of landings in Virginia was 44.4 M, a marked increase from the previous 5-year mean of 35.3 M but only slightly different from the 21 year mean of 43.3 M.

Principal explanations for fluctuations in landings may lie in variations of any one or a combination of the following factors: 1) fishing effort; 2) the parent-progeny relationship; and 3) the atmospheric and oceanographic environment. Progress in uncovering relationships for any of these three has been hindered by the unavailability or incompleteness of biological, environmental and landings data, and particularly the uncertainty of their reliability.

Daily landings of crabs are determined by 1) their availability, that is, the portion of the total population susceptible to capture, and 2) the intensity of fishing. While availability is a function of the effects of varying environmental conditions on crab distribution, significant deviations from the normal environment are not likely to occur in the short term, and changes in catch are likely to be of short duration. Similarly, while cessation of fishing may be caused by an oversupply of crabs or crab meat, strikes by watermen demanding higher prices for their catch, unusual weather conditions that have destroyed fishing gear and/or vessels or have prevented normal fishing effort, these conditions

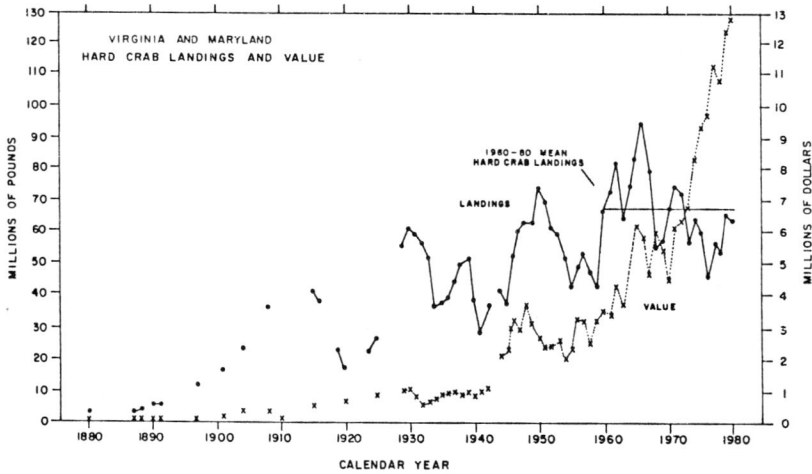

FIGURE 1. Virginia and Maryland hard crab landings and value.

have almost always been of such short duration that the losses in catch are negligible.

Landings over the long term depend primarily on the "fishable" portion in the crab population, which is derived (recruited) by growth from the younger portion. Since "stock density", which is a measure of the proportion of different age or size groups in the population, varies seasonally and interannually, estimates of abundance of each portion of the population are important in understanding the effects of environmental variables and of fishing on the population and on recruitment.

Since the blue crab fisheries are confined to state territorial waters, responsibility for fisheries management rests with the states. Regulatory authority concerning seasonal, geographic, size, sex and catch limits, gear restrictions, and other licensing controls over harvesting is generally retained by the state general assemblies, but may be delegated to commissions that have various degrees of authority to invoke management action at the local level as the need arises.

Despite the stated intention that regulations are for conservation of the blue crab population, most, if not all, regulations of the blue crab fisheries in the United States have been based on local or regional economic, political or sociological demands. Little or no regard has been given for scientific investigations of the status of the population and the fishery before regulations are put into effect nor has regard been given for observing and analyzing conditions in the fishery at later intervals. Such regulations are trial and error efforts to manipulate the fishery to effect whatever objective the management agency is seeking. Untested regulations usually result in the underutilization of the resource.

The historical and popular conception of the life history of the blue crab in the Chesapeake Bay is that the population is self contained within the Bay, that zoeae and megalopae are produced and retained in the highly saline portions of the southern end of the Bay and that their further development to juveniles within the Bay precedes migration to the nursery grounds in the lower salinity regions of the tributaries and the upper Bay. Regulations reflecting this viewpoint have been enacted by Virginia and Maryland over the last 70 years.

"Stock density" implies that there is a relationship between the number of crabs of different age groups in the population. Usually only the relationship between the parents (spawners) and their progeny is considered important enough for study. Seldom mentioned is the competition between crabs, which may be age specific, for food and space.

Most often, the parent-progeny relationship is thought of in the classical sense, that is, the effects that the density (numbers) of adult females may have on the density of very early life-history stages (zoeae and megalopae). Obtaining even relative estimates of the numbers of these early stages is a newly developing area of study and difficult at best. Numbers of these early stages will be less closely correlated than 2 to 12-month old juveniles with the resultant adult stock, probably because of relatively large and fluctuating rates of mortality (affected by food, space and water quality requirements, predation by plankton feeders, and dispersion through water transport). In contrast to the parent-progeny relationship, the relationship between the early stages or the juveniles and the resultant stock is more properly termed a "recruitment" or "spawner-recruit"curve. At present, too little is known of the mode and time of transport of the transition phase between zoeae and megalopae from the open waters of the lower Bay and the adjacent continental shelf waters to the juvenile stage on the nursery grounds to justify speculation on the factors that could affect their survival.

There are three approaches to modeling populations: 1) an empirical method in which trends in the fishery are associated with trends in environmental variables, 2) the surplus-yield "conceptual" model which requires an input of catch and effort data, and 3) the dynamic pool "conceptual" model for which we need details of the biological characteristics of the stock, such as growth, mortality and recruitment rates.

An examination of the life history pattern of the Chesapeake Bay blue crab and its population parameters suggests that a spawner-recruit model would not be useful in setting a management policy, as it is not density-dependent. The species is characterized by the annual production of a large number of young, large interannual fluctuations in production, rapid growth, early attainment of maturity, high mortality rates and a short life span. Those are the characteristics of a density-independent species, exposed to a variable environment in which the population's resources are spent mostly on reproductive (r) functions! In short, the blue crab appears to be an r selected strategist. Because of these characteristics, the blue crab can be fished at high levels of fishing effort, and, because of the short life span and rapid succession of year classes, would have a quick recovery if overfishing occurred. Species with such characteristics are strongly affected by physical, chemical and biological environmental factors! It is axiomatic that for populations fluctuating widely as a response to environment variation that the maximum sustained yield cannot be realistically estimated.

Although the exact mechanisms through which environmental factors affect yearclass strength are only partially known, it is believed that they occur at critical times early in the life cycle of the blue crab.

BIOLOGY, LIFE HISTORY AND DISTRIBUTION

Spatial, seasonal and size distributions of the pre-recruit stages are evidence that chemical, physical and biological factors effect survival and growth.

Female blue crabs mate only once, in the soft shell state at the terminal molt, sometime between early May and the end of October, and shortly thereafter migrate from the brackish waters of the tributaries and the upper Bay to the saltier waters of the southern end of the Bay. Most females of the same age group (year class) reach the terminal molt stage and mate in late summer, July through September, the second summer of life, and do not extrude eggs until the following spring: they may produce two or more egg masses at least two months apart. A varying number of females of the same year class do not reach the terminal molt stage and mate until May or June in the third year and may not extrude eggs until August in the third summer. During the third winter of life, aged females carrying egg remnants on the swimmerets from previous extrusions comprise about 5% of the stock of adult females in the southern end of the Bay[2] and, conceivably, some of them may extrude eggs the fourth summer, although that has not been verified. Adult females are commonly called "sooks", and those with extruded eggs are usually called "sponge crabs" but occasionally called "busted sooks".

Zoeae hatch within 10-14 days and progress through seven, occasionally eight, stages in one month, then metamorphose to the megalopal stage. In the Bay, Stage 1 zoeae are found predominantly in the surface waters along with some later stages, but most later stages, including the megalopa, are in subsurface layers or bottom waters.

In summary, zoeae appear in the Bay plankton June through September, are most numerous in July and August and rarely occur in May, October or November. They are most numerous at the baymouth and occur in rapidly decreasing numbers upestuary. In some years they may be found in large numbers at least 75 km inside the baymouth. Stages I-VII and the megalopa are present in the water column, but most zoeae, predominantly Stage I, are found in the upper 10-15 cm of the surface waters. Evidence of vertical migration was not found by Goy[3] from day and night samples in Hampton Roads, although the results are inconclusive because of the small numbers found. Grant and Olney[4] suggested diurnal migration of the larvae, based upon the large proportion of megalopae and zoeae in subsurface samples in the daytime, in contrast to their relative scarcity in the neuston. They concluded that "this difference in depth distribution of developmental stages may reflect a behavioral adaptation of the species that aids in recruitment of populations to the estuary" (p. 48). Provenzano et al.[5] found peaks of abundance of zoeae at 3 m depth after late night or morning high slack tide at stations occupied just inside the baymouth in 1979. They suggested that the peaks were associated with a night time ebbing and could be due to the input of new zoeae into the water column due to an apparent synchronized hatch. Support for vertical migration was not observed. McConaugha et al.[6,7] reported large numbers of zoeae of all stages and megalopae collected in 1980 from stations located inside and outside the baymouth.

Collection of large numbers of blue crab zoeae and megalopae in continental shelf waters in the last 15 years has revived interest in dispersal and retention mechanisms. Specifics of the distribution and abundance of zoeae and megalopae outside the baymouth and on the continental shelf, and of the site-specific hydrographic conditions where samples are collected, are clues to the abilities of these stages to survive and exist under those conditions, and to the extent of the transport of those stages. These collections do not tell us where the stages originated, nor of the transport mechanisms involved.

Comparison of temperature-salinity conditions in the Bay and on the shelf with tolerances of zoeae and megalopae determined in the laboratory may lead to unwarranted conclusions that in some seasons the natural environmental conditions are too rigorous to sustain life or permit molting. Possibly, tolerances may have been too narrowly defined when temperature-salinity ranges were selected to determine optimal responses, e.g., 15-30°C, 5-19 ppt.[8] Responses to combinations of temperature and salinity of 10-20°C/10-20 ppt and 10-20°C/30-35 ppt are needed to predict survival and growth in the Bay and on the shelf.

Following metamorphosis of the megalopa to the first true crab stage, young-of-the-year crabs move into the tributaries of the southern end of the Bay and into Maryland waters, occasionally as far north as Eastern Bay, Maryland. Information on the distribution and abundance of megalopae and the earliest true crab stages, 1st to 3rd instars, 2-5 mm wide, is scanty. The time of their first appearance as 6-25 mm wide crabs in the lower Bay's tributaries is usually early September, but varies considerably from year to year, rarely mid-August or as late as mid-November. These pre-recruits are seldom as numerous in late summer and fall trawl net catches, taken in monitoring surveys, as in the following spring and summer. Maximum numbers are recorded in June, July and August, the second summer. In June, the usual width range is 25-60 mm.

Growth is rapid. Crabs hatched in late May or early June become five inches in width or larger by mid-August or early September the following year, but those hatched in late August will not reach five inches in width until May or June the third year. Sexual maturity of males is attained a year after hatching and before the crab is fully grown, and during its last three growth stages may mate with one or more females. Females hatched in June become sexually mature in 14 months and mate in the soft shell stage in the terminal molt.

The influence of environmental variables on yield has been demonstrated for a number of crustacean fisheries. In each instance long time series of data were used and regression analyses, consisting of current landings as the dependent variable, were run against combinations of previous years' landings, fishing effort and lagged environmental variables as the independent variables.

Temperature, salinity, substrate and food have been found to be the primary factors affecting growth, survival and distribution of many crustaceans. That these factors are optimal in the Chesapeake Bay is suggested by the fact that hardshell blue crab landings by Virginia and Maryland watermen accounted for almost 48% of the total of East and Gulf coast landings in 1985.[9] In a multiple correlation analysis, 86% ($r^2 = 0.86$) of the variation in commercial hard crab landings from 1964 through 1975 (in the Biological Year, September through August) was explained by fluctuations in Norfolk, VA cooling degree days in May, Delaware Bay meridional wind stress in January and the log transformation of the York River juvenile crab catch per tow in the year of the hatch and through August the following year.[10]

Physical and chemical pollutants, predators, and plants and animals symbiotic with the blue crab are part of the environment that must be acknowledged as actual or potential factors affecting the rates of reproduction, growth and survival, and the behavior and distribution of the blue crab population. Research to identify and describe those factors that could have an impact on the blue crab population has attained marked progress since the mid-1960's.

ESTIMATES OF MORTALITY

Known sources of total mortality of the blue crab, that occur at any stage of its life history, do not account for the low survival rate obtained from estimates of the reproductive and fishing mortality rates. Estimates of the mortality rate from the egg to the adult stage can be made by applying a number of initial assumptions: 1) commercial fishing removes 75 to 90% of the adult stock; 2) the commercial fishery removed 46 to 90 million (M) pounds of crabs a year over the 21-year period from 1960-1980; 3) fecundity rate of adult female blue crabs, which are one-half the survivors, varies from 0.725×10^6 to 2.0×10^6; and 4) there are three crabs in each pound of the commercial catch.

Estimates of the mortality rate range from approximately 0.9999725 to 0.9999959. Commercial fishing removes an additional 0.0000031 to 0.0000251, leaving 0.0000024 to 0.000001 as the rates of removal by other sources. If it is assumed that the commercial catch is grossly underestimated, that the landings could be as much as 150 M pounds, the estimates of the total mortality rate are unchanged: 0.9999725 to 0.9999959.

A second estimate of total mortality can be based on a tenet held by ecologists, that for a population to maintain its position in a community, i.e., to continue to occupy the same niche, on the average over several generations no more than two individuals from each brood may survive to adulthood. If more than two individuals survive, the population will soon outgrow its niche; if less than two individuals survive, the population will be threatened by extinction or must adopt a smaller role in the composition of the community. If reproductive potential averages 1×10^6 eggs per female, the survival rate is estimated as 2×10^{-6} and the mortality rate up to the adult stage is estimated as 0.999998.

SYMBIONTS

At least 84 kinds of viruses, algae, diatoms, bacteria and invertebrates, some identified only to genus, are symbionts of the blue crab on the Atlantic or Gulf coasts (Table 1). Twenty-two (26.5%) of them occur as commensals that may be found on or attached to the external shell or in the gill chamber of the crab as part of the normal flora and fauna, such as algae, bacteria, sponges and corals, coelenterates, mollusks, annelids, barnacles and amphipods, bryozoans, and tunicates. None of these fouling organisms is usually considered a threat to the crab's existence.

The other 62 species occur as parasites or diseases, ten of which have been implicated with mortality of crabs: four viruses, a bacterium, four photozoans, and a fungus. In addition, a nemertean, *Carcinonemertes carcinophila* and a ciliated protozoan, *Lagenophrys callinectes* may occur on the gills in such large numbers as to interfere with gas diffusion across the gill membranes and diminish the capacity of the crab to respire. Mortality caused by parasites and disease is believed to be small, except in crabs infected with the protozoan *Paramoeba perniciosa,* and crab eggs that are infested with the fungus *Lagenidium callinectes. Paramoeba* infections are usually limited to high salinity environments and mortality may reach 14-18%.[11] A large winter mortality of blue crabs in Chincoteague Bay in the late 1960's may have resulted from a combination of cold temperature and *Paramoeba* infection.[11] Fungi may infest from 25-50% of the extruded eggs on as many as 87% of sponge crabs in the Chesapeake Bay in a wide range of salinities, and some or all of the larvae grown in culture.[12]

Effects of the remaining 50 parasites and diseases are not now known to threaten the existence of the blue crab population: 18 bacteria, five viruses, one fungus, eight protozoans, one bryozoan, two nematodes, seven trematodes, one cestode, three annelids, three barnacles, and one isopod.

Bacteria have been found inside the body and in the hemolymph of wild crabs.[13,14,15,16,17,18,19,20,21,22,23,24,25] Stress on crabs, such as injuries, exposure to temperature extremes and dehydration, which may occur during commercial fishing, probably results in decreased effectiveness of resistance mechanisms, and is related to increases in hemolymph bacterial numbers.[24] Infections by viruses of natural, "wild" populations of crabs are relatively unknown. Accumulations of viruses by crabs could occur by feeding on infected tissues, by direct transmission from the water or through wounds, but concentration of viruses by the host has been considered unlikely because of the crab's feeding habits. Mortality of peeler and soft crabs in shedding tanks may result from preexisting viral (or bacterial or amoebic) infections, occurring in crabs before they were caught, or those acquired by contact in the tanks with infected crabs, aggravated by the stress of capture and confinement. Human illness from eating crabs that have accumulated viruses has not been reported.

Several bacteria associated with crabs that have been in contact with fecal pollution, are potential human pathogens, causing gastroenteritis when crabs that are improperly cooked are eaten.[13,14,15,16,17,20,21,26,27,28,29,30,31,32,33] Wound infections on humans can occur when live crabs are carelessly handled.[16,17]

TABLE 1

Symbionts of the Blue Crab

FOULING ORGANISMS
Algae.
    1. (34)
Bacteria
    2. coliforms. (21,35)
Porifera
    3. zoanthid. *Epizoanthus americanus.* (36)
    4. soft coral. *Leptogordia vingulata.* (36)
    5. stony coral. *Astrangia danae.* (36)
    6. sponge. (36, Van Engel unpubl. data)
Coelenterates
    7. *Obelia bidentata.* (34,37)
    8. *Bougainvillia* sp. (34,37)
Mollusks
    9. *Crassostrea virginica.* (34,37, Van Engel unpubl. data)
    10. *Mytilus edulis.* (34,38, Van Engel unpubl. data)
Annelids
    11. polychaetes. (34, Van Engel unpubl. data)
Barnacles
    12. *Balanus eburneus.* (Van Engel unpubl. data)
    13. *Balanus improvisus.* (Van Engel unpubl. data)
    14. *Balanus venustus niveus.* (32,34,35)
Amphipods
    15. (34,36)
Bryozoans
    16. *Acanthodesia tenuis.* (39, Van Engel unpubl. data)
    17. *Alcyonidium mytili* (=*polyoum*). (36,39, Van Engel unpubl. data)
    18. *Alcyonidium verrilli.* (39, Van Engel unpubl. data)
    19. *Membranipora crustulenta.* (39, Van Engel unpubl. data)
    20. *Membranipora tenuis.* (34)
    21. *Conopeum tenuissium.* (34)
Tunicates
    22. *Molgula manhattensis.* (36, Van Engel unpubl. data)
PARASITES (Species that are lethal to the host are marked *)
Viruses. Viruses marked with brackets [ ] were seeded to water for accumulation studies, and were not found naturally occurring in blue crabs.
    Baculovirus. (32,40,41,42). No overt disease.
    23. *Baculovirus* A. (19,41,43)
    24. *Baculovirus* B. (19,43)
        [Coxsachievirus. (44-seeded)]
        [echovirus. (44-seeded)]
    25. enterovirus. (35)
    26. enveloped helical virus. (19,45)
    27. *herpes-like virus. (19,32,42,43,46,47,48). Lethal.
    28. *picorna-like virus. (19,32,42,43 [49 is incorrect]. Lethal.
        [poliovirus. (40, 44-seeded)]

TABLE 1 (continued)
Symbionts of the Blue Crab

    29. *reo-like virus. (19,32,42,43,47,48,49,50). Lethal combination with rhabdo-like virus.
        rhabdo-like virus (2 or 3 kinds).
    30. *rhabdo-like virus A. (42,43,45,51,52,53). Lethal combination with reo-like virus
    31. rhabdo-like virus B. (45)
        [simian rotavirus. (40, 44-seeded)]

Bacteria. (18,23,54)
        chitonoclastic. (19,54,55,56,57,58)
    32. *Acinetobacter* sp. (15,19,21)
    33. *Aeromonas* sp. *(hydrophila?)* (17,21)
    34. *Bacillus* sp. (15,21)
    35. *Benekea* type 1. (32,58,59)
    36. *Clostridium perfringens* (?). (17)
    37. *Clostridium botulinum.* (25,60)
    38. *Enterobacter aerogenes* (?) (35)
    39. *Escherichia coli.* (21)
    40. *Flavobacterium* sp. (15,19,21)
    41. gram-negative, unidentified. (19,61)
    42. *Leucothrix mucor.* (filamentous). (19,65). (Probably only a fouling organism).
    43. *Pseudomonas* sp. *(bathycetes ?).* (21,59)
    44. *Staphylococcus aureus.* (17)
    45. *Vibrio* spp. (14-in water, 15,16,17,21,22,24,35,59,62).
    46. *Vibrio alginolyticus.* (17).
    47. *Vibrio cholerae.* (14-in water, 16,17,22,30)
    48. *Vibrio fischeri.* (21)
    49. **Vibrio parahaemolyticus.* (13, 14-in water, 15,16,19,20,21,22,23,26,27,28,29,31,32,33)
    50. *Vibrio vulnificus.* (16,17)

Fungi
    51. **Lagenedium callinectes.* (19,32,63,64,65,66,67,68,69,70)
    52. *Thraustochytrium* sp. (65)

Protozoans
  Dinoflagellate
        Hematodinium sp. (71)
    53. **Hematodinium perezi.* (11,32,72). May be lethal.
  Amoebae
    54. **Paramoeba perniciosa.* (11,12,32,72,73,74,75,76,77,78,79,80,81). Lethal.
  Microsporidians
        Ameson sp.
    55. **Ameson (Nosema) michaelis.* (11,32,72,78,82,83,84,85,86). Lethal.
    56. *Ameson sapedi.* (11,78,84,87,88,89)
    57. *Ameson* sp. (new) (84)
    58. *Pleistophora cargoi.* (11,72,78,84,90)
    59. *Thelohania* sp. (91,92)
  Haplosporidians
    60. **Haplosporidium* sp. (*Minchinia*-like). (11,32,72,93). Lethal
    61. *Urosporidium crescens.* (11,32,72,78,84,87,88,89,94)
  Suctorian
    62. *Acineta* sp. (32,84)
  Peritrichs
    63. **Lagenophrys callinectes.* (11,32,72,95,96,97). May be lethal in shedding floats.
    64. *Epistylis* sp. (32)
  Holotrich ?
    65. (32)

TABLE 1 (continued)

Symbionts of the Blue Crab

Bryozoans
    66. *Triticella elongata.* (34,39, Van Engel unpubl. data)
Nematodes
    67. other. (34,89)
    68. *Hysterothylacium* sp. (34)
Trematodes
    69-70. digeneans. (2 others) (34)
    71. *Levenseniella capitanea.* (32,34,37,98,99)
    72. *Microphallus basodactylophallus.* (32,34,37,98.100)
    73. *Microphallus nicolli.* (34,101)
    74. *Microphallus pygmaeus.* (34,102)
    75. *Megalophallus diodontis.* (34)
Cestode
    76. (32)
Nemertean
    77. *Carcinonemertes carcinophila.* (32,34,36,37,103,104,105,106)
Annelid
    78. *Myzobdella lugubris.* (32,34,37,107,108,109)
    79. *Calliobdella vivida.* (34)
    80. *Cambrincola vitreus (= mesochoreus?)* (32,34,37)
Barnacle
    81. *Chelonibia patula.* (32,34,36,37,103,110,111,112, Van Engel unpubl. data)
    82. *Octolasmis mulleri.* (32,34,36,37,103,110,113,114,115,116,117, Van Engel unpubl. data)
    83. *Loxothylacus texanus.* (32,34,37,118,119,120)
Isopods
    84. In gill chamber of blue crabs, carapace width approx. 8-25 mm, York River (Gloucester Banks area) and Chisman creek, VA, 1948-1949. W.A. Van Engel collection.

PREDATION

Predation intensity on blue crabs varies with the species of predator, its size, life history stage, physical characteristics and feeding habits, whether it is a resident or migratory species, and the chemical and physical conditions of the environment, but overall appears to be slight or insignificant, judging from gut content analyses. Generally, fish predators < 250 mm standard length (SL) preferentially feed on micro- and macroplankton, small fishes and epibenthos; however, changes in diet may occur with increase in predator size. The frequency of occurrence of blue crabs in the diet is small in lie-in-wait predators and in those fishes whose mouths are not morphologically equipped to capture small blue crabs. Some migratory predators consume blue crabs only in fresh to brackish water, but seek other prey in saltier waters. Throughout a migrant's range, its food may include any abundant organism. Resident fishes are less frequent consumers of blue crabs than migratory megapredators.

Whether predation varies with the life history stage and size of the blue crab cannot be determined from published reports, since the prey stage or size is almost never reported. Predation on blue crab zoeae and megalopae is largely unknown: remains of early-stage brachyurans in fish stomachs are seldom if ever identified other than as "crab zoeae", brachyura zoea" or "megalopae" (see e.g., 121,122,123,124,125,126). However, blue crab megalopae were found in the gut contents of a few juvenile weakfish (*Cynoscion regalis*) on the York River nursery grounds.[127] The role of predators that could be consumers of blue crab megalopae in brackish and salt marshes, in particular the ubiquitous mummichog, *Fundulus heteroclitus,* has been partially investigated. In tidal brackish marshes from Massachusetts to Georgia, larval, juvenile and adult mummichogs forage at high tide in daylight on the marsh surface, primarily on detritus, polychaetes and small crustaceans such as amphipods.[21] Consumption of "crab zoea" by small, 9.5-23.4 mm SL mummichogs collected in the wild has been observed, but consumption of blue crab megalopae has not been confirmed; some sites of study were beyond the natural range of the megalopae, and because the rate of digestion of food by mummichogs is so rapid, fish guts are often empty or the contents unidentifiable.[21] Other ubiquitous small-size fishes and palaemonid crustaceans, which also occur in tidal marshes, have not been investigated as potential predators of megalopae.

Post-megalopal-stage blue crabs seldom appear in the gut contents of fishes that are residents or are coastal species utilizing the inshore, mid- to low-salinity (5-20 ppt) crab and fish nursery grounds along the eastern Atlantic coast and on the Gulf coast. For numerous species, less than 15% of the individuals (= frequency of occurrence) occasionally consume blue crabs (Table 2).

TABLE 2

Species and common name.	Size, mm	Reference
Carcharhinus leucas bull shark	780-805	(128)
Raja eglanteria clearnose skate		(129)
Ictalurus catus white catfish		(127)
Ictalurus furcatus blue catfish	60-229	(128)
Ictalurus punctatus channel catfish		(128*)
Arius felis sea catfish	170-229	(128)
	240-360	(128*)
Bagre marinus gafftopsail catfish		(128*)
Opsanus tau oyster toadfish	58-337	(130)
	32-252	(131)
		(132)
Morone americana white perch	204	(133)
Micropterus salmoides northern largemouth bass	175-209	(128)
Pomatomus saltatrix bluefish	122-870	(133)
	450 mean SL	(134)

Archosargus probatocephalus sheepshead	218-410	(128)
	190-365	(128*)
	145-449 SL	(135)
Lagodon rhomboides pinfish	65-74	(128)
Aplodinotus grunniens freshwater drum	211-347	(128)
Bairdiella chrysura silver perch	70-143	(128)
	20-150	(133)
		(136)
Cynoscion arenarius sand seatrout	59-320 SL	(135)
Cynoscion nebulosus spotted seatrout	275-555	(128*)
	400-495	(138)
	73-532 SL	(135)
Cynoscion regalis weakfish	320-580	(133)
	121-180	(136)
		(137)
	403 mean SL	(134)
		(127)
Leiostomus xanthurus spot	20-150	(133)
Micropogonias undulatus Atlantic croaker	100-199	(128)
	<39- >180 SL	(124)
	95-350 SL	(138)
	135-142	(136)
		(137)
Pogonias cromis black drum	205-460	(128*)
		(135)
		(127)
		(136)
Sciaenops ocellata red drum		(128*)
Ancylopsetta quadrocellata ocellated flounder	25-199 SL	(139)
Citharichthys spilopterus bay whiff	50-149 SL	(139)
Paralichthys lethostigma southern flounder	113-380	(128)
	168-410 SL	(135)

Footnote. Body lengths are shown as designated by the author; lengths that are missing were not provided by the author. Citations designated with an * include reviews of other authors.

There are a few heavy consumers of blue crabs in estuaries (Table 3); the percentage of fishes sampled that had eaten blue crabs (frequency of occurrence) is shown in parenthesis.

Sandbar sharks in the Chesapeake Bay preferred soft blue crabs but occasionally consumed hard crabs, 35-115 mm wide;[133] however, at the mouth of the Chesapeake Bay and on the adjacent continental shelf, between May and October, large sandbar sharks (545-2070 mm TL) fed mostly on epibenthic fishes and some pagurid and brachyuran crustaceans, but no blue crabs.[142] Feeding preferences of the red drum appear similar in Mississippi Sound (usually 6-15 ppt salinity), Lake Pontchartrain (1.2-18.6 ppt),[128] the bay side of the eastern shore of Chesapeake Bay (14-21 ppt),[133] and in waters of Texas, Louisiana and Florida.[141] In Lake Pontchartrain, feeding by red drum on blue crabs varied seasonally, more in spring and summer than fall and winter, with a preference

TABLE 3

Species and common name	Size, mm	Frequency of occurrence, %	Reference
Carcharhinus plumbeus			
sandbar shark	435-852 SL	(57)	(133)
	600-1125	(41.3)	(140)
Lepisosteus oculatus			
spotted gar	405-555	(70)	(128)
			(128*)
Lepisosteus spatula			
alligator gar	903-1472	(65)	(128)
		(87)	(128*)
Ictalurus furcatus			
blue catfish	230-411	(24)	(128)
Opsanus tau			
oyster toadfish	58-337	(36)	(130)
Strongylura marina			
needlefish	380-403 SL	(90)	(133)
Tylosurus acus			
agujon	970-1140	(71.4)	(133)
Roccus mississippiensis			
yellow bass	130-195	(18.5)	(128)
Centropristes striatus			
black sea bass	72-158 SL	(25)	(133)
Cynoscion regalis			
weakfish	403 mean SL	(38)	(134)
Micropogonias undulatus			
Atlantic croaker	200-325	(21.7)	(128)
Sciaenops ocellata			
red drum	384-765 SL	(100)	(133)
	184-625	(62)	(128)
			(128*)
	190-780 SL	(17.3)	(141)

Footnote. Body lengths are shown as designated by the author. Citations designated with an * include reviews of other authors.

for crabs in the lower salinity waters near Lake Pontchartrain and for shrimp in waters in or near Gulf of Mexico waters.[141] In Mississippi Sound, red drum (190-780 mm SL) were frequent feeders on the blue crab but more drum were consuming the "lesser blue crab", *Callinectes similis*.[141]

Frequently the gut contents are described by the common name "crabs", either because the remains cannot be further identified or because the investigators are unfamiliar with decapod crustacean species types. Probably, but not certainly, the "crabs" are xanthid or grapsoid crabs, since such reports relate more often to in-faunal or epibenthic feeders. Fishes consuming "crabs", "zoeae and megalopae" are listed in Table 4.

TABLE 4

Species and common name	Size, mm	Reference
Carcharhinus leucas bull shark		(128*)
Squalus acanthias spiny dogfish		(129)
Gymnura micrura butterfly ray		(129*)
Lepisosteus osseus northern longnose gar	706-1180	(128)
Elops saurus ladyfish		(128*)
Synodus foetens lizard fish		(129*)
Arius felis sea catfish	240-360	(128*)
Bagre marinus gafftopsail catfish		(129*)
Urophycis regia spotted hake		(129)
Roccus mississippiensis yellow bass		(128*)
Micropterus salmoides northern largemouth bass	203-432	(128*)
Caranx hippos crevalle jackfish		(128*)
Lagodon rhomboides pinfish	51-100	(128*)
Bairdiella chrysura silver perch		(128*)
		(143)
Cynoscion arenarius sand seatrout	100-406	(128)
		(128*)
Leiostomus xanthurus spot		(128*)
Menticirrhus saxatilis northern kingfish	19-72	(126)
Menticirrhus americanus southern kingfish	80-250	(126)
Micropogonias undulatus Atlantic croaker		(128*)
		(144)
		(145)
Pogonias cromis black drum	80-200	(128*)
Stellifer lanceolatus star drum	21-80	(126)
Prionotus carolinus northern searobin		(129)
Paralichthys dentatus summer flounder		(129)

Footnote. Body lengths are shown as designated by the author; lengths that are missing were not provided by the author. Citations designated with an * include reviews of other authors.

Some fish species in confinement may prey on blue crabs, whereas in an open environment such predation by the species is rare or has been unreported: the northern puffer (*Sphaeroides maculatus*) in confinement preys on blue crabs and Tangier, Virginia, watermen have used blue crabs in crab pots as bait for puffers (Van Engel unpubl. data). Cannibalism by blue crabs has been reported from stomach content analyses of crabs > 30 mm width,[128,146] although in approximately 3200 stomachs crab remains represented only nine % of the dry weight, almost all consumed by crabs > 60 mm wide.[146] Blue crabs have been observed to feed on crab scraps or injured or dead crabs discarded at piers or from boats or found in peeler crab shedding tanks, since they are opportunistic feeders and scavengers (147,148, Van Engel unpubl. data), and all sizes of paper-shell crabs, which are those that recently molted, regularly consume their own molted exoskeletons.

Sub-adult loggerhead (*Caretta caretta*) and Atlantic ridley (*Lepidochelys kempii*) turtles forage in the lower Chesapeake Bay from May through October. The stocks of these turtles in the Chesapeake Bay regions is not large. Loggerheads consume mostly horseshoe crabs and rock crabs, some blue crabs, spider crabs and clams. Destruction by loggerheads of wire pots set to catch blue crabs in the south end of Chesapeake Bay has been frequently reported by local watermen. Atlantic ridleys feed exclusively on blue crabs.[149,150,151]

River otters (*Lutra canadensis*) in small intertidal ponds and streams in salt marshes and freshwater swamps consume mammals, birds, fishes and invertebrates. In eastern North Carolina, in a 4-year fall and winter collection, 39% of the otters consumed crayfish, blue crabs and shrimp, in that order of preference.[152,153] In southern Louisiana, almost 20% of the otters in salt marshes in a 3-year winter collection consumed blue crabs, second in abundance to cyprinodont fishes, but in swamps 4% of the otters ate blue crabs and 34% ate crayfishes.[154] Little seasonal difference in predation is believed to occur.[154] Mink *(Mustela vison)* in eastern North Carolina marshes preferred fishes, and fewer mink consumed mammals, arthropods, birds, amphibians, and reptiles, in that order; no blue crabs were observed in stomach contents.[153]

CHEMICAL AND PHYSICAL FACTORS

Numerous chemical and physical pollutants occur in the Chesapeake Bay and its tributaries. Sediments, fertilizers, pesticides, herbicides, heavy metals, sewage and petroleum products arrive in the marine environment from both point and non-point sources. Their effects on estuarine biota are not well known. Chronic and acute effects of most of these parameters of water quality on the various life history stages of the blue crab, and their secondary effects on a crab's behavior, caused by the destruction or alteration of the habitat or the depletion of the food supply, are even less well known. Field evidence is difficult to obtain since only large departures from normal water quality parameters may influence changes in the seasonal and spacial distributions of organisms, which otherwise are primarily influenced by temperature, salinity and food availability. Laboratory studies provide data on the effects that specific pollutants may have in enclosed, controlled environments, but field evidence is required to substantiate any lethal or behavioral responses. For example, storm-water runoff from deforested land may contain tannic, folvic and lignic acid components that lower the pH and raise water color. Laboratory studies of blue crabs of all sizes in low salinity water, 0-2 ppt, adjusted artificially with dilute hydrochloric acid between pH 4.6 to 8.2, have shown significant, maximal avoidance to pH < 6.0.[155,156] However, field collections of blue crabs from creeks in East Bay, part of the Apalachicola Bay system, Florida, which exhibit a range of pH from 4.0 to 8.8, have shown greater numbers of juvenile crabs, < 60 mm width, and lesser numbers of larger crabs in regions where pH was < 6.0. The extremely low pH values were usually preceded by moderate to heavy rainfall. Furthermore, seasonal migrations of small blue crabs, January-February and July-August, have been shown to occur in Apalachicola Bay, Florida, when large pH fluctuations occur in East Bay.[155,156] Supportive evidence comes from other laboratory studies which have demonstrated that during upstream migration, as blue crabs acclimate to increasingly lower salinities and a more acidic medium, when salinity falls below 27 ppt they change from osmoconformity to osmoregulation.[157]

The differences between "pollution" and "inorganic or organic enrichment" are in the degree of departure from normal values and the possibility for potential harm. While temperature and salinity of the environment are not usually considered pollutants, large departures from normal can have deleterious effects on the biota. For example, a few adult female blue crabs with extruded eggs (sponge crabs) may first appear in the southern end of the Chesapeake Bay in late April, but intensive activity does not begin until mid-June, and usually ceases by early September. Hatching of crab eggs has been observed to occur in 8-11 days at 25-26°C (77-79°F),[158] at water temperatures that normally occur in the southern end of Chesapeake Bay from mid-June through mid-September, but may not be attained until mid-July and may drop below 25°C in late August, sometimes in the same year.[159] However, at 16°C (60°F) hatching occurs in approximately 45 days:[158] a mean water temperature of 16°C is normally attained in the southern end of the Chesapeake Bay in early May, but may occur anytime from late April to mid-May, and may drop below 16°C in mid-October.[159] Eggs extruded on May 1 may not hatch until June 15, while those extruded on June 15 may hatch on June 25. If mortality due to predation and disease on the eggs and embryos during the early 45-day hatching period is larger than that occurring in the later 10-day period, and if the rates of development of the vectors in the two periods are not substantially different, timing egg extrusion to the warmer temperature would have extreme survival value to the blue crab population. The rate of warming in the spring, after late April, and the time of cooling in the fall may have a marked effect on the success or failure of a year class of crabs. Cooling degree days in May in the year of the hatch, observed at Norfolk, VA, was one of three variables used in a multiple correlation analysis, in which 86% ($r^2 = 0.86$) of the variation in commercial hard crab landings one and a half years later, from 1964 through 1975 was explained.[160]

In the lore of Chesapeake Bay watermen, sub-freezing winter temperatures are associated with the deaths of adult female crabs in the southern end of the Bay. Dead females are often seen in the winter dredge catch, varying geographically and with decreasing salinity from the southern end of the Bay to the Virginia-Maryland border.[161]

Whenever below-average rainfall in late summer and fall and concomitant increases in salinity in the Bay's tributaries occur, juvenile blue crabs have been found more numerous in higher reaches of the tributaries (Van Engel unpubl.

data). In those areas, since space and food are less than occur in the normal nursery areas, competition between individuals promotes stress and thus potentially limits the success of the year class.

Although sediment is the most significant contribution from agriculture to the marine environment, particularly during storm-water runoff, the only known direct effects, reported by local watermen, are losses of soft crabs in shedding houses and a decrease in peeler crab catch following heavy rainfall; the latter is probably the result of a change in crab behavior in the wild in water containing suspended sediment. Secondary effects of sediment transport and deposition are the physical alteration of habitat and the increase in turbidity, resulting in decreased light penetration. Decreased photosynthesis is considered a factor in the decline in abundance of eelgrass (*Zostera*), although excessive nutrients, nitrogen and phosphorus, which promote fouling of the fronds, are probably the major cause.[162]

Less well known are winter mortalities occurring when frequent low pressure centers pass over the southern end of the Bay and on the adjacent ocean coast and create wind tides. Subsurface currents, enhanced in a storm in the relatively shallow Bay and coastal area, sweep crabs along the bottom, where their exoskeletons are severely abraded, resulting in uncounted deaths of crabs along the ocean shore and up to 20% of the dredge catch in the Bay.[161]

Perhaps the most insidious parameter of water quality that can affect blue crabs is a deficiency of dissolved oxygen (DO). The duration and extent of anoxic water in the Bay have accelerated since 1950, primarily in the portion of the Bay from the Patapsco River, MD, south to the vicinity of Reedville, VA, and at depths between the bottom and the halocline (8-14 m in depth). Volume of water with DO concentrations equal to or less than 0.7 mg L^{-1} (defined as "low" value) is estimated to be 15 times more in 1980 than in 1950,[162] and the duration now extends from early May into September. Reduction in DO is primarily due to decomposition of organic material, but there are seasonal differences related to temperature and salinity effects on oxygen solubility. Oxygen deficiencies or barriers may be encountered by juvenile blue crabs migrating to nursery grounds in the estuary. Migration routes, and the depth of water in which migration occurs, are not well understood. If normal migration primarily occurs on flood tide at depth, the presence of anoxic water could be a barrier or a trap, either preventing further migration or forcing crabs to move to shallower waters.[163]

Municipal wastewater discharge, atmospheric fallout and surface runoff contribute trace metals to the marine environment, but the amounts contributed vary with potential sources in each geographic site. Some of the trace metals are essential, in moderation, in nutrition and growth (e.g., copper, zinc, manganese and cobalt) and others potentially toxic (e.g., cadmium, silver, mercury, lead and chromium).[164] A substance in the hepatopancreas and in the gills of the blue crab, a metallothionein-like protein, is involved with trace metal metabolism and with metal-detoxification.[165,166] Bioconcentration of cadmium from seawater by the blue crab has been found to occur primarily at the gills, and lesser amounts occur in the hepatopancreas and hemolymph. Although cadmium has been found associated with hemocyanin in the hemolymph, the low levels of cadmium found are estimated not to affect hemocyanin-mediated oxygen transport.[167] Copper, zinc and nickel have been found in gills and hepatopancreas from environmentally-exposed blue crabs.[168]

Unsubstituted polynuclear aromatic hydrocarbons (PAHs) enter the marine environment from atmospheric fallout of vehicle exhaust, industrial smoke and wood fires. Localized concentrations of hydrocarbons are contributed by effluents from domestic and industrial treatment facilities, oil spillages and urban run-off. PAHs are toxic, and some are believed to be carcinogenic or mutagenic or both when activated through metabolism.[169] Blue crabs have been shown to take up PAHs directly from the water and from food, pass some of it directly through the intestinal tract and rapidly metabolize most of the remainder, converting lipid-soluble to water-soluble compounds.[170] Metabolism of the lipophylic compounds is believed to be carried out by mixed-function oxygenases (MFO), which are enzymes. The main site of metabolism is either in the hepatopancreas[171] or in the green gland, which is essentially an excretory gland.[172] Concentrations of these oxygenases vary with the molt stage of the blue crab, being highest during the intermolt and least during the molt.[172] The y-gland molt-promoting hormone crustecdysone, a steroid hormone, exhibits a cycle opposite that of the oxygenases. Oxygenases are also required in the synthesis and breakdown of crustecdysone. It has been proposed that in instances of exposure to high levels of PAHs, oxygenase activity could be concentrated on detoxifying the PAHs instead of the production of the ecdysone, resulting in a breakdown in the molting sequence and prolonging the intermolt state.[172,173,174] PAH in lipid reserves may be relatively inert. Benzene, an aromatic hydrocarbon, has been shown to delay limb-bud growth and limb regeneration and to cause an increase in the intermolt stage in the blue crab.[175]

Chlorination of wastewater to improve its hygienic quality, may be either insufficient to minimize water-borne disease or produce substances that have an adverse environmental impact. Several compounds have been identified with highly chlorinated wastewater, such as chloramines in freshwater and bromamines in seawater, and their oxidation products. Few studies have been conducted of the effects of any of these substances on marine crustaceans of commercial value. In laboratory studies of acute-toxicity, adult blue crabs have been found more tolerant of chlorine-produced oxidants than most other marine species tested.[176] Respiration rate, ventilation rate and most blood serum constituents were unaffected by the levels of oxidants tested.[176]

Halogenated substances in the marine environment, such as pesticides, solvents and other industrial compounds, are contributed from industrial and domestic wastewaters, including runoff, but some may have been derived from the chlorination of wastewater. Among the toxicants are Kepone, chlorinated phenols, polychlorinated biphenyls, dieldrin, mirex, heptachlor, methoxychlor and DDT and its derivatives. A blue crab primarily accumulates most of its body burden of pesticides through its dietary habit as an opportunistic scavenger. Laboratory studies of blue crabs, involving the bioconcentration of DDT from water treated with low levels of 0.25 and 0.50 ppb to higher levels of 0.01 to 1.0 ppm, exposed from nine months to 12 hours, respectively, have shown no differences in mortality rates between control and treated crabs.[177,178] In one study, uptake appeared to occur primarily at the gills, but DDT and its metabolites were found sequestered in the gills, gonads, claw and backfin muscles, the heart and the hepatopancreas, principally in the latter organ.[177] The number of molts of control crabs and those exposed to low levels of DDT was not found to be significantly different, but crabs exposed to 0.50 ppb DDT experienced fewer molts.[178] In field studies, moribund crabs contaminated with DDT exhibit lack of equilibrium and tremors, and large scale mortality, up to 90% of the population, eventually ensues (179,180,181, Van Engel unpubl. data).

Relatively small amounts of Kepone and low levels of toxicity have been found to occur in juvenile blue crabs that have bioconcentrated Kepone from water.[182] However, crabs that accumulated Kepone through feeding underwent fewer molts and suffered over 80% mortality, and depuration of Kepone was found to be relatively slow.[183] In another study, no statistical differences were found in mortality and molting rates between crabs provided food either uncontaminated or contaminated with Kepone.[184] Concentrations of Kepone have been found to differ in male and female blue crabs, in amount and site of concentration: male crabs concentrated more Kepone in the backfin muscle, and adult females had more Kepone in the ovary and secondarily in the hepatopancreas.[185] Depuration of Kepone in the female blue crab, through the extrusion of Kepone-laden eggs, has been suggested.[186] No effects of contamination in the eggs on hatching or survival have been found.[187] The potential impact of Kepone contamination on the wild blue crab stock has been estimated to be minimal, based on the estimate that the natural diet of crabs contains lower levels of Kepone than used in laboratory studies.[188]

The sorption of some pesticides by the chitin of crustacean exoskeletons has been documented, but no statements have been made of the potential impact on the host animal. Dicamba, 2, 4-D and DDT have been found sorbed by chitin, but not atrazine, propanil or paraquat.[189]

REFERENCES

1. Adams, P.B. 1980. Life history patterns in marine fishes and their consequences for fishery management. *Fish. Bull.* 78(1):1-12.
2. Sette, O.E. and R.H. Fiedler. 1925. A survey of the condition of the crab fisheries of Chesapeake Bay, a preliminary report. *U.S. Bur. Fish. Spec. Mem.* 1607-14, 36 pp.

3. Goy, J.W. 1976. Seasonal distribution and retention of some decapod crustacean larvae within the Chesapeake Bay, Virginia. MA thesis, Old Dominion Univ., Norfolk, VA. 334 pp.
4. Grant, G.C. and J.E. Olney. 1983. Lower Bay zooplankton monitoring program: the August 1978 survey. Virginia Inst. Mar. Sci., School Mar. Sci., Coll. William and Mary, Spec. Sci. Rept. 115, Gloucester Pt. 81 pp.
5. Provenzano, A.J., J.R. McConaugha, K.B. Phillips, D.F. Johnson and J. Clark. 1983. Vertical distribution of the first stage larvae of the blue crab, *Callinectes sapidus,* at the mouth of Chesapeake Bay.. *Estuar. Coastal Shelf Sci.* 16:489-499.
6. McConaugha, J.R., D.F. Johnson, A.J. Provenzano, and R.C. Maris. 1983. Seasonal distribution of larvae of *Callinectes sapidus (Crustacea: Decapoda)* in the waters adjacent to Chesapeake Bay. *Jour. Crust. Biol.* 3(4):582-591.
7. McConaugha, J.R., A.J. Provenzano, D.F. Johnson, J. Clark and P. Sadler. 1981. Offshore displacement and reinvasion of *Callinectes sapidus* larvae in Chesapeake Bay. Abstract *(Estuaries* 4(3):477).
8. Costlow, J.D., Jr. 1967. The effect of salinity and temperature on survival and metamorphosis of megalops of the blue crab, *Callinectes sapidus. Helgolander wiss. Meres.* 15:84-97.
9. U.S. Department of Commerce. 1986. Fisheries of the United States, 1985. U.S. Dept. Comm., NOAA/NMFS Current Fish. Stat. 8368, Washington, D.C.
10. Van Engel, W.A. and R.E. Harris, Jr. 1980. Climatology and blue crab abundance in the Chesapeake Bay. Abstract (Proc. Fourth Ann. Meet., Potomac Chapt. Amer. Fish. Soc.). p. 112.
11. Couch, J.A. and S. Martin. 1982. Protozoan symbionts and related diseases of the blue crab, *Callinectes sapidus* Rathbun from the Atlantic and Gulf coasts of the United States, pp. 71-80. *In:* H.M. Perry and W.A. Van Engel, (Eds.) *Proc. Blue Crab Colloquium, Oct. 18-19, 1979.* Biloxi, MS.
12. Couch, J.A. and H. Tubiash. 1967. A report on the preliminary investigations of blue crab mortalities in South Carolina. U.S. Bur. Comm. Fish. Interlab. Rept. Oxford, MD. 3 pp.
13. Colwell, R.R., T. Kaneko, and T. Staley. 1972. *Vibrio parahaemolyticus* — an estuarine bacterium resident in Chesapeake Bay. Abstracts of papers presented at the Third Food-Drugs from the Sea Conference, Univ. Rhode Island, Kingston. 9 pp.
14. Colwell, R.R., J. Kaper and S.W. Joseph. 1977. *Vibrio cholerae, Vibrio parahaemolyticus,* and other vibrios: occurrence and distribution in Chesapeake Bay. *Science* 198:394-396.
15. Colwell, R.R., T.C. Wicks and H.S. Tubiash. 1975. A comparative study of the bacterial flora of the haemolymph of *Callinectes sapidus. Mar.*

Fish. Rev. 37(5-6):29-33.
16. Davis, J.W. and R.K. Sizemore. 1982. Incidence of *Vibrio parahaemolyticus* associated with blue crabs *(Callinectes sapidus)* collected from Galveston Bay, Texas. *Appl. Environ. Microbiol.* 43(5):1092-1097.
17. Elliot, E.L. 1984. A microbiological study of Chesapeake Bay blue crabs. PhD. Dissertation, Univ. Maryland, 186 pp.
18. Johnson, P.T. 1976c. Bacterial infection in the blue crab, *Callinectes sapidus:* course of infection and histopathology. *Jour. Invert. Pathol.* 28(1):25-36.
19. Johnson, P.T. 1983. Diseases caused by viruses, rickettsiae, bacteria, and fungi, pp. 1-78. *In:* A.J. Provenzano (Ed.) *The biology of Crustacea, Vol. 6, Pathobiology.* Academic Press, New York. 290 pp.
20. Kaneko, T., and R.R. Colwell. 1973. Ecology and *Vibrio parahaemolyticus* in Chesapeake Bay. *Jour. Bacteriol.* 113:24-32.
21. Sizemore, R.K., R.R. Colwell, H.S. Tubiash, and T.E. Lovelace. 1975. Bacterial flora of the hemolymph of the blue crab, *Callinectes sapidus:* numerical taxonomy. *Appl. Microbiol.* 29(3):393-399.
22. Sizemore, R.K. and J.W. Davis. 1982. Source of *Vibrio* spp. found in blue crab hemolymph. Abstract (Ann. Meet. Amer. Soc. Microbiol. 1982). p. 195.
23. Tubiash, H.S., R.K. Sizemore and R.R. Colwell. 1975. Bacterial flora of the hemolymph of the blue crab, *Callinectes sapidus:* most probable numbers. *Appl. Microbiol.* 29(3): 388-392.
24. Welsh, P.C. and R.K. Sizemore. 1985. Incidence of bacteremia in stressed and unstressed populations of the blue crab, *Callinectes sapidus. Appl. Environ. Microbiol.* 50(2):420-425.
25. Williams-Walls, N.J. 1968. *Clostridium botulinum* Type F: isolation from crabs. *Science* 162:375-376.
26. Fishbein, M., I.J. Mehlman, and J. Pitcher. 1970. Isolation of *Vibrio parahaemolyticus* from the processed meat of Chesapeake Bay blue crabs. *Appl. Microbiol.* 20:176-178.
27. Johnson, P.T. 1976d. Gas-bubble disease in the blue crab, *Callinectes sapidus. Jour. Invert. Pathol.* 27:247-253.
28. Krantz, G.E., R.R. Colwell and E. Lovelace. 1969. *Vibrio parahaemolyticus* from the blue crab. *Science.* 164:186-187.
29. Molenda, J.R., W.G. Johnson, M. Fishbein, B. Wentz, I.J. Mehlman, and T.A. Dadisman, Jr. 1972. *Vibrio parahaemolyticus* gastroenteritis in Maryland: laboratory aspects. *Appl. Microbiol.* 24:444-448.
30. Moody, M.W. 1982. Zoonotic diseases, pp. 65-69. *In:* H.M. Perry and W.A. Van Engel, (Eds.). *Proc. Blue Crab Colloquium, Oct. 18-19, 1979.* Biloxi, MS.
31. Peffers, A.S.R., J. Bailey, G.I. Barrow, and B.C. Hobbs. 1973. *Vibrio*

parahaemolyticus gastroenteritis and international air travel. *Lancet* i: 143-145.
32. Overstreet, R. 1978. Marine maladies? Worms, germs, and other symbionts from the northern Gulf of Mexico. Mississippi-Alabama Sea Grant Consortium, MASGP-78-021, Ocean Springs, MS. 140 pp.
33. Sumner, W.A., S.J. Moore, M.A. Bush, B. Nelson, J.R. Molenda, W. Johnson, H.J. Garber, and B. Wentz. *Vibrio parahaemolyticus* gastroenteritis—Maryland. *Morbid. Mortal. Week. Rept.* 20:356.
34. Overstreet, R. 1982. Metazoan symbionts of the blue crab, pp. 81-87. *In:* H.M. Perry and W.A. Van Engel, (Eds.). *Proc. Blue Crab Colloquium, Oct. 18-19, 1979.* Biloxi, MS.
35. Babinchak, J.A., D. Goldmintz, and G. Richards. 1982. A comparative study of autochthonous bacterial flora on the gills of the blue crab, *Callinectes sapidus,* and its environment. *Fish. Bull.* 80(4):884-890.
36. Pearse, A.S. 1947. On the occurrence of ectoconsortes on marine animals at Beaufort, N.C. *Jour. Parasitol.* 33(6):453-458.
37. Overstreet, R. 1983. Metazoan symbionts of crustaceans, pp. 155-250. *In:* A.J. Provenzano (Ed.). *The biology of Crustacea, Vol. 6, Pathobiology.* Academic Press, New York, 290 pp.
38. Cargo, D.G. 1959. Mussels muscling in. *Maryland Tidewater News* 15(2):7.
39. Osburn. R.C. 1944. A survey of the Bryozoa of Chesapeake Bay. *Chesapeake Biol. Lab. Publ.* 63:1-59.
40. Hejkal, T.W. and C.P. Gerba. 1981. Uptake and survival of enteric viruses in the blue crab, *Callinectes sapidus. Appl. Environ. Microbiol.* 41(1):207-211.
41. Johnson, P.T. 1976a. A baculovirus from the blue crab, *Callinectes sapidus. Proc. 1st Int. Colloq. Invert. Pathol.,* p. 24.
42. Johnson, P.T. 1978. Viral diseases of the blue crab, *Callinectes sapidus. Mar. Fish. Rev.* 40(10):13-15.
43. Johnson, P.T. 1985. Blue crab (*Callinectes sapidus* Rathbun) viruses and the diseases they cause, pp. 13-19. *In:* H.M. Perry and R.H. Malone (Eds.) *Proceedings of the national symposium on the soft-shelled blue crab fishery.* Gulf Coast Res. Lab., Biloxi, MS.
45. Johnson, P.T. and C.A. Farley. 1980. A new enveloped helical virus from the blue crab, *Callinectes sapidus. Jour. Invert. Pathol.* 35:90-92.
46. Johnson, P.T. 1976b. A herpeslike virus from the blue crab, *Callinectes sapidus. Jour. Invert. Pathol.* 27:419-420.
47. Johnson, P.T. 1977a. A viral disease of the blue crab, *Callinectes sapidus:* histopathology and differential diagnosis. *Jour. Invert. Pathol.* 29:201-209.
48. Johnson, P.T. 1984. Viral diseases of marine invertebrates. *Helgolander wiss. Meeres.* 37:65-98.
49. Johnson, P.T. and J.E. Bodammer. 1975. A disease of the blue crab, *Callinectes sapidus,* of possible viral etiology. *Jour. Invert. Pathol.*

26(1):141-143.
50. Johnson, P.T. 1977c. Reovirus-like virus of blue crabs, pp. 103-105. *In:* C.J. Sindermann (Ed.) *Disease diagnosis and control in North American marine aquaculture.* Elsevier Sci. Publ. Co., New York.
51. Jahromi, S.S. 1977. Occurrence of rhabdovirus-like particles in the blue crab, *Callinectes sapidus. Jour. Gen. Virol.* 36:485-494.
52. Yudin, A.I. and W.H. Clark, Jr. 1978. Two virus-like particles found in the ecdysial gland of the blue crab, *Callinectes sapidus. Jour. Invert. Pathol.* 32:219-221.
53. Yudin, A.I. and W.H. Clark, Jr. 1979. A description of rhabdovirus-like particles in the mandibular gland of the blue crab, *Callinectes sapidus. Jour. Invert. Pathol.* 33:133-147.
54. Rosen, B. 1967. Shell disease of the blue crab, *Callinectes sapidus. Jour. Invert. Pathol.* 9:348-353.
55. Hood, M.A. and S.P. Meyers. 1973. The biology of aquatic chitinoclastic bacteria and their chitinolytic activities. *La mer (Bull. Soc. franco-japonaise d'oceanogr.)* 11(4):213-229.
56. Overstreet, R.M. and H.D. House. 1973. Some parasites and diseases of estuarine fishes in polluted habitats of Mississippi, pp. 427-462. *In:* C.S. Dawe, J.C. Harshbarger and R.G. Tardiff (Eds.) *Aquatic pollutants and biologic effects with emphasis on Neoplasia.* Ann. New York Acad. Sci. 298.
57. Rosen, B. 1970. Shell disease of aquatic crustaceans, pp. 409-415. *In:* S.F. Snieszko (Ed.) *A symposium of the American Fisheries Society on diseases of fishes and shellfishes. Amer. Fish. Soc. Spec. Publ.* 5. 526 pp.
58. Sandifer, P.A. and P.J. Eldridge. 1974. Observations on the incidence of shell disease in South Carolina blue crabs, *Callinectes sapidus,* pp. 161-184. *In:* R.L. Amborski, M.A. Hood and R.R. Miller (Eds.). *1974 Proc. Gulf Coast Regional Symposium of Aquatic Animals.* LSU-SG-74-05. Baton Rouge: Center for Wetland Resources, Louisiana State Univ.
59. Cook, D.W. and S.R. Lofton. 1973. Chitinoclastic bacteria associated with shell disease in *Penaeus* shrimp and the blue crab *(Callinectes sapidus). Jour. Wildl. Diseases* 9(2):154-159.
60. Kautter, D.A., T. Lilly, Jr., A.J. LeBlanc and R.K. Lynt. 1974. Incidence of *Clostridium botulinum* in crabmeat from the blue crab. *Appl. Microbiol.* 28(4):722.
61. Johnson, P.T. 1976e. An unusual microorganism from the blue crab. *Proc. 1st Int. Colloq. Invert. Pathol.,* Kingston, Ontario, p. 316.
62. Sizemore, R.K. 1985. Involvement of *Vibrio* spp. in soft crab mortality, pp. 21-22. *In:* H.M. Perry and R.F. Malone (Eds.) *Proceedings of the national symposium on the soft-shelled blue crab fishery.* Gulf Coast Res. Lab., Biloxi, Ms.

63. Bland, C.E. 1975. Fungal diseases of marine Crustacea. *Proc. Third U.S.-Japan meeting on Aquaculture at Tokyo, Japan October 15-16, 1974.* Spec. Publ. Fish. Agency, Japanese Govt. and Japan Sea Regional Fish. Res. Lab., Niigata, Japan: 41-48.
64. Bland, C.E. and H.V. Amerson. 1973. Observations on *Lagenidium callinectes:* isolation and sporangial development. *Mycologia* 65(2):310-320.
65. Bland, C.E. and H.V. Amerson. 1974. Occurrence and distribution in North Carolina waters of *Lagenidium callinectes* Couch, a fungal parasite of blue crabs. *Chesapeake Sci.* 15(4):232-235.
66. Couch, J.N. 1942. A new fungus on crab eggs. *Jour. Elisha Mitchell Sci. Soc.* 58(2):158-162.
67. Rogers, M.R. 1945. The occurrence and distribution of the fungus *Lagenidium callinectes* Couch, on the eggs of the blue crab, *Callinectes sapidus* Rathbun. MA thesis, Coll. William and Mary, Williamsburg, 31 pp.
68. Rogers-Talbert, R. 1948. The fungus *Lagenidium callinectes* Couch (1942) on eggs of the blue crab in Chesapeake Bay. *Biol. Bull.* 95(2):214-228.
69. Sandoz, M., R. Rogers, and C.L. Newcombe. 1944. Fungus infection of the eggs of the blue crab, *Callinectes sapidus* Rathbun. *Science.* 99:124-125.
70. Scott, W.W. 1962. The aquatic phycomycetous flora of marine and brackish waters in the vicinity of Gloucester Point, Virginia. Virginia Inst. Mar. Sci. Rept. 36. 12 pp.
71. Newman, M.W. and C.A. Johnson. 1975. A disease of blue crabs (*Callinectes sapidus*) caused by a parasitic dinoflagellate, *Hematodinium* sp. *Jour. Parasitol.* 63:554-557.
72. Couch, J.A. 1983. Diseases caused by Protozoa, pp. 79-111. *In:* A.J. Provenzano (Ed.) *The biology of Crustacea, Vol. 6, Pathobiology.* Academic Press, New York. 290 pp.
73. Johnson, P.T. 1977b. Paramoebiasis in the blue crab *Callinectes sapidus. Jour. Invert. Pathol.* 29(3):308-320.
74. Newman, M.W. and G.W. Ward, Jr. 1973. An epizootic of blue crabs, (*Callinectes sapidus*), caused by *Paramoeba perniciosa. Jour. Invert. Pathol.* 22:329-334.
75. Sawyer, T.K. 1969. Preliminary study on the epizootiology and host-parasite relationship of *Paramoeba* sp. in the blue crab, *Callinectes sapidus. Proc. Nat. Shellfish. Assoc.* 59:60-64.
76. Sawyer, T.K., R. Cox and M. Higginbottom. 1970. Hemocyte values in healthy blue crabs, *Callinectes sapidus,* and crabs infected with the amoeba, *Paramoeba perniciosa. Jour. Invert. Pathol.* 15:440-446.
77. Sawyer, T.K. and S.A. MacLean. 1978. Some protozoan diseases of decapod crustaceans. *Mar. Fish. Rev.* 4(10):32-35.
78. Sprague, V. 1970. Some protozoan parasites and hyperparasites in marine

decapod Crustacea, pp. 416-430. *In:* S.F. Sniezsko, (Ed.), *A symposium on diseases of fishes and shellfishes. Amer. Fish. Soc., Spec. Publ.* 5, Washington.

79. Sprague, V. and R.L Beckett. 1966. A disease of blue crabs (*Callinectes sapidus*) in Maryland and Virginia. *Jour. Invert. Pathol.* 8(2):287-289.

80. Sprague, V. and R.L. Beckett. 1968. The nature of the etiological agent of "gray crab" disease. *Jour. Invert. Pathol.* 11(3):503.

81. Sprague, V., R.L. Beckett and T.K. Sawyer. 1969. A new species of *Paramoeba* (Amoebida, Paramoebidae*)* parasitic in the crab *Callinectes sapidus. Jour. Invert. Pathol.* 14:167-174.

82. Sprague, V. 1965. *Nosema* sp. (Microsporida, Nosematidae) in the musculature of the crab *Callinectes sapidus. Jour. Protozool.* 12(1):66-70.

83. Sprague, V. 1977. Systematics of the Microsporidia. *In:* L. A. Bulla, Jr. and T.C. Cheng (Eds.). *Comparative pathobiology,* Vol. 2. 510 pp.

84. Sprague, V. and J. Couch. 1971. An annotated list of protozoan parasites, hyperparasites, and commensals of decapod Crustacea. *Jour. Protozool.,* 18(3):526-537.

85. Sprague, V., S.H. Vernick and B.J. Lloyd, Jr. 1968. The fine structure of *Nosema* sp. Sprague 1965 (Microsporida, Nosematidae) with particular reference to stages in sporogony. *Jour. Invert. Pathol.* 12(1):105-117.

86. Weidner, F. 1970. Ultrastructural study of microsporidian development. I. *Nosema* sp. Sprague, 1965 in *Callinectes sapidus* Rathbun. *Zeitschrift fur Zellforschund und Mikroskopische Anatomie* 105(1):33-54.

87. Anonymous [De Turk, W.E.]. 1940. The occurrence and development of a hyperparasite, *Urosporidium crescens* (Sporozoa, Haplosporidia) which infests the metacercaria of *Spelotrema nicolli,* parasitic in *Callinectes sapidus. Jour. Tenn. Acad. Sci.* 15:418-419 (Abstract).

88. DeTurk, W.E. 1940a. The occurrence and development of a hyperparasite, *Urosporidium crescens* (Sporozoa, Haplosporidia) which infests the metacercaria of *Spelotrema nicolli,* parasitic in *Callinectes sapidus. J. Elisha Mitchell Sci. Soc.* 56:231-232.

89. DeTurk, W.E. 1940b. The parasites and commensals of some crabs of Beaufort, North Carolina. Thesis, Duke Univ.

90. Sprague, V. 1966. Two new species of *Pleistophora (Microsporida, Nosematidae)* in decapods with particular reference to one in the blue crab. *Jour. Protozool.* 13(2):196-199.

91. Johnson, C.A., III. 1977. A preliminary report on diseases of North Carolina coastal crabs with emphasis on the blue crab, *Callinectes sapidus. ASB (Assoc. Southeast. Biol.) Bull.* 19:77

92. Sprague, V. 1978. Comments on trends in research on parasitic diseases of shellfish and fish. Mar. Fish. Rev. 40(10):26-30.

93. Newman, M.W., C.A. Johnson, III and G.B. Pauley. 1976. A *Minchinia*-like haplosporidan parasitizing blue crabs, *Callinectes sapidus. Jour.*

Invert. Pathol. 27:311-315.
94. Couch, J.A. 1974. Pathological effects of *Urosporiaium* (Haplosporida) infection in microphallid metacercariae. *Jour. Invert. Pathol.* 23(3): 309-396.
95. Couch, J.A. 1966. Two peritrichous ciliates from the gills of the blue crab. *Chesapeake Sci.* 7:171-173.
96. Couch, J.A. 1967. A new species of *Lagenophrys* (Ciliatea: Peritrichida: Lagenophryidae) from a marine crab, *Callinectes sapidus*. *Trans. Amer. Microsc. Soc.* 86(2):204-211.
97. Couch, J.A. 1971. Form, morphogenesis, and host-ciliate relationship of *Lagenophyrs callinectes* (Ciliatea: Peritrichida. PhD dissertation. Florida Sta. Univ. 153 pp.
98. Heard, R.W., III. 1976. Microphallid trematode metacercariae in fiddler crabs of the genus *Uca* Leach, 1814 from the northern Gulf of Mexico. PhD dissertation, Univ. Southern Mississippi, Hattiesburg. 179 pp.
99. Overstreet, R.M. and H.M. Perry. 1972. A new microphallid trematode from the blue crab in the northern Gulf of Mexico. *Trans. Amer. Microscop. Soc.* 91(3):436-440.
100. Bridgeman, J.F. 1969. Life cycles of *Carneophallus choanophallus* n. sp. and *C. basodactylophallus* n. sp. (Trematoda: Microphallidae). *Tulane Studies in Zoology and Botany* 15(3):81-105.
101. Cable, R.M. and A.V. Hunninen. 1940. Studies on the life history of *Spelotrema nicolli* (Trematoda: Microphallidae) with the description of a new microphallid cercaria. *Biol. Bull.* 78:136-157.
102. Hutton, R.F. 1964. A second list of parasites from marine and coastal animals of Florida. *Trans. Amer. Microsc. Soc.* 83:439-447.
103. Causey, D. 1959. Lagniappe. *Turtox News* 37(1):10-11.
104. Davis, C.C. 1965. A study of the hatching process in aquatic invertebrates: XX. The blue crab, *Callinectes sapidus,* Rathbun. XXI. The nemertean, *Carcinonemertes carcinophila (*Kolliker). *Chesapeake Sci.* 6:201-208.
105. Hopkins, S.H. 1947. The nemertean *Carcinonemertes* as an indicator of the spawning history of the host, *Callinectes sapidus*. *Jour. Parasitol.* 33:146-150.
106. Humes, A.G. 1942. The morphology, taxonomy and bionomics of the nemertean genus *Carcinonemertes*. *Illinois Biol. Monogr.* 18(4):1-105.
107. Daniels, B.A. and R.T. Sawyer. 1975. The biology of the leech *Myzobdella lugubris* infesting blue crabs and catfish. *Biol. Bull.* 148(2):193-198.
108. Hutton, R.F. and F. Sogandares-Bernal. 1959. Notes on the distribution of the leech, *Myzobdella lugubris* Leidy, and its association with mortality of the blue crab, *Callinectes sapidus* Rathbun. *Jour. Parasitol.* 45:384, 404, 430.
109. Sawyer, R.T., A.R. Lawler and R.M. Overstreet. 1975. Marine leeches of the eastern United States and the Gulf of Mexico with a key to the species.

Jour. Nat. Hist. 9:633-667.
110. Newman, W.A. 1967. Shallow-water versus deep-sea *Octolasmis* (Cirripedia Thoracica). *Crustaceana.* 12:13-32.
111. Pilsbry, H.A. 1916. The sessile barnacles (Cirripedia) contained in the collections of the U.S. National Museum; including a monograph of the American species. *U.S. Nat. Mus. Bull.* 93:1-366, 76 pp.
112. Williams, A.B. and H.J. Porter. 1964. An unusually large turtle barnacle *(Chelonibia p. patula)* on a blue crab from Delaware Bay. *Chesapeake Sci.* 5:150-151.
113. Coker, R.E. 1902. Notes on a species of barnacle (*Dichelaspis*) parasitic on the gills of edible crabs. *U.S. Fish Comm. Bull.* 1901:399-412.
114. Humes, A.G. 1941. Notes on *Octolasmis mulleri* (Coker), a barnacle commensal on crabs. *Trans. Amer. Microscop. Soc.* 60(1):101-103.
115. Jeffries, W.B. and H.K. Voris. 1983. The distribution, age, and reproduction of the pedunculate barnacle, *Octolasmis mulleri* (Coker, 1902) on the blue crab, *Callinectes sapidus* (Rathbun, 1896). *Fieldiana Zool.,* New Ser. 16: v + 1-10.
116. Pilsbry, H.A. 1907. The barnacles (Cirripedia) contained in the collections of the U.S. National Museum. *U.S. Nat. Mus. Bull.* 60:1-121, pl. I-XI.
117. Walker, G. 1974. The occurrence, distribution and attachment of the pedunculate barnacle *Octolasmis mulleri* (Coker) on the gills of crabs, particularly the blue crab, *Callinectes sapidus* Rathbun. *Biol. Bull.* 147:678-689.
118. Christmas, J.Y. 1969. Parasitic barnacles in Mississippi estuaries with special references to *Loxothylacus texanus* Boschma in the blue crab *(Callinectes sapidus).* Proc. Twenty-Second Ann. Conf. Southeast. Assoc. Game Fish Comm., pp. 272-275.
119. Ragan, J.G. and B.A. Matherne. 1974. Studies of *Loxothylacus texanus,* pp. 185-203. *In:* R.L. Amborski, M.A. Hood and R.R. Miller (Eds.). *1974 Proc. Gulf Coast Regional Symposium of Aquatic Animals.* LSU-SG-74-05. Baton Rouge: Center for Wetland Resources, Louisiana State Univ.
120. Reinhard, E.G. 1950. An analysis of the effects of a sacculinid parasite on the external morphology of *Callinectes sapidus* Rathbun. *Biol. Bull.* 98:277-288.
121. Kneib, R.T. 1986. The role of *Fundulus heteroclitus* in salt marsh trophic dynamics. *Amer. Zool.* 26(1):259-269.
122. Sheridan, P.F. 1979. Trophic resource utilization by three species of sciaenid fishes in a northeast Florida estuary. *Northeast Gulf Sci.* 3:1-14.
123. Smith, S.M., J.G. Hoff, S.P. O'Neil, and M.P. Weinstein. 1984. Community and trophic organization of nekton utilizing shallow marsh habitats, York River, Virginia. *Fish. Bull.* 82(3):455-467.
124. Stickney, R.R., G.T. Taylor and D.B. White. 1975. Food habits of five species

of young southeastern United States estuarine Sciaenidae. *Chesapeake Sci.* 16(2):104-114.
125. Stoner, A.W. 1980. Feeding ecology of *Lagodon rhomboides* (Pisces: Sparidae): Variation and functional responses. *Fish. Bull.* 78(2): 337-352.
126. Welsh, W.W. and C.M. Breder, Jr. 1923. Contributions to life histories of Sciaenidae of the eastern United States Coast. *Bull. U.S. Bur. Fish.* 39:141-201.
127. Van Engel, W.A. and E.B. Joseph. 1968. Characterization of coastal and estuarine fish nursery grounds as natural communities. Virginia Inst. Mar. Sci. Final Rept. Nov. 1965 - Aug. 1967. Contract 14-17-007-531, Comm. Fish. Res. Develop. Act. Mimeo. 43 pp., 58 tables, 56 figs.
128. Darnell, R.M. 1958. Food habits of fishes and larger invertebrates of Lake Pontchartrain, Louisiana, an estuarine community. *Inst. Mar. Sci.* 5:355-416.
129. Hildebrand, S.F. and W.C. Schroeder. 1928. Fishes of Chesapeake Bay. *Bull. U.S. Bur. Fish.* 43(1927) - Part 1. 366 pp.
130. Schwartz, F.J. and B.W. Dutcher. 1963. Age, growth, and food of the oyster toad near Solomons, Maryland. *Trans. Amer. Fish. Soc.* 92(2):170-173.
131. Chrobot, R.J. 1951. The feeding habits of the toadfish (*Opsanus tau*) based on an analysis of the contents of the stomach and intestine. MA thesis, Univ. Maryland, College Park. 35 pp.
132. Gudger, E.W. 1910. Habits and life history of the toadfish (*Opsanus tau*). *Bull. U.S. Bur. Fish.* 28(2):1071-1109.
133. Brooks, H.A., J.V. Merriner, C.E. Meyers, J.E. Olney, G.W. Boehlert, J.V. Lascara, A.D. Estes, T.A. Munroe. 1982. Higher level consumer interactions, Chapter 4, pp. 1-196. *In: Final report. Structural and functional aspects of the biology of submerged aquatic macrophyte communities in the lower Chesapeake Bay.* Virginia Inst. Mar. Sci. School Mar. Sci., Coll. William and Mary, Special Rept. 267, Appl. Mar. Sci. Ocean Engineer.
134. Lascara, J.V. 1981. Fish predator-prey interactions in areas of eelgrass *(Zostera marina)*. MA. thesis, Coll. William and Mary, Williamsburg, 81 pp.
135. Overstreet, R.M. and R.W. Heard. 1982. Food contents of six commercial fishes from Mississippi Sound. *Gulf Res. Repts.* 7(2):137-149.
136. Thomas, D.L. 1971. The early life history and ecology of six species of drum (Sciaenidae) in the lower Delaware River, a brackish tidal estuary. *Ichthyol. Assoc. Bull.* 3. 247 pp.
137. Merriner, J.V. 1975. Food habits of the weakfish, *Cynoscion regalis,* in North Carolina waters. *Chesapeake Sci.* 16(1):74-76.
138. Overstreet, R.M. and R.W. Heard. 1978b. Food of the Atlantic croaker, *Micropogonias undulatus,* from Mississippi Sound and the Gulf of Mexico. *Gulf Res. Repts.* 6(2):145-152.

139. Stickney, R.R., G.T. Taylor and R.W. Heard, III. 1974. Food habits of Georgia estuarine fishes. I. Four species of flounders (Pleuronectiformes: Bothidae). *Fish. Bull.* 72(2):515-525.
140. Medved. R.J. and J.A. Marshall. 1981. Feeding behavior and biology of young sandbar sharks, *Carcharhinus plumbeus* (Pisces, Carcharhinidae), in Chincoteague Bay, Virginia. *Fish. Bull.* 79(3):441-447.
141. Overstreet, R.M. and R.W. Heard. 1978a. Food of the red drum, *Sciaenops ocellata,* from Mississippi Sound. *Gulf Res. Repts.* 6(2): 131-135.
142. Lawler, E.F., Jr. 1976. The biology of the sandbar shark *Carcharhinus plumbeus* (Nardo, 1827) in the lower Chesapeake Bay and adjacent waters. MA thesis, Coll. William and Mary, Williamsburg. 49 pp.
143. Reid, G.K. Jr. 1954. An ecological study of the Gulf of Mexico fishes, in the vicinity of Cedar Key, Florida. *Bull. Mar. Sci. Gulf Caribb.* 4:1-94.
144. Chao, L.N. 1976. Aspects of systematics, morphology, life history and feeding of western Atlantic Sciaenidae (Pisces: Perciformes). PhD dissertation, Coll. William and Mary, Williamsburg. 342 pp.
145. Powell, A.B. and F.J. Schwartz. 1979. Food of *Paralichthys dentatus* and *P. lethostigma* (Pisces: Bothidae) in North Carolina estuaries. *Estuaries* 2(4):276-279.
146. Laughlin, R.A. 1979. Trophic ecological and population distribution of the blue crab, *Callinectes sapidus* Rathbun, in the Apalachicola estuary (north Florida, U.S.A.) Ph.D. dissertation, Florida Sta. Univ., Tallahassee. 143 pp.
147. Hay, W.P. 1905. The life history of the blue crab, *Callinectes sapidus. U.S. Bur. Fish. Rept. 1904.* pp. 395-414 + 4 Pl.
148. Tagatz, M.E. 1969. Some relations of temperature acclimation and salinity to thermal tolerance of the blue crab, *Callinectes sapidus. Trans. Amer. Fish. Soc.* 98(4):713-716.
149. Lutcavage, M.E. 1981. The status of marine turtles in Chesapeake Bay and Virginia coastal waters. MA thesis, Coll. William and Mary, Williamsburg, 127 pp.
150. Musick, J.A., R. Byles, R. Klinger and S. Bellmund. 1984. Mortality and behavior of sea turtles in the Chesapeake Bay. Summary report for 1979 through 1983, Contract NA80FAC00004. Submitted to the National Marine Fisheries Service Northeast Region. 56 pp. + Appendices A-C.
151. Lutcavage, M. and J.A. Musick. 1985. Aspects of the biology of sea turtles in Virginia. *Copeia* 1985 (2):449-456.
152. Wilson, K.A. 1955. The role of mink and otter as muscrat predators in northeastern North Carolina. *Jour. Wildl. Mgmt.* 18(2):199-207.
153. Wilson, K.A. 1959. The otter in North Carolina. *Proc. Southeast Assoc. Fish Game Comm.* 13:267-277.
154. Chabreck, R.H., J.E. Holcombe, R.G. Linscombe and N.E. Kinler. 1982. Winter foods of river otters from saline and fresh environments in Loui-

siana. *Proc. Ann. Conf. Southeast Assoc. Fish and Wildl. Agencies* 36:473-484.
155. Laughlin, R.A., C.R. Cripe and R.J. Livingston. 1978. Field and laboratory avoidance reactions by blue crabs *(Callinectes sapidus)* to storm water runoff. *Trans. Amer. Fish. Soc.* 107(1):78-86.
156. Livingston, R.J., C.R. Cripe, R.A. Laughlin and F.G. Lewis, III. 1976. Avoidance responses of estuarine organisms to storm water runoff and pulp mill effluents, pp. 313-331. *In:* Martin Wiley (Ed.) *Estuarine processes. Vol. I. Uses, stresses, and adaptations to the estuary.* Academic Press, New York.
157. Mangum, C.P. 1976. The function of respiratory pigments in estuarine animals, pp. 356-380. *In:* M. Wiley (Ed.) *Estuarine processes. Vol. 1. Uses, stresses and adaptation to the estuary.* 541 pp.
158. Amsler, M. O'L. and R.Y. George. 1984. The effects of temperature on the oxygen consumption and developmental rate of the embryos of *Callinectes sapidus* Rathbun. *Jour. Exper. Mar. Biol. Ecol.* 82:221-229.
159. Hseih, B.B. 1979. Variation and prediction of water temperature in York River Estuary at Gloucester Point, Virginia. MA. thesis, Coll. William and Mary, Williamsburg, 171 pp.
160. Van Engel, W.A. and R.E. Harris, Jr. 1983. The blue crab fisheries of Virginia and Maryland. A preliminary review of landings and fishing effort, 1929-1981. Virginia Inst. Mar. Sci. Res. Rept. 83-9. Unnumbered.
161. Van Engel, W.A. 1982. Blue crab mortalities associated with pesticides, herbicides, temperature, salinity, and dissolved oxygen, pp. 81-92. *In:* H.M. Perry and W.A. Van Engel, (Eds.). *Proc. Blue Crab Colloquium, Oct. 18-19, 1979.* Biloxi, MS.
162. U.S. Environmental Protection Agency. 1983. Chesapeake Bay: a profile of environmental change. Technical coordinators D.A. Flemer, G.B. Mackiernan, W. Nehlsen, and V.K. Tippie. Vol. 1, pp. i-xxi, 1-200. Vol. 2, Appendices A to D.
163. Carpenter, J.H. and D.G. Cargo. 1957. Oxygen requirement and mortality of the blue crab in the Chesapeake Bay. Chesapeake Bay Inst., The Johns Hopkins Univ., Tech. Rept. 13:1-22.
164. Engel, D.W. and M. Brouwer. 1984. Trace metal-binding proteins in marine molluscs and crustaceans. *Mar. Environ. Res.* 13(1984): 177-194.
165. Roesijadi, G. 1981. The significance of low molecular weight, metallothionein-like proteins in marine invertebrates: current status. *Mar. Environ. Res.* 4(1980-1981):167-179.
166. Wiedow, M.A., T.J. Kneip and S.J. Garte. 1982. Cadmium-binding proteins from blue crabs *(Callinectes sapidus)* environmentally exposed to cadmium. *Environ. Res.* 28:164-170.
167. Brouwer, M., T. Brouwer-Hoexum and D.W. Engel. 1984. Cadmium accumulation by the blue crab, *Callinectes sapidus:* involvement of hemo-

cyanin and characterization of cadmium-binding proteins. *Mar. Environ. Res.* 14:71-88.
168. Engel, D.W. and M. Brouwer. 1984. Cadmium-binding proteins in the blue crab, *Callinectes sapidus:* laboratory-field comparison. *Mar. Environ. Res.* 14(1984):139-151.
169. Mix, M.C. 1984. Polycyclic aromatic hydrocarbons in the aquatic environment: occurrence and biological monitoring, pp. 51-102. *In:* E. Hodgson (Ed.) *Reviews in environmental toxicology,* I. Elsevier Sci. Publ., Amsterdam. 337 pp.
170. Jackim, E. and C. Lake. 1978. Polynuclear aromatic hydrocarbons in estuarine and nearshore environments, pp. 415-442. *In:* M.L. Wiley (Ed.) *Estuarine interactions.* Academic Press, New York. 603 pp.
171. Lee, R.F., C. Ryan and M.L. Neuhauser. 1976. Fate of petroleum hydrocarbons taken up from food and water by the blue crab, *Callinectes sapidus. Mar. Biol.* 37:363-370.
172. Singer, S.C. and R.F. Lee. 1977. Mixed function oxygenase activity in blue crab, *Callinectes sapidus:* tissue distribution and correlation with changes during molting and development. *Biol. Bull.* 153:377-385.
173. Hale, R.C. 1983. Accumulation of toxic organic pollutants in the blue crab, *Callinectes sapidus.* PhD Dissertation, Coll. William and Mary, Williamsburg. 122 pp.
174. Hale, R.C. 1983. Mixed-function-oxygenase enzyme systems: purpose and possible deleterious interactions with organic pollutants in the blue crab. *Jour. Shellfish Res.* 3(1):92. Abstract.
175. Cantelmo, A.C., R.J. Lazell and L.H. Mantel. 1981. The effects of benzene on molting and limb regeneration in juvenile *Callinectes sapidus* Rathbun. *Mar. Biol. Let.* 2:333-343.
176. Laird, C.E. and M. Roberts, Jr. 1980. Effects of chlorinated seawater on the blue crab, pp. 569-579. *In:* R. L. Jolley, W.A. Brungs, and R.B. Cumming (Eds.) *Water chlorination. Environmental impact and health effects.* Vol. 3. Ann Arbor Sci., MI. 1171 pp.
177. Sheridan, P.F. 1973. Uptake, metabolism, and distribution of DDT in organs of the blue crab, *Callinectes sapidus.* MA thesis, Univ. Virginia, Charlottesville, 30 pp.
178. Lowe, J.I. 1965. Chronic exposure of blue crabs, *Callinectes sapidus*, to sublethal concentrations of DDT. *Ecology* 46(6):899-900.
179. Koenig, C.C., R.J. Livingston and C.R. Cripe. 1976. Blue crab mortality: interaction of temperature and DDT residues. *Arch. Environ. Contam. Toxicol.* 4:119-128.
180. Sandholzer, L.A. 1945. The effect of DDT upon the Chesapeake Bay blue crab *(Callinectes sapidus). Fish. Market News* 7(11):2-4.
181. Springer, P.F. and J.R. Webster. 1951. Biological effects of DDT applications on tidal salt marshes. *Mosquito News,* 11(2):67-74.

182. Schimmel, S.C., J.W. Patrick, Jr., L.F. Faas, J.L. Oglesby and A.J. Wilson, Jr. 1979. Kepone: toxicity to and bioconcentration by blue crabs. *Estuaries.* 2(1):9-15.
183. Schimmel, S.C. and A.J. Wilson. 1977. Acute toxicity of Kepone to four estuarine animals. *Chesapeake Sci.* 18(2):224-227.
184. Fisher, D.J. 1980. Effects of ingestion of kepone contaminated food by juvenile blue crabs *(Callinectes sapidus* Rathbun). MA thesis, Coll. William and Mary, Williamsburg. 80 pp.
185. Roberts, M.H., Jr. 1981. Kepone distribution in selected tissues of blue crabs, *Callinectes sapidus,* collected from the James River and lower Chesapeake Bay. *Estuaries* 4(4):313-320.
186. Roberts, M.H., Jr. and A.T. Leggett, Jr. 1980. Egg extrusion as a Kepone-clearance route in the blue crab, *Callinectes sapidus. Estuaries* 3(3):192-199.
187. Leggett, A.T., Jr. 1979. The development of blue crabs, *Callinectes sapidus,* from kepone-contaminated eggs. Master's thesis, Coll. William and Mary, Williamsburg. 85 pp.
188. Bender, M.E. and R.J. Huggett. 1984. Fate and effects of Kepone in the James River, pp. 5-50. *In:* E. Hodgson (Ed.) *Reviews in environmental toxicology,* I. Elsevier Sci. Publ., Amsterdam. 337 pp.
189. Kemp, M.V. and J.P. Wightman. 1981. Interaction of 2, 4-D and dicamba with chitin and chitosan. *Virginia Jour. Sci.* 32(2):34-37.

Contaminant Problems and Management of Living Chesapeake Bay Resources. Edited by
S. K. Majumdar, L. W. Hall, Jr. and H. M. Austin. © 1987, The Pennsylvania Academy of Science.

Chapter Eleven
MOLLUSK CULTURE FOR THE CHESAPEAKE BAY
MICHAEL CASTAGNA
Virginia Institute of Marine Science
School of Marine Science
College of William and Mary
Wachapreague, Virginia 23480

INTRODUCTION

The water quality of the Chesapeake Bay has suffered a decline over the last 5 decades due to anthropomorphic activities. Insidious additions of industrial and farm pollutants to the Bay have created a situation where in many areas there are periodic sublethal levels of chemicals. Although the juveniles and adults seem to survive these levels, they are obviously interfering with some early life stages of the living organisms that make up the bay fauna. Species whose early life history take place out of the Bay (i.e. *Callinectes sapidus*) are less affected by this problem than those species whose eggs, embryos and larvae are found in the Bay.

Over 70,000 chemicals are being manufactured in the United States today. Of these approximately 50,000 are being produced in excess of 1.3 billion pounds annually.[6]

Population shifts and expansion are pushing more people toward this nation's coastlines. By 1990, it has been estimated that 75% of the population will live within 50 miles of the oceans.[10] This population shift is bound to exacerbate the already frightening statistics on land-based pollution of the coastal zone; i.e., there is no toxicity data for 90% of the common chemicals in use which ultimately find their way to the estuaries or oceans. There is no similar data on 65% of the pesticide formulas in common use.[10] To further compound the problem, the fresh water input into the Bay has declined due to increased water demands on the major drainages of the Bay.[20]

These effects can be compared by observing the fouling organisms found on pilings and bulkheads along the coast. In badly fouled areas such as busy harbors, common fouling organisms are absent.

Unfortunately, the environmental quality of the Bay has not been judged by subtle changes in fauna. Instead, in past history, it is judged by the annual fishery production of key species for which the Bay is noted. This has severe drawbacks. Most commercial species are harvested as adults, which means in most cases a time lag of several years. Most harvestable species have regulations and laws governing them which sometimes cloud the picture. Most important, if a decline in a fishery is used as a measure of degradation, then only an increased yield per unit effort would indicate an improvement in the environment. Unfortunately, this is sometimes not effected in a reasonable length of time, and in some cases not in a lifetime.

MOLLUSK FISHERY

The mollusk fisheries of the Bay, especially oysters, were at one time the most important fisheries in Virginia. Virginia until the late 1950's was the leading shellfish producer in the U.S.[11] This fishery has declined sharply, probably due to the decline of the environmental quality of the Bay due to anthropomorphic additions. These additions can both directly and indirectly affect certain life history stages of mollusks. The other two commercial mollusk species in the Bay are hard clams and soft clams. These too have shown a decline. Other species occur in the Bay such as *Rangia cuneata* which are commercially harvested in other areas. Since there is no commercial fishery for this species, I have not included them as a commercial mollusk.

REMEDIAL ACTION

The mollusk harvest has shown a continued decline despite efforts to clean up the Bay and to manage the resource.[11]

Unless the water quality of the Bay is improved, it is doubtful that management or replenishment programs will solve or even have much impact on the problem. History of the fisheries has indicated the commercial mollusk fishery has continued to decline despite regulatory changes or infusions of money for replenishment of stocks or improvement of substrate.[12]

A possible method of increasing production is aquaculture.

At the present time there are a number of commercial firms using intensive aquaculture methods to produce seed oysters and clams. These seed, after being grown through hatchery and nursery phases, are then planted in protected beds or predator exclusion devices such as trays. Although these methods have proven profitable for private firms, it does not mean that the same methods would be successful for repletion programs. However, the yield from this type of culture could certainly supplement the acute seed shortage the entire industry is facing. Further, the harvest of private firms will contribute to the yield of mollusks.

In the Bay the hatchery rearing and remote setting of *Crassostrea virginica* is presently being explored. This is being carried out in the following manner. Adult oysters are brought into the hatchery prior to spawning. One or two can be opened and the condition or ripeness of the gonads ascertained. If the gonads do not appear to be ripe, they can be brought to a ripe condition by holding them in warm flowing seawater (22 to 24°C) with ample unicellular algae for food for one or two weeks.[17] If they are collected from nature in a ripe condition, this step can be eliminated. Once they are ripe, the adult spawners can be held in cooled seawater between 16 and 19°C to prevent spawning and to maintain the gonadal ripeness until needed. Rations of unicellular algae should be fed daily to maintain the stock and prevent reabsorption of the gonads. This technique can be used to hold oysters for several months.

The ripe spawners can then be induced to spawn either individually or as pooled or mass spawn. Mass spawning is usually used in commercial operations. Approximately 25 to 200 spawners are placed in a trough or tank with filtered (10 μ) seawater at about 22°C. Then by raising (28°C) and lowering (24°C) the water temperature in about 45-minute cycles, the animals can be induced to spawn.[3]

Other stimuli are used besides temperature shock or temperature cycling. The addition of gonadal products to the water is usually the best stimulus to trigger spawning.[15,16] The addition of food[2] or the injection of serotonin[9] can also be used as a spawning stimulus. When the bivalves spawn, they simply eject the eggs and sperm into the water where fertilization and development take place. Within about 24 hours the fertilized eggs will have developed through the embryo stages into larvae. The larval stages have a relatively short duration but are important to the life cycle. They contribute to the recruitment, distribution and genetic exchange of adult populations.[7,8,19] The larvae, called veligers, are then grown in a hatchery for the next 8 to 30 days (depending on the species, water temperature and amount of food in the culture tanks). The water is changed 3 or 4 times a week by siphoning the water through an appropriate-sized fine mesh nylon sieve to collect the veliger larvae. The larvae are then placed into a clean container of filtered water with the appropriate unicellular algal food to continue their growth. The larvae grow from about 50 μm to approximately 230 μm. They are then ready to metamorphose. At this point the oyster larvae develop a red spot that can be seen through the shell. The oyster larvae usually metamorphose within 72 hours of its appearance. Early investigators believed this to be a light receptor and identified it as an eyespot. Shortly after the appearance of the eyespot, the larvae develop a foot. At this stage, called pediveliger, the larvae spend part of the time swimming and increasing periods of crawling as if seeking a suitable substrate to attach and cement shell to and start the sessile phase of its existence.

When oyster larvae develop an eyespot, they are referred to as eyed larvae. At this point the larvae can be concentrated by draining the culture through an appropriate-sized mesh sieve. The concentrated larvae can then be rinsed through a stack of sieves stacked in descending order to sort the larvae by size. The larger eyed larvae can then be stored in a moist cold condition. The smaller larvae are put back into culture to continue to grow.

The eyed larvae can be rinsed onto a small piece of nylon mesh cloth. This is folded around the mass of larvae and then wrapped in seawater-soaked paper towels. This packet can then be stored in a cold styrofoam ice chest (with refrigerant packs) for about 5 days. Two and a half million larvae will pack to about the size of a ping pong ball. During this storage period more eyed larvae can be collected, packed and stored, or the eyed larvae can be transported in this condition.[13]

The ability of larvae to successfully withstand this treatment allows the remote setting technique to work. An oyster grower can arrange to receive a given number of eyed larvae transported in this condition for use in a remote setting operation.

The oyster grower will establish a tank or a portable swimming pool near the bay or estuary. He can then fill this container with clean ready to use cultch. Most planters use an extruded plastic mesh about 70 cm long and 30 cm in diameter with a 2 cm stretch mesh made into bags that are then filled with oyster or clam shells. Other material can be used for cultch such as plastic tubing or pieces of rubber tires, but there are limited choices which are as cost effective as shells.

The tank is equipped with an aeration system usually made of $\frac{1}{2}''$ PVC pipe fitted together along the bottom edge of the tank. A series of $\frac{1}{16}''$ holes is drilled in the pipe for air escape. Cross pipes are used in larger tanks. This aeration pipe is attached to a regenerative blower or other air source to produce adequate air to vigorously bubble the water. This ensures the distribution of the larvae in the water.

Before the larvae arrive, the cultch or bags of clean shells are stacked in the tank like cord wood. The tank is filled with filtered seawater. The water can be filtered using a sand filter or a convenient 25 μ bag filter (GAF®). The shells are soaked in the aerated water for 24 to 48 hours to ensure that a bacterial film has coated the shell surfaces. The tank is drained and refilled just before the arrival of the eyed larvae. If necessary, the water in the tank can be heated to about 24-28°C with immersion heaters. The packets of eyed larvae are unpacked and rinsed into the tank of filtered water containing cultch and aerated for about 72 hours. At the end of that period most of the eyed larvae will have metamorphosed and set to become oyster spat. An excellent description of this method has been written by Bruce and Gordon Jones.[13]

After setting has taken place, the bags of cultch are usually moved to a nursery or a pier or float where they are suspended off the bottom, or they are placed on an intertidal area to grow to a larger, heavier shell size that can survive better. They can then be transferred to a growing ground. The bags are cut open and the cultch with seed are spread onto the bottom to grow.

The above describes remote setting in a simple form. If eyed larvae production is adequate to support a larger operation, larger tanks can be used and the cultch handling can be containerized and the entire system mechanized to make it less labor intensive.

In 1986 a west coast oyster company produced 18 billion eyed larvae from their hatchery. If they had a setting success of 20% (well within their average rate), they would produce over 3 and one half billion spat or the equivalent of over 3 and one half million bushels of seed (in Virginia a heavily set bushel of shell cultch will have from 1 to 3 thousand spat per bushel).

As the industry converts to cultured spat, the traditional oyster growing methods will undoubtably change. For instance, since the spat are in containers of some type for setting and nursery growth, it would probably be advantageous both biologically and economically to continue off-bottom culture through the first one or two growing seasons of the oyster. This would greatly increase survival and ultimately the yield at harvest.

Aquaculture of clams is carried out in much the same manner as with oysters. Clams are spawned in the same manner and the larvae are grown using almost identical methods.[5] Since clams require no cultch to attach to when they metamorphose, they are normally allowed to set in a hatchery. After setting, the seed is grown in a nursery in either flowing water troughs or in an upweller system.[4] Upwellers are screen bottom cylinders or square containers in which clams are placed. Flowing water enters the bottom, passing through the clams creating a semi-fluidized bed and then out an opening near the top of the container.[18,1]

When the seed clams attain a length of 8 to 10 mm, they are planted in protected beds or trays, placed in the bay or estuary and grown to market size.[5,14] This culture method has been used primarily for the hard clam, *Mercenaria mercenaria.* The soft clam, *Mya arenaria,* has not been commercially cultured due to its relatively poor market value and its success as a wild fishery.

The cost of culturing bivalves is high when compared to the cost of wild natural seed. However, its dependability makes this method attractive to the commercial industry. This method has been studied as a repletion method for public fishing areas. However, only small token plantings have been made with questionable results. Commercial use of these methods will increase the yield of these valuable mollusks and may relieve some fishing pressures from the natural seed areas.

Since the quality of the Bay has been measured by the yield of its fisheries, this method, by increasing the harvestable yield, may be interpreted as an improvement. This is not necessarily true, but it may mitigate the damage to the mollusk fishery by the degradation of the water quality in the Bay.

LITERATURE CITED

1. Bayes, J.C. 1981. Forced upwelling nurseries for oysters and clams using impounded water systems. In: C. Claus, N. DePauw and E. Jaspers (Eds.) Nursery Culturing of Bivalve Molluscs. European Mariculture Society Special Publication No. 7, pp. 73-83.
2. Breese, W.P. and A. Robinson. 1981. Razor clams *Siliqua patula* (Dixon): Gonadal development, induced spawning and larval rearing. Aquaculture 22: 23-33.
3. Castagna, M. 1983. Review of recent bivalve culture methods. J. World Maricul. Soc. 14: 567-575.
4. Castagna, M. 1984. Methods of growing *Mercenaria mercenaria* from postlarval- to preferred-size seed for field planting. Aquaculture 39: 355-359.
5. Castagna, M. and J.N. Kraeuter. 1981. Manual for growing the hard clam *Mercenaria*. VIMS Special Report in Applied Marine Science and Ocean Engineering, No. 249, 110 pp.
6. Congress of the U.S., Office of Technology Assessment. 1979. Environmental contaminants in food. Library of Congress cat. card No. 79-600207, p. 15.
7. Crisp, D.J. 1974. Factors influencing the settlement of marine invertebrate larvae. In: P.I. Grant and A.M. Mackie (Eds.) Chemoreception in Marine Organisms. Academic Press, London, New York, pp. 177-265.
8. Crisp, D.J. 1976. The role of pelagic larvae. In: P. Spencer-Davis (Ed.) Perspectives in Experimental Biology, Vol. 1, Zoology. Pergamon Press, Oxford-New York, pp. 145-155.
9. Gibbons, M.C. and M. Castagna. 1984. Serotonin as an inducer of spawning in six bivalve species. Aquaculture 40: 189-191.
10. Graham, D.M. 1986. Resource management—getting it together. Sea Technology, Vol. 27, No. 8, August 1986.
11. Haven, D.S., W.J. Hargis and P.C. Kendall. 1978. The oyster industry of Virginia: Its status, problems and promise. VIMS Special Papers in Marine Science, No. 4.
12. Haven, D.S. and J.P. Whitcomb. 1986. The public oyster bottoms in Virginia: An overview of their size, location, and productivity. American Malacological Bulletin, Special Ed. No. 3(1986): 17-23.
13. Jones, Bruce and Gordon Jones. 1983. Methods for setting hatchery produced oyster larvae. Information Report No. 4, Marine Resources Branch, Ministry of Environment, Province of British Columbia. Queens Printer for British Columbia, Victoria, Canada, 1984.
14. Kraeuter, J.N. and M. Castagna. 1977. An analysis of gravel, pens, crab traps and current baffles as protection for juvenile hard clams, *Mercenaria mercenaria*. Proc. of the 8th Annual Meeting of the World Mariculture Society, pp. 581-592.

15. Loosanoff, V.L. and H.C. Davis. 1949. Gonad development and spawning of oysters at several constant temperatures. Anat. Rec. 105: 112.
16. Loosanoff, V.L. and H.C. Davis. 1950. Spawning of oysters at low temperatures. Science 111(2889): 521-2.
17. Loosanoff, V.L. and H.C. Davis. 1963. Rearing of bivalve mollusks. Advances in Marine Biology 1: 1-136.
18. Manzi, J.J., N.H. Hadley and M.B. Maddox. 1986. Seed clam, *Mercenaria mercenaria,* culture in an experimental-scale upflow nursery system. Aquaculture 54: 301-311.
19. Obrebski, S. 1979. Larval colonizing strategies in marine invertebrates. Mar. Ecol. Prog. Ser. 1: 293-300.
20. Williams, J. 1979. Introduction to Marine Pollution Control. John Wiley & Sons, New York, 173 p.

Contaminant Problems and Management of Living Chesapeake Bay Resources. Edited by S. K. Majumdar, L. W. Hall, Jr. and H. M. Austin. © 1987, The Pennsylvania Academy of Science.

Chapter Twelve

MESOZOOPLANKTON AND MICROZOOPLANKTON IN THE CHESAPEAKE BAY

DAVID C. BROWNLEE[1] and FRED JACOBS[2]

[1]The Academy of Natural Sciences
Benedict Estuarine Research Laboratory
Benedict, Maryland, USA 20612
and
[2]Coastal Environmental Services, Inc.
2829 Old North Point Rd.
Baltimore, Maryland, USA 21222

ABSTRACT

The Chesapeake Bay is a plankton based ecosystem in which the zooplankton act as trophic intermediaries between the very productive phytoplankton and bacteria and the higher trophic levels including many of the economically important fish and shellfish species. The distributions of the mesozooplankton from both the Maryland and Virginia waters of the Chesapeake Bay are presented in terms of abundance, biomass and species composition. The dominant species were the calanoid copepods *Acartia tonsa, Eurytemora affinis* and *A. hudsonica.* At certain times of the year and in particular salinity regimes, cladocerans, barnacle nauplii, and polychaete larvae became important. Ctenophores appeared to be major predators on the crustacean zooplankton in summer months. The highest values of mesozooplankton abundance and biomass occurred in the freshwater regions except for the freshwater station at the head of the main stem of the Bay which was relatively depopulate of these organisms. Seasonal distributions of microzooplankton greater than 44 μm in diameter are presented for the Maryland portion of the Bay. Average abundance and biomass decreased with increasing salinity, though a large bloom of the tintinnine ciliate *Tintinnopsis fimbriata* occurred at most of the oligohaline zone stations in May. A comparison of the 44 μm mesh net sample to 20 μm mesh and whole water is also given. This comparison plus a review of the literature suggests the need for future study of the total microzooplankton component based on whole water samples. The distributions presented suggest that the mesozooplankton could be important predators on the microzooplankton especially in late winter in the oligohaline zone and in late summer in the mesohaline zone. The relationship between phytoplankton and zooplankton distributions are also discussed.

INTRODUCTION

Zooplankton, those animals that live suspended in the water column and move passively in relation to water currents, are an integral component of most aquatic ecosystems. They provide an important trophic pathway in which biomass is moved up the food web from the very productive phytoplankton and bacteria to higher trophic levels including many commercially valuable fish and shellfish species. Zooplankton help maintain the high production of these microbial populations by grazing the bacteria and phytoplankton to levels below which competition becomes important and by nutrient regeneration through excretion. The production of fecal pellets by some of the zooplankters results in rapid transport of nutrients and carbon from the surface layers to the benthos. Because of the pivotal trophic position of the zooplankton, they could also be important in concentrating and in transporting toxic compounds through the food web.

The assemblage of organisms that make up the zooplankton are diverse in size, morphology and taxonomic position. Representatives span the size range from 10-20 μm for some ciliates to over a meter for some jellyfish. The Ciliophora, Rotifera, Arthropoda and Schyphozoa are among the diverse phyla that comprise the zooplankton. The plankton have been divided into groupings based on size[1] and those groups applicable to the zooplankton are: nanoplankton (2-20 μm); microplankton (20-200 μm); mesoplankton (0.2-20 mm); macroplankton (2-20 cm); and megaplankton (20-200 cm). In this chapter, we will use a less restrictive definition and will separate the Chesapeake Bay zooplankton into two groups, the microzooplankton (those less than 200 μm) and the mesozooplankton (those greater than 200 μm).

In the Chesapeake Bay, the microzooplankton are made up predominantly of the protozoan phyla Sarcodina and Ciliophora, the Rotifera, and nauplii of the crustacean order Copepoda. The heterotrophic flagellates, though functionally zooplankton, are not considered in this chapter. The taxa of the mesozooplankton include the adults and juveniles of the crustacean orders Copepoda and Cladocera, the Chaetognatha, the gelatinous zooplankton especially the Ctenophora and Schyphozoa, and the merozooplankton-planktonic larval stages of benthos such as oysters, polychaetes, and barnacles. The distribution and ecology of these two groups of zooplankton will be discussed separately below.

MESOZOOPLANKTON

Mesozooplankton are defined as those organisms which are retained on mesh sizes of 200 mm.[2,3] These organisms which feed on phytoplankton, certain microzooplankters, and detritus, provide a vital carbon source to higher trophic levels. The importance of the mesozooplankton community as an essential food source to higher trophic levels is well documented. Most estuarine fish rely on zooplankton for nourishment during some stage of their development. Certain species, like striped bass and white perch consume small zooplankton such as copepods and cladocerans as larvae, switch to mysids, amphipods, etc. as young-of-year and become increasingly piscivorous as they get larger.[4,5,6,7] Others such as bay anchovy and silversides remain plankton feeders (copepods and cladocerans) throughout their lives.[8,9] Species such as menhaden consume phytoplankton, zooplankton and suspended material in varying amounts at different stages of their development.[10,11] Furthermore, zooplanktonic material that manages to avoid predation will die, settle to the bottom and either be consumed by benthos, or be broken down by microorganisms.

Zooplankton studies have been conducted in certain Chesapeake Bay tributaries[12,13] and several taxonomic groups are extremely well studied; however, comprehensive baywide monitoring programs have been generally lacking. Cowles[14] probably provided the most definitive zooplankton data prior to 1970. His study included detailed physical and chemical data for collections made in 1915, 1916, 1917, 1920, 1921, and 1922. In addition, comprehensive seasonal occurrences of a large number of invertebrate groups were presented. However, these results were not quantitative and cruises were not conducted on a regular basis between sampling years.

In August 1971, a zooplankton monitoring program was initiated by Virginia Institute of Marine Sciences in the Virginia portion of the Chesapeake Bay. This continued monthly until July 1973; less frequent sampling was conducted in subsequent years. In more recent times, the Chesapeake Bay Initiative prompted the undertaking of a zooplankton monitoring program (funded by the State of Maryland) in the Maryland portion of the Bay. This monitoring program has continued from 1984 to the present. Zooplankton monitoring efforts have also been initiated in Virginia. Much of the information that follows is drawn from the Maryland program and the 1971-1973 Virginia study.

METHODS

Maryland Study

Mesozooplankton samples were collected monthly at 16 stations (Figure 1) beginning in August 1984. Five stations in the mainstem Bay were sampled.

The northernmost station was near the mouth of the Susquehanna River and the southernmost station was east of Point No Point and north of the mouth of the Potomac River. On the western shore of the Bay, seven tributary stations were sampled. Three of these stations were located in the Patuxent River and three were in the Potomac River. They ranged in salinity from relatively

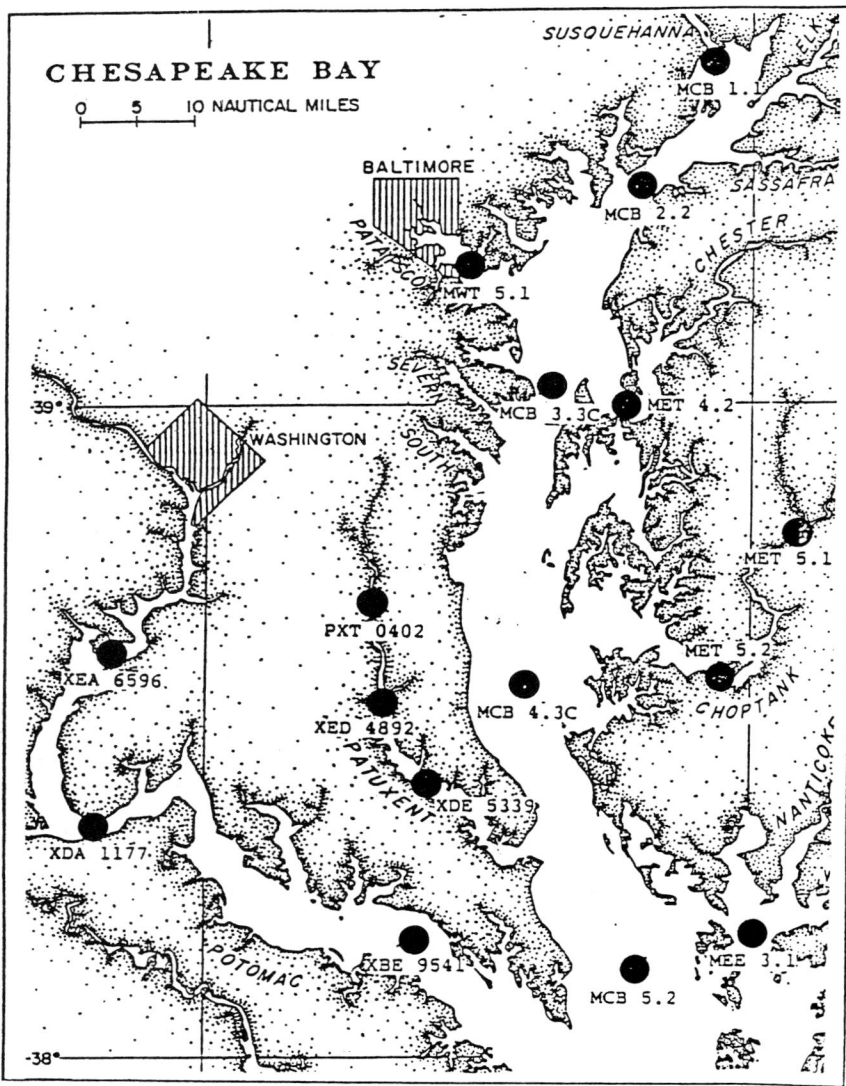

FIGURE 1. Location of mesozooplankton and microzooplankton sampling stations in the Maryland portion of Chesapeake Bay.

freshwater upriver habitats to truly estuarine downriver habitats. The remaining western shore station was located in the Patapsco River near the Key Bridge. Eastern Shore tributary collections were taken at the mouth of the Chester River, at two stations in the Choptank River, and at one station in Tangier Sound.

Mesozooplankton samples were obtained by towing a 20-cm bongo net (202 μm mesh net) in a stepped oblique fashion for each replicate tow. The entire water column was sampled by first deploying the gear a few meters from the bottom and raising the net in progressive time steps, usually 0.5 to 1.5 minutes/step. For stations <8 m in depth, 1-m step intervals were used; steps were 2 m for stations in the depth range of 9-20 m; at stations >20 m in depth, 4-m step intervals were taken. The duration of the tow was 5 minutes during periods of high zooplankton or ctenophore density and 10 minutes during periods of low zooplankton density. The actual volume of water each net filtered was calculated by using flow data from a General Oceanics flowmeter mounted in the mouth of one side of the bongo net.

One taxonomic sample (preserved in 5-10% formalin) and one biomass sample

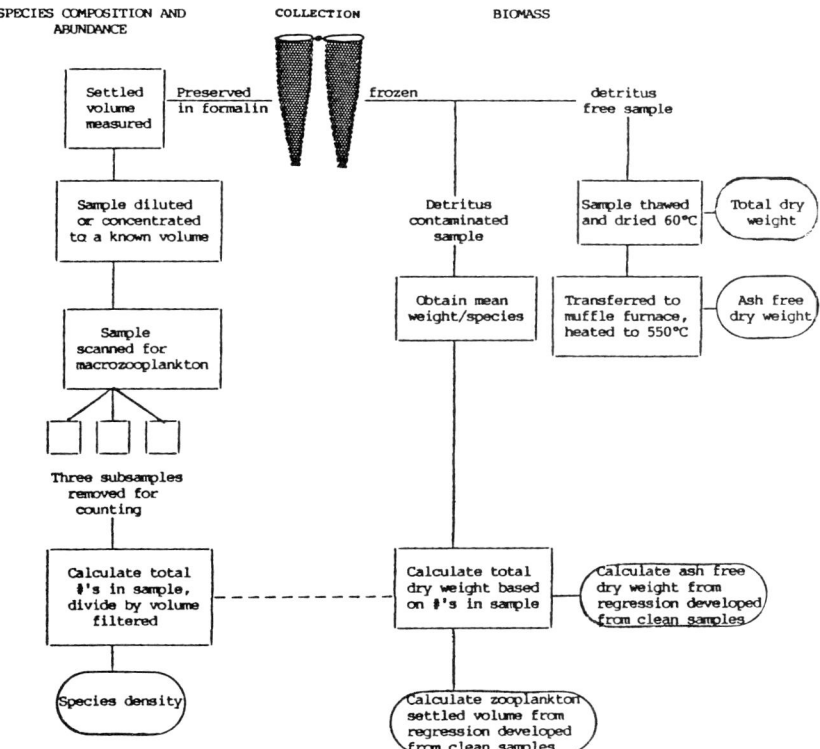

FIGURE 2. Flow chart illustrating processing procedures for mesozooplankton samples taken in Maryland.

(frozen in the field) were collected from each bongo tow. Ctenophores, when they were found, were removed from samples in the field, and their numbers and biomass (settled volume) were recorded from the net that was used as the taxonomic sample. Figure 2 summarizes mesozooplankton processing procedures for taxonomic and biomass samples.

In the laboratory each sample was either concentrated or diluted to a volume that facilitated counting of subsamples taken with a Hensen-Stempel pipette. A modified version of a hierarchical counting technique[15] was employed to obtain reliable density estimates for less abundant as well as dominant species. This procedure consisted of first counting at least 60 individuals of the most dominant forms (e.g., *Acartia tonsa*) in a small subsample (usually 1-2 ml), followed by 5- and 10ml subsamples, from which all species that had counts <60 in the previous subsample were counted.

The relative abundance of each species of mesozooplankton (number/m³) was determined from subsample counts by using the following equation:

$$\text{relative abundance } \#/m^3 = A \times \frac{B}{CD}$$

where:

A = the number of individuals counted in the subsample
B = dilution volume (ml)
C = subsample volume (ml)
D = volume of water filtered (m^3).

The density of macrozooplankton was calculated by enumerating the number of organisms in the entire sample and dividing by the volume of water filtered.

Detritus-free biomass samples were thawed and dried at 60°C to a constant weight. After total dry weights (mg/m³) were recorded, the sample was then transferred to a muffle furnace and heated to 550°C, and the ash-free dry weight was calculated. Ash-free dry weight (g) was determined as [total dry weight (g)]-[ash weight (g)]. Dry weight and ash-free dry weight values were then converted to mg/m³ of water sampled. Greater details of methods and results of the Maryland study are provided in Burton et al.[16]

Virginia Study

Zooplankton were collected monthly in eight subareas of the Virginia portion of the Bay from August 1971-July 1973 (Figure 3). Three to five randomly selected stations were sampled per subarea, insuring a minimum of 24 samples per month. As in the Maryland study, zooplankton were sampled by 20 cm 202 μm bongo nets, obliquely towed from bottom to surface. Laboratory procedures utilized a split-sort technique with a sample splitter.[17] Zooplankton dry weight was obtained through lyophilization rather than heating to constant weight. Greater details of field and laboratory procedures can be found in Jacobs.[18]

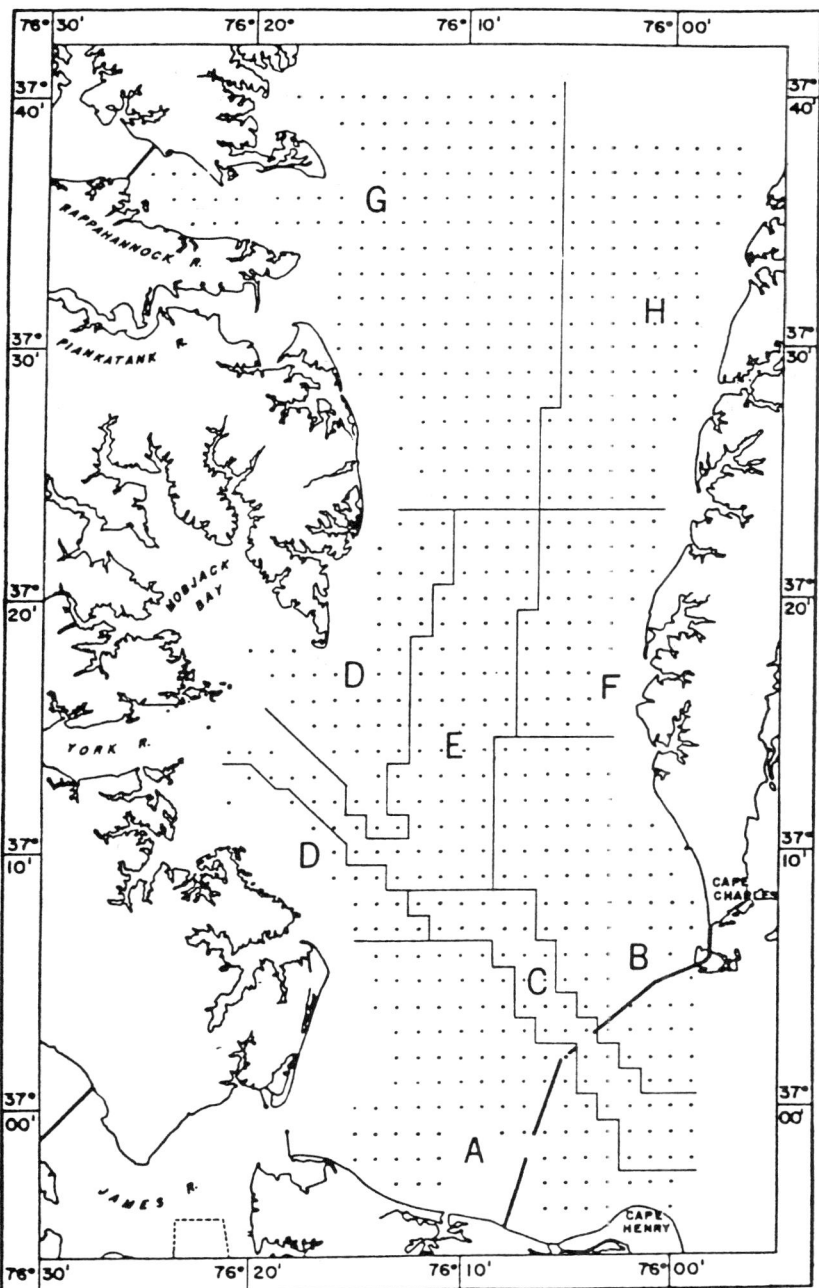

FIGURE 3. Lower Chesapeake Bay study area, indicating subareas sampled monthly from August 1971-July 1973.

TABLE 1

Mean abundance, % composition, and ubiquity of mesozooplankton collected in the Maryland study. Values are averaged over all stations and months (A. clausi = A. hudsonica).

Taxon	Density (5s/m^2)	Percent of Total	Ubiquity (% of Total Samples in which Taxon Occurred)
Acartia tonsa	5503.82	54.13	91.25
Eurytemora affinis	2032.20	19.99	53.61
Bosmina longirostris	715.45	7.04	32.89
Polychaete larvae	506.94	4.99	59.51
Barnacle nauplii	336.12	3.31	73.57
Acartia clausi	254.32	2.50	9.70
Diaphanosoma leuchtenbergianum	153.73	1.51	14.07
Moina micrura	86.48	0.85	13.88
Podon polyphemoides	82.21	0.81	24.71
Acartia sp.	81.54	0.80	7.22
Mesocyclops edax	80.57	0.79	14.26
Copepod nauplii	76.93	0.76	61.60
Cyclops vernalis	43.18	0.42	25.10
Ostracoda	27.71	0.27	57.03
Oithona colcarva	26.03	0.26	28.71
Daphnia retrocurva	23.06	0.23	18.63
Scapholeberis kingi	17.80	0.18	6.46
Gastropod larvae	16.41	0.16	12.93
Centropages hamatus	16.13	0.16	9.32
Pseudodiaptomus coronatus	13.41	0.13	22.62
Harpacticoida	10.99	0.11	30.80
Halicyclops magnaceps	9.82	0.10	4.75
Mollusca	8.49	0.08	13.12
Alonella sp.	4.68	0.05	8.17
Alona spp.	4.33	0.04	1.33
Cyclops bicuspidatus	4.29	0.04	10.08
Diaptomus sp.	3.97	0.04	15.59
Mysid	3.17	0.03	46.20
Ergasilus sp.	2.98	0.03	13.31
Polychaeta	2.96	0.03	3.04
Unid. fish eggs	2.69	0.03	11.79
Ilyocryptus spinifer	1.76	0.02	10.65
Sida crystallina	1.37	0.01	6.84
Brachyurian zoea	1.33	0.01	12.17
Paracyclops fimbriatus poppei	1.09	0.01	5.32
Sagitta sp.	0.97	0.01	14.83
Chydorius sp.	0.84	0.01	9.89
Camptocercus rectirostris	0.82	0.01	3.61
Hydracarina	0.79	0.01	4.75
Leptodora kindtii	0.75	0.01	2.66
Ilyocryptus sp.	0.74	0.01	2.28
Eucyclops agilis	0.70	0.01	9.89
Palaemonetes sp.	0.62	0.01	12.36
Unid. crab zoea	0.49	0.00	1.71

TABLE 1 (continued)

Taxon	Density (5s/m²)	Percent of Total	Ubiquity (% of Total Samples in which Taxon Occurred)
Unid. fish larvae	0.46	0.00	18.25
Alona affinis	0.43	0.00	8.56
Temora turbinata	0.41	0.00	1.33
Gammarus fasciatus	0.40	0.00	17.30
Centropages furcatus	0.38	0.00	0.57
Paracalanus crassirostris	0.32	0.00	1.14
Cyclopoida	0.16	0.00	0.76
Oligochaeta	0.13	0.00	0.19
Morone americana	0.13	0.00	1.14
Corophium lacustre	0.07	0.00	8.17
Alona costata	0.06	0.00	3.42
Argulus sp.	0.06	0.00	13.50
Morone sp.	0.04	0.00	0.57
Alona sp.	0.04	0.00	0.57
Sapherella sp.	0.04	0.00	0.76
Bosmina sp.	0.03	0.00	0.38
Clupeidae	0.02	0.00	0.95
Morone saxatilis	o0.02	0.00	0.76
Monoculodes edwardsi	0.02	0.00	4.94
Eubosmina coregoni	0.01	0.00	0.95
Leptocheirus plumulosus	0.01	0.00	1.33
Chaoborus sp.	0.01	0.00	3.23
Chironomid larvae	0.01	0.00	6.46
Isopoda	0.01	0.00	7.22
Parathemisto compressa	0.01	0.00	1.90
Micropogon undulatus	0.00	0.00	2.28
Anchoa mitchilli	0.00	0.00	1.33
Brachyurian megalops	0.00	0.00	0.95
Pseudopleuronectes americanus	0.00	0.00	0.38
Menidia	0.00	0.00	0.76
Gobiosoma bosci	0.00	0.00	0.76
Piscicolidae	0.00	0.00	0.38
Lucifer faxoni	0.00	0.00	0.57
Dipteran larvae	0.00	0.00	0.38
Euceramus praelongus	0.00	0.00	0.19

RESULTS

Table 1 shows that the estuarine copepod *Acartia tonsa* was the most abundant (5,504/m³) and ubiquitous (observed in 91% of all samples) organisms sampled in the Maryland portion of the Bay. *Eurytemora affinis* was the next most abundant with an average density of 2,032/m³. Together, these two calanoid species made up greater than 65% of the total zooplankton collected. The recurring pattern of a dominance shift between the two species is clearly seen in Figure

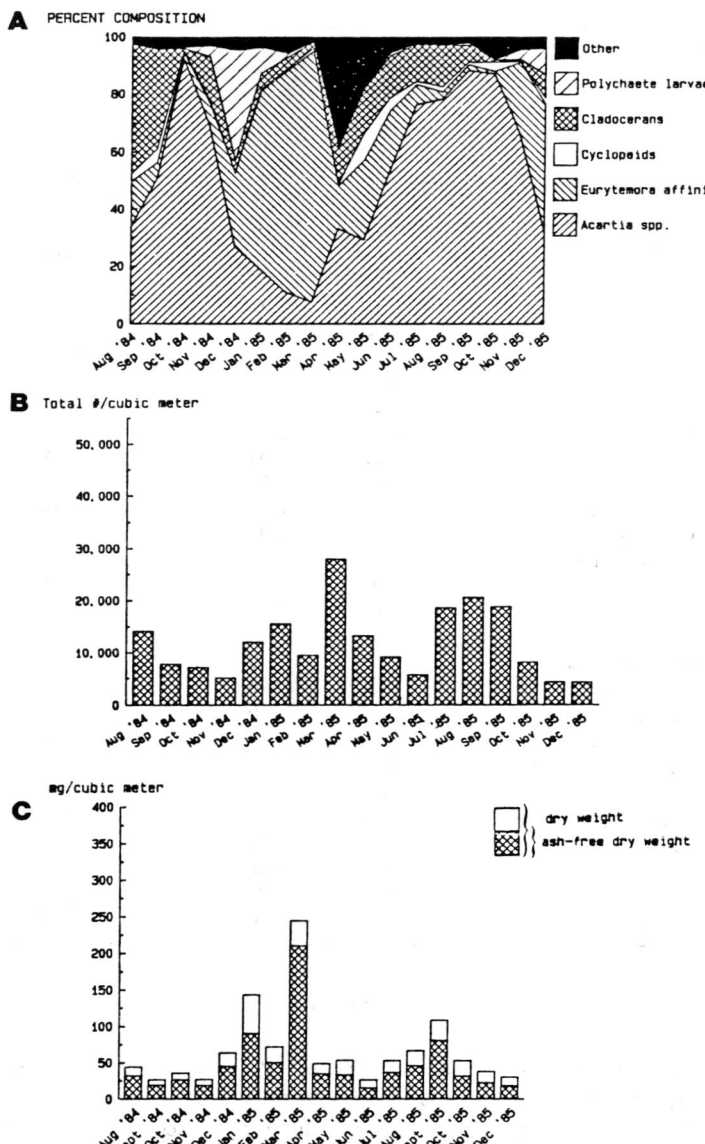

FIGURE 4. Zooplankton species composition (A), total abundance (B), and biomass (C) for all regions of the Maryland sampling area for the August 1984 through December 1985 period.

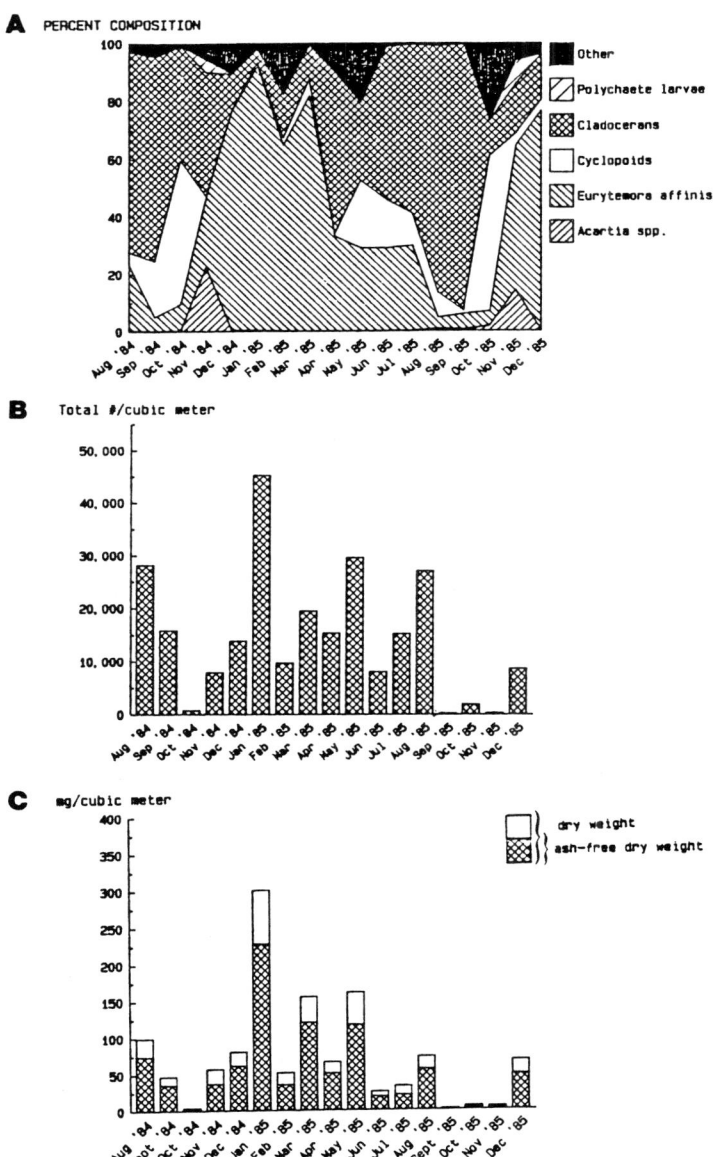

FIGURE 5. Zooplankton species composition (A), total abundance (B), and biomass (C) for Maryland freshwater regions for the August 1984 through December 1985 period.

4. During warmer months, *A. tonsa* made up the largest fraction of organisms in most samples. During the winter sampling period, *E. affinis* was most abundant. Cladocerans, polychaete larvae, and barnacle nauplii (seen in other category in Figure 4), were at times substantial contributors to the plankton community.

Cladocerans comprised >50% of the zooplankton collected at freshwater stations from August to November, 1984 and >60% from April to September, 1985 (Figure 5). Common cladoceran species included *Bosmina longirostris, Diaphanosoma leuchtenbergianum,* and *Moina micrura*. During colder months, *E. affinis* made up most of the freshwater zooplankton community. Cyclopoid copepods made up a significant proportion of collections from August through October 1984 (>20%) and May through December, 1985 (>10%). *Cyclops vernalis, Mesocyclops edax, Halicyclops magniceps,* and *C. bicuspidatus,* were the most commonly observed freshwater cyclopoids. *Acartia* when observed, usually occurred in low numbers in freshwater areas. Freshwater stations in tributaries usually supported extremely diverse communities of considerable abundance and biomass. This can be contrasted with the upper-most main Bay station (usually fresh) which consistently yielded extremely low density and biomass values.

The shift in dominance from *A. tonsa* in summer to *E. affinis* in winter was most evident in oligohaline (0.5-5%) collections (Figure 6). From December 1984 through April 1985, *E. affinis* completely dominated samples. Polychaete larvae appeared abundantly in December 1984 and 1985 making up 30% and 5% of all organisms collected, respectively. Cladocerans (*M. micrura, D. leuctenbergianum, B. longirostris*) were common in summer months, and cyclopoids appeared in large numbers in fall 1984. Total densities and biomass were high during summer (35,000/m^3 in August 1985) and winter (> 50,000/m^3 in February 1985) with relatively lower values occurring during transition seasons. Oligohaline regions contained both freshwater and estuarine species and were therefore usually diverse.

Mesohaline regions (5-18%) generally yielded fewer species than did freshwater and oligohaline zones. The seasonal shift between *E. affinis* and *A. tonsa* as dominants was evident, with the period for dominance of *E. affinis* more restricted in time in the mesohaline region (Figure 7). In the mesohaline region, polychaete and barnacle larvae were extremely abundant on certain occasions. *Oithona colcarva* was the most abundant cyclopoid at higher salinity stations. The seasonal density and biomass pattern observed in the oligohaline region was not as pronounced in either the mesohaline or polyhaline regions (Figure 8) of the Maryland Bay. *A. tonsa* usually dominated in the Maryland polyhaline environment. *E. affinis,* polychaete larvae, and the cladoceran *Podon polyphemoides* at times made substantial contributions to the polyhaline zooplankton community. Plankton diversity in this region was generally lower than in lower salinity areas.

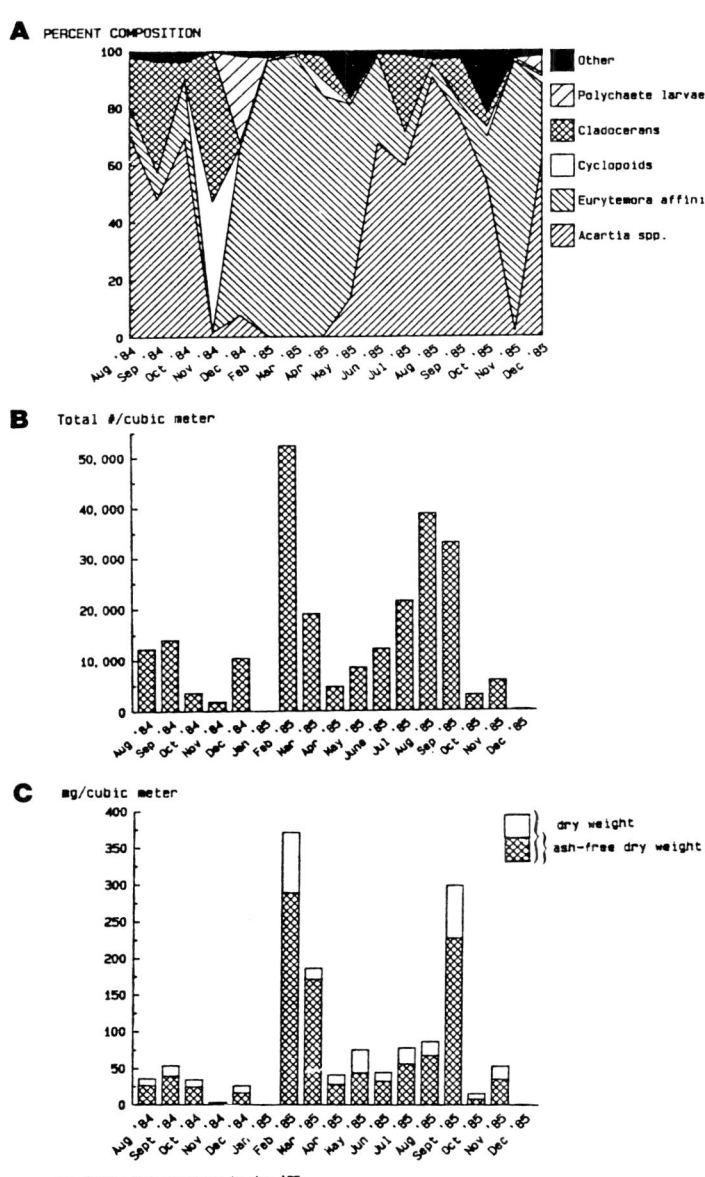

FIGURE 6. Zooplankton species composition (A), total abundance (B), and biomass (C) for Maryland oligohaline regions for the August 1984 through December 1985 period.

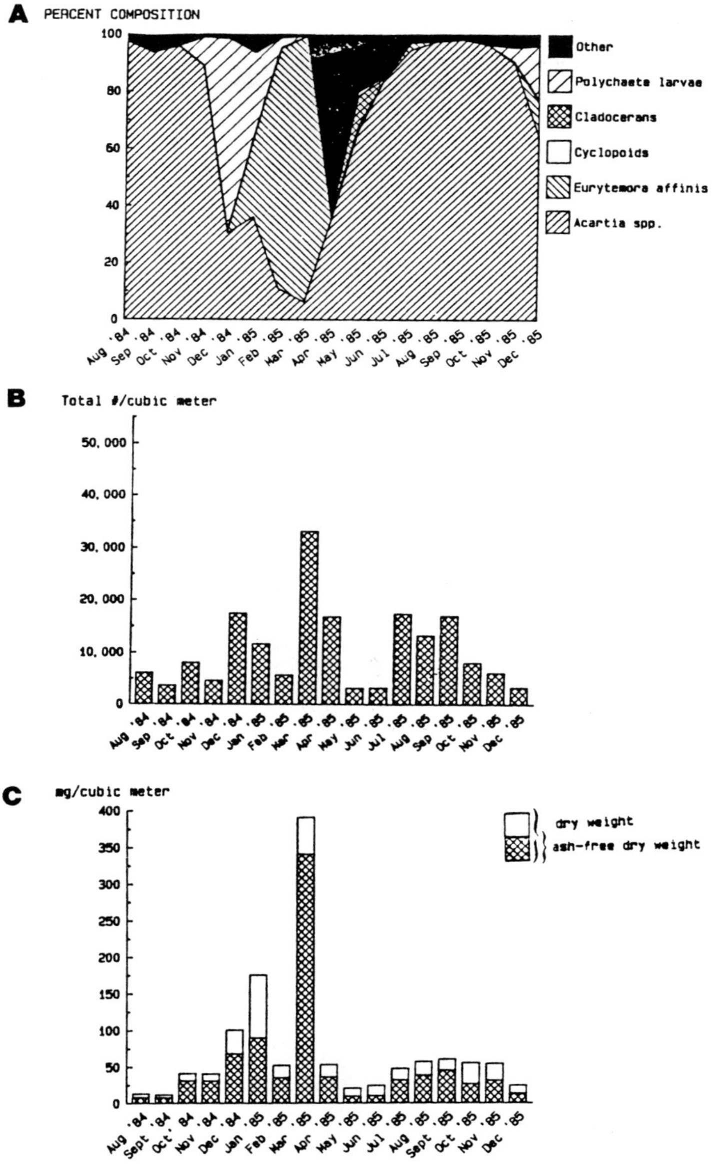

FIGURE 7. Zooplankton species composition (A), total abundance (B), and biomass (C) for Maryland mesohaline regions for the August 1984 through December 1985 period.

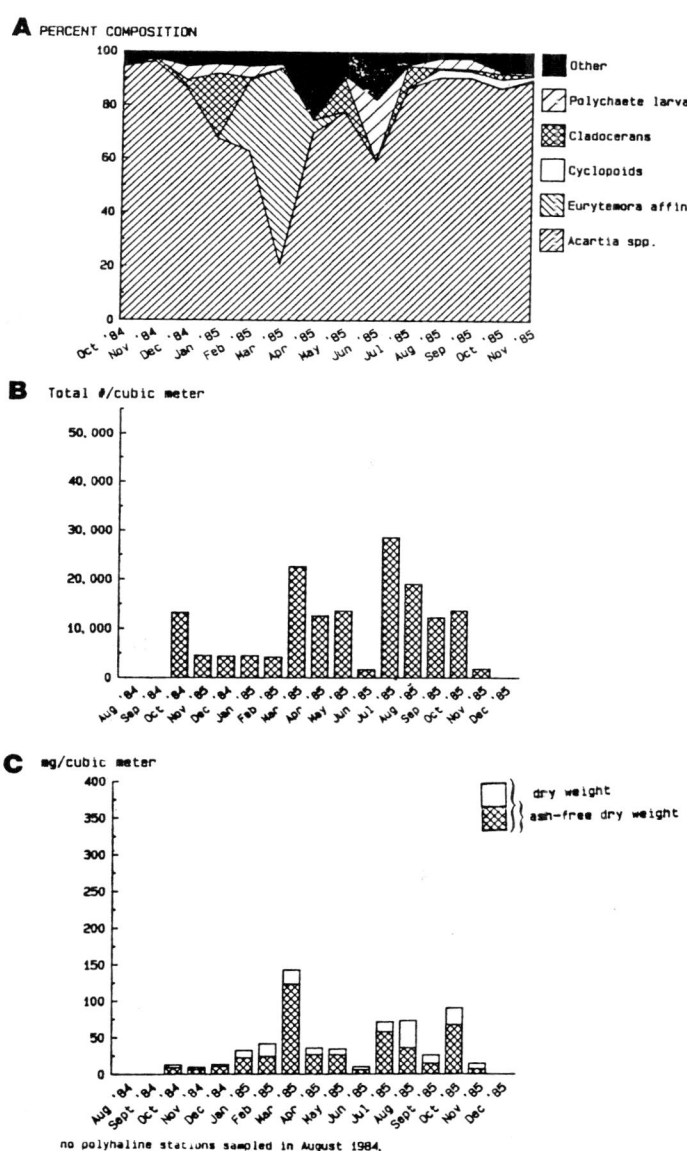

FIGURE 8. Zooplankton species composition (A), total abundance (B), and biomass (C) for Maryland polyhaline regions for the august 1984 through December 1985 period.

Most of the Virginia portion of the Bay can be characterized as polyhaline, with the northern most subareas G & H being either high mesohaline or low polyhaline (Table 2). Salinities, especially surface values were greatly reduced by Tropical Storm Agnes, which passed over the study area in June 1972 (Figure 9).

Figure 10 describes the temporal pattern of zooplankton biomass in the Virginia portion of the Bay between August 1971 and July 1973. As in regions of the Maryland Bay, biomass peaks appeared to occur in summer and winter. August 1971 was a major summer peak for both dry weight (DW) and ash-free dry weight (AFDW), averaging 258 and 211 mg/m^3, respectively. There was a steady decline in DW and AFDW from the August peak through December, when lowest values occurred with DW averaging 9.5 mg/m^3 and AFDW 5.2 mg/m^3. Biomass increased by an order of magnitude in January 1972, with DW averaging 64 mg/m^3. This overall increase continued through February and attained a winter peak in March 1972, with DW and AFDW values averaging 199

TABLE 2

Mean salinity of all sampled depths by subarea and month in the Virginia Chesapeake Bay

Regular Cruises	Subarea								Total Study
Month	A	B	C	D	E	F	G	H	Ares
Aug 1971	25.02	25.45	24.75	18.91	21.05	22.98	17.31	20.16	21.79
Sept	24.62	25.50	27.19	21.33	23.78	22.75	18.13	21.33	22.81
Oct	21.15	25.15	24.33	18.92	20.24	23.26	17.45	21.09	21.61
Nov	22.34	27.05	25.64	17.86	21.99	22.03	—	22.50	23.25
Dec	—	—	—	20.01	21.26	—	17.30	21.54	20.08
Jan 1972	22.37	22.72	24.34	19.49	19.75	20.98	17.50	18.64	20.81
Feb	23.00	23.66	23.61	20.74	21.14	21.64	17.10	20.71	21.43
Mar	23.79	27.37	26.94	17.29	18.35	21.36	15.71	19.72	21.84
Apr	20.86	19.81	21.54	17.55	18.98	18.34	14.27	18.83	18.95
May	19.67	22.05	23.89	16.05	18.44	20.98	14.03	16.53	19.19
June	25.59	22.64	26.35	17.38	19.37	20.03	—	—	22.37
July	15.33	19.15	24.13	12.49	17.81	17.77	11.55	13.93	16.47
Aug	22.43	20.18	24.09	16.48	19.64	18.58	14.35	15.98	18.91
Sept	18.50	21.71	21.80	18.19	19.24	20.16	16.30	15.52	19.08
Oct	21.91	—	22.52	19.69	20.53	23.41	17.66	22.49	21.14
Nov	26.92	23.04	27.83	19.62	20.62	21.66	—	—	23.64
Dec	20.10	21.35	21.30	17.96	19.34	19.26	14.77	18.92	19.65
Jan 1973	22.81	23.10	26.65	15.40	19.22	18.60	12.00	16.92	18.82
Feb	17.42	21.70	22.36	15.23	18.69	21.38	13.80	18.19	18.75
Mar	21.14	20.82	24.79	17.67	16.97	17.90	—	16.59	19.57
Apr	21.30	22.14	22.06	17.04	17.76	17.34	14.41	15.68	18.34
May	23.20	23.90	24.38	16.19	18.55	19.97	12.85	16.83	19.52
June	17.33	23.16	23.63	16.60	18.44	20.07	15.06	19.20	19.37
July	25.30	26.94	26.69	19.25	20.36	22.44	16.19	19.15	22.40

—indicates not sampled.

mg/m³ and 146 mg/m³, respectively. Values declined somewhat in the following months until a summer peak was recorded in July and August 1972. DW values averaged 135 mg/m³ and 114 mg/m³ during these months while AFDW measurements averaged 90 mg/m³ and 100 mg/m³. As in 1971, the fall months of 1972 saw a progressive reduction in biomass through November. Average values began increasing in December 1972 and the second year's winter peak occurred in February 1973, DW averaging 61 mg/m³ and AFDW 52 mg/m³.

The organic fraction (i.e., AFDW) generally comprised the major component of dry weight, exceeding 65% in most instances. An obvious deviation from this pattern occurred in June 1973, when AFDW contributed less than 35% to the DW. In that month molluscan larvae averaged 1,700/m³, which was about 70% of zooplankton total abundance. The bulk of these small organisms was shelled, inert material which accounted for the low AFDW/DW relationship.

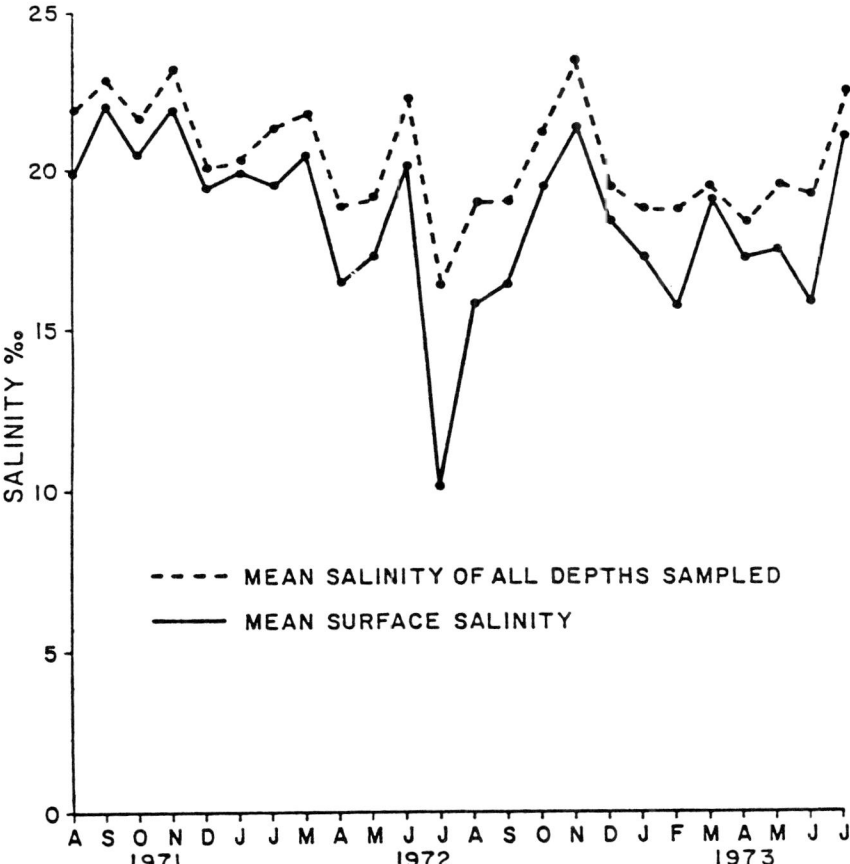

FIGURE 9. Mean monthly salinity in the Virginia Chesapeake Bay.

The zooplankton community in the Virginia portion of the Bay reflected the higher salinities characteristic of the region; both estuarine and marine organisms were observed.[8] The composition of the community during representative summer and winter periods is depicted in Figure 11. *Acartia tonsa* was generally the dominant copepod in warmer months with *A. hudsonica* being the most abundant form in winter. Other copepods that were observed included: *Labidocera aestiva, Paracalanus crassirostris, Paracalanus* spp., *Temora turbinata, T. longicornis, Centropages furcatus, C. hamatus, C. typicus, Pseudodiaptomus coronatus, Eucalanus pileatus* and *Calanus finmarchicus, Eurytemora affinis, Oithona* sp., and *Corycaeus* sp. Many of these species are marine. Other cyclopoids, parasitic and harpacticoid forms (such as *Euterpina acutifrons)* were also occasionally found. *E. affinis* abundances were greatly reduced relative to the Maryland portion of the Bay, and were generally only a minor contributor to the zooplankton community in the Virginia mainstem Bay. Cladocerans that were abundant included *Evadne tergestina, Penilia avirostris,* and *Podon polyphemoides.* Chaetognaths, most notably *Sagitta tenuis,* were at time observed over much of the study area, but were most abundant in the subareas of highest salinity. These organisms reached peak densities in September of 1972 and 1973. Decapod larvae were abundant and diverse in summer months with dominant species including *Upogebia affinis,*

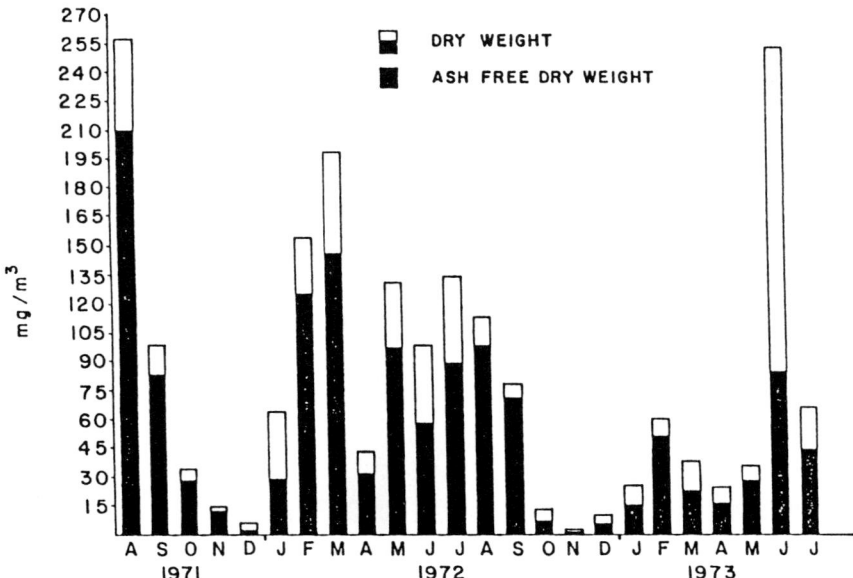

FIGURE 10. Monthly means of dry weight and ash-free dry weight in the Virginia Chesapeake Bay study area.

Callinectes sapidus, Pinnixia chaetopterana and *Hexapanopeus augustifrons*. *Crangon septemspinosa* was the only decopod larvae commonly collected in winter months. Other groups such as barnacle, polychaete, molluscan larvae, mysids, hyromedosae, ctenophores, fish eggs and larvae were also collected.

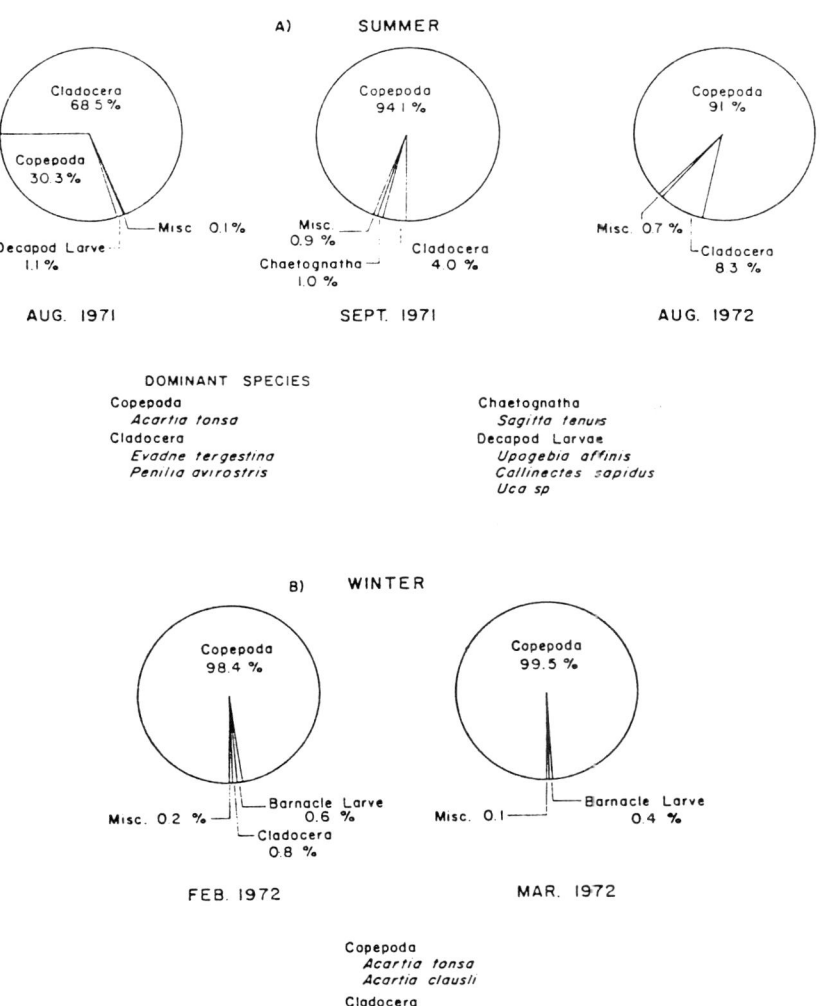

FIGURE 11. Dominant major groups in the Virginia portion of the Chesapeake Bay, expressed as percentage of total zooplankton (*A. clausi* = A. hudsonica).

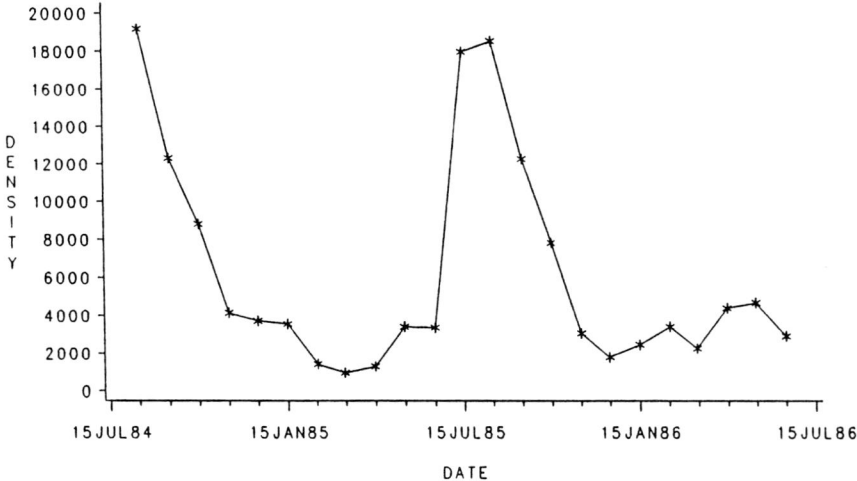

FIGURE 12. Estimated mean densities of *A. tonsa* at Maryland stations (freshwater regions excluded)

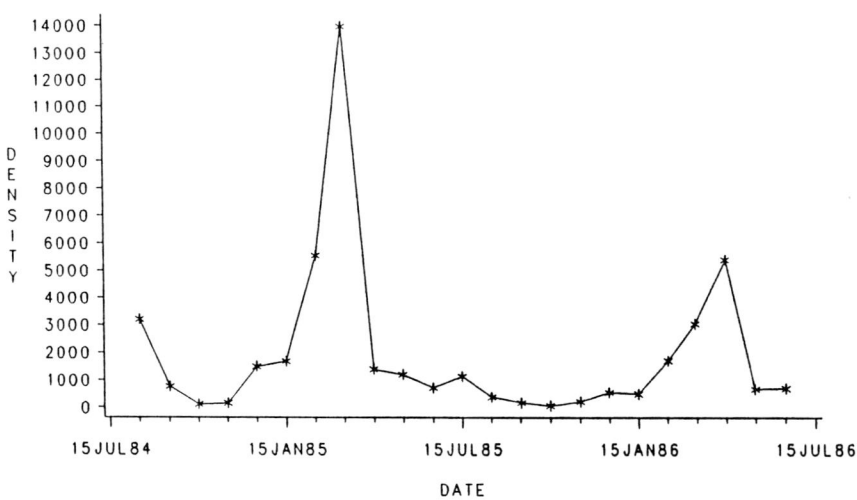

FIGURE 13. Estimated mean densities of *Eurytemora affinis* at Maryland stations.

ABUNDANT TAXA

Copepods—*Acartia tonsa, Acartia hudsonica, Eurytemora affinis*

For the most part, *A. tonsa* was the most ubiquitous and abundant zooplanktonic organisms in the Chesapeake Bay (Table 1, Figure 11). this was supported by Grant and Olney[19] who found *A. tonsa* and *Pseudodiaptomus coronatus* to be the dominant species in August 1978. *A. tonsa* commonly occurs in most temperate estuaries from oligohaline to marine systems.[20,21] In Maryland, *A. tonsa* exhibited peak abundances in summer, and was never completely absent in brackish waters (Figure 12). *E. affinis* assumed the dominant role in winter, this pattern being consistent in 1985 and 1986 (Figure 13). At higher salinity stations, the contribution of *Acartia hudsonica* (formerly *Acartia clausi* became more important in winter (Figure 14). This contribution

TABLE 3

Mean monthly abundance of A. tonsa *and* A. hudsonica *in the Virginia Chesapeake Bay, August 1971-July 1973.*

Date	*A. tonsa*/m^3	*A. hudsonica*/m^3
Aug 1971	27,115	0
Sept	15,405	0
Oct	5,078	0
Nov	1,720	0.1
Dec	1,342	60
Jan 1972	580	475
Feb	1,102	9,835
Mar	410	21,500
Apr	60	6,795
May	342	4,287
June	139	0
July	16,497	0
Aug	18,527	0
Sept	18,416	0
Oct	615	0
Nov	168	0.3
Dec	375	49
Jan 1973	753	245
Feb	1,164	3,461
Mar	95	2,603
Apr	7	2,074
May	275	2,156
June	565	0
July	2,640	0

became far more marked in Virginia, where *A. hudsonica* replaced *A. tonsa* as the dominant copepod during winter (Table 3). Densities of *A. hudsonica* commonly exceeded 2,000/m^3 in the lower Bay.[18,22]

E. affinis is a dominant copepod in lower salinity waters of many North American and European estuaries during winter months. Although found between 0 and 30 o/oo, it appears most abundantly in salinities less than 15 o/oo. This species appears to be particularly well suited to surviving in turbidity maximum regions, characterized by high levels of suspended particulate materials, low dissolved oxygen, and generally low biological diversity. Laboratory studies have indicated best survival in low temperature waters (less than 10°C), between salinities of 3 and 10o/oo, and poor survival above 20o/oo.[23,24,25] In salinities of 12 to 28 o/oo, Grant and Olney[22] found *E. affinis* to be widespread with low densities over the Virginia portion on the Bay in March, 1978.

Conover[26] studied the seasonal distribution of *A. hudsonica* and *A. tonsa* in Long Island Sound and found cycles somewhat similar to those in lower Chesapeake Bay. He reported that *A. hudsonica* first appeared in late November or early in December, reached maximum levels in May during two successive years, and disappeared in July or early August. *A. tonsa* appeared in June of each year, rose to a mid-summer maximum in August and then decreased more or less steadily throughout fall and winter. Since Long Island Sound is a more northern estuary, *A. hudsonica* should peak later in the spring there than in Chesapeake Bay. Similarly, an earlier seasonal recovery of *A. tonsa* in the Chesapeake could be anticipated. Conover[26] indicated June to be a "transition" month for *Acartia* species in Long Island Sound; in the lower Chesapeake Bay

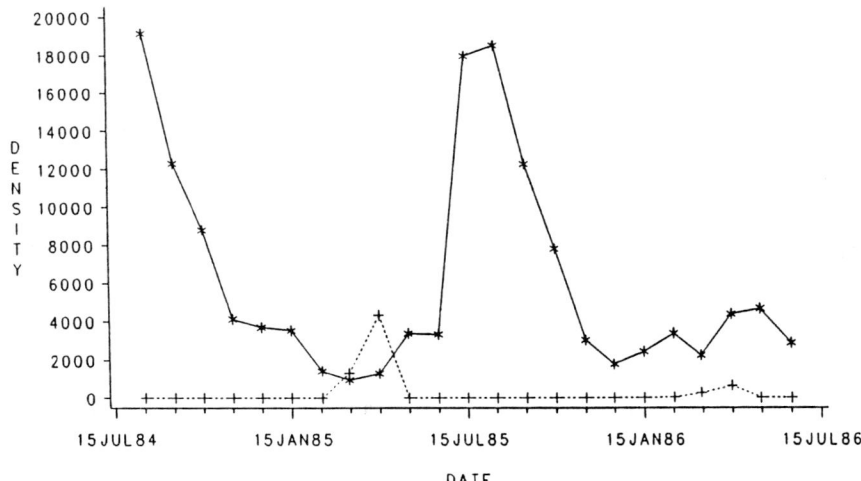

FIGURE 14. Estimated mean densities of *A. tonsa (*)* and *A. hudsonica (+)* at Maryland stations (mesohaline and polyhaline regions only).

the transition occurred in May. Similar seasonal cycles have been observed in other east coast estuaries[27,28,29,30] and as far away as Onagawa Bay, Japan.[31]

Jeffries[32] found that *A. tonsa* would initially replace *A. hudsonica* at the head of the estuary, since *A. tonsa* was better able to propagate at reduced salinities. In the present study, May (both 1972 and 1973) was considered the "transition" month, where *A. tonsa* was beginning to establish itself and replace *A. hudsonica*. Grouping the subareas according to salinity patterns, and establishing an *A. hudsonica/A. tonsa* abundance ratio, yielded results that supported Jeffries findings (Table 4). In both years, *A. tonsa* population replaced *A. hudsonica* to a greater extent in the lower salinity (A, B and C) regions.

The fact that *A. hudsonica* was completely absent from the lower Bay (and other temperature estuaries) during summer months, raises certain questions about the mechanism of seasonal succession of this species. Kasahara et al.[33] found resting eggs of several copepod species including *A. hudsonica* to occur in shallow sediments. Kasahara et al.[34] determined that highest numbers of *A. hudsonica* eggs were in the sediments in summer months. Zillioux and Gonzalez[35] have shown that there can be a period of temperature-dependent dormancy in the eggs of *A. tonsa*. Dormant resting eggs of *A. hudsonica* that matured in late fall or early winter could account for the large increases of this species in winter months. In Naragansett Bay, Sullivan and McMannus[30] presented evidence that showed timing of the seasonal appearance and disappearance of *A. hudsonica* is controlled by temperature and its effect on hatching success of resting eggs. However, other researchers[36,37,31] found other relationships between temperature and hatching.

TABLE 4

Spatial variations in the A. hudsonica/A. tonsa *ratio in the Virginia Chesapeake Bay, May 1972 and 1973*

Date		Subarea Groups		
		A,B,C,	D,E,F,	G,H
May 1972	Mean salinity of all depths sampled	21.9	18.5	15.3
	A. hudsonica/A. tonsa abundance ratio	19:1	20:1	4:1
May 1973	Mean salinity of all depths sampled	23.8	18.2	14.8
	A. hudsonica/A. tonsa abundance ratio	7:1	11:1	2:1

Cladocerans

Cladocerans appeared to be most abundant in warmer months, at the extreme geographic ranges of the sampling areas (i.e., freshwater tributaries in Maryland and high salinity stations of the Virginia area of the Bay). The temporal distribution of the most abundant freshwater species, *B. longirostris* and *D. leuchtenbergianum* is depicted in Figure 15. Greatest numbers of these two species occurred in summer at the upstream stations of the Potomac and Choptank Rivers. A number of other cladoceran species also occurred in freshwater and oligohaline environments in Maryland (See Table 1).

The estuarine cladoceran *Podon polyphemoides* occurred commonly at higher salinity Maryland stations making up 20% and 10% of the zooplankton catch in January and May, 1985, respectively. In the Virginia portion of the bay, greater than 90% of all cladocerans sampled between October 1971-June 1972 and September 1972-June 1973 were *P. polyphemoides*,[38] with May being a peak period of abundance for this species. Bosch and Taylor[39] found this species to be both euryhaline and eurythermal in its distribution, with distinct population maxima in the central portion of the Bay. These maxima were attained between 11°C and 26°C, and salinities between 8 and 18o/oo. These results are consistent with findings from both the Maryland and Virginia studies.

In the Virginia program, the greatest number of cladocerans occurred in August, 1971, averaging greater than 64,000/m³ over the study area. The subareas which contained the greatest numbers were B (averaging 153,000/m³), F (114,000/m³), H (64,000/m³) and C (53,000/m³). These "eastern" subareas of greatest abundances were dominated by *Evadne tergestina* and *Penilia avirostris*. Although both species have been shown to penetrate well into the Chesapeake Bay,[40] greatest numbers occur in higher salinity areas.

Cladoceran abundances for the second summer exhibited significantly lower numbers when compared with the previous year (Table 5). Cladocerans in August 1972 averaged only 1822/m³ and were present in significant numbers only in

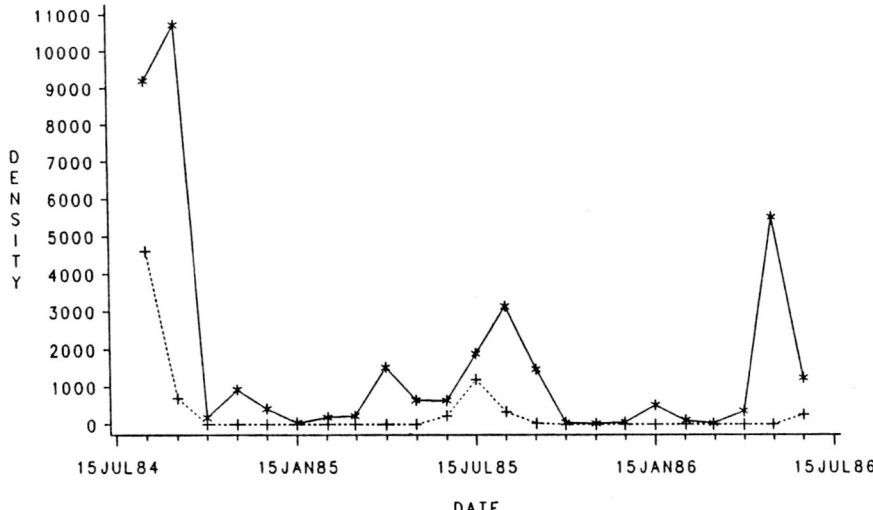

FIGURE 15. Estimated mean densities of *B. longirostris* (*)and *D. leuchtenbergianum* (+) at Maryland stations (fresh and oligohaline regions only).

TABLE 5

Mean density (#/m³) of major zooplankton taxa averaged over the Virginia study area

Regular Cruises Month	Copepods	Cladocerans	Polychaete Larvae	Barnacle Larvae	Decapod Larvae	Chaetognaths	Molluscan Larvae
Aug 1971	27,493	64,239	3	9	1,008	17	19
Sept	15,870	670	13	8	43	162	36
Oct	5,238	4	0	23	2	16	3
Nov	1,954	19	1	5	0	4	0
Dec	1,561	113	6	23	0	0	1
Jan 1972	1,711	186	2	56	1	0	100
Feb	13,352	115	3	79	11	0	5
Mar	25,395	2	4	97	8	0	8
Apr	6,882	8	6	260	7	0	1
May	6,847	4,740	31	463	27	0	120
June	173	1,070	7	238	39	0	35
July	16,700	5	1	8	9	1	6
Aug	19,953	1,822	6	4	83	13	30
Sept	19,308	3	1	3	22	78	13
Oct	672	0	1	52	0	4	4
Nov	208	1	3	7	0	20	2
Dec	574	7	6	5	1	0	1
Jan 1973	1,452	259	2	25	0	0	1
Feb	7,829	1	44	9	1	1	11
Mar	4,192	1	2	25	5	0	3
Apr	3,608	11	10	227	5	0	1
May	4,533	2,970	2	535	12	0	4
June	596	173	0	4	16	0	1,719
July	2,842	1,390	3	92	52	2	551

subareas A, B and C (occurring in very low numbers, if at all, in the other subareas). This followed a July average of 5 cladocerans/m³, which represented less than 0.05% of the total zooplankton. A year later, in July 1973, cladocerans averaged 1390/m³ over the study area, which amounted to greater than 27% of the total zooplankton abundance. It appears that depressed salinities resulting from the passage of Tropical Storm Agnes over the study area may have been a factor in reducing cladoceran populations in the summer of 1972. Grant et al.[41] indicated that Tropical Storm Agnes had its greatest impact on the more saline species *E. tergestina* and *P. avirostris*. Bosch and Taylor[40] reported that the minimum salinity tolerance of these species in Chesapeake Bay was 16o/oo and 18o/oo, respectively. In July 1972, average salinities for the entire water column in six of the eight subareas were all below 18o/oo. The continued depressed salinities in several subareas in August clearly approached or were under levels considered habitable for *E. tergestina* and *P. avirostris*. In the summer of 1978, Grant and Olney[19] found the distribution of these two species to be restricted to the bay mouth. Paffenhofer and Orcutt[42] found *P. avirostris* to be well adapted for survival at low resource levels and suggested the species to be less limited by food than by predators. In addition to the above mentioned species, Bryan,[38] and Bryan and Grant[43] also recorded *Podon intermedius, Podon leuckarti* and *E. nordmanni* from samples obtained from lower Chesapeake Bay.

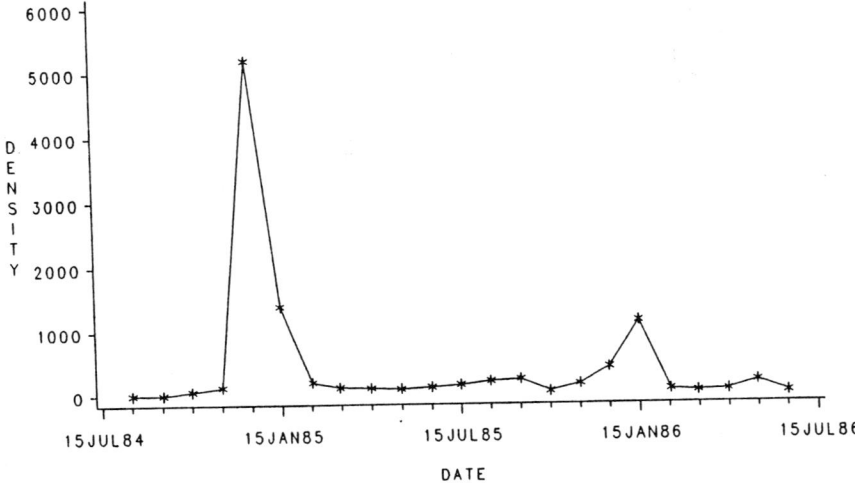

FIGURE 16. Estimated mean densities of *Polychaete larvae* at Maryland stations (freshwater regions excluded).

Polychaete Larvae

During most periods abundances of polychaete larvae were extremely low, with occasional large contributions to the plankton. In the Maryland portion of the bay, winter and summer pulses were evident (Figure 16). In December 1984, 70% of the zooplankton samples taken in mesohaline areas consisted of a spionid polychaete larvae. A similar, less marked peak, occurred in January 1986. Based on sets of benthic larvae taken in successive months it is likely that the spionid larvae was *Scolecolepides viridis*.[44] Polychaete larvae were also found at certain stations in considerable numbers in summer months.

Polychaete abundances in the Virginia portion of the bay were also highly variable (Table 5), and did not strongly indicate a consistent yearly trend. Perhaps monthly sampling is insufficient for describing abundance patterns of this meroplanktonic taxonomic group.

Barnacle nauplii

These organisms occurred most commonly in the mesohaline and polyhaline regions of the Maryland Bay, and were widespread over the Virginia portion.

Highest densities in both regions occurred in spring (April-May) (Figure 17, Table 5). Densities occurring in April 1985 in Maryland regions were the highest observed over the bay, an average approaching at 6,000/m^3. At the lower Patuxent and Choptank River stations barnacle nauplii densities exceeded 40,000/m^3 and 14,500/m^3 respectively. April 1985 densities at other meso- and polyhaline stations in Maryland ranged between 60 and 5,000/m^3. Highest densities during periods of peak barnacle nauplii densities in Virginia, ranged from 1,000-2,000/m^3.

Other Groups

Several species of cyclopoids occurred in the Maryland portion of the bay. Other groups such as decapod larvae, and chaetognoths occurred infrequently in the Maryland portion of the bay but were far more important in the Virginia portion of the bay. The ctenophore, *Mnemeopsis leidyi* appeared abundantly in warmer months.

Decapod larvae were particularly speciose in summers of 1971 and 1972 in the lower bay.[45,18] Grant and Olney[19] identified forty one decapod taxa in their 1978 summer collections. In August 1971, densities averaged over 1,000/m^3 (Table 5), with highest abundances recorded in subareas A, B and C.[18] *Upogebia affinia* were taken in greater than 90% of all stations sampled in August 1971 and 1972.

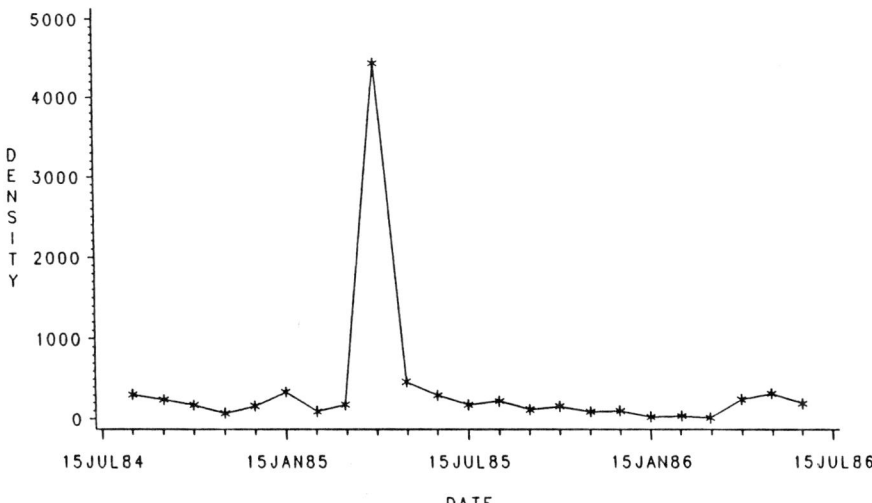

FIGURE 17. Estimated mean densities of barnacle nauplii at Maryland stations (freshwater regions excluded).

Crangon septemspinosa occurred throughout the year and was the only larval decapod species taken during winter months in the 1971-1973 survey. Grant and Olney[22] also found the decapod community to consist largely of this species in winter months, with larvae being restricted to the lower part of the study area. *C. septemspinosa*'s importance as a winter spawner was reported by Sandifer[46,47] who found it to be responsible for winter decapod peaks occurring at his bay stations. Furthermore, his study also observed the characteristic summer decapod peak, and indicated a steady decrease in species diversity proceeding from the bay mouth to the York and Pamunkey rivers.

The distribution of chaetognaths (arrow worms) into the bay suggests that they are clearly limited by salinity.[48,19] Chaetognaths were sometimes found over the entire Virginia study area, only occasionally collected in Maryland in substantial numbers, and most abundant in highest salinity areas of the lower bay. Peak abundances occurred in September 1971 and 1972 (Table 5). Grant[48] identified five species of chaetognaths from the study area, namely: *Sagitta tenuis, S. enflata, S. elegans, S. hispida* and *S. tasmanica*. Of these, *S. tenuis* was the most dominant, accounting for nearly 99% of all chaetognaths captured. Grant[48] and Grant and Olney[22,19] found two major seasonal groups, a summer-fall group including *S. tenuis, S. enflata* and *S. hispida* and a winter-spring group including *S. tenuis, S. enflata* and *S. tasmanica* (very rare). Furthermore, *S. tenuis* was observed to breed continuously throughout summer and early fall while only one generation of *S. elegans* was produced in the bay during the colder part of the year.

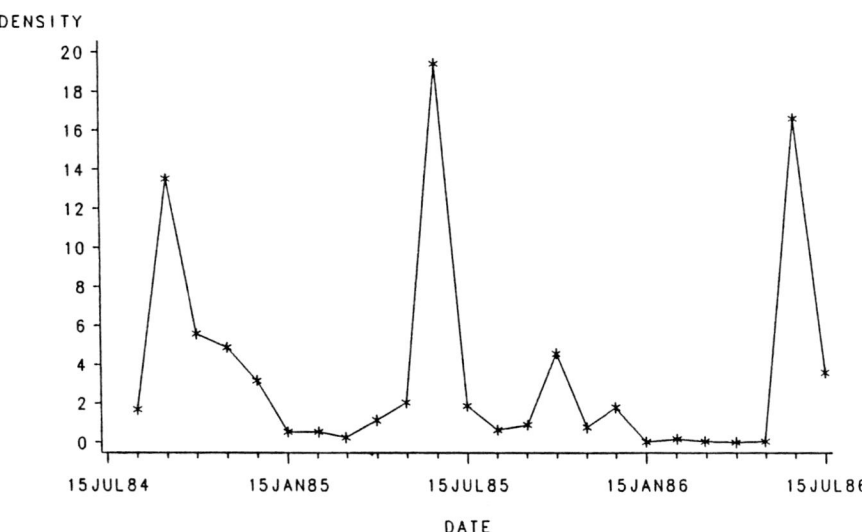

FIGURE 18. Estimated mean densities of ctenophores at Maryland stations (mesohaline and polyhaline regions only).

The monthly abundance pattern of the ctenophore *Mnemeopsis leidyi* (Figure 18) in Maryland indicates peak periods during summer in mesohaline and polyhaline waters. These jelly-like organisms have been shown to be voracious feeders, often creating substantial grazing pressure on zooplankton communities. In Narragansett Bay, *M. leidyi* cropped 5-10% of daily zooplankton standing crop throughout the bay and up to 30% in localized areas.[49] Bishop[50] estimated the cropping rate on *Acartia* to be as much as 30% in the Patuxent River, while Burrell[51] attributed 73% of total zooplankton mortality in the York River to ctenophore predation. Miller[52] showed that ctenophores removed up to 48% of the copepod biomass per day in a North Carolina estuary. In a four-year study, Deason and Smayda[53] showed that the summer pulse of ctenophores was accompanied by declines in zooplankton. Furthermore, they suggested that ctenophores not only control zooplankton population abundance, but in an indirect manner, phytoplankton abundances as well.

The present study also suggests an inverse relationship between ctenophore abundance (Figure 18) and copepod density during warmer months. Peak periods of *Mnemeopsis* that occurred in June 1985 and 1985, corresponded to low numbers of *A. tonsa* for those months (Figure 12). Furthermore, ctenophore abundances increased substantially between August and September 1984, while *A. tonsa* densities showed a decrease between the same months. Examination of the relationship between ctenophore biomass and zooplankton density at selected stations in the summer of 1984 and 1985, also yielded similar findings.[16]

MICROZOOPLANKTON

Historical Review

The first distributional survey of microzooplankton in the Chesapeake Bay was conducted by Wolfe et al.[54] The sampling scheme consisted of cross-bay transects from the mouth of the Bay to the mouth of the Patapsco River near Baltimore. The sampling scheme covered the time periods from October 1915 to September 1916 and from January 1920 to June 1921. Unfortunately, the data, based on whole water samples concentrated by centrifugation, were presented as total protozoans and the dinoflagellates were included in this group. As the dinoflagellates would make up a large though unknown portion of the protozoans, comparison to more recent data on microzooplankton (excluding dinoflagellates) is not possible. Non-loricate ciliates were present among the species list provided in this paper though members of the ciliate suborder Oligotrichina were surprisingly absent. Oligotrichines are among the dominant ciliates found in whole water samples from the Chesapeake Bay (see below).

Since the study by Wolfe et al., most studies which considered microzooplankton distributions were based on net samples which underestimate certain fractions of this group and did not consider the non-loricate ciliates. Morse[55] gave a qualitative analysis of the seasonal distribution of tintinnines, rotifers, copepods and phytoplankton at one station in the Patuxent River. Beginning in September 1943, and continuing for over two years, samples were taken biweekly using a #20 mesh (76 μm) net. Thus, most of the tintinnines and some of the rotifers would not be quantitatively collected due to their small size (63 and results presented below).

Using a continuous underway sampling system for plankton, Whaley & Taylor[56] made transects down the main stem of the Bay from the Susquehanna flats to near the mouth of the Bay. Between October 1955 and October 1956, nine such transects were conducted. A 65 μm net was employed so this study had the same limitations as the survey by Morse discussed above.

A survey of the plankton of the western shore of the Bay in the vicinity of the Calvert Cliffs Nuclear Power Plant was conducted between June 1969 and May 1970 by Mulford.[57] Monthly samples were taken by pump and passed through a 10 μm mesh net. In addition to phytoplankton, tintinnines, rotifers and copepod nauplii were enumerated. Of the microzooplankton considered in this study, only the smallest tintinnids could have passed through the net.

This review revealed several gaps in our knowledge about microzooplankton in the Chesapeake Bay. There was a noticeable lack of specific or even generic identification of the microzooplankton. When species and genera were listed,[54,55] abundances per taxa had not been reported. Taxonomic separation within the rotifers and copepod nauplii was not attempted in any of these studies. In addition, due to collection and handling techniques, the smallest, most fragile, and often most abundant microzooplankters, the non-loricated ciliates, were usually overlooked. The paucity of information on ciliated protozoa and sarcodinids in the Chesapeake Bay was stressed by Small.[58] In light of the above it is concluded that published values of total microzooplankton abundance have not previously existed for the Chesapeake Bay.

METHODS

To promote better estimation of microzooplankton abundances in the future and to allow for data comparisons between studies, a brief discussion of collection and preservation techniques is given here. The fraction of the microzooplankton most often missed in survey studies is the small non-loricate ciliates. Ciliates, especially those of the suborder Oligotrichina, are known to burst when collected with a net and some can squeeze through even a 10 μm net. To properly enumerate all the non-loricate ciliates, whole water samples should be used and the Utermöhl method of counting phytoplankton[59] using

TABLE 6

Microzooplankton (l^{-1}), Upper Choptank River (MET 5.1, Surface layer), September 11, 1985.

	Sampling						Method
	44μm Net			20 μm Net			Whole Water
Organism	A^1	B^2	X	A^3	B^4	X	A^5
Rotifera							
Synchaeta spp.	44.8	45.8	45.3	57.2	76.5	66.8	0
Brachionus angularis	156	179	167	185	207	196	100
Brachionus bidenta	2.4	0	1.2	4.4	0	2.2	0
Unidentified rotifers	0	0	0	17.6	13.5	15.6	0
Total	203	225	214	264	297	281	100
Tintinnina							
Tintinnopsis cratera	46.5^6	71.9^8	59.2	189	212	200	100
T. Fimbriata	274	470	372	356	324	340	400
T. rapa/parva	60.2^7	37.1	48.6	$3,986^9$	$3,375^9$	3,681	7,700
T. turbo	54.3	74.1	64.2	57.2	63.0	60.1	200
T. subacuta	0	6.5	3.25	0	0	0	0
T. minuta	0	0	0	0	27	13.5	1,300
Tintinnidium spp.	163	129	146	1,245	1,354	1,300	6,600
Tintinnidiidae sp. #1	0	0	0	74.8	45	59.9	1,200
Unidentified Tin.	4.8	0	2.4	0	0	0	100
Total	602	788	695	5,909	5,400	5,655	17,600
Non-loricate ciliates							
Didinium sp.	4.8	10.9	7.85	26.4	22.5	24.4	100
Oligotrichina	2.4	0	1.2	4.4	4.5	4.4	3,000
Unidentified ciliates	2.4	4.36	3.38	0	0	0	2,500
Total	9.6	15.3	12.4	30.8	27.0	28.8	5,600
Sarcodinids	61.4	43.6	52.5	110	162	136	800
Nauplii and Meroplankton							
Copepod Nauplii	132	179	156	136	171	154	0
Barnacle nauplii	0	2.2	1.1	0	0	0	0
Pelecypod larvae	0	10.9	5.45	0	0	0	
Unidentified larvae	14.2	2.2	8.2	30.8	9	19.9	200
Total	146	194	170	167	180	174	200
Total microzooplankton		1,144			6,275		24,300
% of whole water sample		5%			26%		

[1] n = #1 $^{-1}$ x 0.42
[2] n = #1 $^{-1}$ x 0.46
[3] n = #1 $^{-1}$ x 0.23
[4] n = #1 $^{-1}$ x 0.22
[5] n = #1 $^{-1}$ x 0.010
[6] n = #1 $^{-1}$ x 1.27
[7] n = #1 $^{-1}$ x 0.85
[8] n = #1 $^{-1}$ x 0.92
[9] n = #1 $^{-1}$ x 0.036

an inverted microscope should be employed. Unfortunately, some of the larger ciliates as well as rotifers and copepod nauplii will not be present in the Utermöhl chambers in sufficient numbers for reliable estimates. Therefore, a combination of net (10-44 μm, 25-50l) samples and whole water (5-25 ml settled) samples would provide the most complete information.

The fixative and method of fixation are also important in obtaining accurate numbers. Dale and Burkill[60] found a reasonable agreement between counts made on living and fixed ciliates except that live counts were higher when detritus levels were high. They settled 50 ml of sample for the fixed counts, while we recommend that only 5-25 ml be settled in areas with high levels of detritus such as the Chesapeake Bay. Bouins fixative[61] and buffered formaldehyde (2% final concentration) are suitable fixatives. For the ciliated protozoa, the way in which the fixative was added was found to be critical. When Bouins fixative was added dropwise to a 1 ml depression containing sea water and freshly collected oligotrichine ciliates, those individuals that were in the immediate area of the drop were fixed very well. Those that were away from the drop and swam toward it experienced a gradient of the fixative concentration, burst and were no longer recognizable as organisms. Thus, it is recommended that the fixative should be mixed with the sample very rapidly. For instance, the sample should be poured onto the fixative resulting in rapid mixing and in the organisms in the sample experiencing only high concentrations of fixative. Pouring of the fixative onto the sample is not recommended as a temporary gradient of fixative may be set up which may result in destruction of cells.

As part of the Chesapeake Bay Water Quality Monitoring Program, a comparison of collection techniques, 44 μm mesh net (currently in use) vs. 20 μm mesh net vs. whole water, was made for microzooplankton.[62] Two stations were sampled, one with high loads of detritus in the upper Choptank River and one from an area having much lower detrital levels near the mouth of the Chester River (stations MET 5.1 and MET 4.2, respectively, Figure 1). The results, presented in Tables 6 and 7 indicate large differences between the methods for certain organisms. The values underlined in the tables point out interesting comparisons between methods for particular groups of organisms. Note that for the whole water samples the number counted is the value given divided by 100. Thus values less than 3000 in this column should not be used for comparisons.

The rotifers showed slightly higher numbers in the 20 vs. the 44 μm net samples for each station. There were too few rotifers in the whole water samples to make reasonable comparisons to the other counting methods. *Synchaeta* is the dominant rotifer genus in the estuarine portion of the Bay and its tributaries. The data suggest that some portion of this population was being missed with the 44 μm net.

There were 8 times as many tintinnine ciliates in the 20 μm than in the 44 μm net samples in the Choptank River sample but this difference was only about 3 fold in the Chester River sample. The whole water samples had even higher

TABLE 7

Microzooplankton (l^{-1}), Chester River (MET 4.2, Surface layer), September 11, 1985.

	Sampling						Method
	44μm Net			20 μm Net			Whole Water
Organism	A^1	B^2	X	A^3	B^4	X	A^5
Rotifera							
Synchaeta spp.	160	145	152	164^8	220	192	400
Brachionus plicatilis	1.86	2.4	2.13	0.64	0.66	0.65	0
Keratella earlinae	0	0	0	0.64	0.66	0.65	0
Kellicottia sp.	0	0	0	0.64	0	0.32	0
Trichocerca sp.	0	0	0	0	0	0	100
Total	162	147	154	166	221	193	500
Tintinnina							
Tintinnopsis dadayi	53.3	34.2	43.8	57.6	21.8	39.7	200
T. tocanteninsis	4.34	3.6	3.97	108	128	118	400
T. subacuta	23.9^6	17.4	20.6	19.8	12.5	16.2	300
T. radix	4.34	1.20	5.54	1.92	5.9	3.91	0
T. rapa/parva	1.24	0	0.62	0.64	0.66	0.65	100
T. fimbriata	8.68	2.4	5.54	3.2	1.98	2.59	0
T. lata	0	0	0	0.64	0.66	0.65	100
T. baltica	0	0	0	1.28	0	0.64	0
T. beroidea	0	0	0	0.64	1.32	0.98	0
T. minuta	0	0	0	0	0	0	2,700
Tintinnidium spp.	1.86	0	0.93	0	0	0	500
Leprotintinnus		1.20	0.60	0	0	0	0
Eutintinnus pectinis	0.62	0.60	0.61	57.6^8	59.4	58.5	300
E. apertus (type)	6.2	2.4	4.3	6.4	6.6	6.5	0
Tintinnidiidae sp.#1	0	0	0	0	0	0	100
Total	104	63	83.5	258	237	247	4,600
Non-loricate ciliates							
Peritrichs	2.48	0	1.24	100^8	93.7	96.8	2,100
Oligotrichina	1.86	2.4	2.13	16.6	21.1	3l7.7	6,400
Mesodinium	0	0	0	0	0	0	200
Unidentified ciliates	0	0	0	2.56	0	1.28	1,300
Total	4.34	2.4	3.37	119	115	117	10,000
Nauplii and Meroplankton							
Copepod nauplii	96.5^7	74.4	85.4	109	75.2	92.1	100
Polychaete larvae	9.92	7.8	8.86	6.4	10.6	8.5	0
Pelecypod larvae	9.92	3.0	6.46	6.4	7.92	7.16	0
Total	16	85.2	101	122	93.7	108	100
Total microzooplankton			342			665	15,200
% of Whole Water Sample			2%			4%	

[1] $n = \#1^{-1} \times 1.61$ [3] $n = \#1^{-1} \times 1.56$ [5] $n = \#1^{-1} \times 0.01$ [7] $n = \#1^{-1} \times 1.94$
[2] $n = \#1^{-1} \times 1.66$ [4] $n = \#1^{-1} \times 1.52$ [6] $n = \#1^{-1} \times 3.22$ [8] $n = \#1^{-1} \times 1.07$

concentrations of these protozoans than the 20 μm net with 3 and 19 fold increases for the Choptank and Chester River samples, respectively. This was 25 and 55 times more tintinnines than caught on the 44 μm net, respectively. It is interesting to note that the tintinnine ciliate that was most abundant in the whole water sample from the Chester River was not even seen in the 44 or 20 μm net samples. The species *T. cratera, T. rapa/parva, T. minuta, Tintinnidium* spp., *E. pectinis,* and Tintinnidiidae sp. #1 were underestimated with the 44 μm net. These same species, with the exception of *T. cratera,* were also underestimated by the 20 μm net. *T. fimbriata, T. turbo, T. subacuta* and *T. dadayi* were among the few species of tintinnines which appeared to be retained quantitatively by the 44 μm mesh net. Note also that many of the tintinnines were not found in the whole water samples in sufficient numbers to provide reliable estimates of their abundance.

The nonloricate ciliates were grossly underestimated by both net samples. Many of these forms are small and some, especially the Oligotrichina, are known to burst on contact with nets during sampling or are very elastic and will squeeze through openings smaller than their smallest body dimensions. The copepod nauplii and merozooplankton, which are generally larger organisms than those discussed above, were caught equally well by both nets and were too sparse in the whole water samples to make any reliable comparisons. However, Beers and Stewart[63] found that some nauplii passed through a 35 μm mesh net.

Considering the whole water sample as representing the "true" total microzooplankton numbers, the percentages retained by the 44 and 20 μm mesh nets are 5 and 26%, respectively, for the Choptank River sample, and 2 and 4%, respectively, for the Chester River sample. Thus, on a numbers basis both net sampling procedures seriously underestimate the total microzooplankton. Similar results, 88% of the microzooplankton numbers passing a 35 μm mesh net, were found by Beers & Stewart.[63]

In terms of biomass, the percent loss is not as great with 49 and 44% retained by the 44 μm net for the Choptank and Chester Rivers, respectively. To estimate biomass for this comparison and for the data presented below, rotifers, tintinnine ciliates, non-loricate ciliates, sarcodines, copepod nauplii, pelecypod larvae and other larvae were assumed to have an average biomass of 8.2×10^{-5}, 3.2×10^{-6}, 5.2×10^{-6}, 1.0×10^{-5}, 7.1×10^{-5}, 2.0×10^{-4}, and 1.0×10^{-3} mg carbon per individual, respectively. In the above comparison, combined tintinnine and non-loricate ciliate biomass is greater than the biomass of all the other microzooplankton combined. It follows that if the turnover rates of these ciliates are greater than those of the other taxa (expected due to their smaller size), then ciliate secondary production would also be greater. Thus the ciliate component of the microzooplankton is an important fraction that is often grossly underestimated.

DISTRIBUTION OF MICROZOOPLANKTON
IN THE CHESAPEAKE BAY

The microzooplankton distributions discussed below are based on samples collected under the supervision of Drs. Kevin Sellner and David Brownlee, Academy of Natural Sciences, Benedict Estuarine Research Laboratory, Benedict, Maryland, as part of the State of Maryland's Office of Environmental Program's Water Quality Monitoring Program on the Chesapeake Bay and its tributaries. Monthly since August 1984, two composite samples, one from five depths below and the other from five depths above the pycnocline, were taken at the same 16 stations sampled for mesozooplankton (Figure 1). The samples were collected on a 44 μm net and therefore only the larger microzooplankton were counted and enumerated (see discussion above).

General Trends

Microzooplankton abundance and biomass averaged over the entire year, decreased with increasing salinity. This was true for the main stem of the Bay as well as the tributaries; Potomac, Patuxent and Choptank Rivers. Month to month variations in this pattern did occur, the most important of which was in May, 1985. At this time, there was a large peak in microzooplankton in the oligohaline stations of the main stem, Potomac River and Choptank River. This peak was caused, for the most part, by a bloom of the tintinnine ciliate *Tintinnopsis fimbriata*. Unfortunately, the salinity structure of the Patuxent River was such that the up-river station (PXT 0402) was fresh water and that of the mid-river station (XED 4892) was mesohaline. Thus, the oligohaline portion of the river, in which a bloom of *T. fimbriata* might have occurred, was not sampled in May.

Averaged over all stations, the total microzooplankton showed late summer and fall peaks in abundance and biomass (Figure 19). There was also abundance and biomass peaks in the spring for many of the individual stations though often occurring in different months (March-May) which resulted in them being smoothed out in the averaged data. In terms of abundance, rotifers were the dominant organism for much of the year except for March, May, and August of 1985 when copepod nauplii, tintinnines, and copepod nauplii were dominant, respectively. Tintinnines were well represented from May through the early fall. Sarcodinids, restricted mostly to the freshwater stations, were never abundant but represented their greatest percent of abundance in the colder months when total microzooplankton abundances were small. Copepod nauplii were most significantly represented in the months of October, 1984, February through April and June through September, 1985. Based on the mesozooplankton data, the spring peak in abundance of copepod nauplii is assumed to be predominantly *Eurytemora affinis* while the summer and fall peaks of nauplii would mostly

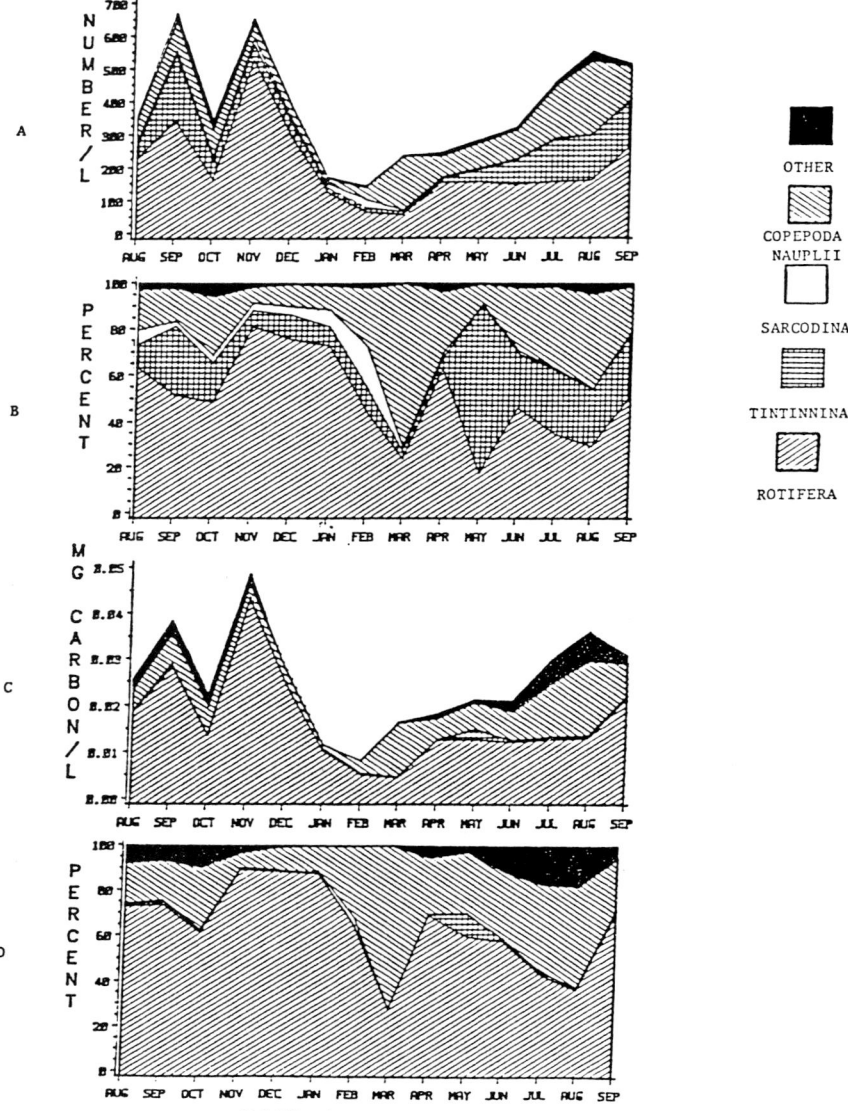

FIGURE 19. Seasonal distribution of microzooplankton averaged over all stations. A. Abundance. B. Relative abundance. C. Biomass. D. Relative biomass.

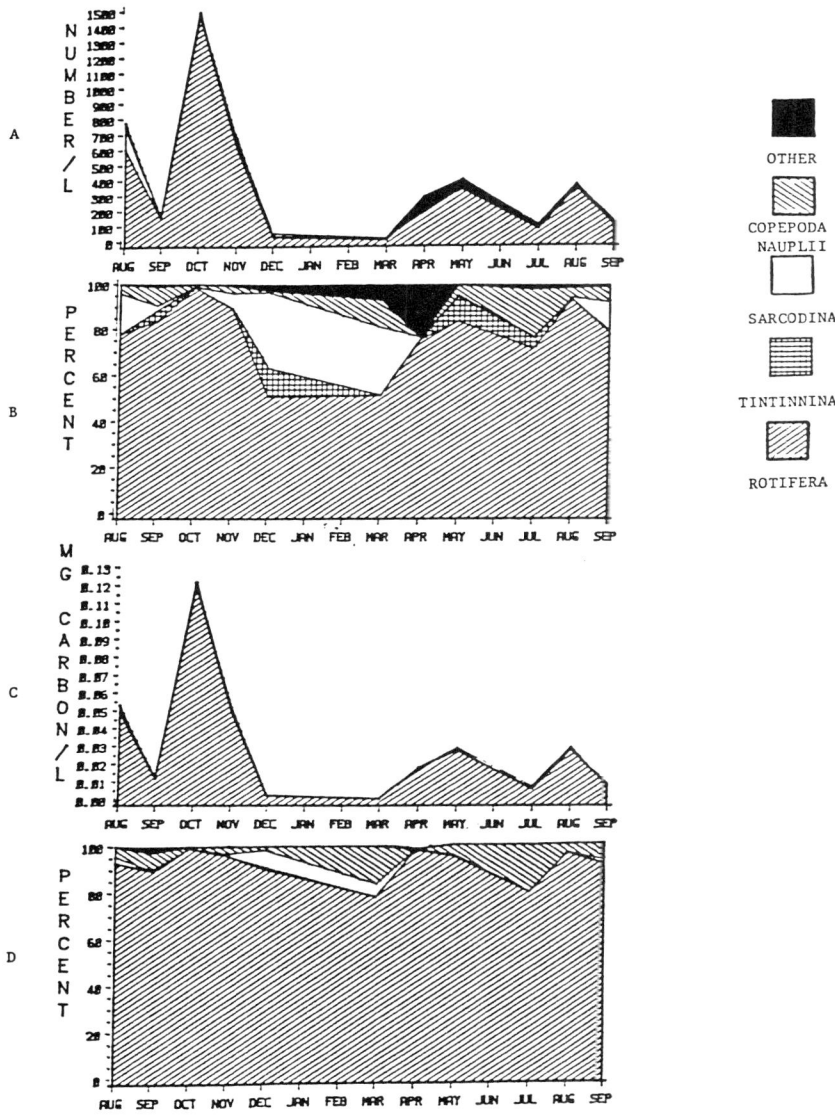

FIGURE 20. Seasonal distribution of microzooplankton. Main Bay. Station MCB 1.1. A. Abundance. B. Relative abundance. C. Biomass. D. Relative biomass. No counts available for January, February, and June.

be comprised of *Acartia tonsa*.

Rotifers and copepod nauplii represented the majority of the biomass throughout the year with rotifers being dominant except in March and August 1985 (Figures 19C & D). The tintinnines (greater than 44 μm) because of their small size and the sarcodinids due to their low numbers did not make up a significant portion of the biomass. The biomass of the "other microzooplankton" category became important in the summer of 1985 due predominantly to the presence of merozooplankton.

Below, the seasonal microzooplankton distributions are described for each of the five stations along the main stem of the Bay. As these stations range from fresh water to polyhaline, they are used to represent the patterns found in the tributaries. The distributions are then summarized and any important exceptions to the main bay pattern are discussed.

Main Stem

The biomass of microzooplankton, over the entire time period considered showed a general decrease from fresh to saline waters (Figures 20-24). At the freshwater station (MCB 1.1, Fig. 20), rotifers were the dominant throughout the year (always greater than 50% by numbers and 80% by biomass) though the sarcodinids represented a significant portion (up to 30%) of the numbers in the colder months when total microzooplankton numbers were low. Copepod nauplii represented 10 to 20% of the biomass in February-March and in June-July, 1985. There was a very large peak in total microzooplankton abundance in October with minor peaks in May and August.

At the oligohaline (transition zone) station (MCB 2.2, Fig. 21), total microzooplankton peaked at over 4000/L in May due mostly to a bloom of the tintinnine, *Tintinnopsis fimbriata;* lesser peaks occurred in September, 1984 and December, 1985. Rotifers were less dominant (generally less than 55% by numbers and 65% by biomass though reaching about 90% of the biomass in December) in the transition zone than in any of the other salinity regimes. Tintinnines, on the other hand, exhibited their greatest numbers and biomass in the oligohaline region. Copepod nauplii represented 50% or more of the biomass in February-March and 30 to 40% in September-October, 1984 and July and August, 1985. Again the late winter-early spring peak should be composed predominantly of the species *Eurytemora affinis* and the summer and fall peaks due to *Acartia tonsa.* Other microzooplankters, especially merozooplankton, were 15 to 50% of the biomass in the fall, 1984 and summer of 1985.

The remaining three mesohaline stations (MCB 3.3c, Fig. 22; MCB 4.3c, Fig. 23; MCB 5.2, Fig. 24) all showed a similar pattern with major peaks in abundance in late summer-early fall and minor peaks in winter and spring. Summed over the entire time period, rotifers were the dominant organisms at all three stations though they were dominant for a lesser portion of the year the further

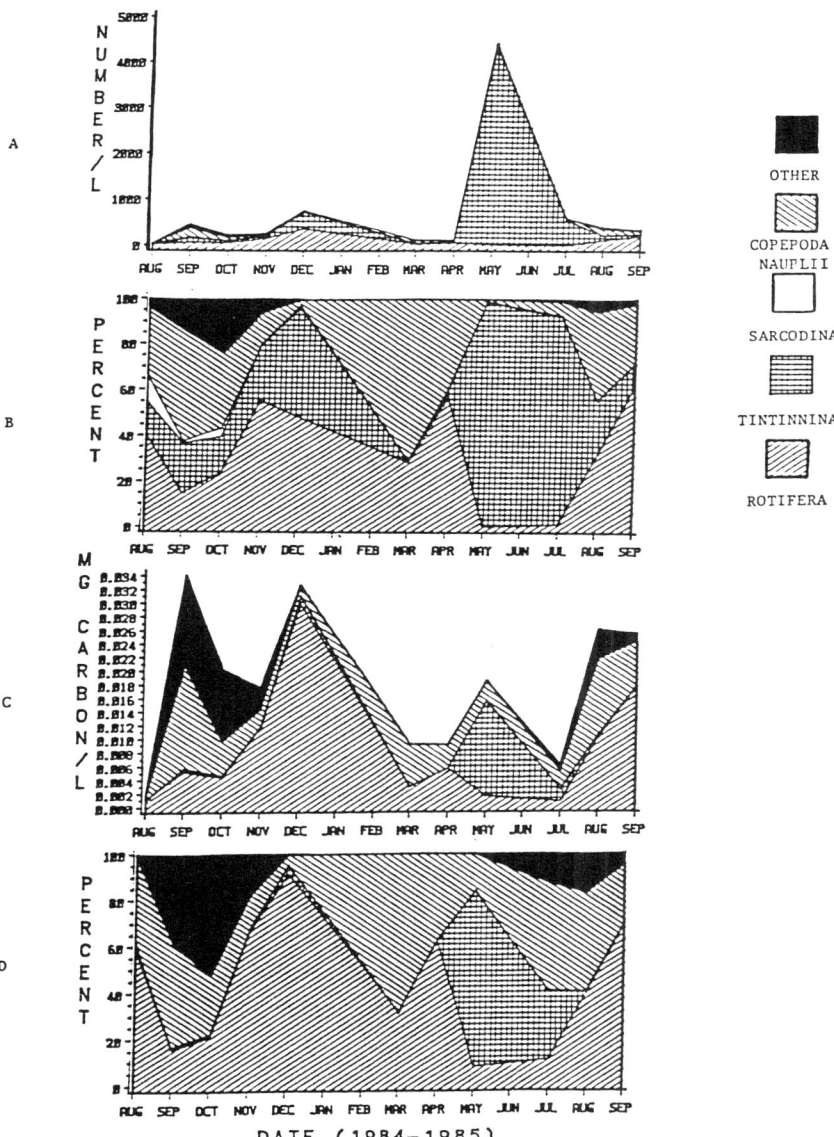

FIGURE 21. Seasonal distribution of microzooplankton. Main Bay. Station MCB 2.2. A. Abundance. B. Relative abundance. C. Biomass. D. Relative biomass. No counts available for January, February and June.

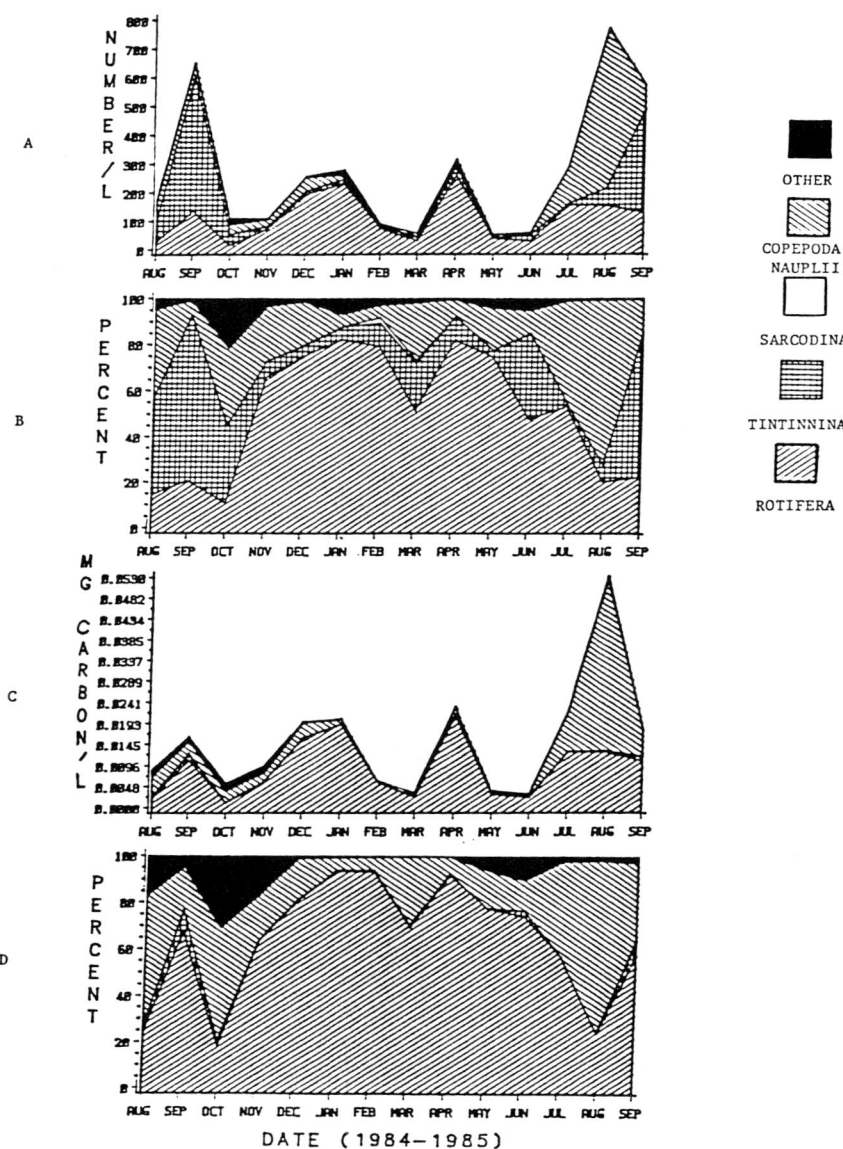

FIGURE 22. Seasonal distribution of microzooplankton. Man Bay. Station MCB 3.3c. A. Abundance. B. Relative abundance. C. Biomass. D. Relative biomass.

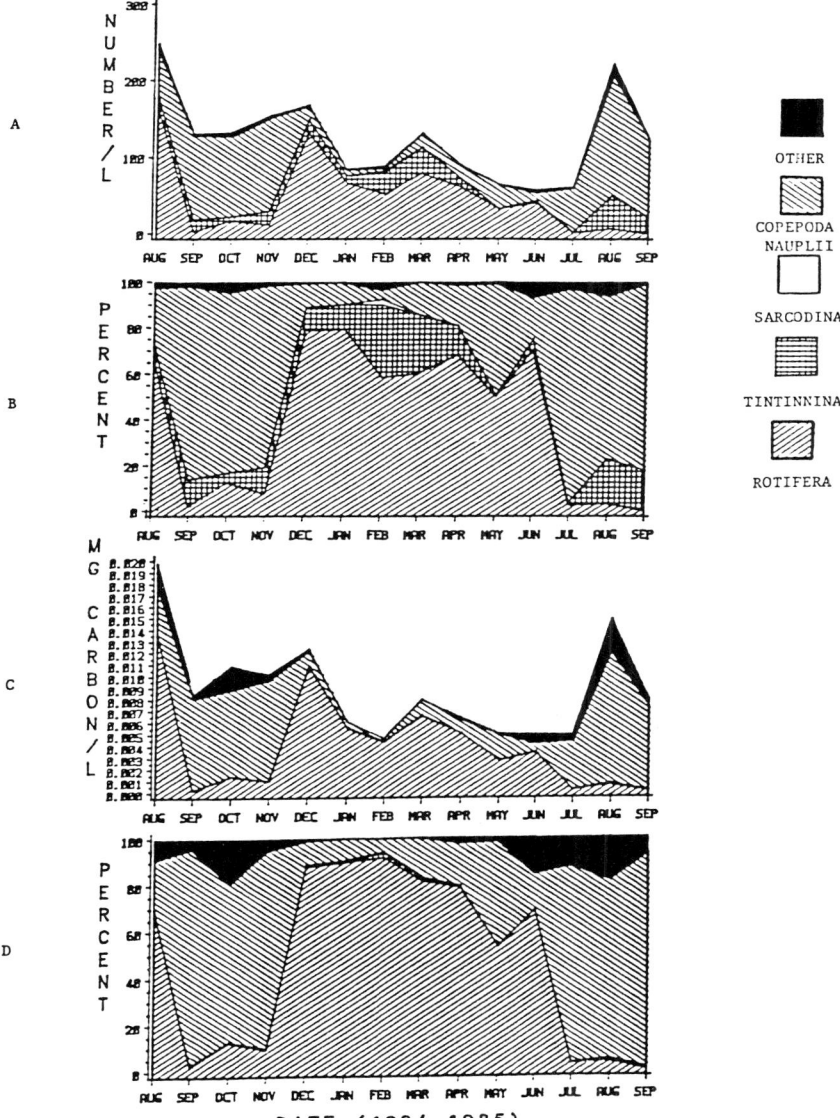

FIGURE 23. Seasonal distribution of microzooplankton. Main Bay. Station MCB 4.3c. A. Abundance. B. Relative abundance. C. Biomass. D. Relative biomass.

down Bay the station. The species composition of rotifers, dominated by *Synchaeta* sp., was different from the very diverse rotifer taxa found at the freshwater station. Inversely, copepod nauplii, especially *Acartia tonsa,* became more important down Bay. Tintinnines, except for relatively high abundances in August-October, 1984 and September 1985 at station MCB 3.3c, were not a very important component of the biota at these stations in either numbers or especially biomass. "Other microzooplankton", predominantly merozooplankton, were generally 10 to 20% of the biomass in summer and fall at the three stations though reached 40-50% in June-July, 1985 at station MCB 5.2.

SUMMARY

The microzooplankton distributions can be generally divided into the freshwater, oligohaline and mesohaline-polyhaline zone patterns. The freshwater pattern was characterized by a large fall peak in biomass and lesser peaks in spring and summer (main stem, MCB 1.1, Fig. 20; Patuxent River, PXT 0402). The Potomac River freshwater station (XEA 6596) was somewhat different with only a small fall peak and then with increasing biomass from the spring through the summer of 1985. This summer increase in biomass was associated with the development of a bloom of the blue green alga, *Microcystis aeruginosa.* Rotifers were strongly dominant at all freshwater stations.

The oligohaline zone stations (main stem, MCB 2.2, Fig. 21; Potomac River, XDA 1177; Patuxent River, XED 4892; Choptank River, MET 5.1; Baltimore Harbor, MWT 5.1) generally showed important late summer-early fall (1984 and 1985) and winter peaks in biomass and a lesser spring peak. The Potomac River transition zone station differed from this pattern by not having a winter peak and by having a large biomass peak in early spring (March). The Patuxent River transition zone station differed by having a very strong fall peak in biomass. During May in the main Bay, Potomac River, and Choptank River transition zone stations, the tintinnine, *T. fimbriata* occurred in bloom proportions. This species also showed an increase in numbers in May at the Baltimore Harbor station though at much lower densities. In general, tintinnines were relatively more abundant and represented a greater proportion of the total at the transition zone stations. In early spring, nauplii of the copepod *Eurytemora affinis* were an important component of the microzooplankton at these stations, while the population levels of all other taxa were suppressed.

The mesohaline zone stations (Main Bay, MCB 3.3c, Fig. 22, MCB 4.3c, Fig. 23, MCB 5.2, Fig. 24; Potomac River, XBE 9541; Patuxent River, XDE 5339; Choptank River, MET 5.2; Chester River, MET 4.2; Tangier Sound, MEE 3.1) were characterized by major biomass peaks in late summer-early fall with lesser winter, spring and summer peaks. The Patuxent River station differed in that

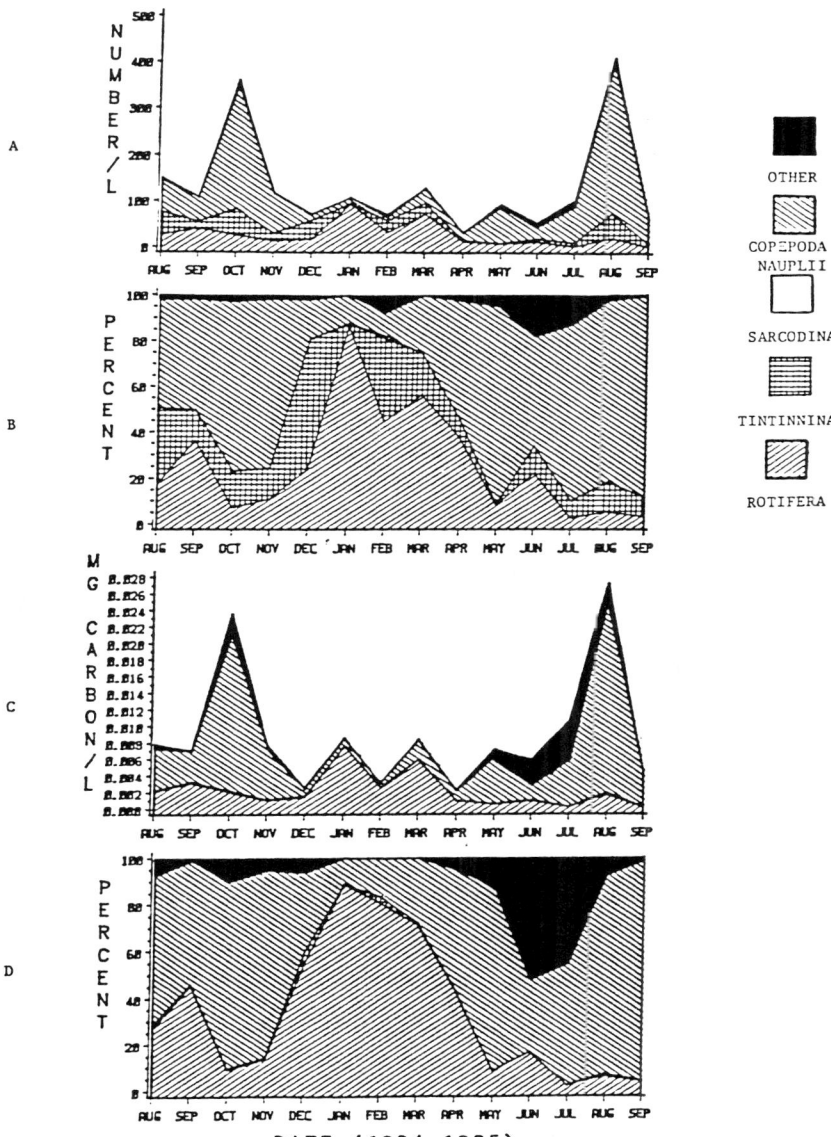

FIGURE 24. Seasonal distribution of microzooplankton. Main Bay. Station MCB. 5.2. A. Abundance. B. Relative abundance. C. Biomass. D. Relative biomass.

the late summer-early fall peaks were not especially strong compared to the peaks at other seasons. The Choptank River station did not show the second (1985) late summer-early fall peak. The Chester River differed in having a strong winter peak and lesser peaks in fall, 1984, and spring and summer, 1985. In general, the dominants were rotifers during the colder months and copepod nauplii (largely *Acartia tonsa)* and invertebrate larvae in the warmer months. Exceptions were found in the Potomac, Patuxent, and Choptank Rivers where rotifers also were important in the late summer-early fall of 1984 and in the Chester River where rotifers were dominant at all times except in March when copepod nauplii represented just over 50% of the biomass.

Microzooplankton Taxa

The thirty most abundant microzooplankton taxa are listed in Table 8 in order of dominance based on the percent of total abundance over all stations and months. Note that *Synchaeta* spp., copepod nauplii, Difflugiidae, Cyphoderiidae and Centropyxidae contain multiple species.

Rotifers were important in all salinity regimes but were less dominant at the transition zone stations. In the fresh water stations, the rotifer taxa were diverse while at the higher salinity stations, species of the genus *Synchaeta* dominated. At the fresh water stations, the peak abundances occurred during the warmer months (summer and early fall) while in the more saline areas the rotifers represented a greater proportion of the total numbers in the cooler months.

Synchaeta species were difficult to separate taxonomically in the preserved samples. *Synchaeta* sp. A and B in Table 8 were morphologically distinct enough to count separately while those included in *Synchaeta* spp. constituted a mix including but not limited to *S. stylata, S. pectinata* and *S. oblonga.* Species identification of *Synchaeta* is based on characteristics observed in live specimens or on the detailed structure of the trophi (hard structure of the feeding apparatus) which requires special processing to extract.

Tintinnine ciliates were most abundant in the transition zone stations and were rarely abundant in freshwater. They were generally present throughout the year with peak abundance occurring in spring, summer or fall depending on the particular station. Large blooms of *T. fimbriata* occurred at the oligohaline zone stations in May.

Several problems exist in the present study of tintinnid distributions. For example, only a small percentage of the tintinnines are included due to sampling techniques. Averaging the data from Tables 6 and 7, it can be seen that 97% of the tintinnine numbers and 64% of the tintinnine species have been missed. An even higher percentage of the non-loricate ciliates are lost. Further study of the smaller than 44 μm fraction of the microzooplankton needs to be conducted. In addition, the taxonomy of the Tintinnina is presently based solely on the morphology of the lorica though this can at times be misleading and the use of cytological criteria has been shown to be important.[64,65]

TABLE 8

Mean, maximum and percent of total abundance over all stations and dates.

Taxon	Taxanomic Group	Density (#/L)	Maximum Density (#/L)	Percent of Total
Synchaeta spp.	Rotifera	117.1	2,513	29.38
Tintinnopsis fimbriata	Tintinnina	106.0	4,334	20.59
Copepod nauplii	Copepoda	39.2	1,006	19.45
Synchaeta sp. A	Rotifera	34.2	609	2.89
Tintinnopsis dadayi	Tintinnina	17.6	486	2.89
Keratella cochlearis cochlearis	Rotifera	27.2	481	2.80
Polyarthra sp.	Rotifera	36.8	378	2.38
Difflugiidae	Sarcodina	14.3	270	2.06
Synchaeta sp. B	Rotifera	65.2	296	1.80
Tintinnopsis subactua	Tintinnina	11.7	442	1.76
Tintinnopsis radix	Tintinnina	22.1	439	1.37
Brachionus angularis	Rotifera	24.9	205	1.21
Trichocerca sp.	Rotifera	14.5	264	1.11
Brachionus calyciflorus	Rotifera	16.2	170	1.05
Pelecypoda larvae	Pelecypoda	7.4	63	1.02
Keratella cochlearis tecta	Rotifera	16.0	232	0.89
Filinia sp.	Rotifera	21.0	223	0.66
Polyarthra vulgaris	Rotifera	17.4	101	0.62
Notholca acuminata	Rotifera	12.3	93	0.48
Brachionus plicatilus	Rotifera	11.4	66	0.33
Hexarthra mira	Rotifera	16.3	132	0.30
Tintinnopsis subacuta	Tintinnina	4.5	28	0.24
Filinia longiseta	Rotifera	10.9	47	0.24
Acineta sp.	Ciliophora	15.6	149	0.22
Brachionus caudatus	Rotifera	10.0	63	0.22
Keratella sp.	Rotifera	15.2	132	0.22
Cyphoderiidae	Sarcodina	1.8	20	0.17
Centrophyxidae	Sarcodina	2.7	30	0.17
Conochilus unicornis	Rotifera	15.6	68	0.16
Arcella sp.	Sarcodina	2.9	46	0.15

The Sarcodina, restricted mostly to the fresh water stations, were found in low numbers throughout the year. In winter when total microzooplankton numbers were low, these amoeboid organisms could make up a significant portion of the microzooplankton community. Most of the species encountered belonged to the family Difflugiidae. Sawyer[66] studied the free-living marine amoebae from the upper Chesapeake Bay.

The distribution of copepod nauplii reflected the distribution of the two dominant adult copepods, *Eurytemora affinis* Figure 13) and *Acartia tonsa* (Figure 14). The major abundance peaks of the adults of these two species were separated both temporally and spacially. *E. affinis* reached it highest numbers in the oligohaline portions of the bay and its tributaries in late winter early spring while *A. tonsa* peaked in summer and early fall in the higher salinity regions. As expected, the nauplii followed the adult distributions.

COMPARISON TO OTHER STUDIES

There have been few studies that investigated the distribution over time and space of microzooplankton in an estuary. Dale and Burkill[60] examined the distribution of microzooplankton along a salinity gradient (less than 0.5 to 31 o/oo) in the River Tamar Estuary, southwest England, at a single time period, April 1980. In agreement with the Chesapeake Bay study, they found that tintinnines and oligotrichines were most abundant at the lower salinity stations. In contrast, they found that the remaining ciliate taxa increased in numbers with increasing salinity up to 27 o/oo and then decreased seawards. In the comparison study discussed above (Tables 6 and 7), tintinnine, dominated in the oligohaline Choptank station and oligotrochines dominated at the mesohaline Chester station which fits this pattern. However, these groups of ciliates were not sampled quantitatively in the regular collection of the Chesapeake Bay Monitoring Study (due to use of nets and the size mesh of the nets, see above) and therefore, no statistically sound comparison can be made.

The seasonal cycle of ciliated protozoa from a single local in the Kiel Bight, River Tamar Estuary, and Long Island Sound have been investigated by Smetacek,[67] Burkill,[68] and Capriulo & Carpenter,[69] respectively. Hargraves[70] examined the seasonal cycle of tintinnines at four stations in Narragansett Bay, Rhode Island. Smetacek found a bimodal seasonal distribution of ciliated protozoa in Kiel Bight with spring and fall maxima. Tintinnines were relatively unimportant compared to the naked ciliates. Burkill found that the naked ciliates peaked in biomass in May whereas the tintinnid ciliates exhibited a bimodal seasonal distribution with peaks in May and in October. Except for the spring peak when naked ciliate biomass was about equivalent to tintinnine biomass, the former showed approximately an order of magnitude lower biomass than the later. Capriulo and Carpenter also found naked ciliates to bloom in spring and to be relatively unimportant relative to tintinnine ciliates at other times of the year. They found that tintinnines peaked in summer rather than showing the spring and fall peaks described by Smetacek and by Burkill. Hargraves found a peak in spring and a lesser peak in late summer for tintinnines. The studies of Capriulo and Carpenter and Hargraves were conducted in high salinity estuaries with salinity always above 25o/oo which is higher than any of the values for the Chesapeake Bay study discussed above.

In the Chesapeake Bay, total microzooplankton averaged over all stations, showed peak abundances in late summer and fall, though spring and winter blooms occurred at particular stations. For instance, the oligohaline zone stations had a large spring peak of tintinnines while the mesohaline zone stations showed late summer and fall peak in abundance.

Relationship to Phytoplankton and Mesozooplankton

The microzooplankton peaks in spring and fall followed by one month or coincided with the peaks in diatom-rich phytoplankton. The summer and winter peaks appeared to be associated with blooms of small flagellates and 1-2 μm diameter coccoid cells (perhaps bluegreen algae) and in summer also with high primary production. Heterotrophic bacteria and heterotrophic flagellates, also a potential food source for microzooplankton, were not investigated in this study.

Low microzooplankton numbers occurred at times when mesozooplankton were abundant. For instance, in February and March in the oligohaline zone when *E. affinis* was abundant, the microzooplankton exclusive of copepod nauplii became very depressed (Figure 21). This also occurred in late summer early fall in the meso- and polyhaline stations when *A. tonsa* was prevalent (Figures 22-24). The peaks in abundance of the mesozooplankton followed by one to two months the peaks in both the microzooplankton and the phytoplankton distributions. The importance or magnitude of the energy flow from microzooplankton to higher trophic levels in nature has not been well documented though the results of the micro- and mesozooplankton distributions in the Bay argue for a strong coupling between these two groups at all times of the year except perhaps during the spring bloom. The link between microzooplankton and mesozooplankton has been established in the laboratory.[71,72,73] The distributional data presented here and these laboratory studies suggest that the flow of energy between the microzooplankton and mesozooplankton in nature deserves further study.

In April during the spring phytoplankton bloom, both micro- and mesozooplankton showed concurrent peaks in abundance suggesting that food was not limiting. In support of this hypothesis, sediment trap data (low C/N ratios and C/chlorophyll levels,[74]) suggested that phytoplankton at this time were sedimenting into the lower water without being reworked (i.e. grazed by zooplankton). This deposition of organic matter may have initiated the beginnings of low oxygen bottom water in the Chesapeake Bay.

ACKNOWLEDGMENTS

The mesozooplankton and microzooplankton programs discussed for the Maryland portion of the Chesapeake Bay were supported by the State of Maryland's Office of Environmental Programs' Water Quality Monitoring Program. William Burton and Ivy Moss assisted in the field collections and in the laboratory analyses and Eric Ross provided analytical assistance for the Maryland mesozooplankton program. Dr. George C. Grant directed the Virginia zooplankton program and provided advice and discussions concerning this study. Field collections of the microzooplankton were conducted by Richard

Lacouture, Charles Parrish, Sharyn Hedrick, Kevin Braun, and Carolyn Watson. Stella Brownlee and Richard Jacobsen provided the laboratory analyses for the microzooplankton and Stella Brownlee, Suzan Gunsalus, Marjorie Marsh, and Richard Osman assisted in the computer analyses and graphics. Dr. Kevin Sellner assisted in the direction of the microzooplankton program and provided valuable discussions concerning the results.

REFERENCES

1. Seiburth, J. McN., V. Smetacek, and J. Lenz. 1978. Pelagic ecosystem structure: Heterotrophic compartments of the plankton and their relationship to plankton size fractions. *Limnol. Oceanogr.* 23(6): 1256-1263.
2. Biological Methods Panel Committee on Oceanography (E. Ahlstrom, Chairman). 1969. Recommended procedures for measuring the productivity of plankton standing stock and related oceanic properties. National Academy of Sciences, Washington, D.C. 59 pp.
3. UNESCO. 1968. Zooplankton sampling. D.J. Tranter (Ed.). Monographs on oceanographic methodology, No. 2. 174 pp.
4. Raney, E.C. 1952. The life history of the striped bass, *Roccus saxatilis* (Walbaum). *Bull. Bingham Oceanogr. Coll.* 14:5-97.
5. Markle, D.F., and G.C. Grant. 1970. The summer food habits of the young-of-the-year striped bass in three Virginia rivers. *Chesapeake Sci.* 11:50-54.
6. Setzler, E.M., W.R. Boynton, K.V. Wood, H.H. Zion, L. Tucker, and J.A. Mihursky. 1979. Synopsis of biological data on striped bass, *Morone saxatilis* (Walbaum) Cont. No. 802 Center for Environmental Studies, University of Maryland, Solomons, MD.
7. Setzler-Hamilton, E.M., P.W. Jones, G.E. Drewry, F.D. Martin, K.L. Ripple, M. Beaven and J.A. Mihursky. 1982. A comparison of larval feeding habits among striped bass, white perch and clupeidae in the Potomac estuary. Final Report to Maryland Power Plant Siting Program, University of Maryland, Solomons, MD.
8. Carr, W.E.S. and C.A. Adams. 1973. Food habits of juvenile marine fishes occupying seagrass beds in the estuarine zone near Crystal River, Florida. *Trans. Am. Fish Soc.* 102:511-540.
9. Gilmurray, M.C. and G.R. Daborn. 1981. Feeding relations of the Atlantic silverside *Menidia menidia* in the Minas Basin, Bay of Fundy. *Mar. Ecol. Prog. Ser.* 6:231-235.
10. June, F.C. and F.T. Carlson. 1971. Food of the young Atlantic menhaden, *Brevoortia tyrannus,* in relation to metamorphosis. *Fish Bull. U.S. Fish Wildl. Ser.* 68:493-512.
11. Durbin, A.G. 1979. Food selection by plankton feeding fishes. *In:* Predatory-prey systems in fisheries management, pp. 203-218. H. Clepper, Ed. Sport

Fishing Institute, Washington, D.C.
12. Herman, S.S., J.A. Mihursky and A.J. McErlean. 1968. Zooplankton and environmental characteristics of the Patuxent River Estuary. *Chesapeake Sci.* 9:67-82.
13. Lippson, A.J., M.S. Haire, A.F. Holland, F. Jacobs, J. Jensen, R.L. Moran-Johnson, T.T. Polgar, and W.A. Richkus. 1979. Environmental Atlas of the Potomac Estuary. Prepared by Martin Marietta Environmental Center for Maryland Department of Natural Resources, Power Plant Siting Program. 279 p.
14. Cowles, R.P. 1930. A biological study of the offshore waters of Chesapeake Bay. *Bull. U.S. Bur. Fish* 46:277-381.
15. Alden, R.W., III, R.C. Dahiya & R.J. Young, Jr. 1982. A method for the enumeration of zooplankton subsamples. *J. Exp. Mar. Biol. Ecol.* 59:185-206.
16. Burton, W.H., I. Moss and F. Jacobs. 1986. Chesapeake Bay water quality monitoring program mesozooplankton component: August 1984-December 1985. Prepared for Maryland Department of Health and Mental Hygiene, Office of Environmental Programs, by Martin Marietta Environmental Systems.
17. Burrell, V.G., W.A. Van Engel and S.G. Hummel. 1974. A new device for subsampling plankton samples. *J. Cons. perm. int. Explor. mer.* 35:364-366.
18. Jacobs, F. 1978. Zooplankton distribution, biomass, biochemical composition and seasonal community structure in lower Chesapeake Bay, Ph.D. Dissertation, University of Virginia, Charlottesville. 104 p.
19. Grant, G.C. and J.E. Olney. 1983. Lower Bay zooplankton monitoring program. The August 1978 survey. Spec. Sci. Rept. No 115. Virginia Institute of Marine Science. 81 p.
20. Sharpe, K.W. 1910. Notes on the marine Copepoda and Cladocera of Woods Hole and adjacent regions, including a synopsis of the genera of the Harpacticoida. *Proc. U.S. Nat. Mus.* 38:405-436.
21. Wilson, C.B. 1932. The copepods of the Woods Hole region Massachusetts. *Bull. U.S. Nat. Mus.* 158:1-635.
22. Grant, G.C. and J.E. Olney. 1979. Lower Bay monitoring program: an introduction to the program and results of the initial survey of March 1978. Spec. Sci. Rept. No. 93. Virginia Institute of Marine Science. 92 p.
23. Roddie, B.D., R.J. Leakey, and A.J. Berry. 1984. Salinity-temperature tolerance and osmoregulation in *Eurytemora affinis* (Poppe) (Copepoda: Calanoida) in relation to its distribution in the zooplankton of the upper reaches of the Forth estuary. *J. Exp. Mar. Biol. Ecol.* 79:191-211.
24. Heinle, D.R. 1969. Temperature and zooplankton.*Chesapeake Sci.* 10:186-209.
25. Bradley, B.P. 1975. The anomalous influence of salinity on temperature tolerances of summer and winter populations of the copepod *Eurytemora*

affinis. Biol. Bull. 148:26-34.
26. Conover, R.J. 1956. Oceanography of Long Island Sound, 1952-1954. VI. Biology of *Acartia clausi* and *A. tonsa. Bull. Bingham Oceanogr. Coll.* 15:156-233.
27. Lee, W.Y. and B.J. McAlice. 1979. Seasonal succession and breeding cycles of three species of *Acartia (*Copepoda: Calanoida) in a Maine estuary. *Estuaries* 2:228-235.
28. Hulsier, E.E. 1976. Zooplankton of Lower Narragansett Bay, 1972-1973. *Chesapeake Sci.* 17:260-270.
29. Cronin, L.E., C. Daiber and E.M. Hulbert. 1962. Quantitative seasonal aspects of zooplankton in the Delaware River estuary. *Chesapeake Sci.* 3:63-93.
30. Sullivan, B.K. and L.T. McManus. 1986. Factors controlling seasonal succession of the copepods *Acartia hudsonica* and *A. tonsa* in Narragansett Bay, Rhode Island: Temperature and resting egg production. *Mar. Ecol. Prog. Ser.* 28:121-128.
31. Uye, S. 1980. Development of neritic copepods *Acartia clausii* (Giesbrecht) (Copepoda: Calanoida) in inlet waters. *J. Exp. Mar. Ecol.* 57:55-83.
32. Jeffries, H.P. 1962. Succession of two *Acartia* species in estuaries. *Limnol. Oceanogr.* 7:354-364.
33. Kasahara, S., S. Uye and T. Onbe. 1974. Calanoid copepod eggs in sea-bottom muds. *Mar. Biol.* 26:167-171.
34. Kasahara, S., S. Uye and T. Onbe. 1975. Calanoid copepod eggs in sea-bottom muds. II. Seasonal cycles of abundance in the populations of several species of copepods and their eggs in Inland Sea of Japan. *Mar. Biol.* 31:25-29.
35. Zillioux, E.J. and J.G. Gonzalez. 1972. Egg dormancy in a neritic calanoid copepod and its implications to over-wintering in boreal waters. *In:* Fifth European marine Biology Symposium. Ed. by B. Battaglis. pp. 217-230.
36. Landry, M.R. 1975. Seasonal temperature effects and predicting development rates of marine copepod eggs. *Limnol. Oceanogr.* 20:434-440.
37. Landry, M.R. 1978. Population dynamics and production of a planktonic marine copepod, *Acartia clausi,* in a small temperate lagoon on San Juan Island, Washington, *Int. Revue. ges. Hydrobiol.* 63:77-119.
38. Bryan, B.B. 1977. The ecology of the marine Cladocera of lower Chesapeake Bay. Ph.D. dissertation. University of Virginia. 101 p.
39. Bosch, H.F. & R.W. Taylor. 1973. Distribution of the cladoceran *Podon polyphemoides* in the Chesapeake Bay. *Mar. Biol.* 19:161-171.
40. Bosch, H.F. & R.W. Taylor. 1968. Marine cladocerans in the Chesapeake Bay estuary. *Crustaceana* 15:161-164.
41. Grant, G.C., B.B. Bryan, F. Jacobs, and J.E. Olney. 1977. Effects of Tropical Storm Agnes on zooplankton in the lower Chesapeake Bay. pp. 425-442. *In:* The Effects of Tropical Storm Agnes on the Chesapeake Bay Estuarine

System. Chesapeake Res. Consortium Publ. No. 54. 639pp.
42. Paffenhofer, G.A. and J.D. Orcutt, Jr. 1986. Feeding growth and food conversion of the marine cladoceran *Penilia avirostris*. *J. Plankton Res.* 8:741-754.
43. Bryan, B.B. and G.C. Grant. 1974. The occurrence of *Podon intermedius* in Chesapeake Bay. *Chesapeake Sci.* 15:120-121.
44. Holland, A.F. 1987. Personal communication Versar, Inc.
45. Goy, J. 1976. Seasonal distribution and the retention of some decapod larvae within the Chesapeake Bay. Masters Thesis. Old Dominion University, Norfolk, VA.
46. Sandifer, P.A. 1972. Morphology and ecology of Chesapeake Bay decapod crustacean larvae. Ph.D. Dissertation, University of Virginia, Charlottesville, Virginia. 532 p.
47. Sandifer, P.A. 1975. The role of pelagic larvae in recruitment to populations of adult decapod crustaceans in the York River estuary and adjacent lower Chesapeake Bay, VA. *Estuarine Coastal Mar. Sci.* 3:269-380.
48. Grant, G.C. 1977. Seasonal distribution and abundance of the Chaetognaths in the lower Chesapeake Bay. *Estuarine Coastal Mar. Sci.* 5:809-824.
49. Kremer, P. 1979. Predation by the ctenophore *Mnemeopsis leidyi* in Narragansett Bay Rhode Island. *Estuaries* 2:97-105.
50. Bishop, J.W. 1967. Feeding rates of the ctenophore *Mnemeopsis leidyi*. *Chesapeake Sci.* 8:259-264.
51. Burrell, V.G. 1968. The ecological significance of a ctenophore *Mnemeopsis leidyi* in a fish nursery ground. M. A. Thesis. The College of William and Mary, Williamsburg, VA.
52. Miller, R.J. 1974. Distribution and biomass of an estuarine ctenophore population *Mnemeopsis leidyi*. *Chesapeake Sci.* 15:1-8.
53. Deason, E.E. and T.J. Smayda. 1982. Ctenophore-zooplankton-phytoplankton interactions in Narragansett Bay, Rhode Island, USA, during 1972-1977. *J. Plankton Res.* 4:203-217.
54. Wolfe, J.J., B. Cunningham, N. Wilkerson and J. Barnes. 1926. An investigation of the microplankton of Chesapeake Bay. *J. Elisha Mitchell Sci. Sco.* 42:25-54.
55. Morse, D.C. 1947. Some observations on seasonal variation in plankton population, Patuxent River, Maryland, 1943-1945. *Chesapeake Biol. Lab.,* 65:1-31.
56. Whaley, R.C. and W.R. Taylor. 1968. A plankton survey of the Chesapeake Bay using a continuous underway sampling system. Chesapeake Bay Institute, Technical Report 36:89 pp.
57. Mulford, R.A. 1972. An annual plankton cycle on the Chesapeake Bay in the vicinity of Calvert Cliffs, Maryland, June 1969-May 1970. *Proc. Acad. Nat. Sci. Phili.* 124(3):17-40.

58. Small, E.B. 1972. Free living protozoa of the Chesapeake Bay exclusive of Foraminifera and the flagellates. *Chesapeake Sci.* 13:596-597.
59. Lund, J.W.G., C. Kipling and E.D. LeCren. 1958. The inverted microscope method of estimating algal numbers and the statistical basis of estimations by counting. *Hydrobiologia* 11:143-170.
60. Dale, T. and H. Burkill. 1982. "Live counting" — a quick and simple technique for enumerating pelagic ciliates. *Ann. Inst. Oceanogr.* 58(S): 267-276.
61. Coats, D.W. and J.F. Heinbokel. 1982. A study of reproduction and other life cycle phenomena in planktonic protists using an acridine orange fluorescence technique. *Mar. Biol.* 67(1):71-79.
62. Brownlee, D.C., S.G. Brownlee, K.G. Sellner. 1985. A comparison and analysis of microzooplankton collection techniques: 40 μm mesh net vs. 20 μm mesh net vs. whole water. Report to the State of Maryland's Office of Environmental Programs, Water Quality Monitoring Program.
63. Beers, J.R. and G.L. Stewart. 1967. Microzooplankton in the euphotic zone at five locations across the California Current. *J. Fish Res. Bd. Canada* 24(10):2053-2068.
64. Brownlee, D.C. 1977. The Significance of Cytological Characteristics as Revealed by Protargol Silver Staining in Evaluating the Systematics of the Ciliate Suborder Tintinnina. Masters Thesis, Department of Zoology, University of Maryland, 146 pp.
65. Laval-Peuto, M. and D.C. Brownlee. 1986. Identification and systematics of the Tintinnina (Ciliophora): Evaluation and suggestions for improvement. *Ann. Inst. Oceanogr.* 62(1):69-84.
66. Sawyer, T.K. 1971. Isolation and identification of free-living marine amoebae from upper Chesapeake Bay, Maryland. *Trans. Amer. Microsc. Soc.* 90(1):43-51.
67. Smetacek, V. 1981. The annual cycle of protozooplankton in the Kiel Bight. *Mar. Biol.* 63:1-11.
68. Burkill, P.H. 1982. Ciliates and other microplankton components of a nearshore food-web: standing stocks and production processes. *Ann. Inst. Oceanogr.* 58:335-349.
69. Capriulo, G.M. and E.J. Carpenter. 1983. Abundance, species composition and feeding impact of tintinnid micro-zooplankton in central Long Island Sound. *Mar. Ecol. Prog. Ser.* 10:277-288.
70. Hargraves, P.E. 1981. Seasonal variations of tintinnids (Ciliophora: Oligotrichida) in Narragansett Bay, Rhode Island, U.S.A. *J. Plank. Res.* 3(1): 81-91.
71. Berk, S.G., D.C. Brownlee, D.R. Heinle, H.J. Kling and R.R. Colwell. 1977. Ciliates as a food source for marine planktonic copepods. *Microb. Ecol.* 4:27-40.
72. Robertson, J.R. 1983. Predation by estuarine zooplankton on tintinnid ciliates. *Estuarine Coast. Shelf Sci.* 16:27-36.

73. Stoecker, D.K. and N.K. Sanders. 1985. Differential grazing by *Acartia tonsa* on a dinoflagellate and a tininnid. *J. Plank. Res.* 7(1):85-100.
74. Boynton, W., M. Kemp, J. Garber & J. Barnes. 1986. Nutrient flux. Supplement to Maryland Office of Environmental programs Chesapeake Bay Water Quality Monitoring Program, Ecosystems processes component. Chesapeake Biological Laboratory, University of Maryland, Solomons, Md.

Chapter Thirteen
CONTAMINANTS IN CHESAPEAKE BAY: THE REGIONAL PERSPECTIVE

GEORGE R. HELZ[1] and ROBERT J. HUGGETT[2]

[1]Department of Chemistry and Biochemistry
University of Maryland
College Park, MD 20742
and
[2]Virginia Institute of Marine Sciences
College of William and Mary
Gloucester Point, VA 23062

ABSTRACT

Industrial and municipal point sources of contaminants are scattered along the shores of Chesapeake Bay and its tributaries, but reach especially high density at Norfolk, Va., and Baltimore, Md. Sedimentation and various chemical processes in many cases conspire to restrict the water-borne transport of contaminants away from point sources. Kepone, residual chlorine, volatile halogenated hydrocarbons, and anthropogenic trace metals are well-studied examples of point-source contaminants. For the most part, their concentrations in water and sediments drop to nearly immeasurable values within a distance of a few kilometers, or sometimes a few tens of kilometers, from their sources.

On the other hand, certain contaminants have now been shown to be truly regionally dispersed. Included are polychlorinated biphenyls, phthalate esters, anthropogenic trace metals (Cu, Zn, Pb), polycyclic aromatic hydrocarbons, herbicides and weapon derived radionuclides. Most of these enter the Bay in significant amounts from the atmosphere. Thus their dispersion throughout the Bay is not dependent on aquatic transport processes. Although it is tempting to link the existence of this regional contamination with well publicized regional biological problems, no link has yet been proven.

INTRODUCTION

There is a widespread popular perception that Chesapeake Bay is dying. Just what is meant by death in seldom defined, but the term is used in connection with declines that are occurring in the quality of the Bay as a recreational resource and as a source of food for humans. In discussing this issue recently, Schubel[1] cited the following six symptoms of illness: a) declines in harvest of anadromous fish, b) declines of oyster harvests and poor spat set, c) retreats in submerged aquatic vegetation, d) blooms of blue-green algae and dinoflagellates, e) increases in nutrient levels, and f) increases in extent and duration of summer anoxia in the bottom waters. Other chapters in this book explore the current status of knowledge regarding some of these problems. In this chapter, we want to investigate what role toxic substances may play in these problems, if any.

At the outset, it should be noted that many of the problems cited by Schubel[1] are regional in nature and are not restricted to one tributary or local embayment. Possibly the best documented example is the retreat of aquatic vegetation. The Environmental Protection Agency's Chesapeake Bay Program produced a detailed survey of the Bay-wide extent of this retreat during the decade, 1965 to 1975.[2] In this period, extensive losses of submerged aquatic plants occurred on Susquehanna Flats, in the lower reaches of the Patuxent, Potomac, Rappahannock, York, and indeed in innumerable other places throughout the Bay. The phenomenon occurred in waters ranging widely in salinity. It was not confined to a single species, so it seems unlikely to have been caused by disease. Similar retreats were not observed elsewhere along the Atlantic Coast, so a climatic control is also unlikely.

If an ecological phenomenon such as this were caused by chemical contaminants, it would be necessary that these contaminants be pervasive throughout the Bay. Therefore it is of great interest to evaluate the extent to which the various toxic substances found in the Bay have become pervasive in their distribution. Further, for those that are pervasively distributed, we would like to know how they came to be dispersed throughout this far reaching estuary, which is a geographic feature larger than either Rhode Island or Delaware.

CONTAMINATION IN HIGHLY IMPACTED AREAS

Industrial activity along the shores of Chesapeake Bay is concentrated in two major centers, Baltimore, and Norfolk (see map, Figure 1). These two urban areas have gradually engulfed the shores of the Patapsco and Elizabeth Rivers, respectively. Evidence of serious environmental contamination is not hard to find in these tributary estuaries, and similarities between the two are striking.

To illustrate the physical distribution of contaminants in these tributaries, we will make use of the data base now available on trace metals (summarized in Table 1). For no other toxic materials are our data as extensive. However, since trace metals have natural as well as anthropogenic sources, we will use enrichment factors, rather than raw concentration data, to identify contaminated

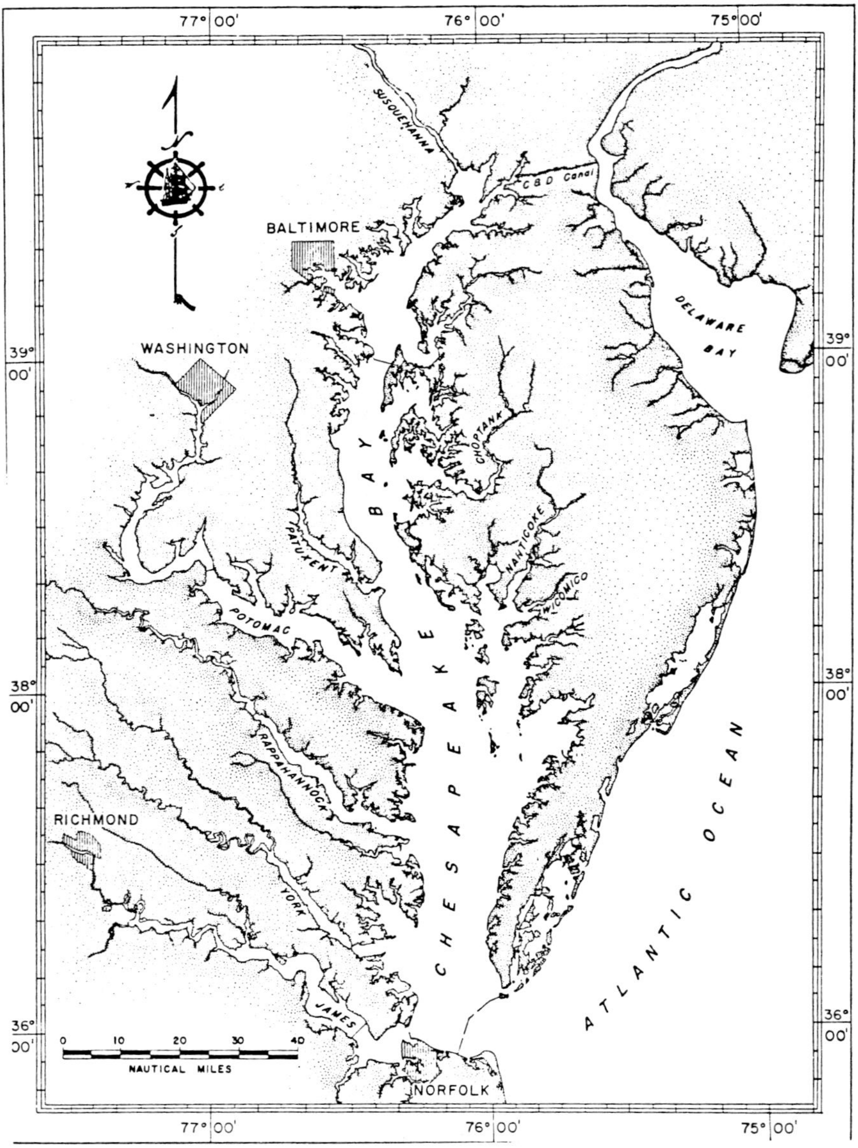

FIGURE 1. Map of Chesapeake Bay showing locations of Baltimore and Norfolk, the two major industrial centers.

zones. The enrichment factor, EF, is an index of the enhancement of the concentration of a metal over that expected naturally, and is defined by the following equation:

(1) $EF = (M/Al)_{sample} / (M/Al)_{shale}$

Where M designates a trace metal and Al represents aluminum. A sediment containing a metal in the same ratio to aluminum as found in shale would have an enrichment factor of unity, indicating no enrichment. High enrichment factors imply that the metal is present in anomalously high concentrations in the sample relative to shale, which is being taken here as representative of uncontaminated estuarine sediments. In the deepest parts of most sediment cores from Chesapeake Bay, enrichment factors approach unity, supporting the choice of shale as a useful reference composition.

TABLE 1

Summary of Data Sources Concerning the Elemental Composition of Chesapeake Bay Sediments.

Area	Elements	Reference
MAIN STEM	Al, C, Ca, Co, Cr, Cu, Fe, H, K, Mg, Mn, Na, Ni, Pb, S, Si, Ti, V	3
	Cr, Cu, Ni, Pb	4
	Cd, Cu, Fe, Mn, Ni, Zn	5
	Ag, Al, Cd, Co, Cr, Cu, Fe, Mn, Ni, Pb, V, Zn	6
	Al, Co, Cr, Cu, Fe, Ga, Mn, Ni, Org C, Org N, Pb, Si, Ti, V, Zn	7-13
	As, Cd, Cu, Fe, Hg, Mn, Ni, Pb, Sn, Zn	14
BACK RIVER	Cd, Cu, Pb, Zn	15
PATAPSCO	Cd, Cr, Cu, Hg, Mn, Ni, Pb, Zn	16
	As, Cd, Cr, Cu, Hg, Mn, Ni, Pb, Zn	17,18
	Al, Co, Cr, Fe, Mn, Ni, Si, Ti, V, Zn	19
RHODE RIVER	Cd, Cr, Fe, Mn, Zn	20
PATUXENT	Cd, Co, Cr, Cu, Fe, Mn, Ni, Pb, Zn	21
POTOMAC	Ag, Ba, Cd, Co, Cr, Cu, Fe, Li, Mn, Ni, Pb, Sr, V, Zn	22
	Ba, Co, Cr, Cu, Fe, Mn, Pb, Sr, Ti, V, Zi, Zn	23
RAPPAHANNOCK, YORK	Cu, Zn	24,25
ELIZABETH	Al, Cd, Cr, Cu, Fe, Hg, Pb, Zn	26
	Al, Co, Cr, Fe, Mn, Ni, Si, Ti, V, Zn	13
	Cd, Co, Cr, Cu, Fe, Mn, Ni, Pb, Zn	64

In Figure 2, zinc has been chosen to illustrate the nature of contamination in the sediments of Baltimore Harbor. Zinc is used in a wide range of industrial activities and is thus a good general marker of man's effect on the environment. Zinc has also been measured by most workers studying trace metals in Bay sediments (Table 1).

Zinc itself is relatively non-toxic. Extremely high concentrations are necessary to produce toxic responses in most organisms. However, its usefulness as a tracer stems from the fact that the distribution of zinc tends to be controlled by the same processes that control the distribution of considerably more toxic metals, such as mercury and cadmium (which occur in the same group as zinc in the periodic table) as well as copper and lead. The marine chemistry of these metals is dominated by their affinity for particles. Similarly, they all are immobilized in anoxic environments by precipitation as sulfides.

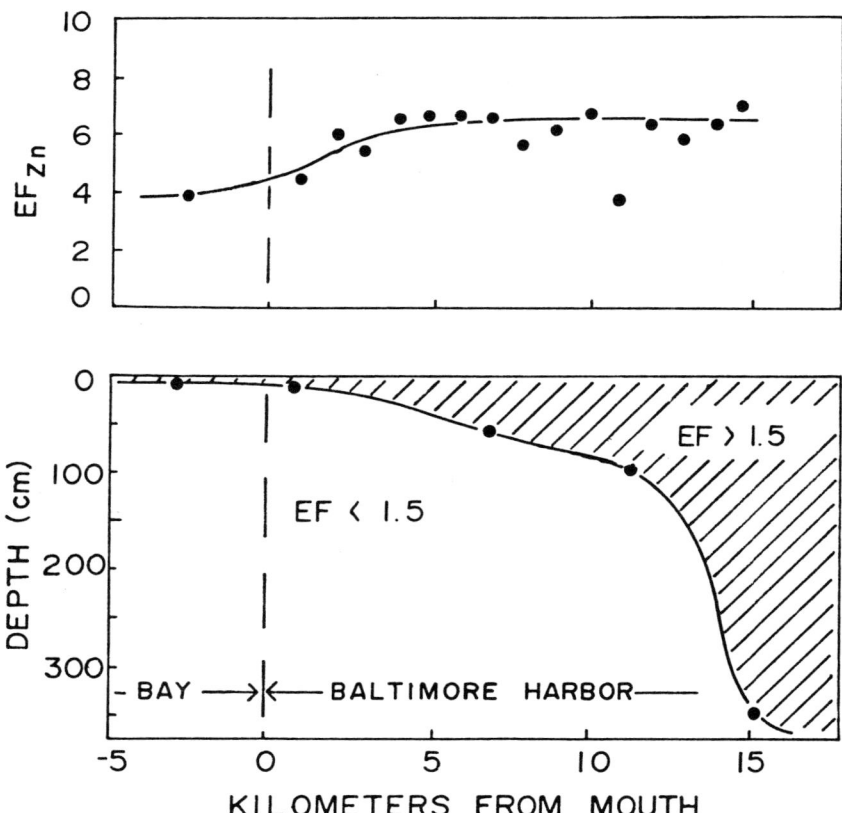

FIGURE 2. Enrichment of zinc in sediments of Baltimore Harbor as a function of distance from the mouth. Upper: Enrichment Factor (see text) with respect to shale in surface sediments. Lower: Thickness of sediments with Enrichment Factor greater than 1.5. (Data from refs. 13 and 18).

The upper part of Figure 2 shows the enrichment factor for zinc in the surface sediments of Baltimore Harbor. As shown, these sediments tend to be enriched relative to shale throughout the Harbor by roughly a factor of seven, although this factor drops to four around the mouth. Vertically downward in the sediments, the enrichment factor declines. The lower part of Figure 2 indicates that the layer of enriched sediment is only about 20 cm thick near the mouth of the Harbor. However, it thickens to over 300 cm toward the upper end of the Patapsco estuary, where much of the anthropogenic zinc discharged historically to the Harbor may still reside.[19]

Figure 3 shows vertical profiles of the zinc enrichment factor at selected coring sites in the Elizabeth River. In the surface sediments, enrichments are much more intense than in the Patapsco estuary, but the thickness of the contaminated zone is much thinner, generally less than 50 cm.

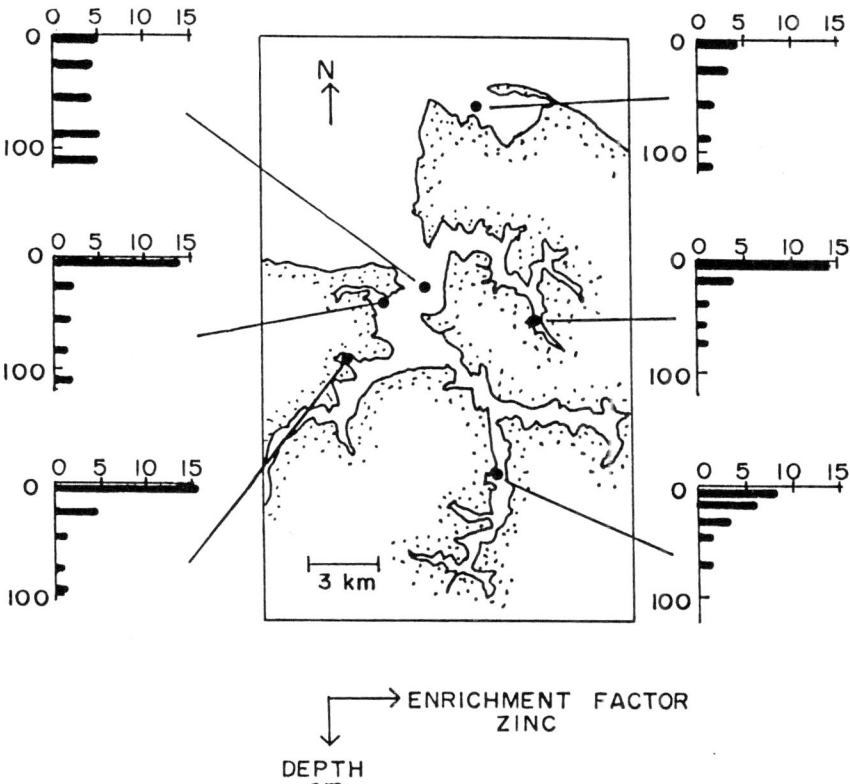

FIGURE 3. Enrichment Factor for zinc in selected cores from the Elizabeth River, Va. Note extreme enrichments in surface layers in several cases. Also note that the depth of zinc penetration is generally less than in the upper part of Baltimore Harbor, shown in the previous figure. (Data from ref. 13).

276 Contaminant Problems and Management of Living Chesapeake Bay Resources

Sinex and Helz[19] point out that the great thickness of the contaminated zone in the upper part of the Patapsco exceeds the historic net accumulation of sediment in this part of the harbor. Therefore, zinc and other contaminants must have been mixed downward into older sediment deposited before the time that anthropogenic contaminants were introduced. The mixing processes are probably related to the nearly continual dredging that has occurred in Baltimore Harbor since the middle of the 19th century. Dredging not only plows contaminants downward, but also dilutes contaminated surface sediment with clean, underlying sediment. Possibly much of the difference between the Elizabeth River and the Patapsco estuary, that is, greater zinc enrichment in a thinner

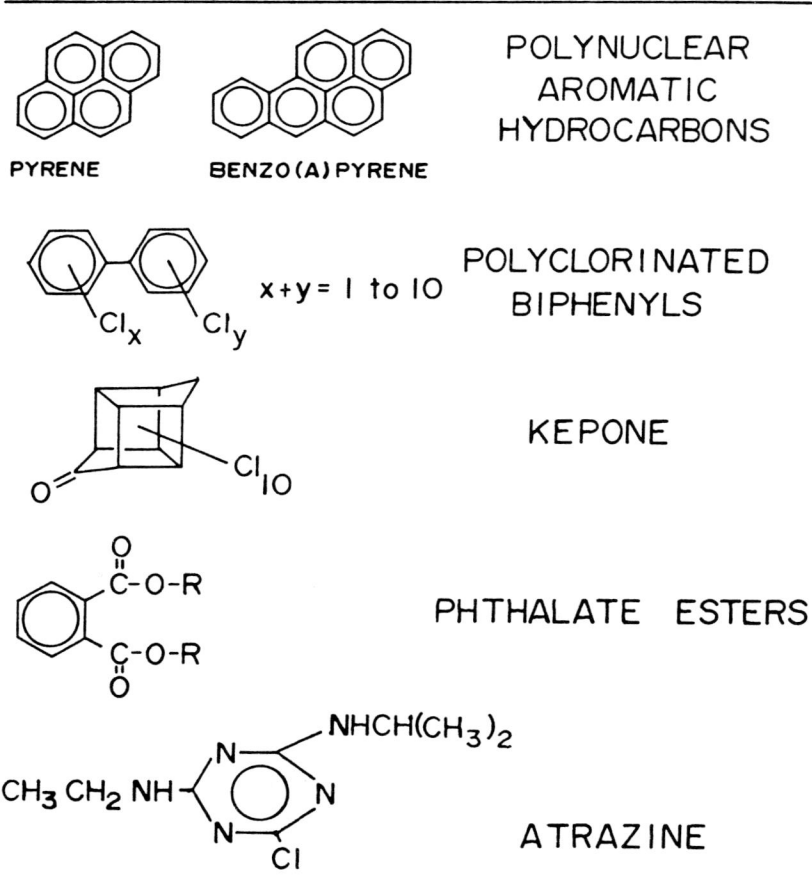

FIGURE 4. Structures of some organic compounds discussed in this chapter.

surface layer in the Elizabeth, is a reflection of less vertical mixing.

The locations of highly zinc-enriched sediments in these industrialized tributaries delineate zones of general contamination. Innumerable other toxic materials, only a few of which have been carefully investigated so far, occur in these zones. Most of these toxic materials are organic compounds.

Detailed chemical surveys of the organic chemicals in bottom sediments of the Chesapeake Bay are relatively recent. Less than ten years have passed since the first was undertaken. Several reasons are apparent for this new interest: a) the awareness that compounds other than those which are regulated can cause toxicological problems; b) the increase in quantities of synthetic organic compounds which are produced, used and disposed of in this country; c) technological advancements in analytical instrumentation which have allowed more comprehensive analyses and d) national and regional pollution episodes which have caused catastrophic ecological and/or economic impacts.

Until the late seventies, most of the data on organic chemicals in the environment concerned chlorinated hydrocarbons, such as DDT, DDE, Dieldrin, etc. Analysts were mostly unaware of another group of compounds, the polychlorinated biphenyls (PCB's; see Figure 4 for structures of these and other organic compounds discussed in this chapter). Often PCB's interfered with the analyses for pesticides. Some of these compounds are of toxicological concern because they are bioconcentrated, and in high concentrations they are thought to be teratogens and carcinogens. Some of the PCB congeners eluted from gas chromatographic columns at the same time as individual pesticide compounds, giving rise to overestimates in the early data for pesticides.

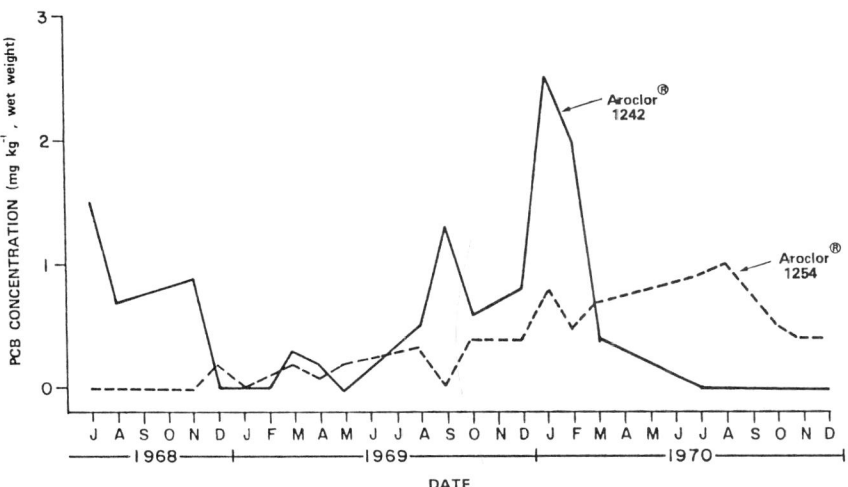

FIGURE 5. Historical data on PCB concentrations in oysters from the Elizabeth River. (Arochlor is a Monsanto trade name for PCB's).

A re-examination of archived gas chromatograms clearly show that PCB's were present in Chesapeake Bay as early as the 1960's. Retrospective analyses can be performed using these gas chromatograms with more recently derived ones of PCB and DDT standards. Figure 5 shows the concentrations of the PCB's, Aroclor 1254 and 1242, in oysters collected from the Elizabeth River in the late 1960's and early 1970's. It should be noted that such retrospective calculations do not yield data of the quality that would have been obtained had the PCB's been quantified originally, but the trends should be valid and the concentrations approximate.

Although analogous data on PCB's in oysters from Baltimore Harbor do not exist, the Harbor is nonetheless clearly contaminated with PCB's. Tsai, et al.[18] report concentrations exceeding 2 mg/Kg in surface sediments from the Harbor. Similar results were obtained in a Westinghouse survey.[27] Figure 6 shows the number of PCB congeners reported at various sites in the Harbor according to the latest survey.[31]

The most abundant anthropogenic organic compounds in the Chesapeake Bay fall into a class called polynuclear aromatic hydrocarbons (PAH's). These,

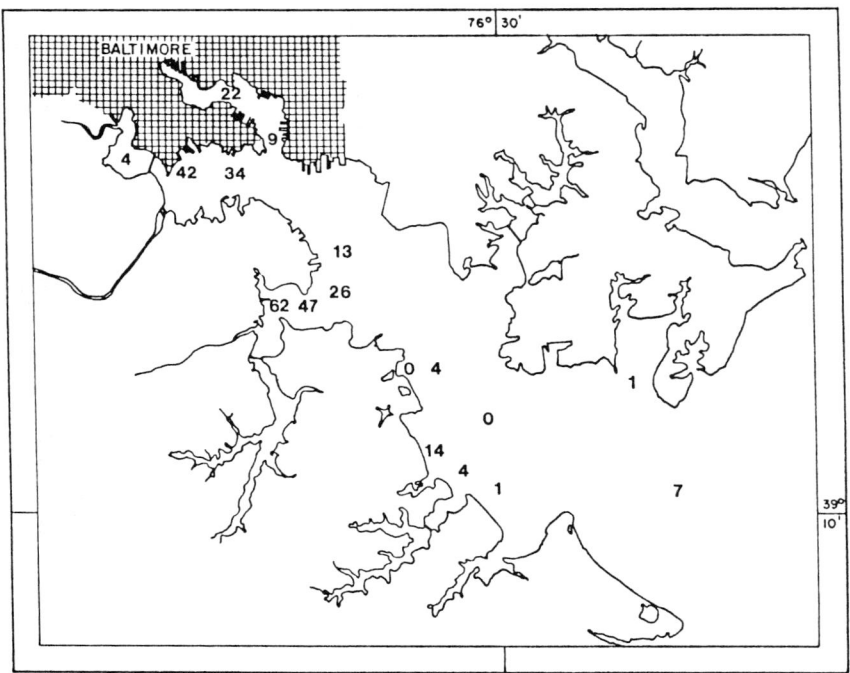

FIGURE 6. Map of Baltimore Harbor showing numbers of PCB congeners identified at selected sites. Note that the number of congeners identified declines in the seaward direction.

like the PCB's, are highly hydrophobic and thus display a strong particle affinity when placed in natural waters. The PAH's occur naturally in fossil fuels and also are produced during the burning of carbonaceous fuels such as wood, coal and refined petroleum products. Therefore they are emitted from smoke stacks, internal combustion engines, wood stoves, etc. Once emitted, they can be transported through the air to deposit on the Bay's surface, or they can settle on land, only to be subsequently eroded and transported in the Bay.

Many compounds in this class are known to be mammalian carcinogens and therefore are suspected to be human carcinogens. Additionally, they have been implicated in chemically induced tumors in the English sole, *Parophrys vetulus*, inhabiting polluted portions of Puget Sound.[28]

The highly industrialized Baltimore Harbor and Elizabeth River tributaries are markedly contaminated with PAH's. Figure 7 shows the concentration of one PAH, benzo(a)pyrene, in the sediments of Baltimore Harbor.[29] Levels are highest in the vicinity of domestic and industrial outfalls.

Since the turn of the century, there have been five wood treatment facilities along the shores of the Elizabeth River,[30] only one of which remains. These operations used creosote, a mixture of PAH's, to treat wood for protection against fungi and worms. There have been documented creosote spills from

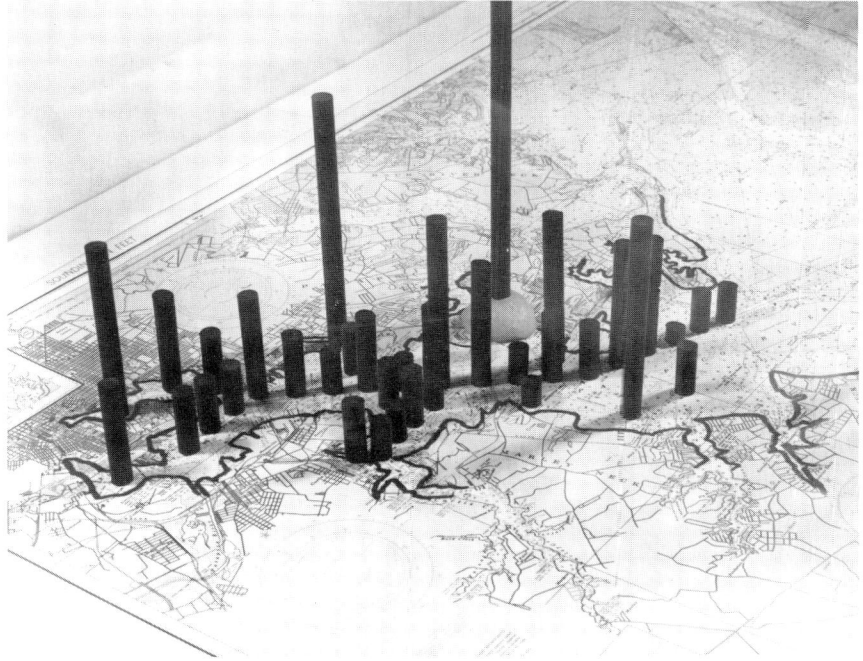

FIGURE 7. PAH concentrations in Baltimore Harbor. Concentrations are indicated by height of the bars.

these facilities, and there are still chronic, relatively low level inputs from the operations. The bottom sediments in the River contain a record of these events. Figure 8 shows the distribution of pyrene in sediment collected from the channel of the river. Concentrations increase in the upstream direction, reaching a peak near the sites of the wood treatment plants.[31,32,33]

Perhaps one of the most studied chemical pollution episodes in an estuary is the Kepone contamination of the James River. Kepone is a halogenated organic compound, intended as a pesticide, and is extremely resistant to both chemical and biological breakdown. It was produced at Hopewell, Virginia. From the late 1960's to the mid 1970's, thousands of pounds entered the James.[35,36,37,38] As is the case with PCB's and PAH's, Kepone rapidly associates with suspended sediments. In the James, it therefore accumulated in regions of high sediment deposition, especially near the freshwater-saltwater interface, where the turbidity maximum occurs.

Since the source of Kepone was eliminated in the mid-seventies, uncontaminated sediments have been slowly burying the pollutant. In areas of the river where the sedimentation rates are relatively high, maximum concentrations are found fifty or more centimeters below the surface. However, near Hopewell, where sedimentation is low, the highest levels are still found near the surface. Figure 9 presents examples of both situations. Baileys Creek mouth is an area of low sedimentation while at Tar Bay, the sedimentation rate is higher.

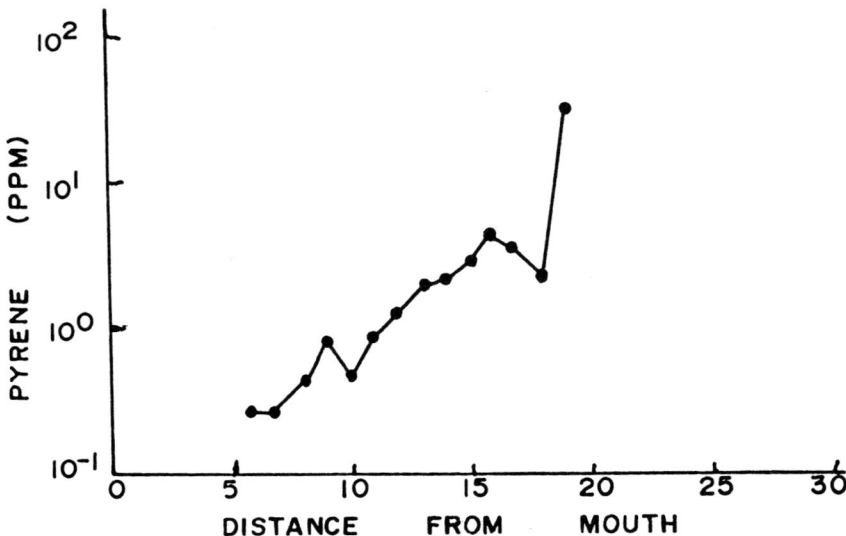

FIGURE 8. Decline of pyrene (a specific PAH compound) in the seaward direction in the Elizabeth River sediments.

BIOLOGICAL EFFECTS IN HIGHLY CONTAMINATED AREAS

In the preceding pages, we have cited evidence that toxic substances occur in high concentrations and are ubiquitous in parts of the Bay system that are impacted by large numbers of nearby discharge sources. However, we have as yet presented no evidence that these substances are producing measurable harm. The reason is that evidence of harm is in fact sparse, and much of the existing evidence is indirect.

Proof of environmental damage by toxic substances is difficult ot obtain because in nature, individual organisms that are weakened by exposure to toxic materials are apt to become targets of predators. Therefore marine biologists rarely have the advantage enjoyed by medical doctors of being able to examine ill patients in order to diagnose the cause of illness. Of course acute exposures

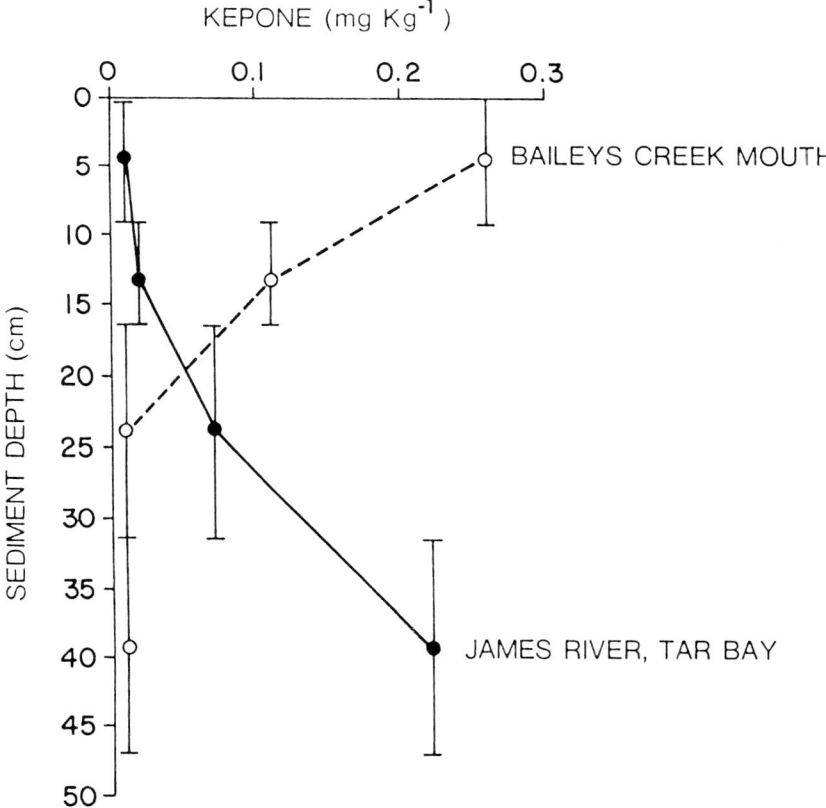

FIGURE 9. Kepone concentrations in sediment cores from the James River. Bars indicate depth interval of sediment analyzed. The sedimentation rate at Tar Bay is much greater than at Baileys Creek.

to toxic materials can produce massive kills that are likely to come to the attention of biologists. However, chronic, sublethal exposures, which might in the long run have very serious consequences for the viability of species, may be observable by biologists only through population declines of the type mentioned in the introduction. Unfortunately population declines are usually difficult to document, except for commercially important species. Furthermore, even well documented population declines commonly can not be unequivocally tied to one particular cause.

Despite the difficulties, there is evidence that the extreme contamination levels found in the Elizabeth River and in Baltimore Harbor are harmful to organisms that attempt to live there. Finfish collected from the most contaminated areas of the Elizabeth River have been found to suffer from a variety of maladies. These include lesions of the skin, liver and gill, fin erosion and cataracts.[33] Laboratory experiments which exposed finfish to bottom sediments from the river duplicated the symptoms found in feral populations.[34] Although it has not been clearly established which toxic materials are responsible for these problems, the correlation of disease with PAH abundance in the River tends to implicate creosote.

Tsai, et al.[18] exposed two species of finfish and one species of clam to suspensions of sediment from Baltimore Harbor. The Harbor sediments were lethal to the finfish at concentrations as much as two orders of magnitude lower than control suspensions. There was a considerable range in the toxicity of sediments from different regions of the Harbor. Toxicity correlated with a species diversity index obtained from earlier, quarterly field samplings of benthic invertebrates, crabs, fish eggs and larvae, and adult fish. Greater toxicity was associated with lower diversity. This relationship testifies that the toxic materials in Harbor sediments indeed are affecting organisms that live in the Harbor. However, toxicity measured by bioassays, could not be correlated with the concentration of any one toxic material, because of high covariance among the toxic materials measured. This is a general problem in contaminated environments; where one toxic material is found, there will usually be a whole suite of toxic materials. Clams were comparatively tolerant of toxic materials in the Harbor sediments over a time period of up to 96 h.

There has been no demonstrated biological effect of Kepone on the James River biota. However, since concentrations in edible portions of species were above FDA action levels, commercial fishing was restricted. More details on the toxicities and concentrations of Kepone in the biota can be found in other chapters of this volume.

EVANESCENT CONTAMINATION

Those contaminants which are found in high concentrations in the sediments

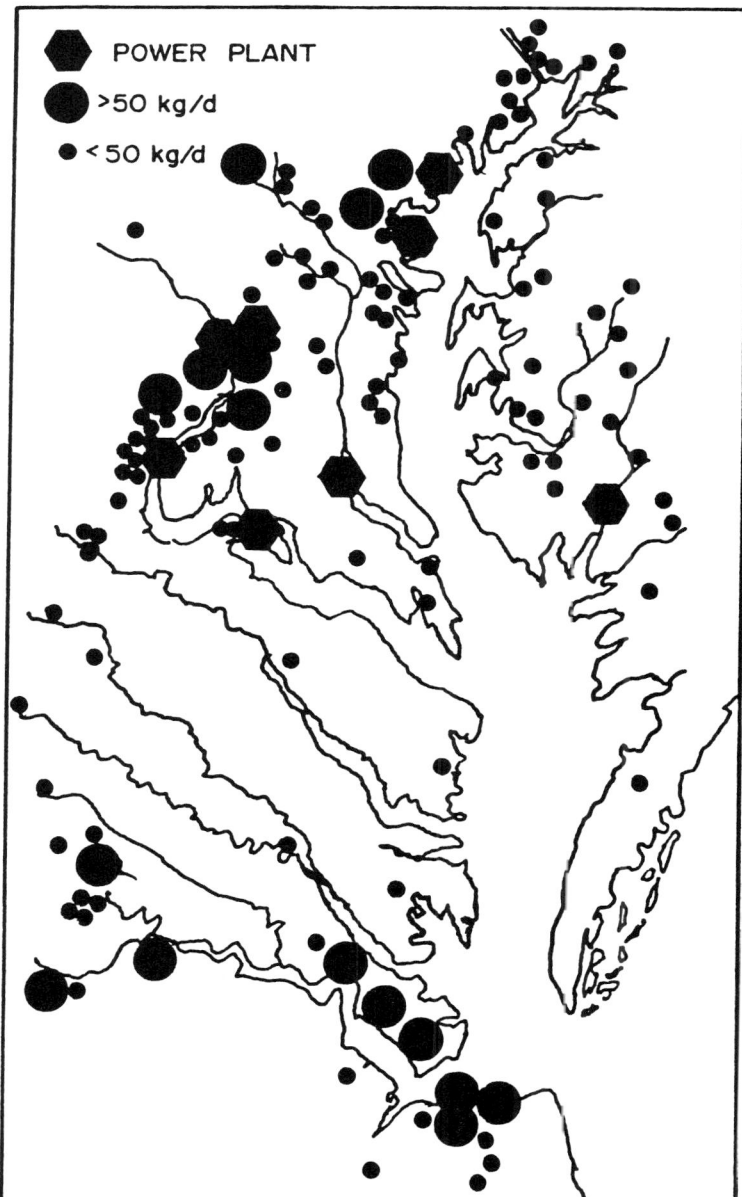

FIGURE 10. Map showing wide distribution of chlorine discharge sources around Chesapeake Bay. Small circles indicate wastewater treatment plants discharging less that 50 kg of chlorine per day, while large circles indicate treatment plants discharging more than this. Hexagons designate some of the large electric power plants that use chlorine for fouling control. Some of these power plants discharge thousands of kilograms of chlorine per day during warmer months. (Modified from ref. 63).

of highly impacted areas like the Elizabeth River and Baltimore Harbor share two characteristics. They are materials that readily sorb to sediment particles, and they are materials that persist for long periods of time in the environment. Because of these characteristics, the sediment column preserves a historical record of their introduction into the Bay. On the other hand, there are some evanescent contaminants, which we know to be injected in large amounts into the Bay, but which leave no record in sediments because they are either degraded rapidly or removed from the Bay's waters by mechanisms other than sorption to sediments.

Chlorine is one such contaminant. Figure 10 is a map showing points where discharges of chlorinated water are known to exist. Chlorine is added to effluents from wastewater treatment plants for disinfection and to power plant cooling waters to prevent biofouling of heat exchangers. Both applications require a chemical that is highly toxic to a broad spectrum of organisms, and chlorine has proven ideal for this purpose. Unfortunately, chlorine in high enough concentrations poses a threat to all aquatic organisms, including those that are not intended targets of its use. The literature on its toxicity to aquatic vertebrates, molluscs, algae and other groups of organisms is extensive.[39]

Inspection of Figure 10, showing the immense number of widely distributed chlorine sources along the Bay's tributaries, suggests that chlorine ought to be a truly regional contaminant. However, this proves not to be the case because of rapid decay of residual chlorine compounds in receiving waters. Field investigations at the Chalk Point electric power plant, one of the largest chlorine sources in the Chesapeake system, indicated that 90% of the chlorine dose typically was consumed by redox processes before the cooling water emerged from the plant's conduits.[40,41] The remainder decayed according to a rate law approximately first order with respect to chlorine. The half life was a few hours at most. No residual chlorine was detectable in the Patuxent River outside the discharge canal, although brominated macromolecules, produced by the action of chlorine on dissolved humic materials in salt water,[42] were detectable as far away as 6 km.[40] Experience at a number of wastewater treatment plants, including two of the largest (Blue Plains in Washington, D.C., and Back River, in Baltimore) indicates that active residual chlorine, even at levels as low as a few parts per billion, is very hard to find more than a hundred meters or so from a wastewater outfall.

The rapid decay of chlorine in natural waters raises serious doubts about its significance as a hazardous material in the aquatic environment despite the large number and wide distribution of its sources. However, this statement needs to be qualified in several important ways. First, there can be no doubt that excessive applications of chlorine can produce serious consequences. Possibly the best documented example occurred in the James River, in 1973, where control failure at a wastewater treatment plant apparently resulted in a massive fish kill.[43] Chlorine concentrations as high as 2.2 mg/L were found around the outfall. Aquarium tests with this water showed it to be highly toxic, but also showed that the toxicity could be largely removed by addition of sodium thiosulfate. By reducing the chlorine feed rate at the plant by a factor of two, further kills were averted.

A second qualification is that in small tributaries where wastewater discharges can contaminate the entire cross section of a stream in the immediate vicinity of an outfall, chlorine residuals might seriously interfere with migratory patterns of fish. Thus it is at least plausible that the decline in shad harvests has been caused by creation of chemical barriers to upstream migration in spawning season. However, this must be treated as an unproven hypothesis at the present time.

A third qualification is that in the case of chlorination by power plants, where massive quantities of water are pumped into the plant and directly treated with chlorine, there will be significant losses in standing stocks of pelagic organisms, such as algae and fish eggs and larvae. Decreases in primary productivity around power plants using chlorine is well documented.[44,45,46,47] However, Goldman, et al.[48] argue that this effect is unlikely to seriously reduce the viability of an estuarine ecosystem because of the rapid recovery time of algae.

There are other examples of evanescent contaminants that may be widespread in the Chesapeake Bay but which leave no permanent record of their presence. Halogenated organic solvents, particularly various chlorinated methanes, ethanes and ethenes, are widely used in dry cleaning, degreasing and similar activities, and these compounds sometimes can be found in municipal and industrial wastewaters. Helz and Hsu[49] found several of these compounds in the finished wastewater of the Back River treatment plant in Baltimore. However, in summer, they were unable to find the same compounds in Back River itself, apparently owing to the rapid rate at which the solvents are known to volatilize from receiving waters. On the other hand, when they returned to Back River in winter, during a time when volatilization was restricted by ice cover, they found readily detectable concentrations as far as 10 km downstream from the outfall of the treatment plant. Several compounds in this class are carcinogenic to laboratory animals. What impact, if any, they have in Chesapeake Bay is entirely unknown. However, because of their evanescent character, effects would almost certainly be confined to areas near sources.

PERVASIVE LOW-LEVEL CONTAMINATION

We have now discussed the behavior in Chesapeake Bay of two kinds of contaminants with very different properties: long lived materials which are strongly sorbed to sediments (e.g. PAH's, PCB's and trace metals) and short lived materials (e.g. chlorine, halogenated solvents). In the cases described, the geographic dispersion of the contaminants is restricted to the vicinity of their input to the Bay by various processes such as entrapment on sediments, chemical or biological decomposition, and volatilization.

No matter where we look in the Bay, we find evidence of some chemical contamination. For example, although Figure 2 showed that the layer of zinc contaminated sediments in Baltimore Harbor thinned towards the mouth, and that the enrichment factors declined, nonetheless evidence of zinc contamination did not disappear at the mouth of the Harbor. Indeed, some enrichment of surface sediments with zinc is a feature found everywhere in the Bay, except near the mouth.[1] This is illustrated in Figure 11 which presents vertical profiles of copper, zinc, lead and a natural radioisotope, ^{210}Pb, at a site about 10 km south of the mouth of the Patuxent River. This locality is nearly as remote as

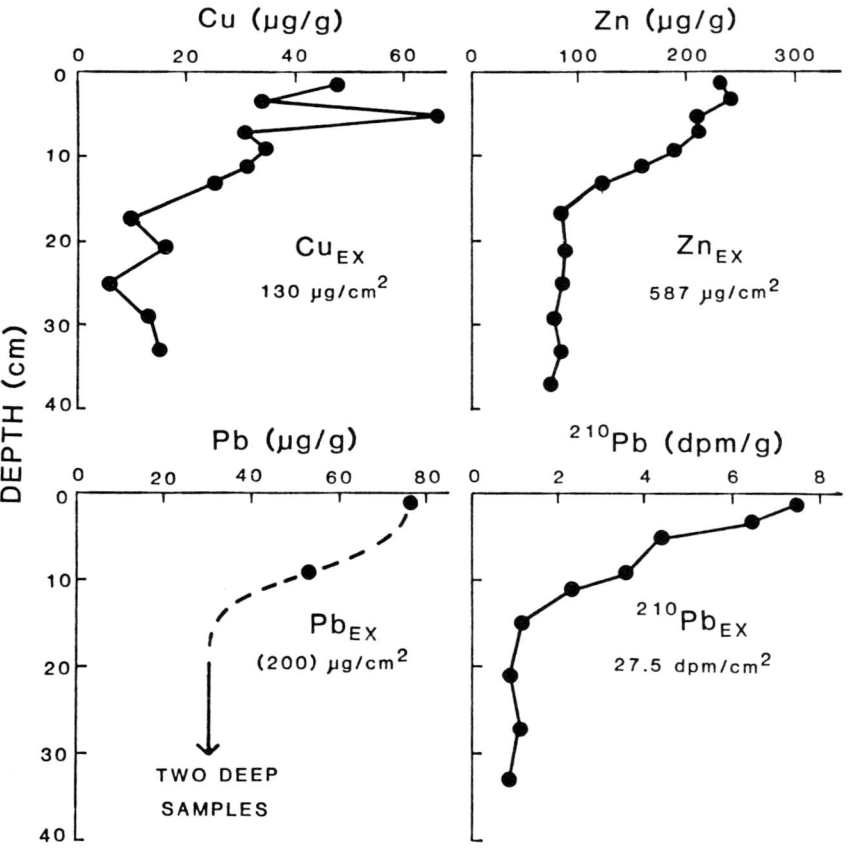

FIGURE 11. Vertical profiles of Cu, Zn, Pb and the natural radioisotope, ^{210}Pb in a sediment core from a site roughly equidistant between Baltimore and Norfolk. these data indicate that, over the past century, metals have contaminated even localities remote from shoreline sources.

possible from shoreline industrial sources in Chesapeake Bay, being roughly equidistant from Baltimore and Norfolk. Yet all four components in Figure 11 are enriched at the surface compared to their concentrations in older sediments at depth.

In the case of ^{210}Pb, the downward decreasing profile is due to a natural process, radioactive decay. Lead-210 is produced in the atmosphere by decay of ^{222}Rn which has escaped from soils containing natural ^{238}U. The ^{210}Pb is deposited from the atmosphere into the Bay's surface waters, from which it is rapidly scavenged to sediments. So far as is known, the flux of ^{210}Pb into the Bay has not changed with time. In sediments, ^{210}Pb decays until an equilibrium activity, supported by radium in the sediment, is reached. Roughly a century (i.e. five half lives) is required to attain equilibrium. Thus the ^{210}Pb profile provides a time scale for the core in Figure 11.

In contrast, the downward decreasing concentrations of the stable elements, copper, zinc and lead, in this figure can only be explained by a lower delivery of these elements to the deposition site in the past. This implies that the present deposition rates of these metals exceed pre-industrial rates. Based on comparison with the ^{210}Pb profile, the enrichment of near-surface sediments with Cu, Zn and Pb has occurred over roughly the past century, corresponding to the period of industrialization in the United States. The only alternative hypothesis, that the enrichment is due to diagenetic mobilization processes that somehow concentrate these metals near the surface, is untenable in this case because sediments in the mid-Bay are permanently anoxic and sulfidic.[50] These are conditions under which these particular metals would not be mobile. Because sites like the one represented in Figure 11 are remote from shoreline sources and because the flux of Cu, Zn and Pb needed to account for the surface enrichments is similar to fluxes measured from the atmosphere, it has been argued that the source of this pervasive trace metal enrichment in Chesapeake Bay is the atmosphere.[10,11] Recent studies have established that the northeastern United States is blanketed by contaminated air masses that deposit anthropogenic trace metals, acidity, and other contaminants to both land and water surfaces.[51,52]

Many of the contaminants found in highly impacted areas are also now found in remote areas, but at much lower concentrations. There are probably no pristine, truly uncontaminated sites left in Chesapeake Bay. In the case of contaminants that display a strong sediment affinity and that therefore are restricted in their mobility in the aquatic environment, transport through the atmosphere may be the chief route of delivery to remote sites.

Over three hundred different PAH compounds have been detected at one place or another in the sediments of the Bay. The aerial distribution of PAH's in the top 2 cm of the sediments is given in Figure 12. It is apparent that PAH's are more abundant in the northern Bay and at the mouths of the major tributaries.

There are several possible reasons for the higher contamination in the Northern Bay. One is that the Northern Bay is more exposed to the contaminated air masses that travel from the Ohio River Valley towards New England. Since these air masses are known to be contaminated with sulfur dioxide originating from coal combustion, it is a reasonable inference that they are also contaminated with PAH's, which are combustion by-products. Another possible reason that

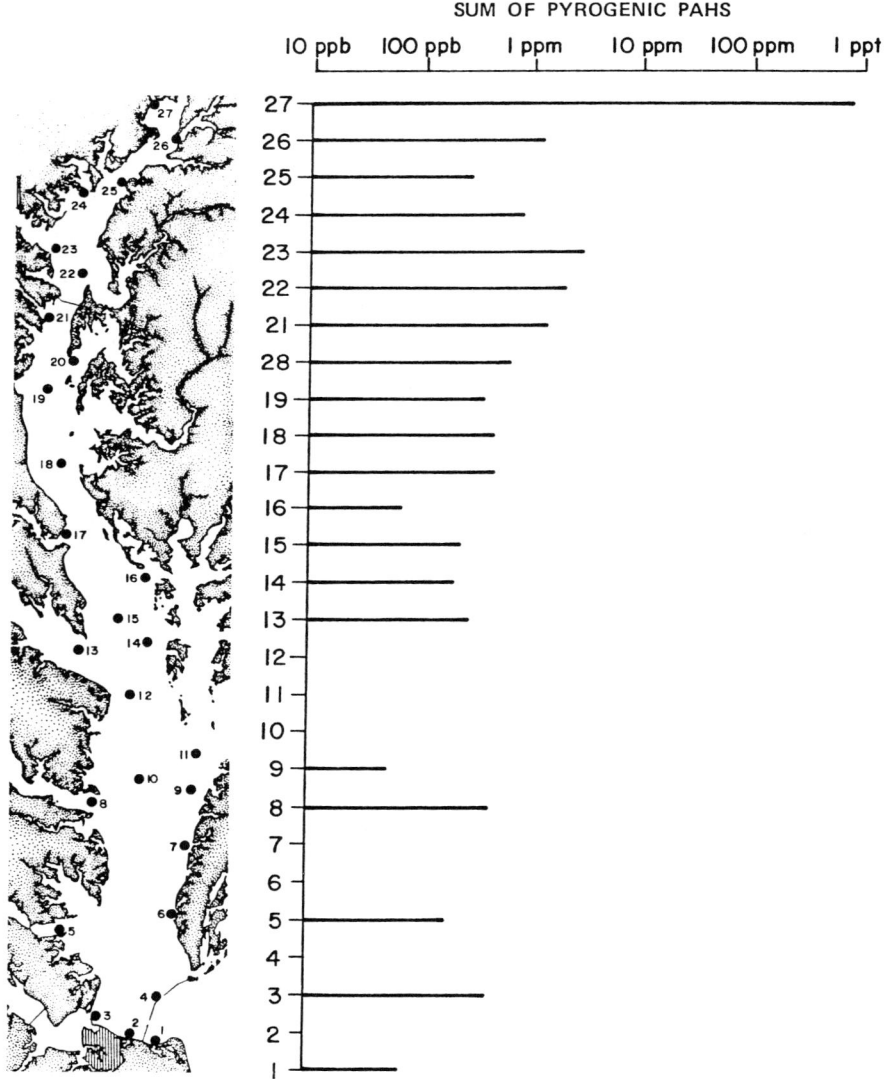

FIGURE 12. Total pyrogenic PAH's in surface sediments from Chesapeake Bay. Note log scale on abscissa. Concentrations are consistently higher in the northern Bay.

higher PAH concentrations occur in the Northern Bay is that the sediments there have a higher silt and clay content than those farther south. Since finer grained sediments have a higher surface area per unit mass and usually contain more natural organic matter, compounds which sorb to surfaces or partition into organic phases will be more concentrated in finer grained sediments than in

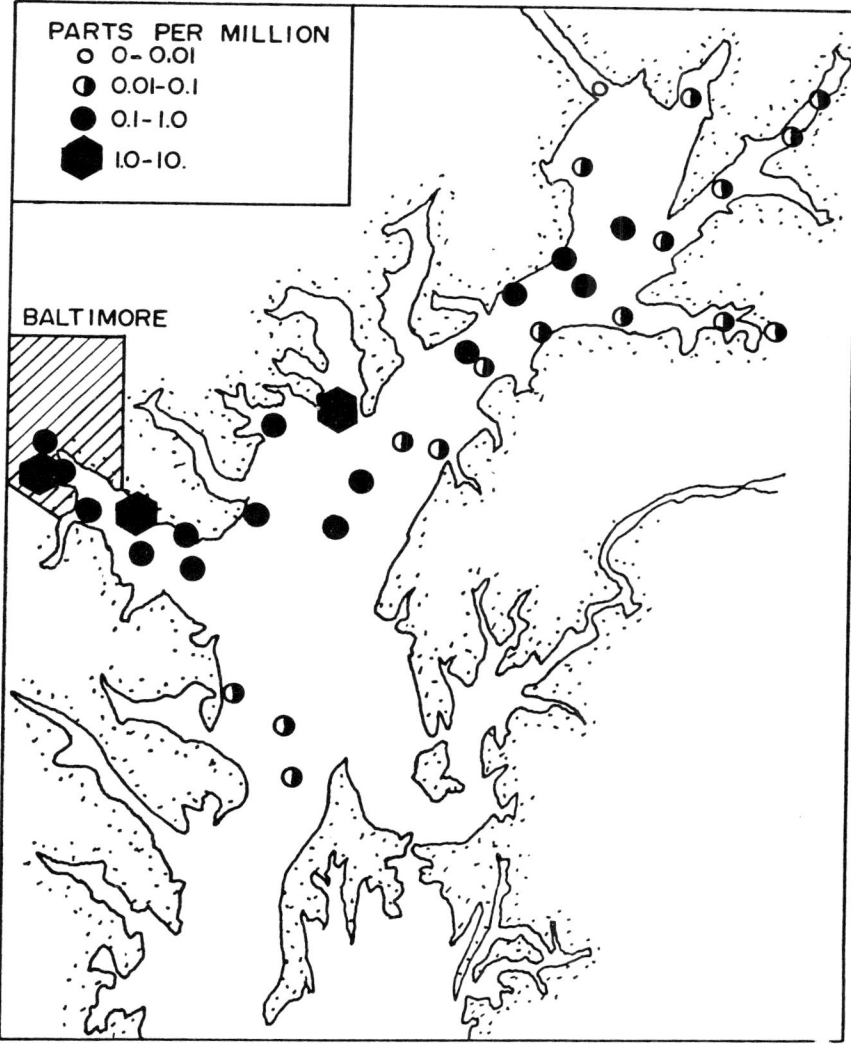

FIGURE 13. PCB's in surface sediments of the upper Chesapeake Bay. Note high concentrations in Baltimore Harbor. (After ref. 27).

coarser grained ones. Still another possibility is that there is a higher human population density around the northern Bay, leading to a higher density of both local aquatic and atmospheric discharges.

Figure 13 shows the distribution of polychlorinated biphenyls (PCB's) in sediments of the upper Bay, based on a Westinghouse study.[27] In only two cases were the concentrations near the detection limit. Everywhere else, readily measurable concentrations were found, although as with the PAH's, the concentrations at remote stations were markedly lower than in Baltimore Harbor. These compounds have no known natural sources; thus their widespread distribution throughout the Bay clearly indicates human contamination on a very large scale. In the same Westinghouse study, DDT residues and chlordane were investigated. The DDT residues were widely distributed, but present at levels closer to the detection limit than the PCB's. Chlordane was found in relatively few sites.

The Maryland Department of Health has monitored PCB's in fish and shellfish for a number of years.[53] Table 2 reproduces some of these data. It is clear that PCB's are also widely found in fish, but the concentrations appear to be about an order of magnitude lower than FDA limits (i.e. 2 mg/L). Thus while this class of compounds has become widely distributed, it is not certain that it is creating a serious problem, at least for human consumers. Whether PCB's have long term adverse effects on aquatic life, is a very difficult question for which no satisfactory answer is available.

An ongoing program at the Virginia Institute of Marine Science is designed to quantify chlorinated hydrocarbons in seafood collected from Virginia. The data show that, in general, the PCB concentrations are low compared to the FDA action level, as has been found in Maryland. A general trend of increasing concentrations with time is apparent for some species collected from near the highly populated areas around Hampton Roads. Finfish enter the Bay from the ocean in spring with lesser PCB burdens than when they migrate out in fall.[54]

Polychlorinated biphenyls are known to have been transported throughout

TABLE 2

Polychlorinated Biphenyl (Arochlor 1254) in Fish and Shellfish from Maryland Waters (Data from 53).

Species	No. Samples	Concentration in mg/L Mean	Range
Rockfish	44	0.23	0-0.58
Seatrout	12	0.05	0.02-0.13
White Perch	13	0.21	0-0.42
All Finfish	80	0.20	0-0.58
Oysters	115	0.02	0-0.07
Softshell Clams	13	0.02	0-0.06

the world and an atmospheric route for their dispersion is now accepted.[55] Initially, the idea that PCB's could enter the atmosphere in appreciable concentrations was viewed with a great deal of skepticism, because the vapor pressure of pure PCB's is very low. However, it is now understood that even compounds with very low vapor pressures can be volatilized from water at appreciable rates if their solubility in water is very low.[56]

Several other types of toxic materials are known to be regionally distributed in the Bay, even though they have not yet been studied in detail on a regional basis. For example, the man-made plutonium radioisotopes (Pu-238, 239, and 240) are found everywhere in surface sediments.[6,11] The level of radioactivity that they contribute to Bay sediments is negligible, however, compared to the activity from natural uranium and thorium series isotopes. Plutonium has entered the Bay as fallout resulting from atmospheric testing of nuclear weapons and from atmospheric burnup of man made satellites equipped with nuclear power reactors.

Phthalate esters are another class of organic contaminants that have been found wherever they have been looked for in the Bay's sediments,[57,58] although sampling density is still sparse for these compounds. Phthalate esters are added to synthetic plastics to give them flexibility and to modify their physical properties. They are moderately toxic to aquatic invertebrates; reproductive impairment of 60% has been reported for *Daphnia magna* exposed to only 3 ug/L of di-2-ethyhexyl phthalate.[59] There has been speculation that these compounds were involved in oyster mortalities in the Chester River in the 1970's, but this

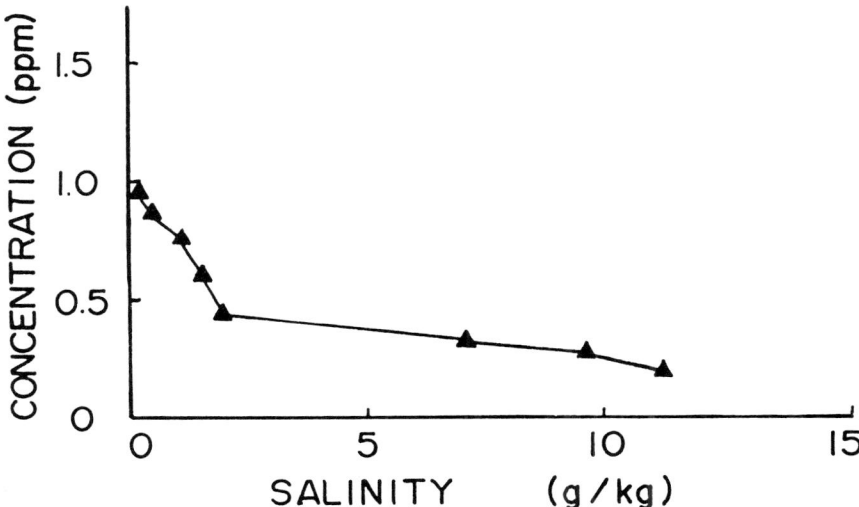

FIGURE 14. Atrazine concentration as a function of salinity in the main stem of the Chesapeake bay. (After ref. 2).

hypothesis is difficult to test and has not been proven. Phthalates are volatile, as analytical chemists have discovered to their grief. Phthalates occur in most plastic labware and therefore permeate laboratory air. They can readily contaminate samples being analyzed unless extreme precautions are taken.

Herbicides are yet another class of chemical contaminants that may be attaining a Bay-wide distribution, at least on a seasonal basis. Herbicide use has greatly expanded in the last decade or so because of the rising popularity of no till agriculture. As a consequence of their use on fields to be planted with very common crops, such as corn and soybeans, herbicide sources are numerous, widespread and diffuse. Figure 14 shows concentrations of atrazine, the most extensively used of the herbicides in Maryland and Virginia.[2] Low concentrations are found in the water over a wide salinity range. Glotfelty, et al.[60] report a detailed investigation of atrazine and simazine in the Wye River tributary over a 3 year period. Atrazine concentrations as high as 300 ug/L were observed in runoff shortly after application to fields. However concentrations in the Wye River, itself, were more than an order of magnitude lower and declined with increasing salinity, apparently due to conservative dilution. The inventory of atrazine in the Wye decreased with a 30 day half life over the growing season.

The observed concentrations of dissolved atrazine appear to be too low by about one order of magnitude to account for the loss of submerged aquatic vegetation in the Bay.[2] However the evidence of atrazine in the main stem of the Bay is of interest because this is one of the few known examples of contamination of a large area of the Bay by a compound that is believed to be transported mainly through the water. For all of the other examples of regionally distributed contaminants that we have discussed, atmospheric deposition is known to be a significant, and possibly the dominant source. The characteristics of atrazine that make it amenable to transport over considerable distances in water are a moderately high solubility (33 mg/L), which means that it has limited susceptibility to sorption on sediments,[61] a low vapor pressure, and a relatively high resistance to chemical degradation, especially in the absence of high concentrations of humic materials.[62]

SUMMARY

In this chapter, we have seen that highly industrialized areas, such as Baltimore Harbor and the Elizabeth River, are highly contaminated. Extremely high concentrations of many toxic materials are found in sediments. However, the concentrations in sediments decline seaward from these highly impacted zones suggesting that sorption to sediments greatly restricts transport of many of the toxic materials into the main Bay. Kepone was transported tens of kilometers seaward of its source, but it too was trapped in the turbidity maximum zone of the James.

Redox reactions, volatilization, microbial degradation and other processes also limit transport of certain types of contaminants, which we have called evanescent contaminants. Only a compound like atrazine, which has a moderate water solubility, low vapor pressure and moderate resistance to chemical breakdown, would be clearly capable of escaping via water from the immediate vicinity of its release point and of affecting the main Bay over a large area.

Despite the evidence of limited water-borne transport, many toxic materials, including materials known to be strongly particle-associated, are found in low concentrations in surface sediments throughout the Bay. This is true even at sites remote from shoreline sources of contaminants. There are probably no uncontaminated localities left in the Bay. For many of these substances, there is evidence that the pathway of delivery to the site of deposition has been primarily via the atmosphere. The role of atmospheric processes in contamination of Chesapeake Bay needs much more attention from researchers in the future than it has received so far.

In highly impacted areas, such as the Elizabeth River and Baltimore Harbor, evidence of adverse impact upon aquatic organisms and reduced biological diversity exists. It is likely that toxic materials are responsible for these effects. However, the pervasive low level contamination occurring in the main stem of the Bay has not been unequivocally linked to any biological deterioration.

REFERENCES

1. Schubel, J.R. 1986. The Life and Death of the Chesapeake Bay. Maryland Sea Grant. University of Maryland, College Park. 45 pp.
2. Environmental Protection Agency. 1982. Chesapeake Bay Program Technical Studies: A Synthesis. U.S. Environmental Protection Agency, Washington, D.C. 635 pp.
3. Sommer, S.E., and A.J. Pyzik. 1974. Geochemistry of middle Chesapeake Bay sediments from Upper Cretaceous to present. Chesapeake Science. 15:39-44.
4. Schubel, J.R. and D.J. Hirschberg. 1977. Pb^{210}-Determined sedimentation rate, and accumulation of metals in sediments at a station in Chesapeake Bay. Chesapeake Science. 18:379-382.
5. Cronin, L.E., D.W. Pritchard, J.R. Schubel, and J.A. Sherk. 1974. Metals in Baltimore Harbor and the Upper Chesapeake Bay and their accumulation in oysters. Phase I. Chesapeake Bay Institute and Chesapeake Biological Laboratory Joint Report. 90 pp.
6. Goldberg, E.D., V. Hodge, M. Koide, J. Griffin, E. Gamble, O.P. Bricker, G. Matisoff, G.R. Holdren, Jr., and R. Braun. 1978. A pollution history of Chesapeake Bay. Geochim. Cosmochim. Acta 42:1413-1425.
7. Sinex, S.A. 1981. Trace element geochemistry of modern sediments from Chesapeake Bay. Ph.D. Thesis, University of Maryland, College Park, MD. 190 pp.

8. Cantillo, A.Y. 1982. Trace element deposition histories in the Chesapeake Bay. Ph.D. Thesis, University of Maryland, College Park, MD.
9. Sinex, S.A. and G.R. Helz. 1981. Regional geochemistry of trace elements in Chesapeake Bay sediments. Environ. Geol. 3:315-323.
10. Helz, G.R., S.A. Sinex, K.L. Ferri and M. Nichols. 1985. Processes controlling Fe, Mn and Zn in sediments of northern Chesapeake Bay. Estuarine, Coastal and Shelf Sci. 21:1-16.
11. Helz, G.R., G.H. Setlock, A.Y. Cantillo, and W.S. Moore. 1985. Processes controlling the regional distribution of ^{210}Pb, ^{226}Ra and anthropogenic zinc in estuarine sediments. Earth and Planet. Sci. 76:23-34.
12. Helz, G.R. and S.A. Sinex. 1986. Influence of infrequent floods on the trace metal composition of estuarine sediments. Mar. Chem. 20:1-11.
13. Helz, G.R., S.A. Sinex, G.H. Setlock and A.Y. Cantillo. 1983. Chesapeake Bay Trace Elements. U.S. Environmental Protection Agency EPA-600/53-83-012 (NITS No. PB 83-207621).
14. Harris, R., M. Nichols, G. Thompson, J. Banacki, and G. Vadas. 1980. Heavy metal inventory of suspended sediment and fluid mud in Chesapeake Bay. Special Scientific Report 99. Virginia Institute of Marine Science, Gloucester Point, VA. 111 pp.
15. Helz, G.R., R.J. Huggett, and J.M. Hill. 1975. Behavior of Mn, Fe, Cu, Zn, Cd and Pb discharged from a waste-water treatment plant into an estuarine environment. Water Res. 9:631-636.
16. Villa, O. Jr., and P.G. Johnson. 1974. Distribution of metals in Baltimore Harbor sediments. Annapolis Field Office, Region III, U.S. Environmental Protection Agency. Tech. Rept. 59 (EPA 903/9-74-012).
17. Trident Engineering Associates, Inc. 1977. Evaluation of the problem posed by in-place pollutants in Baltimore Harbor and recommendation of corrective action. U.S. Environmental Protection Agency, Washington, D.C. EPA 440/5-77-015b. 78 pp.
18. Tsai, C.F., J. Welch, C. Kwei-yang, J. Shaeffer, and L.E. Cronin. 1979. Bioassay of Baltimore Harbor sediments. Estuaries. 2:141-153.
19. Sinex, S.A. and G.R. Helz. 1982. Entrapment of zinc and other trace elements in a rapidly flushed, industrial harbor. Environ. Sci. Technol. 16:820-825.
20. Frazier, J.H. 1976. The dynamics of metals in the American Oyster. *Crassostrea virginica*. II. Environmental effects. Chesapeake Sci. 17:188-197.
21. Ferri, K.L. 1977. Input of trace metals to Mid-Chesapeake Bay from shore erosion. M.S. Thesis, University of Maryland, College Park, MD. 80 pp.
22. Pheiffer, T.H. 1972. Heavy metals analyses of bottom sediment in the Potomac River Estuary. Annapolis Field Office, Region III, U.S. Environmental Protection Agency. Technical Report 49. 22 pp.
23. Mielke, J.E. 1974. Geochemical study of the sediment of the Potomac River Estuary. Ph.D. Thesis, George Washington University, Washington, D.C.

133 pp.
24. Huggett, R.J. and M.E. Bender. 1974. The effects of tropical storm Agnes on the copper and zinc budgets of the Rappahannock River. In: The effects of Tropical Storm Agnes on the Chesapeake Bay estuarine system. Chesapeake Bay Research Consortium Pub. No. 34. p. B31-B43.
25. Huggett, R.J., F.A. Cross and M.E. Bender. 1975. Distribution of copper and zinc in oysters and sediments from three Coastal Plain estuaries. Proceedings of a Symposium on Mineral Cycling in Southeastern Ecosystems, Augusta, GA, 1974. ERDA Symposium Series, CONF-740513, 224-238.
26. Johnson, P.G., and Villa, O. Jr. 1976. Distribution of metals in Elizabeth River sediments. Annapolic Field Office, Region III, U.S. Environmental Protection Agency, Technical Report 61. (EPA 903/9-76-023).
27. Munson, T.O. 1976. Biochemistry. In: Upper Bay Survey, Final Report to the Maryland Department of Natural Resources. Westinghouse Electric Corp. Ocean Sciences Division. Annapolis, MD. Vol. 2. 30 pp.
28. Malins, D.C., M.M. Krahn, E.W. Brown, L.D. Rhodes, M.S. Myers, B.B. McCain and S. Chan. 1985. Toxic chemicals in marine sediment and biota from Mukilteo, Washington: Relationships with hepatic neoplasms and other hepatic lesions in English Sole (*Parophrys vetulus*). JNCI 74:487-494.
29. Huggett, R., R. Bieri, P. deFur, W. MacIntyre, P. Shou, C. Smith, and C. Su. 1981. The present state of organic xenobiotics in the Chesapeake Bay—A synthesis paper. Special Report. Virginia Institute of Marine Science and School of Marine Science, College of William and Mary, Gloucester Point, Va.
30. Lu, M.Z. 1982. Organic compound levels in a sediment core from the Elizabeth River of Virginia. Thesis, School of Marine Science, College of William and Mary, Williamsburg, Va. 157 pp.
31. Bieri, R., C. Hein, R. Huggett, P. Shou, H. Slone, C. Smith and C. Su. 1982. Toxic organic compounds in surface sediments from the Elizabeth and Patapsco Rivers and Estuaries. Final Report to U.S. E.P.A. Grant No. R806012-01.
32. Bieri, R., C. Hein, R.J. Huggett, P. Shou, H. Slone, C. Smith and C. Su. 1986. Polycyclic aromatic hydrocarbons in surface sediments from the Elizabeth River subestuary. Internat. J. Environ. Anal. Chem. 26:97-113.
33. Huggett, R.J., M.E. Bender, and M.A. Unger. 1987. Polynuclear aromatic hydrocarbons in the Elizabeth River, Virginia. In: K. Dixon, A. Maki, and W. Brungs (Eds.) Fate and Effects of Sediment Bound Chemical in Aquatic Systems. Pergamon Press, in press.
34. Hargis, W.J., M.H. Roberts and D.E. Zwerner. 1984. Effects of contaminated sediments and sediment-exposed effluent water on an estuarine fish: Acute toxicity. Marine Environ. Res. 14:327-335.
35. Huggett, R.J., M.M. Nichols and M.E. Bender. 1980. Kepone contamination in the James River estuary. In: Contaminants and Sediments. R.A.

Baker (Ed.), Ann Arbor Science Publishers, Inc. 1:33-52.
36. Huggett, R.J. and M.E. Bender. 1980. Kepone in the James River. Environ. Sci. Tech. 14:918-923.
37. Strobel, C.J., R.E. Croonenberghs, and R.J. Huggett. 1981. The suspended sediment-water partitioning coefficient for Kepone in the James River, Virginia. Environ. Pollution 2:367-372.
38. Bender, M.E. and R.J. Huggett. 1984. Fate and effects of Kepone in the James River estuary. In: Reviews in Environmental Toxicology. E. Hodgson (Ed.) Elsevier Science Publishers. 5-50.
39. Hall, L.W. Jr., G.R. Helz, and D.T. Burton. 1981. Power Plant Chlorination—A Biological and Chemical Assessment. Ann Arbor Science Publishers. 234 pp.
40. Helz, G.R., R. Sugam, and A.C. Sigleo. 1984. Chemical modifications of estuarine water by a power plant using continuous chlorination. Environ. Sci. Technol. 18:192-199.
41. Helz, G.R., R. Sugam, and R.Y. Hsu. 1978. Chlorine degradation and halocarbon production in estuarine waters. In: R.L. Jolley, H. Gorchev, and D.H. Hamilton, Jr. (Eds.) Water Chlorination, Environmental Impact and Health Effects. Ann Arbor Science Publishers, P. 209-222.
42. Sigleo, A.C., G.R. Helz, and W.H. Zoller. 1980. Organic-rich colloidal material in estuaries and its alteration by chlorination. Environ. Sc. Technol. 14:676-679.
43. Bellanca, M.A. and D.S. Bailey. 1977. Effects of chlorinated effluents on aquatic ecosystems in the lower James River. J. Water Pollut. Control Fed. 49:639-645.
44. Morgan, R.P. and R.G. Stross. 1969. Destruction of phytoplankton in the cooling water supply of a steam electric station. Ches. Sci. 10:165-171.
45. Hamilton, D.H., D.A. Flemer, C.W. Keefe, and J.A. Mihursky. 1970. Power plants: effects of chlorination of estuarine primary production. Science. 169:197-198.
46. Carpenter, E.J., B.B. Peck and S.J. Anderson. 1972. Cooling water chlorination and productivity of entrained phytoplankton. Mar. Biol. 16:37-40.
47. Coughlan, J. and M.H. Davis. 1983. Effect of chlorination on entrained plankton at several United Kingdom coastal power stations. In: R.L. Jolley, et al. (Eds.) Water Chlorination, Environmental Impact and Health Effects. 4:1053-1063.
48. Goldman, J.C., M. Capuzzo, and G.T.F. Wong. 1978. Biological and chemical effects of chlorination at coastal power plants. In: R.L. Jolley, et al. (Eds.) Water Chlorination, Environmental Impact and Health Effects. 2:291-305.
49. Helz, G.R. and R.Y. Hsu. 1978. Volatile chloro- and bromocarbons in coastal waters. Limnol. Oceanogr. 23:858-869.
50. Hill, J.M. 1984. Identification of sedimentary biogeochemical reservoirs

in Chesapeake Bay. Unpublished Ph.D. dissertation, Northwestern University, Evanston, IL. 354 pp.
51. Galloway, J.N., J.D. Thornton, S.A. Norton, M.L. Volchok, and R.A.N. McLean. 1982. Trace metals in atmospheric deposition: a review and assessment. Atmos. Environ. 16:1677-1700.
52. Munger, J.W. and S.J. Eisenreich. 1983. Continental-scale variations in precipitation chemistry. Environ. Sci. Technol. 17:32A-42A.
53. Eisenberg, M., R. Mallman, and H.S. Tubiash. 1980. Polychlorinated biphenyls in fish and shellfish of the Chesapeake Bay. Mar. Fisheries Rev. February 1980, 21-25.
54. Edstrom, R. 1986. High resolution gas chromatographic-Hall electrolytic detector analysis of PCB congeners in blue fish (*Pomatomaus saltatric*) and grey trout (*Cynoscion regalis*) in Virginia. Submitted to J. Environ. Tox. and Chem.
55. Swackhamer, D.L. and D.E. Armstrong. 1980. Estimation of the atmospheric and non-atmospheric contributions and losses of polychlorinated biphenyls for Lake Michigan on the basis of sediment records at remote lakes. Environ. Sci. Technol. 20: 879-883.
56. Mackay, D. and P.J. Leinonen, 1975. Rate of evaporation of low-solubility contaminants from water bodies to the atmosphere. Environ. Sci. Technol. 9:1178-1180.
57. Peterson, J.C. and D.H. Freeman. 1982. Method Validation of GC-MS-SIM analysis of phthalate esters in sediment. Inern. J. Environ. Anal. Chem. 12:277-291.
58. Peterson, J.C. and D.H. Freeman. 1982. Phthalate ester concentration variations in dated sediment cores from the Chesapeake Bay. Environ. Sci. Technol. 16:464-469.
59. Sanders, H.O., M.L. Foster, Jr., and D.F. Walsh. 1973. Toxicity, residue dynamics, and reproductive effects of phthalate esters in aquatic invertebrates. Environ. Res. 6:84-90.
60. Glotfelty, D.E., A.W. Taylor, A.R. Isensee, J. Jersey, and S. Glen. 1984. Atrazine and simazine movement to Wye River Estuary. J. Environ. Quality. 13:115-121.
61. Chiou, C.T., L.J. Peters, and V.H. Freed. 1979. A physical concept of soil-water equilibria for nonionic organic compounds. Science. 206:831-832.
62. Khan, S.U. 1978. Kinetics of hydrolysis of atrazine in aqueous fulvic acid solution. Pesticide Sci. 9:39-43.
63. Anonymous. 1982. Chlorine—Bane or Benefit? Proceedings of a Conference on the Uses of Chlorine in Estuaries. Chesapeake Bay Foundation. Annapolis, MD. 212 pp.
64. Rule, J.A. 1986. Assessment of trace element geochemistry of Hampton Roads Harbor and lower Chesapeake Bay area sediments. Environ. Geol. Water Sci. 8:209-219.

Contaminant Problems and Management of Living Chesapeake Bay Resources. Edited by S. K. Majumdar, L. W. Hall, Jr. and H. M. Austin. © 1987, The Pennsylvania Academy of Science.

Chapter Fourteen
NUTRIENTS IN CHESAPEAKE BAY
DAVID L. CORRELL
Smithsonian Environmental Research Center
Box 28,
Edgewater, MD 21037

ABSTRACT

Phosphorus, nitrogen, and silicon are the key nutrient elements with respect to phytoplankton production in Chesapeake Bay. Control of this production at healthy levels can only be attained by management of these nutrients. Phytoplankton can assimilate phosphorus only as dissolved orthophosphate and silicon only as dissolved orthosilicate, but may assimilate nitrogen as ammonium, simple organic fractions, or nitrate. However, natural communities recycle all three elements by mineralizing and hydrolyzing more complex nutrient fractions to these biologically available forms. Algal populations undergoing extended rapid growth in conditions under which neither light nor any of the essential nutrients are in short supply have an element composition in which the atomic ratio of N:P is about 15 to 16 and the atomic ratio of Si:N for diatoms is about 1.0 to 1.3. These are called the Redfield ratios. These ratios may change by one or two orders of magnitude if one or more nutrients are presently, or were recently limiting, or if light intensity is limiting. Smaller shifts in these ratios can also result from temperature changes or changes in species composition.

High concentrations of nitrate and orthosilicate and high atomic ratios of N:P are found in the Bay's headwaters and those of its major tributary tidal rivers in the winter and spring. As these waters move down the bay they become depleted in total-N and nitrate more rapidly than in total-P. Available orthophosphate declines rapidly to a level of about 0.1-0.2 μg at 1^{-1}. During the summer and fall, riverine input volumes and their nutrient concentrations are lower and concentrations of total-P and N in the Bay are more dependent upon rates of benthic regeneration. Surface water concentrations of dissolved orthophosphate in the open Bay remain at about 0.1-0.2 μg at 1^{-1}, and nitrate levels are very low, but phosphorus levels in the tributary tidal rivers rise substantially. Total-P concentrations in the open Bay near Annapolis also increase severalfold over spring levels. Concentrations of available silicate are usually high but in the spring in some estuarine reaches, these concentrations become depleted in surface waters by diatom blooms.

Land discharges during an average year provide 65 and 22%, respectively, of the Bay's annual nitrogen and phosphorus inputs as well as all of the available silicate inputs. Coastal plain watersheds are responsible for 13% of the nitrogen and 66% of the phosphorus inputs due to these land discharges. The atomic ratios of N:P in coastal plain land discharges average only 8 while those from the rest of the Bay's watershed average 115. These land inputs of nutrients are several times larger during wet years and much less during dry years. The majority of the land discharges of nitrogen occur as nitrate during the winter and spring while the majority of the phosphorus is discharged in the spring and summer. Atmospheric deposition on the tidal waters accounts for 10% of total-N inputs and 5% of total-P inputs. Point sources account for 25 and 73% of total-N and P inputs, respectively. They are delivered fairly constantly during the year and have an average N:P atomic ratio of about 4.

Recycling of nutrients within the water column and between the water column and the bottom sediments and fringing marshes is a very dynamic process which allows an average nitrogen nutrient molecule to be reused over 100 times per year. Much of the nitrogen and phosphorus recycling occurs within the plankton community in the water column but silicon recycling occurs primarily in the bottom sediments. The tidal marshes recycle particulate N and P into dissolved forms. Benthic regeneration of N and P fractions is also an important process, especially in the summer and fall.

Concentrations of available P and N fractions in surface waters are not reliable indices of phytoplankton nutrient limitation since recycling is often very rapid. If the atomic ratio of nutrients in the phytoplankton is used to assess nutrient limitation, growth rates should also be measured. The best way to determine how a plankton population will respond to altered nutrient inputs is by direct manipulations in large-scale, continuous-culture bioassays.

INTRODUCTION

Nutrients may be defined as elements such as carbon, hydrogen, oxygen, phosphorus, potassium, iodine, nitrogen, sulfur, calcium and magnesium, which are essential constituents of all biota. Nutrients also include elements such as silicon, which are required only by some important groups of organisms. In order for biota to grow they must be able to assimilate sufficient amounts of these elements from their environment and be able to convert them to the various chemical forms necessary for growth and metabolic processes. Animals generally acquire adequate amounts of these nutrients as components of their food, and these animal nutrient interactions will not be considered in this review. In contrast, most autotrophic or photosynthetic biota must assimilate these nutrients directly from solution as some chemical form of the element. For the purpose of this review, the use by heterotrophic biota of organic carbon compounds as an energy source will not be discussed, even though such compounds are often referred to as organic nutrients.

Hydrogen and oxygen are readily available to biota in water and, in the Chesapeake, high concentrations of inorganic carbon, potassium, iodine, sulfur, calcium and magnesium are found in all waters which are significantly brackish. However, the concentrations of phosphorus, nitrogen and silicon in Chesapeake Bay are often in a range at which autotrophs might respond to increases or decreases in concentration. These responses may be in terms of the biomass of autotrophs, their species composition, or both. Not only are the concentrations of these three nutrients normally within this biologically meaningful range, but the chemical forms in which they are found differ temporally and spatially such that their availability for assimilation by the autotrophs is highly variable.

Proper management of Chesapeake Bay must include the control of autotrophic growth rates (primary production). Optimum rates of primary production are those which are high enough to support healthy populations of various species of desired animals which feed either directly or indirectly on these autotrophs. However, excessively high primary production rates must be avoided so as to prevent episodes of depleted dissolved oxygen, brought about by the decay of excessive accumulations of primary producer biomass. Such episodes of depleted dissolved oxygen are already common in the upper reaches of many of the Bay's tributary subestuaries and, during the growing season, in the deeper waters of the Bay's channel. This is widely perceived to be a major problem. Primary production in Chesapeake Bay is principally controlled by light penetration, the supply of mineral nutrients (P, N, Si), grazing and water circulation or exchange. Of these controls, only the supply of nutrients is ammenable to management.

RESULTS AND DISCUSSION

How Do Algae Respond To Nutrients?

Today, Chesapeake Bay primary production is almost totally due to the photosynthetic activity of algae rather than vascular plants. These algae include a taxonomically and biochemically diverse spectrum but all of these algae, including the chlorophytes, differ from vascular plants in their responses to phosphorus. Vascular plants require a rather well defined amount of phosphorus for growth. If they are exposed to a sudden increase in available phosphorus, they may grow faster but their phosphorus content per cell or per biomass doesn't increase very much. Their ability to store excess phosphorus is very limited. However, all algae have the ability to rapidly assimilate phosphorus and to store it as polyphosphate or volutin in their cells. This is especially true after extended exposure to low concentrations of available phosphorus. This rapid assimilation and storage of up to 50-times as much phosphorus per cell is called luxury consumption or the overplus phenomenon. If supplies of available phosphorus again decline, the excess phosphorus can then be utilized during subsequent algal growth. Plants can only assimilate phosphorus as dissolved orthophosphate. Thus, plant growth is sometimes reduced in the presence of high concentrations of total phosphorus due to low concentrations of dissolved orthophosphate. However, plants and other organisms can respond to low levels of available phosphorus by synthesis of a broad spectrum of phosphatases, phosphodiesterases, and polyphosphatases, which may then hydrolyze more complex phosphorus compounds to orthophosphate. One should not assume, therefore, that if dissolved orthophosphate concentrations are very low then algal uptake rates will be low or limiting. Thus, in the presence of less than 0.1 μg at 1^{-1} dissolved phosphate-P, uptake rates were often found to be from 1 to 10 μg at $1^{-1}hr^{-1}$.[2,3] Algal phosphate assimilation is an energy-requiring, active transport process in the cell wall membrane. The smaller an algal cell, the more surface area of cell wall membrane it has per cellular volume. Thus, in the presence of low concentrations of dissolved orthophosphate small phytoplankton may have a competitive advantage over larger forms in meeting their phosphorus nutrient requirements. The importance of this relationship has been demonstrated in Chesapeake Bay phytoplankton by autoradiography studies of plankton exposed to ^{33}P-labeled phosphate.[4,5]

Algae metabolize and respond to nitrogen in a manner rather similar to that of vascular plants. When the supply of available nitrogen is altered, changes in nitrogen content per biomass do occur but it has been known for a long time that these changes are usually not more than two or three-fold (e.g. 6). Plants readily assimilate nitrogen as either dissolved nitrate or ammonium and can also utilize many low-molecular weight dissolved organic nitrogen compounds such as amines and urea. If the supply of all of these available forms of nitrogen is inadequate, the plants can also synthesize hydrolytic enzymes such as proteinases to break down larger more complex organic nitrogen compounds. Finally, if nitrogen still becomes limiting, some microorganisms, especially cyanophytes, are able to fix elemental nitrogen if other conditions are favorable.

The question is often asked, which nitrogen fraction do algae prefer to assimilate? For Chesapeake Bay phytoplankton the answer seems to be ammonium first, then simple organic-N compounds such as urea, and finally nitrate.[7,8] These studies also found that in the presence of low total available nitrogen concentrations, rates of algal assimilation for all fractions were proportional to their concentrations.

Many vascular plants contain significant amounts of silicon, but some algae, especially diatoms, require large amounts of silicon for the synthesis of their cell wall. Thus, when silicon is not available in sufficient amounts, diatoms are unable to compete with other types of algae which don't have this requirement. An example of the occurrence of this condition has been described in Lake Michigan. As the lake became more enriched with phosphorus and nitrogen, algal populations increased. However, the supply of silicon did not increase. Gradually an algal population which had previously been dominated by diatoms changed to one in which diatoms played only a minor role.[9,10] Plants can only assimilate silicon as monomeric orthosilicate, which is released from diatom frustules by diagenesis of polysilicates at mildly alkaline pHs or by hydrolysis of aluminum silicate soil minerals at mildly acidic pHs.

The Redfield Ratio

A series of studies of laboratory cultures of algae and of natural phytoplankton populations lead in the 1930s, 1940s, and 1950s to a concept that algae, under reasonably good growth conditions, will have an elemental composition with relatively defined atomic ratios. For N:P this ratio is about 15-16. These ratios have become known as the Redfield ratios.[11] For marine diatoms the Si:N atomic ratio is about 1.0 to 1.3.[11,12] Natural systems in which the atomic ratio of a nutrient such as phosphorus to other mineral nutrients is less than the Redfield ratio are often assumed to be systems in which algal growth or biomass is ultimately limited by the depleted nutrient's concentrations or at least that algal growth rates in such systems will be reduced. It is well to be aware, however, that these Redfield atomic ratios are approximations and that the assumptions implied are complex and not necessarily valid. Thus, a pure culture of an alga growing in a constant and saturating supply of nutrients will vary its phosphorus content per biomass four-fold depending only upon what stage of the algal cell division cycle is present.[13] Also, the ratio of nitrogen to phosphorus in algae can vary approximately two-fold due only to variation in either light intensity or light quality.[14] The N:P ratio of algal cultures in stationary growth phase varied from 5 to about 20 under N or P limitation, respectively.[15] Laboratory cultures have also been shown to have approximately the Redfield ratio of N:P when their growth is light limited.[16] Likewise, four-fold shifts in N:P ratios were shown to be due to changing only temperature or light[17] and N:P ratios varied three-fold between algal species when all other variables were held constant.[14] Algal cultures subjected to phosphorus deficiency shift from N:P ratios of about 15 to about 100, then upon exposure to phosphate shift to ratios of less than 10.[18] Algal cultures which had been deprived of adequate phosphorus also exhibited a 26-fold increase in cellular phosphorus contents (primarily as polyphosphates) within 1 hour of exposure to phosphate.[19] These are examples of the phosphate "overplus phenomenon". Thus, one must be very cautious indeed in using the Redfield ratios. Nevertheless, the atomic ratios of nutrients in algal cells and in the water are often useful clues when attempting to elucidate algal-nutrient interactions. A good rule of thumb is not to attempt to use the Redfield ratio concept when one is dealing with algal populations recently exposed to significant changes in concentrations of available N and P or with populations whose growth rates are light limited.

If one really wants to determine how the algae present at a given time and place in the Chesapeake will react to altered nutrient concentrations or nutrient ratios in the water column, it is best to ask the algae this question by running bioassays. These bioassays are much more realistic if they involve large-scale continuous culture methods (e.g., 20). Thus, one directly measures how the specific algal populations respond to experimentally increased levels of individual nutrient fractions or combinations of two or more nutrient fractions. In such bioassays it is imperative that care be taken to control light intensity and quality as well as suspended sediment concentrations. It is also advisable that the species composition of the plankton in the bioassay apparatus be representative of the natural population. Algal responses in these bioassay systems may also be cautiously interpreted by the measurement of changes in the N:P ratios within the algal cells.

PATTERNS OF NUTRIENT CONCENTRATIONS IN CHESAPEAKE BAY

Despite the acknowledged importance of nutrients to the overall ecology of Chesapeake Bay, very few extensive surveys of the nutrient patterns of the Bay as a whole have been published. In contrast, many studies have been published of specific subestuaries or isolated sections of the open Bay. I will, therefore, attempt to generalize and describe an overall pattern. The reader should be aware, however, that year to year variations due to weather are significant and that the patterns described here need to be further refined in future studies.

The Susquehanna River provides about one half of the freshwater input to the Bay, while the Potomac and James Rivers together provide another 30 percent. These inputs and the Coriolis forces resulting from the earth's rotation interact with the Ocean at the Bay's mouth. The results include a surface current from the head of the Bay to the mouth, especially on the western shore and a deep water counter current, especially along the eastern shore. This counter current mixes with surface waters, especially at points of channel shoaling, such as occurs in the open Bay just above Annapolis.[21]

Winter and Early Spring

Peak flows from the Susquehanna River in late winter and spring contain high nutrient concentrations. In the mid 1960's, total-P concentrations at the head of the Bay were one to two µg at 1^{-1} and total-N concentrations were 80 to 105 µg at 1^{-1}.[22] Nitrate comprised about 85% of the nitrogen and the atomic ratio of total-N to total-P was about 60. Available silicate concentrations at the head of the Bay were about 60 µg 1^{-1} in the late 1970's.[23] In the early 1970's total-P concentrations in winter and spring discharges from the Susquehanna River were 0.9 to 1.6 µg at 1^{-1} while dissolved phosphate concentrations were 0.1 to 0.4 µg at 1^{-1}.[24]

Nitrate and phosphate levels in the mid 1960's were somewhat higher in the Potomac River above Washington, DC, and the atomic ratio of nitrate-N to phosphate-P varied from 15 to 25.[22] Available silicate concentrations were slightly higher than in the Susquehanna River, and in the James River were about 140 µg at 1^{-1}.[23]

These high nutrient concentrations are altered as these surface waters are swept to the south along the western shore of the Bay. The overall effect of a series of important processes in the water column is a decrease in the concentrations of readily available inorganic forms of phosphorus and nitrogen. These processes include dilution with seawater, biological assimilation and conversion to other nutrient fractions, and sedimentation to deeper waters or to the bottom. In spring (mid 1960's), by the time the surface waters arrived off the western shore near Annapolis they contained about 35 µg at 1^{-1} of nitrate-N and less than 0.2 µg at 1^{-1} of phosphate-P.[22] In the early 1970's these waters contained about one µg at 1^{-1} of total-P and still had less than 0.2 µg at 1^{-1} of phosphate-P.[24] In the spring (late 1970's), these western shore surface waters contained about 30 µg at 1^{-1} of available silicate.[23] Spring concentrations of total-P in these surface waters increased from about one µg at 1^{-1} in 1971 to about two by 1976[25] while dissolved phosphate-P levels were still about 0.2 µg at 1^{-1} in 1976. In the winter and spring of 1973, surface waters near the mouth of the Bay contained from 0.5 to 0.9 µg at 1^{-1} of total-P.[24]

Growing Season

During the summer and fall, freshwater input rates are lower and also contain lower nutrient concentrations. In the 1960's, the Susquehanna River contained about one µg at 1^{-1} of total-P and 40 to 60 µg at 1^{-1} of total-N, while the Potomac River contained from three to four µg at 1^{-1} of phosphate-P and 50 to 70 µg at 1^{-1} of nitrate-N.[22] By September in the mid 1960's, most of the mid and lower bay surface waters contained less than one µg at 1^{-1} of nitrate-N and about the same levels of ammonium-N.[22] In August 1973, total-P concentrations in mid Bay surface waters were reported to vary from 1.2 to 1.8 µg at 1^{-1}.[24] Total-P in surface waters near Annapolis was found to increase from about one or two µg at 1^{-1} in the summer and fall of the early 1970's to about five µg at 1^{-1} in the mid 1970's.[25] Values of about five µg at 1^{-1} have continued to be observed in the early 1980's (Correll, unpublished data). Such a change in the pattern of total phosphorus concentrations in mid Bay surface waters seems to indicate a change from the relatively stable seasonally unchanging pattern of the 1960's to the seasonal behavior typically found in many of the tributary subestuaries. In these tributaries (e.g. the Patuxent River,[26] phosphate concentrations exhibit a large summer peak due to the effects of benthic regeneration processes. N to P atomic ratios shift from very high in the spring to very low in the summer and fall. However, plankton assimilation and recycling appear to maintain a fairly constant open bay orthophosphate concentration of 0.2 μg at l^{-1} or less.

Available orthosilicate concentrations have a different pattern. Diatom blooms during the spring, summer and fall often assimilate nearly all of the silicate in the surface waters of the Rappahannock, James, and York River subestuaries,[27] especially at locations midway between their mouths and the head of the tide. The pattern for available silicate concentrations in the open bay remains to be clearly established. However, a late spring minimum of less than 5 μg at l^{-1} of orthosilicate in the mid Bay region has been observed.[23]

Generally, one expects to find higher concentrations of phosphate, ammonium and available silicate in the deeper waters of channels in the open Bay and in major tributary subestuaries, especially in the summer and fall. Thus, in July, 1965 ammonium-N concentrations in the mid Bay channel were 10 to 12 μg at l^{-1} in the bottom water but only four to six μg at l^{-1} in the surface water.[22] Also in August, 1973 phosphate concentrations were about 1.7 μg at l^{-1} in deep mid Bay waters but only 0.05 μg at l^{-1} at the surface.[24] In July, 1980 at the mouth of the Potomac River, available silicate concentrations were about 12 μg at l^{-1} at the bottom but only about eight μg at l^{-1} near the surface.[23]

Higher nutrient concentrations in deep waters are produced by the dissolution of biological materials which are sedimented from the surface waters to the deeper waters and the bottom sediments. These bottom waters with their higher nutrient contents are moved up the estuary by the deep counter current and eventually are mixed into the surface waters. This process acts as an effective nutrient trapping and recycling mechanism for Chesapeake Bay as a whole.[28]

Sources of N, P, and Si Inputs

Nutrients primarily enter the Bay via atmospheric deposition, land discharge, and sewage treatment plant outfalls. The relative magnitudes of these three nutrient sources differ for each nutrient and also change seasonally and yearly. Different patterns are also found in different portions of Chesapeake Bay. Thus, during the year, discharges from sewage treatment plants tend to be reasonably constant, while land discharges are highly seasonal and very subject to modification by the occurrence of high or low rates of precipitation. The question is often asked, what is the largest source of nutrients to Chesapeake Bay? Such a question is so incomplete that it has no answer. After we have discussed nutrient sources at some length, we will be better prepared to address more specific questions.

Atmospheric deposition, which includes both wet and dry deposition, is not a significant direct source of phosphorus or available silicate directly to the surface of Chesapeake Bay. Deposition of bulk precipitation (Rainfall plus dry material which sediments out of the atmosphere between rain events) 20 km south of Annapolis over a six year period in the 1970's averaged 0.81 kg total-P ha^{-1} year^{-1} of which about 20% was orthophosphate.[29] To put this in perspective, the average concentration of total-P in bulk precipitation was 0.08 μg at l^{-1}. Although there is a significant amount of silicon present as aluminum silicate soil minerals in precipitation, which is transported as dust and then deposited on the surface of the Bay, these minerals are not hydrolyzed to orthosilicate at significant rates in brackish waters. Therefore, little available silicate enters the Bay in precipitation.

However, significant amounts of nitrogen nutrients do enter the Bay in atmospheric deposition. At the same location 20 km south of Annapolis, bulk precipitation over a seven and half year period in the 1970's deposited an average of 10.6 kg total-N ha^{-1} yr^{-1} of which 4.8 Kg were nitrate-N and 2.6 Kg were ammonium-N.[30] Nitrogen deposition was also higher in the spring and summer than in fall and winter. The average concentration of nitrate-N in spring time bulk precipitation was 37 μg at l^{-1} for nitrate-N and 18 μg at l^{-1} for ammonium-N.[30] Average watershed discharge for spring time over a five year period in the same region contained an average of 23 μg at l^{-1} of nitrate-N and only 5 μg at l^{-1} of ammonium-N.[30] In the above study, when hydrologic factors were taken into account, it was concluded that bulk precipitation and land discharge were of about equal importance as an average annual source of readily available nitrogen fractions to the estuary. During an average rainfall year, bulk precipitation was a larger source in the summer and fall. Only during unusually wet periods did land discharges dominate the nitrogen loading.[30] It should be made clear, however, that we are only comparing nitrogen deposition rates on the open water areas of the estuary with land discharge rates derived from many diffuse terrestrial sources. These terrestrial sources include the nitrogen contents of precipitation falling on the estuarine watershed.

Now let us discuss land discharge as a source of Chesapeake Bay nutrients. These discharges are often called diffuse or non-point sources and include both overland flows from the land during storm events and ground water discharges. They also include surface discharges from impervious surfaces such as roads, roofs, and parking lots. They do not include discharge from sewage treatment plants and factories (point sources) even when those releases occur as discharges into freshwater streams and rivers which in turn drain into Chesapeake Bay. Thus, it is not correct to equate the nutrient discharge of a river with land discharge, unless no point sources exist on that river's watershed. These diffuse sources of nutrients are a major source of nitrogen and phosphorus and essentially the total source of available silicon.

Before one can understand the nature and regional characteristics of land discharges into Chesapeake Bay we must discuss some of the geologic and cultural-usage differences among the various subwatersheds of the Bay. The 178,000 km^2 area of Chesapeake Bay watershed is distributed within three geological domains; the Atlantic coastal plain, the piedmont and the ridge and valley region. About 19% or 33,800 km^2 of the total area is within the coastal plain.[31] Land discharges from these three domains differ substantially in their chemistry. The coastal plain is less weathered than the piedmont or the ridge and valley areas and its soils contain much higher concentrations of phosphorus. Its soils contain no calcareous minerals and bedrock lies deeply buried and out of effective contact with both surface and ground water discharges. The piedmont has relatively thin soils with a high proportion of clay. Its bedrock is composed of nonsedimentary formations. The ridge and valley domain is composed of ancient worn-down mountain ridges. It contains a large amount of sedimentary rock, including limestone. Superimposed on these geologically different domains are the effects of man's activities over the last four centuries.

In the coastal plain and some areas of nearby piedmont, farming has caused heavy erosion and the phosphorus-laden sediments have been deposited in the tidal headwaters of the Bay and its tributaries. Extensive row cropping which originally extended as near the water as practical gave way in the eighteen century to farms in which strips of land along stream channels were left in forest, either because they were to steep or to wet to cultivate with ease.

The piedmont and ridge and valley domains were settled somewhat later than the coastal plain but were eventually cleared and in many cases continue today to be intensively farmed. For example, about as much of the Susquehanna River watershed is farmed as the topography allows. Since the land is usually well drained and the stream valleys contain the most level productive soils, relatively little forest has been allowed to grow along the stream banks. Finally, man has built dams on some of the rivers such as the Susquehanna. These impoundments trap suspended soil particles with their high phosphorus content but allow much of the nitrogen to be transported on into the bay.

With the advent of chemical agriculture in the 1940's much higher levels of nitrogen were added to croplands. While the objective of enhanced crop production was achieved, land discharges inevitably were subsequently more enriched with nitrogen.

Now let us look more specifically at land discharges of nutrients from the coastal plain domain. Annual discharges from a watershed with a typical coastal plain land use composition (36% agricultural fields) have been shown to have a total-N: total-P atomic ratio of about 8 and area-weighted mean annual nutrient discharges of 2.6-3.8 and 0.79-0.91 kg ha^{-1}, respectively for total-N and total-P.[32,33] The phosphorus is discharged almost entirely in overland flows during storms and most of the phosphorus is associated with suspended soil particles. Since spring and summer storms are more intense, they result in more erosion and most phosphorus discharges occur during those seasons.[32] Nitrogen discharge is more complex. Some is discharged in overland flows but the majority is discharged as nitrate in ground water. These nitrate discharges occur almost entirely in the winter and spring.[33] Discharges of available silicate also occur primarily as ground water but the concentration of orthosilicate in ground water discharges doesn't fluctuate very much from season to season.[34]

If one considers these discharges on a seasonal basis, the atomic ratio of total-N to total-P was highest (about 12) in the fall, about 11 in the winter, 7.4 in spring and only 4.3 in the summer illustrating the importance of taking seasonal patterns into consideration.[33] Also, the percentage of total-N discharged as nitrate peaked in the winter (63%), followed by spring (39%), fall (28%), and summer (11%).[33]

The effects of land use upon watershed nutrient discharge are dramatic. For example, cropland/riparian forest watersheds discharged five times as much nitrogen and seven times as much phosphorus as forested watersheds, while pasture lands had intermediate discharges.[35] The biological availability of the discharged nutrients also differed among land uses. Nitrate was 53% of total-N discharges from pastureland, 46% from cropland, and only 13% from forest while orthophosphate was 47% of total-P from pastureland, 29% from cropland, and 24% from forest.[35] Annual area yields of total-N (the amount of nitrogen discharge per watershed area) had a correlation of 0.98 with percentage of row crops on the watershed.[36]

While these findings clearly indicate the important and dominating effects of agriculture on land discharges of nutrients, a higher level of analysis has further emphasized the magnitude of this cropland nutrient source. All of the above data were collected on discharges from cropland watersheds which included extensive streamside or riparian forest areas. Detailed studies of such watersheds have now shown that the actual discharges from cropland fields are much higher, but are dramatically modified by the riparian forests along the primary drainage channels. Thus, the total-P content of overland flows were reduced 81% and the total-N content 83% during their passage through riparian forest.[29] Likewise, the nitrate content of groundwater flows were reduced 94% during their passage through riparian forest strips. Changes in nitrate concentrations and pH indicated that high rates of nitrification in the croplands were releasing high concentrations of acid and nitrate while high rates of denitrification in the riparian forest were neutralizing most of the acid and removing most of the nitrate.[37] A systems analysis of the overall landscape indicated that of all the nutrient inputs to the entire watershed only one percent of the nitrogen and seven percent of the phosphorus were discharged to the estuary. Even so, the estuarine receiving waters were highly overenriched.[38] Thus, in the coastal plain the management of riparian vegetation on small feeder streams is critical, since even a small decrease in their efficiency of N and P interception could be catastrophic to the estuary.

Now let us consider land discharges from the piedmont and the ridge and valley domains. Despite the fact that about 81% of the watershed of the Bay lies on the piedmont and the ridge and valley area, relatively little long-term detailed research has been published on nutrient discharges from these domains and on their relationship to land use. The best overall data on total nutrient discharge from wastersheds in these domains have resulted from the national eutrophication survey.[39] Stream concentrations of nutrients were monitored on thirty watersheds which had no point sources and which drained into the Chesapeake in Pennsylvania, West Virginia, Maryland and Virginia. Land use compositions were also determined. The total area of the thirty drainage basins was 925 square kilometers and they had an area weighted mean of 39% of their area in agricultural fields. Area-weighted mean annual nutrient discharges were 5.2 and 0.10 kg ha^{-1} for total-N and total-P, respectively (atomic ratio of N:P equals 115). These area yields for nitrogen and atomic ratios are much higher than those reported from the coastal plain for areas with approximately the same proportion of cropland. These phosphorus area yields are also much lower than those reported for coastal plain watersheds, however, this may in part be due to the fact that the national eutrophication survey sampling involved only spot samples, which tend to underrepresent storm events and thus phosphorus yields.[32] The most useful publications resulting from detailed studies of nutrient discharges from agricultural lands are the results of the Mahantango Watershed study located 40 km north of Harrisburg, PA. Sixty-two percent of the watershed is in agricultural crops. This study reported overall diffuse source discharges of 82 and 115 g dissolved orthophosphate-P ha^{-1} yr^{-1} in 1970 and 1971, respectively.[40] However, these studies did not report area yield land discharges for total-P which would be much higher or for total-N. However, several publications included nitrate-N concentration data for stream discharges. Nitrate-N varied from 100 μg at 1^{-1} during storm crests to 1,000-1,500 μg at 1^{-1} during base flow in the summer and fall of 1969.[41] In 1984 a subwatershed which was 53 percent cropland released 39.3 kg nitrate-N ha^{-1} and the annual average nitrate concentration was 485 μg at 1^{-1}. Only 4.3 kg ha^{-1} of this nitrate-N discharge occurred during the summer and fall. Peak nitrate-N concentrations in the spring were about 850 μg at 1^{-1}.[42] This study also investigated the interception of nitrate-N by riparian vegetation, which typically occupies a much narrower band on these ridge and valley stream banks. Nitrate-N concentrations in ground water were reduced about 50 percent while the ground water was percolating the last 10-20 meters before entering the stream. These decreases were attributed to denitrification and some direct measurements confirmed that significant rates of denitrification were occurring in these soils.[42] Thus, the concentrations of nitrate in the streams in these agricultural areas of the ridge and valley region are approximately an order of magnitude higher than in streams draining similar primarily agricultural areas in the coastal plain. Concentrations of total-P in streams of the agricultural area of the ridge and valley domain are not well documented.

Nitrate concentrations in land discharges from two forested watersheds, Hauver Branch and Hunting Creek in the Catoctin Mountains of central Maryland were about 60 μg at l^{-1} in winter and only 10-15 μg at l^{-1} the rest of the year in 1982 and 1983.[43]

Overall, it seems clear that land discharges from the Bay's watersheds in the piedmont and the ridge and valley domains are richer in total-N and especially nitrate-N than the land discharges from the coastal plain. Likewise, it is even more apparent that the Bay's coastal plain watershed discharges are much richer in phosphorus. Overall the result is very low (c.a. 8) atomic N:P ratios in land discharges from the coastal plain and high ratios (above 100) in land discharges from the rest of the Bay's watersheds.

Nutrients from point sources include both sewage outfalls and industrial facilities. These are major sources of nitrogen and phosphorus loadings to Chesapeake Bay. Generally, much more is known about point sources than about diffuse sources, but much of this information is not published. The locations and discharge volumes of sewage treatment plants were inventoried in 1974.[44] Volumes of both municipal and industrial outfalls on the lower Susquehanna River were also summarized in 1974.[45] However, volumes of discharge change fairly rapidly. Thus, sewage outfalls at Washington, D.C., increased 27% between 1970 and 1979.[46] Total volumes of municipal sewage effluent and industrial sewage effluent in the mid 1970's were estimated to be 1.5×10^9 and 2.1×10^8 $m^3 yr^{-1}$, respectively.[47] The composition of sewage outfalls is not as well known as the volume and is more subject to change with alterations in treatment. Thus, while the Washington, D.C., outfall volume increased during the 1970's, the phosphorus and nitrogen contents decreased 78% and 17% respectively.[46] However, if average nutrient composition values are used from 13 Potomac River municipal outfalls[48] and for lower Susquehanna River industrial outfalls[45], total point source loadings for Chesapeake Bay were 33,000 t yr^{-1} for nitrogen and 18,000 t yr^{-1} for phosphorus with an average atomic ratio of N:P of about 4.[47] Another estimate[46] of point sources for Chesapeake Bay in 1971 was 32,000 t yr^{-1} for nitrogen and 10,500 t yr^{-1} for phosphorus. However, this estimate excluded point source outfalls above the fall line (the transition from the piedmont to the coastal plain), which are substantial. However, the argument can be made that some of the point source phosphorus loadings from above the fall line become bound to suspended sediments and are trapped in reservoirs. Perhaps the correct estimate for total point source phosphorus loadings actually delivered to the Bay lies somewhere between the above estimates.

It becomes apparent that we know relatively little about land discharges of nutrients in the piedmont and the ridge and valley domains, especially about the effects of land use on these discharges. We also have a deficiency of information on the nutrient trapping effects of large reservoirs. While we might wish for better information or more availability and synthesis of information, we are on relatively firm ground when it comes to atmospheric deposition, land discharges in the coastal plain, and point sources. A detailed knowledge of these nutrient sources is very important if we are to attempt to manage nutrient inputs effectively.

Fairly complete data on riverine delivery rates to Chesapeake Bay for various nutrient fractions from all sources combined on various major subwatersheds have been taken in recent years by monitoring flow and nutrient composition of the larger rivers at the fall line. Most of these data are unpublished. These data indicate that inputs from major rivers have very high N:P atomic ratios.[46] Thus, for the Susquehanna the ratio was about 70 during low flows and about 50 during high flow. For the Potomac River this ratio averaged 25, but varied from 15 during low flow to 37 at high flows.[46] However, it must be remembered that these riverine inputs are composed of a mixture of land discharges from the piedmont and ridge and valley regions plus upstream point sources. Inputs of land discharges from the coastal plain which enter the bay below these sites, have N:P atomic ratios of about 8. The majority of point source loadings also enter below these monitoring sites and have N:P atomic ratios of about 4. Thus, these major riverine inputs with relatively high N:P ratios are diluted by major sources with low N:P ratios as they move into the Bay.

The magnitudes of the various nutrient sources are compared in Table 1. Remember that these values are highly dependent upon weather. Thus, during wet years, atmospheric and land discharge loadings will be higher while during periods of drought they will be much lower. Also remember that nitrogen loadings in land discharge occur primarily in the winter and spring while land discharges of phosphorus occur primarily during spring and summer.

Nutrient Recycling In Chesapeake Bay

High concentrations of nutrients from the sources discussed above are swept

TABLE 1

Approximate Annual Nutrient Loadings for Chesapeake Bay Under Average Rainfall Conditions.

Source	Total-N (tonnes)	Total-P (tonnes)	Atomic Ratio (N:P)
Atmospheric deposition on tidal waters	12,600	970	29
Land discharges			
Coastal plain	10,800	2,870	8
Piedmont plus ridge and valley	75,000	1,440	115
Point sources*	32,500	14,200	4
TOTAL	*130,900*	*19,500*	*15*

*mean of estimates in (46) and (47).

into the Chesapeake in late winter and spring. As these waters are gradually transported toward the Atlantic Ocean, plankton proliferate and begin to assimilate and transform the various nutrient fractions. However, if these plankton communities were entirely dependent upon fresh inputs of these nutrients, productivity for the overall bay would still be low due to nutrient limitation. For example, to maintain present rates of primary production in one tributary subestuary of the bay, the Rhode River, requires the assimilation of 3.2 t of nitrogen ha^{-1} yr^{-1} by the phytoplankton while only 0.025 t ha^{-1} yr^{-1} of new nitrogen inputs are available from all external sources combined.[30] In order to maintain high productivity throughout the growing season nutrients must be reused or recycled at least 100 times per year. This recycling involves many processes and occurs in the water column, the bottom sediments, and the tidal marshes along the shoreline as nutrients are transported from their various source areas to either bottom sediments or the ocean (Fig. 1).

Nutrient loadings delivered to the Chesapeake's tidal headwaters as riverine inputs may be significantly altered before they reach the deeper channels of the open bay or its larger tributary tidal rivers. Much of the phosphorus and organic nitrogen is associated with suspended sediments, especially in cases where there are no reservoirs along the channels of the rivers draining into the Bay to intercept these particulates. All but the finest phosphorus-rich particulates usually sediment to the bottom in the open water tidal headwaters or adjacent tidal marshes. The amount of nutrients trapped in the bottom sediments is quite large. Phosphorus removal in one coastal plain headwaters varied from 33% to 200% of direct local inputs due to land discharge and atmospheric deposition. Nitrogen removal varied from 32% to 380% of direct local inputs from land discharge and atmospheric deposition.[33,49] Thus, these headwaters are capable of trapping nutrients from both riverine inputs and the water column

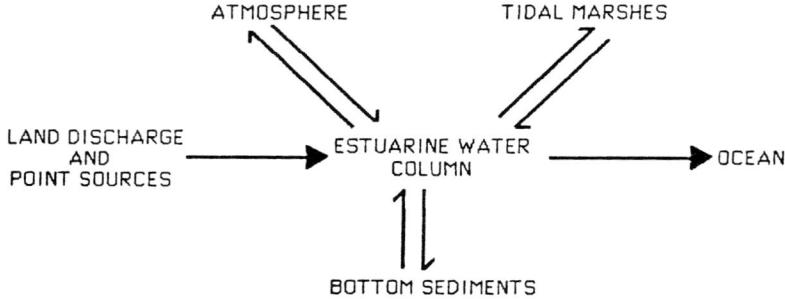

FIGURE 1. Schematic summary of nutrient sources, reservoirs, and sinks for the Chesapeake Bay system.

of the main estuarine basin in drought periods, and may also remove significant amounts of nutrients even during wet periods. The tidal marshes seem to function primarily as nutrient processors by trapping particulate N and P fractions contained in flooding tidal waters and then releasing dissolved fractions to ebbing waters.[49,50,51]

Much of the nutrient recycling in Chesapeake Bay occurs within the plankton community in the water column. One can be assured that this is the case in studies of assimilation in which uptake of isotopically-labeled nutrient fractions are measured, since the water column samples are entirely enclosed in bottles during incubation. Such studies of orthophosphate uptake and recycling found plankton uptake rates of 0.02 to 2.95 μg at l^{-1} hr^{-1} along the axis of the bay for the growing season, which corresponded with dissolved orthophosphate pool turnover times of a few minutes to a few hours.[52] Rates of plankton uptake in the Rhode River subestuary varied from 0.3 to 23 μg at l^{-1} hr^{-1} and were demonstrated to be primarily due to bacterial uptake.[2,3] Just as small phytoplankton have a competitive edge in nutrient assimilation over larger phytoplankton, bacteria have an even larger surface area to cell volume ratio than the small phytoplankton. This results in very rapid bacterial uptake and phosphorus recycling.[4,5] Although bacteria are responsible for most of the phosphate assimilation at any one time, this bacterial phosphorus is rapidly recycled back to dissolved phosphate (Fig. 2). Although the mechanisms for this nutrient recycling from the bacterial biomass are not completely understood, the end result for the nutrients is clear. Since changes in the biomass of bacteria are highly correlated with changes in phytoplankton biomass, and since the biomass of the phytoplankton is usually about ten times that of the bacteria, most of the net long-term removal of total phosphorus from the water column in the deep water areas of the bay is due to phytoplankton assimilation.

A similar pattern exists for nitrogen cycling in the water column (Fig. 3). The differences from phosphorus recycling are primarily due to the possibility for assimilation of several fractions of nitrogen. In general, since ^{15}N experiments

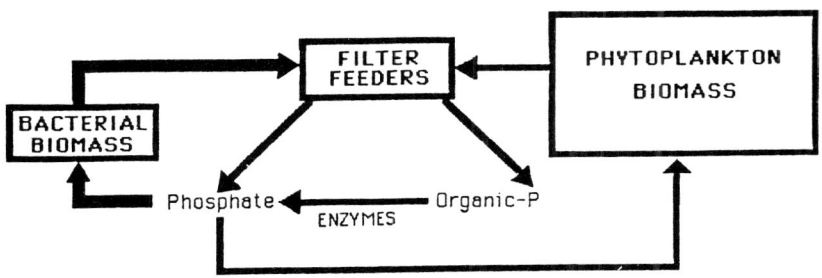

FIGURE 2. Phosphorus recycling pathways in the estuarine water column. The areas of boxes are approximately proportional to pool sizes and the width of arrows are approximately proportional to rates.

are more costly to carry out than radiophosphorus experiments, less is known about rates of nitrogen cycling. However, rates of uptake during the growing season for ammonium and nitrate-N by plankton in the water column varied from 0.01 to 0.6 μg at N l^{-1} hr^{-1} .[7,53] These rates corresponded to turnover times for available-N pools of from one to 90 minutes.[7,53] More recent ^{15}N studies on rates of nitrogen transformation between ammonium, nitrate, nitrite, and nitrous oxide in the water column indicated that nitrogen cycling can be as rapid as phosphorus cycling.[54] Either nitrification or ammonification (Fig. 3) may occur rapidly in the water column depending upon the conditions and both involve small losses of nitrous oxide.[54]

Much less data are available on the pathways of silicon processing in the water column, but available information would seem to indicate that silicon is not rapidly recycled to orthosilicate after assimilation by the plankton.[23,27] Thus, once surface water pools of orthosilicate are assimilated, diatom growth is dependent upon regeneration from the bottom sediments and transport of the released silicate back into the surface waters.

When the water column warms up during the growing season, microbial metabolism in the bottom sediments speeds up and nutrient releases to the water column are accelerated. In the Patuxent River subestuary, rates of ammonium release from bottom sediments peak in July, August, and September at 600-700 μg at m^{-2} hr^{-1}.[55] These rates are sufficient to meet over half of the plankton assimilative demand in the lower tidal river. In the open bay, summer rates of benthic ammonium release range from 13 to 40% of the calculated phytoplankton N requirement while phosphate fluxes were low, (less than 5% of ammonium flux), probably due to the well oxygenated nature of the

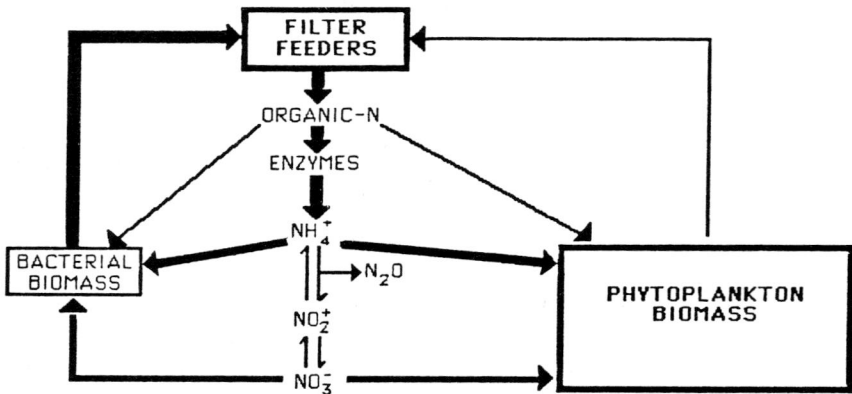

FIGURE 3. Nitrogen recycling pathways in the estuarine water column. Areas of boxes and width of arrows are approximately proportional to pool sizes and rates, respectively.

sediments.[56] In both the open bay and the Patuxent River, significant movement of nitrate from the water column into the bottom sediments was was observed in the spring, probably due to denitrification in the bottom sediments.[57] In cases where the sediments were well-oxygenated, nitrate releases from bottom sediments were observed in the summer and fall. A study of the affects of the spring-neap cycle of tidal stratification-destratification in the York River subestuary in August, clearly demonstrated the rapid accumulation of both ammonium and phosphate and a decrease in dissolved oxygen in the bottom waters during periods of stratification. This was followed by reversals in these concentrations during periods of vertical mixing.[58] These processes and nutrient concentration patterns seem to be common in many areas of the bay and its tidal tributaries. Benthic regeneration rates for orthosilicate have also been estimated to be on the order of five times greater than land discharge input rates.[23]

One might generalize by concluding that pool sizes of dissolved orthophosphate, ammonium, and nitrate are not a reliable indication of nutrient limitations on the growth of phytoplankton since recycling of other forms of N and P into these pools of available nutrients can be very rapid. The total pool of abiotic N and P in the water column, that is all fractions which are not inside of planktonic organisms, may be a better index of nutrient limitations. However, even that doesn't take into account benthic regeneration rates (Fig. 1). Thus, if one wishes to assess the extent to which nutrients are limiting phytoplankton growth rates, one should first directly measure growth, or at least primary production rates, and also the ratios of key nutrient elements within the biomass of phytoplankton. Finally, one should attempt to also conduct direct, large-scale, continuous-culture bioassays in which the plankton community is asked the question, how will it respond to altered nutrient inputs?

SUMMARY

In recent years Chesapeake Bay has experienced a series of undesirable events. These included frequent finfish kills in the summer and the loss of many shellfish beds. The general chain of events begins with excessive nutrient enrichment, which brings about undesirably heavy growths of phytoplankton. Decay of accumulations of dead phytoplankton involves high rates of consumption of dissolved oxygen leading to low dissolved oxygen concentrations during the growing season in the bottom waters of the Bay's deeper channels and diurnally low oxygen concentrations in the headwaters of many tidal river tributaries. Extended periods of low dissolved oxygen in bottom waters results in the suffocation of benthic animals including shellfish. Even short periods of low dissolved oxygen in surface waters, which may occur during the night or during extended periods of heavy cloud cover, result in finfish mortality and crab 'jubilees'.

If we are to improve conditions in the Chesapeake Bay by reversing the trend of increasing nutrient loadings which has occurred over the last several decades, both point and diffuse sources of nitrogen and phosphorus throughout the watershed of the Bay must be reduced. Partial solutions, which are designed to limit phosphorus in one geographic region of the Bay and nitrogen in another will have benefits which are, at best, restricted to a few small areas of the overall Chesapeake system.

Point sources of nutrients, primarily sewage treatment system outfalls, both on the tidal waters and the watershed freshwater streams should be reduced by upgrading or redesigning existing systems to remove most of the phosphorus and nitrogen prior to discharge. Chemical, biological, and land disposal methods have been developed and shown to be effective. The methods selected for a given system should, of course, be determined by the local conditions.

Diffuse sources of nutrients, primarily in land discharges, are inherently more difficult to manage. However, in recent years a better understanding has been emerging of the mechanisms and pathways of nutrient release from disturbed uplands. Most of the phosphorus in land discharges is transported in overland flow during storms and is associated with eroded soil particles. Better control of erosion in croplands and urban areas would significantly reduce diffuse source phosphorus loadings. Most of the nitrogen in land discharges is transported as nitrate in ground water draining agricultural and urban areas. Protection and proper management of riparian forest buffer strips between these source areas and the primary or first order streams which convey this ground water into larger streams would prevent most of this nitrate from being discharged into the Chesapeake. Most of the non-nitrate nitrogen in land discharges is organic-N, which is associated with soil particles in overland flows. The same soil conservation procedures recommended for phosphorus retention would also retain this organic-N.

REFERENCES

1. Keck, K, and H. Stich. 1957. The widespread occurrence of polyphosphate in lower plants. *Annals Bot. N.S.* 21: 612-619.
2. Correll, D.L., M.A. Faust, and D.J. Severn. 1975. Phosphorus flux and cycling in estuaries, pp. 108-135. *In:* L.E. Cronin (ed.) *Estuarine Research Vol. I.* Academic Press, New York, NY, pp. 738.
3. Faust, M.A. and D.L. Correll. 1976. Comparison of bacterial and algal utilization of orthophosphate in an estuarine environment. *Mar. Biol.* 34: 151-162.
4. Faust, M.A. and D.L. Correll. 1977. Autoradiographic study to detect metabolically active phytoplankton and bacteria in the Rhode River Estuary. *Mar. Biol.* 41: 293-305.

5. Friebele, E.S., D.L. Correll, and M.A. Faust. 1978. Relationship between phytoplankton cell size and the rate of orthophosphate uptake: *in situ* observations of an estuarine population. *Mar. Biol.* 45: 39-52.
6. Ketchum, B.H. 1939. The development and restoration of deficiencies in the phosphorus and nitrogen composition of unicellular plants. *J. Cell. Comp. Physiol.* 13: 373-381.
7. McCarthy, J.J., W.R. Taylor, and J.L. Taft. 1975. The dynamics of nitrogen and phosphorus cycling in the open waters of the Chesapeake Bay, pp. 664-681. *In:* T.M. Church (ed.). *Marine Chemistry in the Coastal Environment.* Symp. 18, American Chemical Society, Washington, D.C.
8. McCarthy, J.J., W.R. Taylor and J.L. Taft. 1977. Nitrogenous nutrition of the Chesapeake Bay. 1. Nutrient availability and phytoplankton preferences. *Limnol. Ocean.* 22: 996-1011.
9. Schlelski, C.L. and E.F. Stoermer. 1971. Eutrophication, silica depletion, and predicted changes in algal quality in Lake Michigan. *Science* 173: 423-424.
10. Schlelski, C.L., E.F. Stoermer, D.J. Conley, J.A. Robbins, and R.M. Glover. 1983. Early eutrophication in the lower great lakes: new evidence from biogenic silica in sediments. *Science* 222: 320-322.
11. Redfield, A.C. 1958. The biological control of chemical factors in the environment. *American Scientist* 46: 205-212.
12. Brezezinski, M.A. 1985. The Si:C:N ratio of marine diatoms: interspecific variability and the effect of some environmental variables. *J. Phycol.* 21: 347-357.
13. Correll, D.L. and N.E. Tolbert. 1962. Ribonucleir acid-polyphosphate from algae. I. Isolation and physiology. *Plant Physiol.* 37: 627-636.
14. Wynne, D. and G.Y. Rhe. 1986. Effects of light intensity and quality on the relative N and P requirement (the optimum N:P ratio) of marine planktonic agae. *J. Plankton Res.* 8: 91-103.
15. Terry, K.L., J. Hirata, and E.A. Laws. 1985. Light-, nitrogen-, and phosphorus-limited growth of *Phaeodactylum tricornutum* Bohlin strain TFX-1: chemical composition, carbon partitioning, and the diel periodicity of physiological processes. *J. Exp. Mar. Biol. Ecol.* 86: 85-100.
16. Tett, P., S.I. Heaney, and M.R. Droop. 1985. The Redfield ratio and phytoplankton growth rate. *J. Mar. Biol. Ass. U.K.* 65: 487-504.
17. Jahnke, J., H.J. Rick, and L. Aletsee. 1986. On the light and temperature dependence of the minimum and maximum phosphorus contents in cells of the marine plankton diatom *Thalassiosira rotula* Meunier. *J. Plankton Res.* 8: 549-555.
18. Sakshaug, E. and O. Holm-Hansen. 1977. Chemical composition of *Skeletonema costatum* (Grev.) Cleve and *Pavlova* (Monochrysis) Lutheri (Droop) Green as a function of nitrate- phosphate-, and iron-limited growth. *J. Exp. Mar. Biol. Ecol.* 29: 1-34.

19. Sicko-Goad, L. and T.E. Jensen. 1976. Phosphate metabolism in blue-green algae. II. Changes in phosphate distribution during starvation and the "polyphosphate overplus" phenomenon in *Plectonema boryanum. Amer. J. Bot.* 63: 183-188.
20. D'Elia, C.F., J.G. Sanders, and W.R. Boynton. 1986. Nutrient enrichment studies in a coastal plain estuary: phytoplankton growth in large-scale, continuous cultures. *Can. J. Fish. Aquatic Sci.* 43: 397-406.
21. Biggs, R.B. and L.E. Cronin. 1981. Special characteristics of estuaries, pp. 3-23. *In:* B.J. Neilson and L.E. Cronin (eds.). *Estuaries and Nutrients.* Humana Press, Clifton, N.J., pp. 643.
22. Carpenter, J.H., D.W. Pritchard, and R.C. Whaley. 1969. Observations of eutrophication and nutrient cycles in some coastal plain estuaries, pp. 210-221. *In: Eutrophication: Causes, consequences, correctives.* National Academy of Sciences, Washington, D.C. pp. 661.
23. D'Elia, C.F., D.M. Nelson and W.R. Boynton. 1983. Chesapeake Bay nutrient and plankton dynamics: III. The annual cycle of dissolved silicon. *Geochim. Cosmochim acta* 47: 1945-1955.
24. Taft, J.L. and W.R. Taylor. 1976. Phosphorus distribution in the Chesapeake Bay. *Ches. Sci.* 17: 67-73.
25. Correll, D.L. 1981. Eutrophication trends in the water quality of the Rhode River (1971-1978), pp. 425-435. *In:* B.J. Neilson and L.E. Cronin (eds.) *Estuaries and Nutrients.* Humana Press, Clifton, N.J. pp. 643.
26. D'Elia, C.F. 1985. Nutrient enrichment effects in Chesapeake Bay: the nitrogen vs. phosphorus controversy, pp. 3-16. *In: The fate and effects of pollutants: a symposium.* University of Maryland Sea Grant Program, College Park, MD.
27. Anderson, G.F. 1986. Silica, diatoms and a freshwater productivity maximum in Atlantic Coastal Plain estuaries, Chesapeake Bay. *Estuar. Coast. Shelf Sci.* 22: 183-198.
28. Correll, D.L. 1978. Estuarine productivity. *Bioscience* 28: 646-650.
29. Peterjohn, W.T. and D.L. Correll. 1984. Nutrient dynamics in an agricultural watershed: observations on the role of a riparian forest. *Ecology.* 65: 1466-1475.
30. Correll, D.L. and D. Ford. 1982. Comparison of precipitation and land runoff as sources of estuarine nitrogen. *Estuar. Coast. Shelf Sci.* 15: 45-56.
31. Seitz, R.C. 1971. Drainage area statistics for the Chesapeake Bay freshwater drainage basin. Chesapeake Bay Institute, Johns Hopkins University, Baltimore, MD 21 pp.
32. Correll, D.L., T.L. Wu, E.S. Friebele, and J. Miklas. 1977. Nutrient discharge from Rhode River watersheds and their relationship to land use patterns, pp. 413-437. *In:* D.L. Correll (Ed.) *Watershed research in eastern North America, Vol. I.* Smithsonian Press, Washington, D.C. pp. 469.
33. Correll, D.L. 1981. Nutrient mass balances for the watershed, headwaters

intertidal zone, and basin of the Rhode River estuary. *Limnol. Ocean.* 26: 1142-1149.
34. Correll, D.L., J.J. Miklas, A.H. Hines, and J.J. Schafer, (in press). Chemical and biological trends associated with acidic atmospheric deposition in the Rhode River watershed and estuary. *Water, Air, Soil Pollut.*
35. Correll, D.L. 1983. N and P in soils and runoff of three coastal plain land uses, pp. 207-224. *In:* R. Lowrance, R. Todd, L. Asmussen, and R. Leonard (eds.). *Nutrient cycling in agricultural ecosystems.* University of Georgia, Athens, GA pp. 602.
36. Correll, D.L. and D. Dixon. 1980. Relationship of nitrogen discharge to land use on Rhode River watersheds. *Agro. Ecosyst.* 6: 147-159.
37. Peterjohn, W.T. and D.L. Correll. 1986. The effect of riparian forest on the volume and chemical composition of baseflow in an agricultural watershed, pp. 244-262. *In:* D.L. Correll (ed.). *Watershed Research Perspectives.* Smithsonian Press, Washington, DC pp. 1421.
38. Jordan, T.E., D.L. Correll, W.T. Peterjohn, and D.E. Weller. 1986. Nutrient flux in a landscape: the Rhode River watershed and receiving waters, pp. 57-76. *In:* D.L. Correll, (Ed). *Watershed Research Perspectives.* Smithsonian Press, Washington, D.C. pp. 421.
39. Omernik, J.M. 1976. The influence of land use on stream nutrient levels. U.S. Environmental Protection Agency, Corvallis, Oregon, 106 pp.
40. Gburek, W.J. and W.R. Heald. 1974. Soluble phosphate output of an agricultural watershed in Pennsylvania. *Water Rescur. Res.* 10: 113-118.
41. Gburek, W.J. and W.R. Heald. 1970. Effects of direct runoff from agricultural land on the water quality of small streams, pp. 61-68. *In: Relationship of agriculture to soil and water pollution.* Cornell University, Ithaca, New York.
42. Schnabel, R.R. 1986. Nitrate concentrations in a small stream as affected by chemical and hydrologic interactions in the riparian zone, pp. 263-282. *In:* D.L. Correll (ed.). *Watershed Research Perspectives.* Smithsonian Press, Washington, D.C. pp. 421.
43. Katz, B.G., O.P. Bricker, and M.M. Kennedy. 1985 Geochemical mass-balance relationships for selected ions in precipitation and stream water, Catoctin Mountains, Maryland. *Amer. J. Sci.* 285: 931-962.
44. Brush, L.M., Jr. 1974. Inventory of sewage treatment plants for Chesapeake Bay. Chesapeake Research Consortium, Gloucester Point, Virginia. 62 pp.
45. Clark, L.J., V. Guide, and T.H. Pheiffer. 1974. Summary and conclusions, nutrient transport and accountability in the lower Susquehanna River. U.S. Environmental Protection Agency, Annapolis, M.D.
46. Jaworski, N.A. 1981. Sources of nutrients and the scale of eutrophication problems in estuaries, pp. 83-110. *In:* B.J. Neilson and L.E. Cronin (eds.). *Estuaries and Nutrients.* Humana Press, Clifton, N.J. pp. 643.
47. Correll, D.L. 1976. The relative contributions of point and non-point

sources of nutrients and pathogens to the water quality of the bay, pp. 19-31. *In: Water quality goals for Chesapeake Bay — What are they and how can they be achieved?* Virginia Polytechnic Institute and State University, Blacksbury, VA.
48. Jaworski, N.A. 1969. Nutrients in the upper Potomac River basin. U.S. Environmental Protection Agency, Annapolis, MD.
49. Jordan, T.E., D.L. Correll, and D.F. Whigham. 1983. Nutrient flux in the Rhode River: Tidal exchange of nutrients by brackish marshes. *Estuar. Coast. Shelf Sci.* 17: 651-667.
50. Jordan, T.E. and D.L. Correll. 1985. Nutrient chemistry and hydrology of interstitial water in brackish tidal marshes of Chesapeake Bay. *Estuar. Coast. Shelf Sci.* 21: 45-55.
51. Axelrad, D.M., K.A. Moore, and M.E. Bender. 1976. Nitrogen, phosphorus, and carbon flux in Chesapeake Bay marshes. Virginia Polytechnic Institute and State University, Blacksburg, VA 182 pp.
52. Taft, J.L. and J.J. McCarthy. 1975. Uptake and release of phosphorus by phytoplankton in the Chesapeake Bay estuary, U.S.A. *Mar. Biol.* 33: 21-32.
53. Wheeler, P.A., P.M. Glibert, and J.J. McCarthy. 1982. Ammonium uptake and incorporation by Chesapeake Bay phytoplankton: short-term uptake kinetics. *Limnol. Ocean.* 27: 1113-1128.
54. McCarthy, J.J., N. Kaplan, and J.L. Nevins. 1984. Chesapeake bay nutrient and plankton dynamics. 2. Sources and sinks of nitrite. *Limnol. Ocean* 29: 84-98.
55. Kemp, W.M. and W.R. Boynton. 1984. Spatial and temporal coupling of nutrient inputs to estuarine primary production: the role of particulate transport and decomposition. *Bull. Mar. Sci.* 35: 522-535.
56. Boynton, W.R. and W.M. Kemp. 1985. Nutrient regeneration and oxygen consumption by sediments along an estuarine salinity gradient. *Mar. Ecol. Prog. Ser.* 23: 45-55.
57. Jenkins, M.C. and W.M. Kemp. 1984. The coupling of nitrification and denitrification in two estuarine sediments. *Limnol. Ocean.* 29: 609-619.
58. D'Elia, C.F., K.L. Webb, and R.L. Wetzel. 1981. Time varying hydrodynamics and water quality in an estuary, pp. 597-606. *In:* B.J. Neilson and L.E. Cronin (eds.). *Estuaries and Nutrients.* Humana Press, Clifton, N.J. pp. 643.

Chapter Fifteen
CONTAMINANT EFFECTS ON CHESAPEAKE BAY FINFISHES

RONALD J. KLAUDA[1] and MICHAEL E. BENDER[2]

[1]The Johns Hopkins University
Applied Physics Laboratory
Environmental Sciences Group
Shady Side, Maryland 20764

and

[2]Virginia Institute of Marine Science
School of Marine Science
College of William and Mary
Gloucester Point, Virginia 23062

ABSTRACT

Habitat deterioration is consistent with perceived population declines for several resident and anadromous finfish species in Chesapeake Bay that are subjected to different levels of fishing pressure (e.g., striped bass versus blueback herring). Diminution of habitat quality has natural and anthropogenic roots that are difficult to separate. Recent contaminant effects studies focused on Chesapeake Bay fishes can be grouped as follows: (a) mathematical and statistical modeling studies aimed at elucidating contaminant and stock trend relationships using extant data and theoretical insights, (b) biological and chemical field surveys in selected areas to demonstrate spatio-temporal associations between levels of toxic organic and inorganic chemicals and absence or reduction of sensitive species, (c) measurements of condition factors and tissue residues of chemical contaminants in juvenile and older fishes, (d) laboratory studies of life stage and species sensitivities to an array of toxic contaminants, and (e) in-situ field studies designed to measure the effects of habitat quality on specific life stages of selected species. Contaminant-related research has focused primarily on striped bass, American shad, and river herrings. Two currently intensive areas of investigation are the leaching of tributyltin (TBT) antifouling paints into marina areas and acidic deposition in freshwater coastal plain tributaries. These recent studies collectively support several tentative conclusions that deserve further study: (a) deterioration of spawning and nursery habitats in Chesapeake Bay, via the influx of toxic contaminants, may be contributing to poor recruitment in some anadromous and resident finfishes, (b) larvae and newly transformed juveniles are more sensitive to most contaminants than embryos and older life stages, (c) recent adverse effects (likely the past 20 years) of contaminants on juvenile production in several finfish species may not be related to historical variations in stock abundance, but could be responsible for keeping several species populations at currently low abundance levels, and (d) adverse contaminant effects on finfishes are likely to be highly variable from year to year, among weeks within a year, from river to river or estuary to estuary, among specific spawning and nursery areas within a river or estuary, from species to species, and among life stages within a species. The current state of the art in fish population models limits the extent to which documented contaminant effects on individuals can be used to precisely predict responses of populations. Finfish management decisions must, at least for the foreseeable future, be based on less than accurate scientific predictions of risks associated with the current contaminant levels in Bay habitats. Future studies should continue to identify those species, life stages, and spawning or nursery areas in Chesapeake Bay that are most sensitive to contaminant effects and would most benefit from stringent controls on contaminant inputs.

INTRODUCTION

Finfish spawning and nursery habitats in Chesapeake Bay are typical of most temperate latitude estuaries—highly fluctuating. Unpredictable, temporally and spatially heterogeneous environmental conditions impose mortalities on the early life stages of anadromous and resident species that collectively exceed 99% during the first year of life.[1,2] Successful species must possess life history traits that form a reproductive strategy for persistence in the fluctuating and uncertain environment[3,4,5] that exposes their fragile early life stages to an array of mortality sources (Figure 1).

Despite unpredictability of high mortality of eggs and larvae and substantial variation in year class success, existing fisheries records show that iteroparous Bay fishes such as the anadromous striped bass (*Morone saxatilis*) and American shad (*Alosa sapidissima*), and the resident yellow perch (*Perca flavescens*) and white perch (*Morone americana*) have until recently been reasonably successful. Since the early to mid 1970's, these four species and other anadromous and resident Bay finfish have experienced a series of relatively poor year classes. Below average reproductive success has been reflected in a steady decline in sport and commercial fisheries landings over the past decade.[6,7,8,9,10] The innate abilities of these species populations to persist in the face of environmental uncertainty, developed over evolutionary time, is apparently being threatened in the latter half of the 20th century by one or more stressors in Chesapeake Bay.

Prior to the late 1940's, there was little concern about pollution and habitat degradation, except in localized situations. For example, Galtsoff[11] studied the effects of sulfate pulp mill wastes on oysters in the York River, Virginia, and Davis[12] investigated the effects of copper pollution in the Patapsco River, Maryland. Massman et al. wrote in 1952 that chemical pollution had temporarily affected some fish species in local areas, but had not resulted in long-range losses of economically important species in the Bay as a whole.[3] About a decade later in 1961, Mansueti[14] observed that Chesapeake Bay had been subjected to the

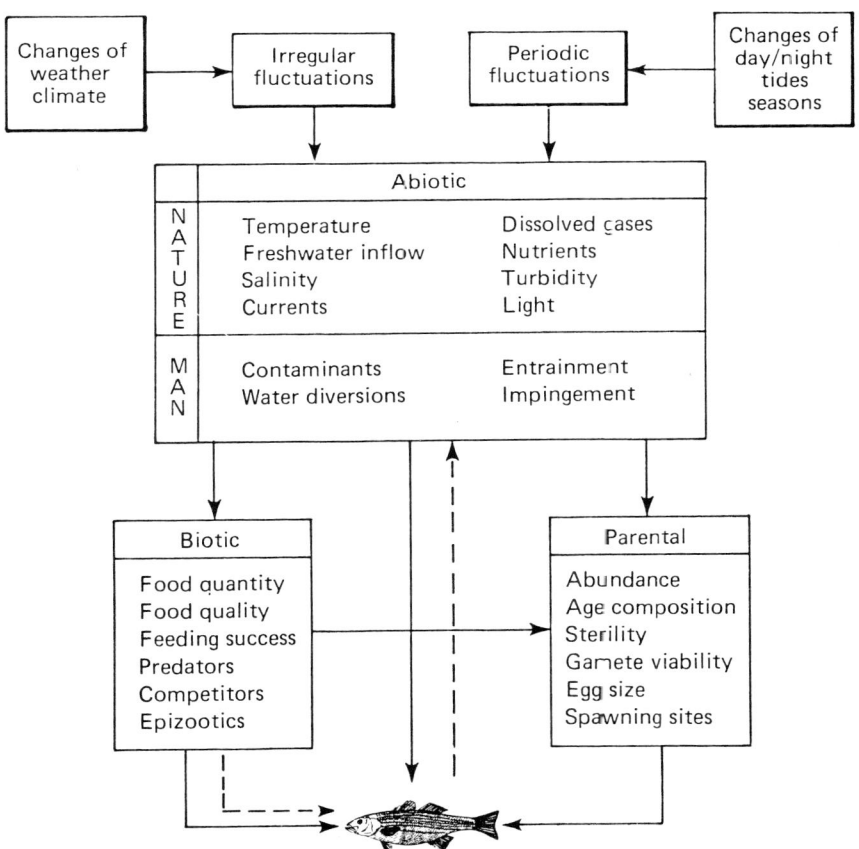

FIGURE 1. Conceptual model of environmental factors which may alter early life stage survival and influence juvenile abundance of Chesapeake Bay finfishes.

effects of civilization. He characterized some effects as catastrophic and others as moderate but sustained.

By 1967, however, L. Eugene Cronin reported that several anadromous and resident Bay finfishes were declining.[15] He suggested that "subtle chemical pollution seems to have high potential for serious and unexpected damage to the estuarine ecosystem", even though chemical contaminants were not reported to be a general problem in the Bay at that time. When the U.S. Environmental Protection Agency initiated the multi-year, $27 million Chesapeake Bay Program (1976-1981), toxic chemical pollution was one of three major study objectives.[16]

Overfishing and deterioration of habitat quality (commonly labelled pollution) are usually the prime causes of dramatic and extended fish stock declines.[17] The relative importance of these two sources of mortality are often hotly debated in the scientific literature.[18] These and other hypotheses have been recently investigated in an attempt to explain the poor recruitment trends for several Bay finfishes.[19,20,21,22,23,24,25,26] Deterioration of spawning and nursery habitats from contaminant inputs is one attractive hypothesis because the Chesapeake Bay Program recently revealed that contaminants entering the Bay are not quickly flushed into the Atlantic Ocean. Rather, because of unique circulation patterns in the Bay, they tend to accumulate.[16] A recent decline in habitat quality is consistent with perceived stock declines for several finfish species that have been subjected to quite different degrees of fishing pressure but use similar habitats for spawning and nursery activities. In contrast to many anadromous and resident species, stocks of oceanic species (e.g., bluefish *Pomatomus saltatrix* and Atlantic menhaden *Brevoortia tyrannus*), which spawn in marine waters and move into the Bay as fully transformed juveniles, have either remained stable or increased during the same recent decade when several anadromous and resident species declined.[6]

Diminution of habitat quality in Chesapeake Bay has natural and anthropogenic roots that are difficult to separate. To illustrate, a large-scale natural event like Tropical Storm Agnes in 1972 caused heavy rainfalls of up to 30 cm in the drainage basin of Chesapeake Bay. The ensuing floods not only transported immense sediment and nutrient loads into estuaries and the mainstem Bay, but also resuspended and deposited unknown amount of contaminants into finfish spawning and nursery habitats.[27,28] The relative impacts of sediment, nutrient, and contaminant inputs associated with this major storm on Bay fishes can probably never be satisfactorily allocated. Nevertheless, because so many fish species of importance to the economy of this region are reproducing so poorly in recent years, researchers have continued to investigate all potential causes, including contaminants.

Other papers in this volume highlighted the kinds of contaminants measured in the water column and sediments of Chesapeake Bay. The objective of this paper is to discuss the role that these contaminants may be playing in the declining population status of several finfishes. Review of a series of recent studies will form the basis for our perspectives on the contaminant effects hypothesis. This paper is not intended to provide an exhaustive literature review of toxic chemical effects on Bay fish species. Rather, we intend to summarize a representative sample of current contaminant effects research and establish a milestone for measuring progress to date and for planning future investigations.

Contaminant effects studies on Chesapeake Bay finfishes can be grouped as follows:
(1) mathematical and statistical modeling exercises aimed at elucidating contaminant versus stock trend relationships using extant data and theoretical insights,
(2) biological and chemical surveys of selected habitats designed to reveal spatio-temporal associations between levels of potentially toxic organic and inorganic contaminants and the absence or reduction of various sensitive species,
(3) measurements of condition factors and contaminant residues in tissues of juvenile and older fishes from selected habitats,
(4) laboratory studies of life stage and species sensitivities to acute and chronic concentrations of toxic contaminants, and
(5) in-situ field studies designed to measure the effects of ambient water quality and contaminants in specific habitats on specific life stages of selected species.

Examples of recent contaminant effects studies discussed here encompass several aquatic habitat quality concerns. These are: (a) biocides, such as chlorine and organotins (tributyltin); (b) polynuclear aromatic hydrocarbons or PAHs; (c) mixtures of inorganic and organic contaminants; (d) acid deposition, which includes the effects of depressed pH and mobilization of toxic metals; and (e) radionuclides. These topics embrace point source and non-point source pathways for a range of inorganic and organic contaminants. Current research on this array of topics is understandably variable in scope and distributed unevenly among the 100 + finfish species that occupy portions of Chesapeake Bay during some segment of their life cycles.

RESULTS

BIOCIDES

Kepone

Restrictions on the commercial harvest of some species of finfish are still in effect in the James River, Virginia, 10 years after the discovery that the pesticide Kepone (decachlorooctahydro-1, 3, 4-metheno-2H-cyclobuta (cd) pentalen-2-one) had contaminated the estuary. Kepone was produced in the City of Hopewell (at river km 120) by two firms between 1967 and 1975. The pesticide entered the tidal river through a variety of primarily point-source routes which included chemical plant discharges, runoff from contaminated land fills, and sewage effluents. Bender and Huggett[29] reviewed the data available through 1982 on the status of Kepone contamination in the James River estuary. This discussion is based primarily on that review. Loesch et al.[30] surveyed several Virginia tributaries to Chesapeake Bay and detected Kepone above the action level (0.3 µg/g) in tissues from juvenile finfishes collected in the James River and its tributary, the Chickahominy River. No Kepone above the action level was detected in tissue samples of juveniles collected in the Mattaponi, Pamunkey, Rappahannock, or Potomac rivers.

Bottom sediments of the James River are contaminated from the source at Hopewell to near the river mouth. Figure 2 shows the mass of Kepone estimated

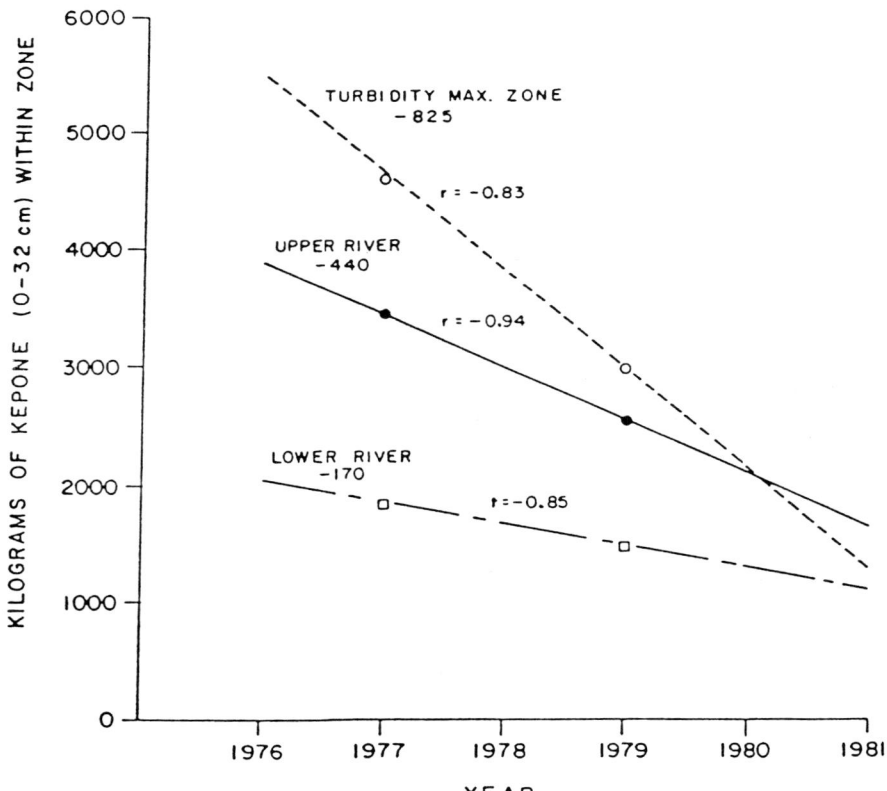

FIGURE 2. Calculated regression lines of Kepone concentrations in the sediments of three zones in the James River.[29]

to be present in the upper 32 cm of river bed sediments as a function of time and location. The rate of burial or dilution (i.e., the slope of the lines in Figure 2) is greatest in the turbidity maximum zone, followed by the upper estuary, and is considerably less in the lower estuary. Since these bed sediments serve as the source of Kepone available to aquatic organisms, the rate at which burial or dilution occurs is extremely important in determining exposure levels.

The relationship between Kepone residues observed in finfish species as a function of the change in Kepone mass in river sediments over time is shown

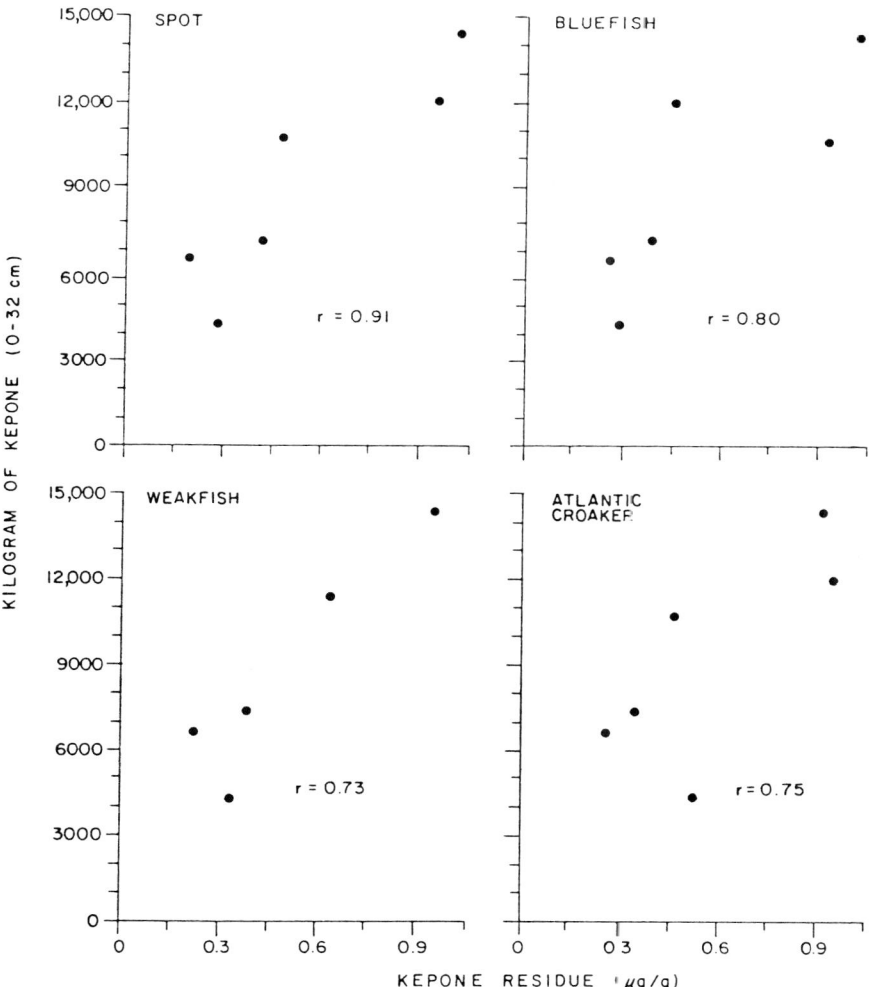

FIGURE 3. Kepone concentrations in James River sediments (0-32cm) from 1976 to 1981 versus Kepone residues in four species.[29]

in Figure 3. Figure 4 depicts the change in third quarter (July-September) residues for Atlantic croaker (*Micropogonias undulatus*) and spot (*Leiostomus xanthurus*) from the lower James River as a function of time (1976-1985). Residues for both species declined through 1980, increased in 1981, and then again declined, but at a much slower rate.

After Kepone contamination of the James River was discovered in 1975, numerous studies were conducted to estimate its impact on aquatic biota. The majority of investigations to establish effects levels were conducted by researchers at the U.S. Environmental Protection Agency laboratory in Gulf Breeze, Florida, and by staff and students of the Virginia Institute of Marine Science. Space precludes a detailed discussion of all technical findings. For more information, the reader is referred to references cited in Bender and Huggett[29] and the recent study by Fisher et al.[31]

Acute, partial chronic, and chronic toxicity tests were used to estimate effects of Kepone on aquatic life. In some cases, bioassays established no effect

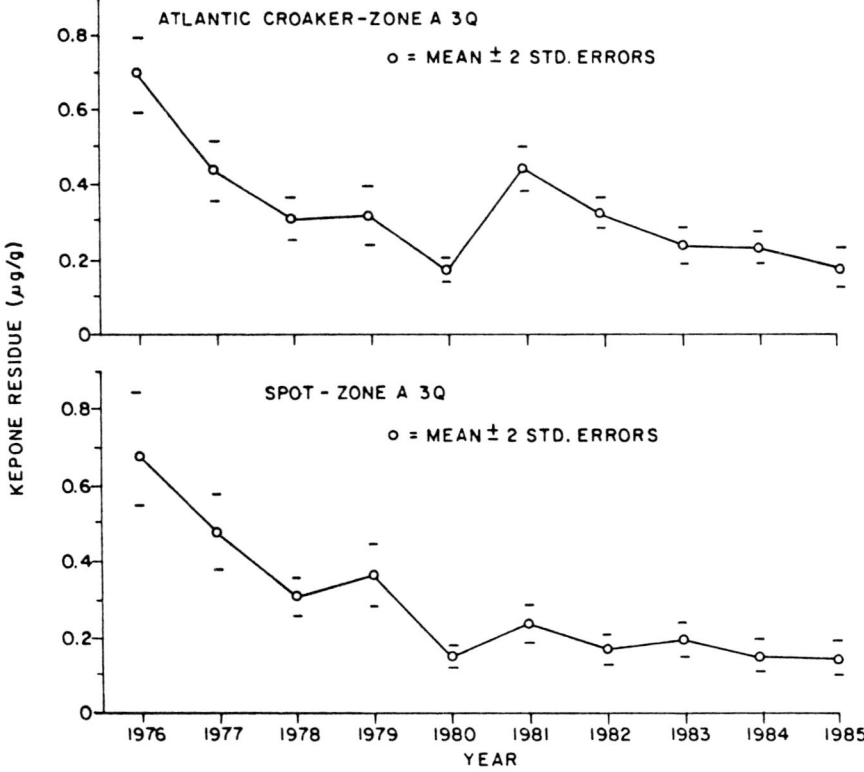

FIGURE 4. Kepone residues in Atlantic croaker and spot, 1976-1985.

levels; i.e., exposure levels at which no significant difference in growth or reproduction were observed compared to the control groups. Other studies estimated the maximum acceptable toxicant concentration (MATC) application factor, defined as the ratio of the chronic no effect level to the 96-h LC50 level (concentration which kills 50% of the test organisms). Figure 5 compares the measured no effect level for several test species to levels of Kepone found in the James River. Exposure levels in the river are well below no effect levels. Figure 6 shows the MATC's for eight finfish species using a very conservative application factor of 0.001.[32]

In summary, laboratory bioassays showed that exposure to the pesticide Kepone can produce measurable acute and chronic effects on marine, estuarine

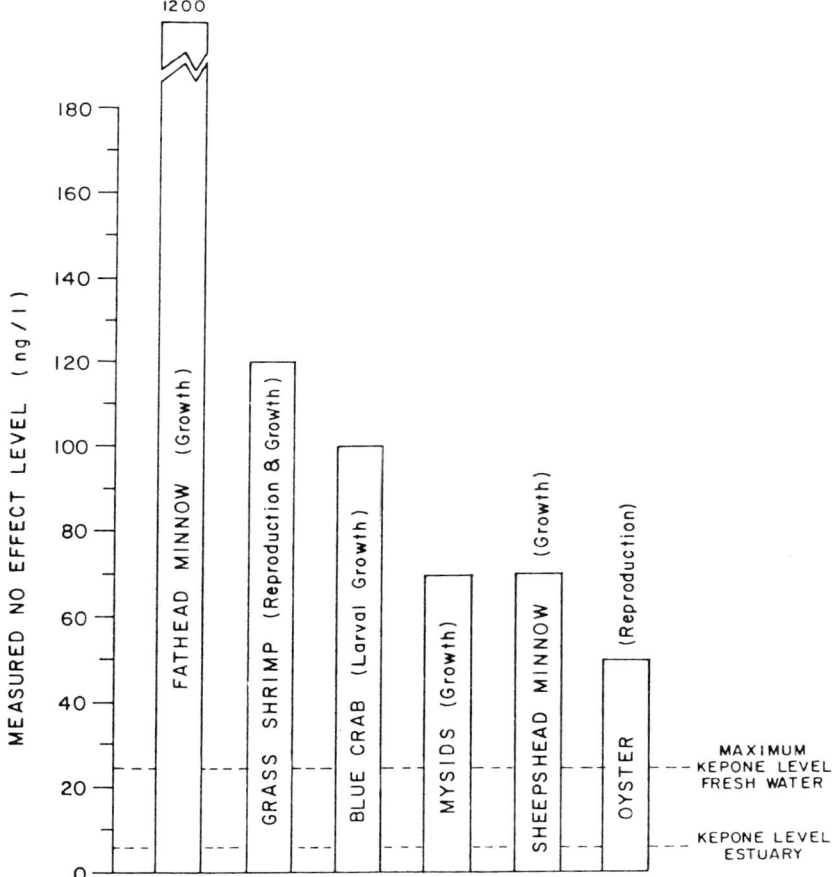

FIGURE 5. Measured no effect levels for Kepone.[29]

and freshwater finfishes that inhabit Chesapeake Bay. However, Kepone concentrations necessary to cause detrimental effects are considerably greater than concentrations measured in the James River. If these conclusions about Kepone effects on finfish are correct, then the major impact of this contaminant in the James River is economic loss due to restrictions on fishing. The James River was closed to all forms of fishing in December 1975 because of Kepone contamination.[30] The ban has since been modified several times. At present, seasonal restrictions limiting commercial fishing for some species are still in effect, and the harvest of striped bass is prohibited throughout the year. Quantitative estimates of economic impacts on the fishery are not available. Many commercial fishermen participated in legal actions against the manufacturing firms to collect damages. All claims were settled out of court.

Chlorine

Chlorine is the fifteenth most abundant element in the earth's surface, and

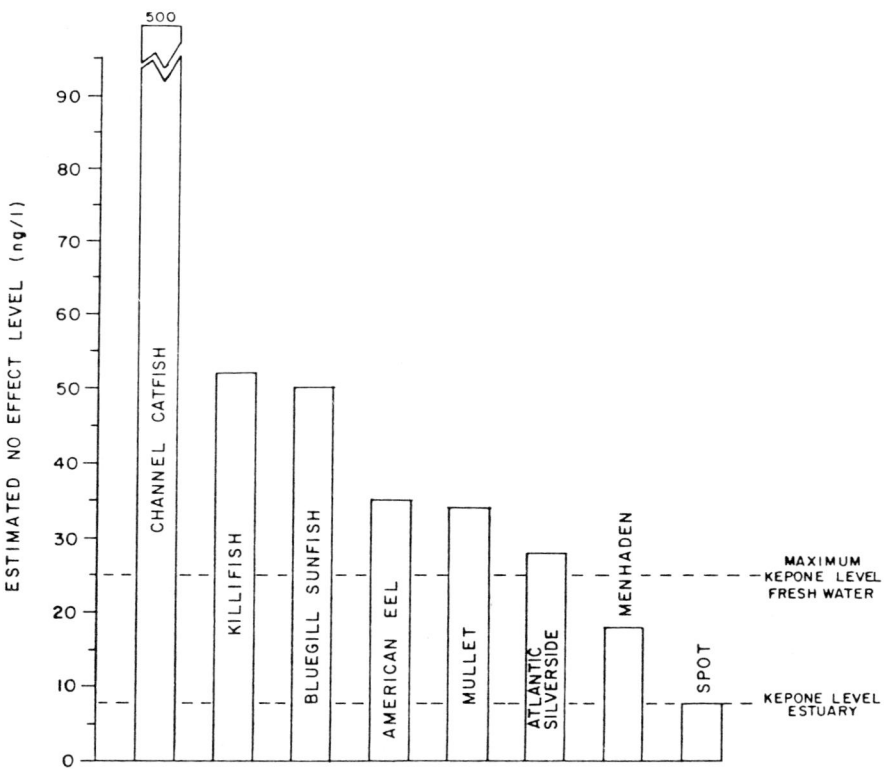

FIGURE 6. Estimated no effect levels for Kepone.[29]

is most commonly present as the Cl^{-1} anion form in estuaries and oceans where concentrations can reach 1.9% by weight.[33] In freshwater habitats, chlorine has an average concentration of about 8 mg/L. Chlorine has been a widely used biocide for disinfection in water and wastewater treatments in both municipal and industrial applications since the late 1800's. Presently, chlorine is the predominant biocide for control of fouling in condenser systems of electrical generating stations (power plants) in the United States.[34,35]

Chlorination of fresh and saline waters may form halogenated organics in water bodies receiving chlorinated discharges. Jolly et al.[36] identified 50 chloroorganic compounds from natural freshwater chlorinated at a power plant. Chlorination of saline waters results in the formation of predominantly brominated rather than chlorinated organics.[37]

During the early 1970's, researchers learned that chlorination of municipal and industrial wastewaters and use of chlorine biocides at power plants produced oxidation compounds considered to be carcinogenic to man[38,39] and toxic to aquatic organisms.[40,41,42,43,44] The potential environmental effects of chlorine releases to freshwater and estuarine habitats have been recognized in Chesapeake Bay since the early 1970's. Tsai[45] related decreases in fish species diversity below sewage treatment plants in Virginia, Maryland, and Pennsylvania to sewage chlorination and turbidity. Bellanca and Bailey[46] demonstrated that high residual chlorine was the causative factor in a major fish kill which occurred in 1974 along the James River, Virginia. Coulter[47] speculated that population declines of striped bass in the Bay could be partially attributable to increased chlorine usage.

As a result of these and other investigations, many acute and chronic bioassay studies have been conducted to define safe limits for chlorine in fresh and estuarine waters (see reviews in[48,49,50,51,52] and recent study results in [53,54,55,56,57,58]). These and other studies were conducted to: (1) better define the toxic concentrations of chlorine; (2) evaluate the toxicity of bromine chloride, a potential alternative biocide[59,60,61,62,63]; (3) determine the decay rates of several chlorine produced oxidants (CPOs) in estuarine waters; (4) determine if dechlorination is effective in reducing the toxicity of chlorinated effluents; and (5) determine organism avoidance responses to CPOs. The results of numerous studies contributed to the formulation of effluent limitations on chlorine discharges in both Maryland and Virginia (personal communication with M.J. Garreis, Maryland Department of Health and Mental Hygiene; K. Buttleman, Virginia Water Control Board.)

Chlorine is toxic to several finfish species which inhabit Chesapeake Bay during some portion of their life cycle (Table 1 and Figure 7). However, species responses to chlorine discharges are affected by many physical, chemical, and biological factors, including temperature, dilution capabilities of the receiving water body, chemical speciation of chlorine, presence of other water quality parameters (e.g., ammonia, dissolved oxygen, pH), and the life stages or age

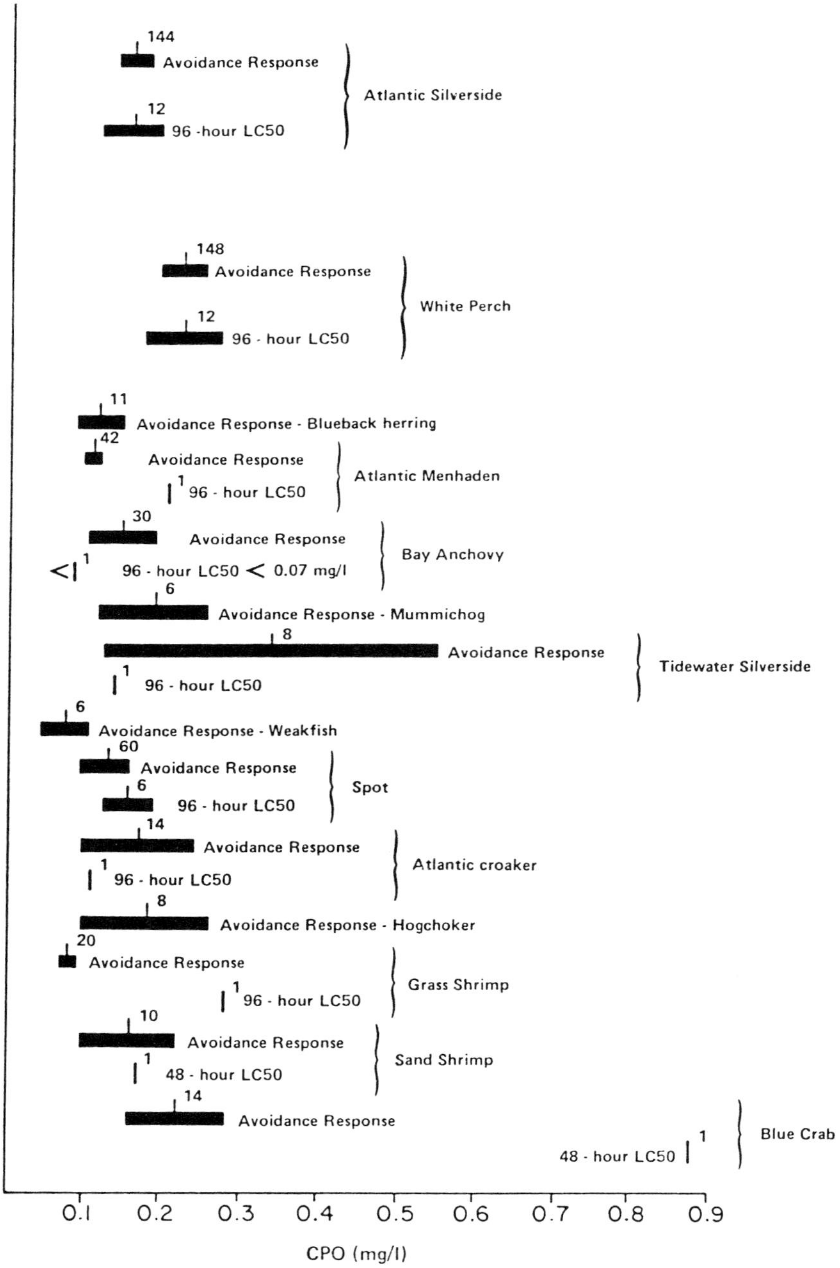

FIGURE 7. Levels of chlorine produced oxidants that are toxic to and avoided by marine and estuarine animals.[68,69]

groups of exposed organisms. Reports of fish kills directly caused by chlorine discharges are rare except where clean water was not accessible to exposed fishes.[70] Mobile life stages can detect and avoid potentially toxic chlorinated effluents;[71] however, only a few Bay species have been intensively studied across a wide range of temperatures and salinities.[72] Since the larval stages tend to be more sensitive to chlorine than older juveniles in laboratory tests,[53] larvae are less mobile, possess less avoidance capability, and are therefore more vulnerable than older life stages to chlorine discharges into freshwater tributaries of Chesapeake Bay.

After the 1974 James River fish kill, a task force was formed in Virginia to recommend measures designed to reduce potential impacts from CPOs in estuarine waters. Of particular concern was the oyster resource in the James River, but protection of finfish species was also important. Roberts et al.[73] showed that oyster and clam larvae were extremely sensitive to CPOs with LC50 concentrations in the low ug/L (parts per billion) range. The James River is the major source of seed oysters in Virginia. State officials were concerned that

TABLE 1

Mean acute toxicity of chlorine (in ug/L of total residual chlorine) to finfish species which inhabit the Chesapeake Bay system during all or a portion of their life cycles.

Species	Mean LC50s (24- to 96-h)	Reference
Channel catfish	90	[64]
Ictalurus punctatus		
Yellow perch	205	[64]
Perca flavescens		
Atlantic silverside	37	[64]
Menidia menidia		
Tidewater silverside (juvenile)	54	[64]
Menidia beryllina		
Naked goby (juvenile)	80	[64]
Gobiosoma bosci		
Spot	90	[64]
Leiostomus xanthurus		
Striped bass (*Morone saxatilis*)		[53]
larva (22-d old)	140 (141-147)[a]	
juvenile (60-d old)	190 (178-209)[a]	
juvenile (388-d old)	230 (226-240)[a]	
Atlantic menhaden		
Brevoortia tyranus		
Alewife	129 (30-227)[b]	[66]
Alosa pseudoharengus		
Blueback herring		
A. aestivalis	250	[67]
larva (48-h old)		

[a]95% confidence interval
[b]Range

chlorine could be responsible in part for decreased spatfalls observed in the river since the early 1960's.[74] Current chlorine discharge limits in the Virginia and Maryland portions of Chesapeake Bay appear to be sufficiently strict to protect finfish spawning and nursery habitats.

In summary, point-source discharge of chlorine and oxidant products into the waters of Chesapeake Bay are not likely to be detrimental to finfish populations except in localized portions of small freshwater tributaries or in the immediate vicinity of major discharges. Areas inhabited by sensitive and relatively non-mobile eggs and larvae are most at risk. Juvenile and older finfish appear capable of avoiding toxic concentrations of CPOs, a behavior that should decrease adverse effects. However, when fish avoid a specific area, spawning activities may be impaired and potential habitat is lost to the population, either temporarily or permanently. The relative importance of lost habitat will influence the ultimate effects of chlorine discharges on Bay fish populations.

Organotins

In recent years, the potential effects of organotin compounds, such as tributyltin (TBT), on Chesapeake Bay biota has become a major environmental issue. Several factors are responsible for the concern: (1) increased use of TBT in antifouling paint on both recreational and commercial watercraft in the Bay; (2) presence of potentially toxic concentrations of TBT in marina areas of the Bay;[75] (3) concentrations of TBT exceeding proposed water quality standards in the United Kingdom (20 ng/L)[76] have been reported in some Chesapeake Bay rivers;[75] (4) a recent proposal by the U.S. Navy to use organotin-based paints on all Naval vessels;[77] and (5) laboratory and field studies in England, France, and the United States which have shown that TBT is highly toxic to several aquatic species.[78,79,80]

The use of organotin paints to prevent growth of fouling organisms on boat hulls has increased since the early 1960's. These paints possess excellent antifouling actions, long lifetimes, and almost no corrosion.[81,82] Tributyltin (TBT), triphenyltin, and tricyclohexyltin compounds are the major biocidal organotins.[83] Presently, there are no effluent guidelines or water quality regulations for organotins in the United States.[83]

Organotin biocides are generally more toxic to aquatic biota than are other major organic contaminants such as polycyclic aromatic hydrocarbons (PAHs), chlorinated pesticides, and polychlorinated biphenyls or PCBs.[83] Organotins can be lethal to fish in the low ug/L concentrations.[84,85] Toxic levels of organotin compounds for several species of Bay finfishes are presented in Table 2. The sheepshead minnow (*Cyprinodon variegatus*) is very sensitive, with a 21-d LC50 for bis (tri-n-butyltin) oxide (TBTO) of 1 ug/L. The limited amount of information on avoidance capabilities of Bay finfishes to organotins (Table 2) suggests that some species (e.g., striped bass) may not avoid low concentrations

TABLE 2

Toxicity of TBTO[a] to finfishes found in Chesapeake Bay.

Species	Concentration (ug/L)	Exposure Time	Type of Test	Test Medium	Type of Response	Life Stage	Reference
Mummichog (*Fundulus heteroclitus*)	24	96 hr	Static	SW	Mortality	Adult	[86]
	1.0 - 13.8	20 min	Flow-through	SW	Avoidance	Adult	[87]
	20.8 - 28.0[b]	96 hr	Flow-through	SW	Mortality	Sub-adult	[91]
	15.2 - 30.4[b]	96 hr	Flow-through	SW	Mortality (LC50)	Larva	[91]
Striped bass (*Morone saxatilis*)	14.7 - 24.9	20 min	Flow-through	FW	Avoidance	Juvenile	[88]
Sheepshead minnow (*Cyprinodon variegatus*)	13 - 17	96 hr	Static	SW	Mortality (LC50)	Juvenile	[89]
	1	21 d	Flow-through	SW	Mortality (LC50)	Juvenile	[85]
	22.8 - 30.1[b]	96 hr	Flow-through	SW	Mortality (LC50)	Sub-adult	[91]
Atlantic menhaden (*Brevoortia tyranus*)	5.5 - 24.9	20 min	Flow-through	SW	Avoidance	Juvenile	[88]
	3.6 - 6.4[b]	96 hr	Flow-through	SW	Mortality (LC50)	Juvenile	[91]
Atlantic silverside (*Menidia menidia*)	6.7 - 11.6[b]	95 hr	Flow-through	SW	Mortality (LC50)	Sub-adult	[91]
Tidewater silverside (*Menidia beryllina*)	2.3 - 4.0[b]	96 hr	Flow-through	SW	Mortality (LC50)	Larva	[91]
Bluegill (*Lepomis macrochirus*)	7.6	96 hr	Static	FW	Mortality (LC50)	Juvenile	[90]
Channel catfish (*Ictalurus punctatus*)	12	96 hr	Static	FW	Mortality (LC50)	Juvenile	[86]

[a]TBTO = bis (tri-n-butyltin) oxide
SW = salt water
FW = fresh water
[b]Concentrations measured in test tanks

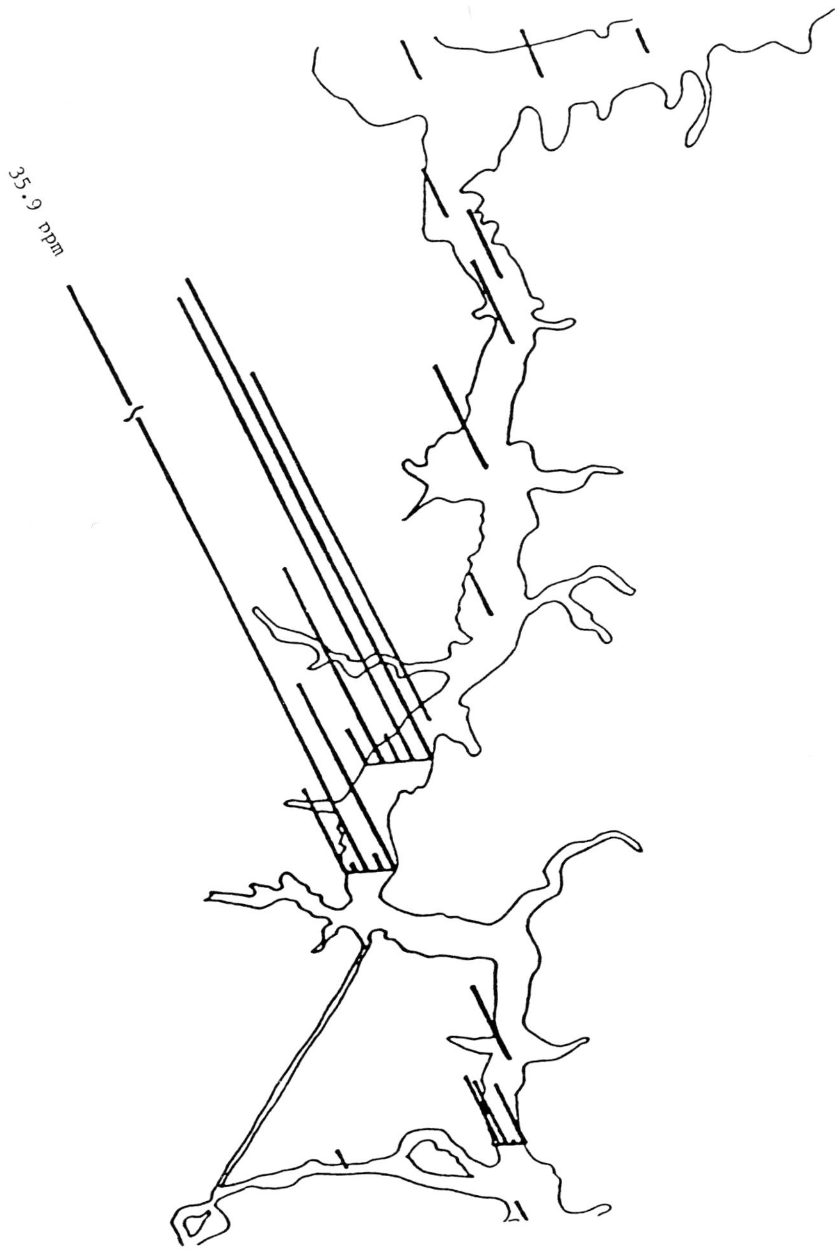

FIGURE 8. Surface sediment concentrations (mg/kg-dry wt) of benzo(a)pyrene along the Elizabeth River, Virginia, in 1985 (0.75 cm = 1 mg/kg or ppm.[94])

in the environment. Adverse effects on such species could occur if their limited avoidance capabilities resulted in extended exposures to TBT-contaminated areas. The sublethal effects of long-term, low-level exposures of organotins to finfishes have not been adequately studied.[83,87] Conversely, the mummichog appears to possess keen avoidance capabilities to low levels of organotins, presumably due to their highly sensitive chemoreceptor system.[87]

In summary, TBT leached into the aquatic environment from antifouling coatings on boat hulls is most concentrated in harbor and marina areas.[83] Therefore, their potential effects on Chesapeake Bay finfishes would presumably be most serious in these localized areas. However, because boats and ships are mobile and organotin compounds bioconcentrate in the food chain, all Bay habitats navigable by TBT-treated boats could be exposed to these contaminants. The effects of TBT on finfish populations in Chesapeake Bay are not yet known. Laboratory toxicity data suggest serious potential problems so research activity on these compounds is currently intense. Given projected increases in use of these antifouling paints[92] and their high toxicity, organotins must be viewed as a major contaminant problem in Chesapeake Bay. Recently, the States of Maryland and Virginia passed legislation to restrict the application of TBT paints on watercraft that use Chesapeake Bay.

Polynuclear Aromatic Hydrocarbons

Polynuclear aromatic hydrocarbons (PAHs) can enter the aquatic environment via several routes, but primarily through the incomplete combustion of

TABLE 3

Percentage of fish showing gross abnormalities from exposure to contaminants in the Elizabeth River, Virginia. Data are means of three samples collected in October, November, and December, 1983.[94]

Abnormality by Species	Kilometers from River Mouth										
	6.5	8.5	10.5	12.5	15.0	17.0	19.0	21.5	23.5	25.5	28.0
Fin Erosion											
Hogchoker[a]	0.7	0	0	0.4	1.4	5.5	4.3	11.2	1.9	0	0.5
Toadfish[b]	0	0	11.0	5.0	0	11.5	30.1	26.3	25.0	0	0
Cataracts											
Spot[c]	0	0	0.1	0	3.0	0.8	9.6	6.0	0.2	0.3	0
Weakfish[d]	0.2	0	0	0.8	1.0	1.8	3.5	14.0	21.0	2.5	7.5
Atlantic Croaker[e]	3.3	1.4	1.5	2.2	4.5	7.9	15.8	15.9	18.1	2.5	5.6

[a] *Trinectes maculatus*
[b] *Opsanus tau*
[c] *Leiostomus xanthurus*
[d] *Cynoscion regalis*
[e] *Micropogonias undulatus*

carbonaceous materials or through industrial processes that convert coal into synthetic fuels.[93] Other sources of PAHs include the manufacture of carbon black, creosote, soot, vehicular emissions (especially diesel), residual oil, and wood smoke. PAHs are of concern to scientists because some can become mutagenic or carcinogenic after being metabolized.

Field observations suggest that fishes in the Elizabeth River, Virginia, are severely stressed because of sediment contamination with PAHs. Figure 8 shows the distribution of one PAH, benzo(a)pyrene, in surface sediments along the Southern Branch of the Elizabeth River.[94] Incidence of abnormalities (e.g., skin lesions, cataracts, fin erosion) in native fishes increased at sampling stations which were heavily contaminated with PAHs (Table 3 and Figure 9).

In laboratory exposures of spot (*Leiostomus xanthurus*) to contaminated sediments from the Elizabeth River, dermal lesions and fin rot similar to those in fish collected from the river were observed.[95] Weeks and Warinner[96] found that the phagocytic efficiency of macrophages from spot and hogchoker (*Trinectes maculatus*) resident in the Elizabeth River was reduced when compared to fish from control stations. The bioavailability of PAHs to oysters in the Elizabeth River was demonstrated using transplant studies.[94]

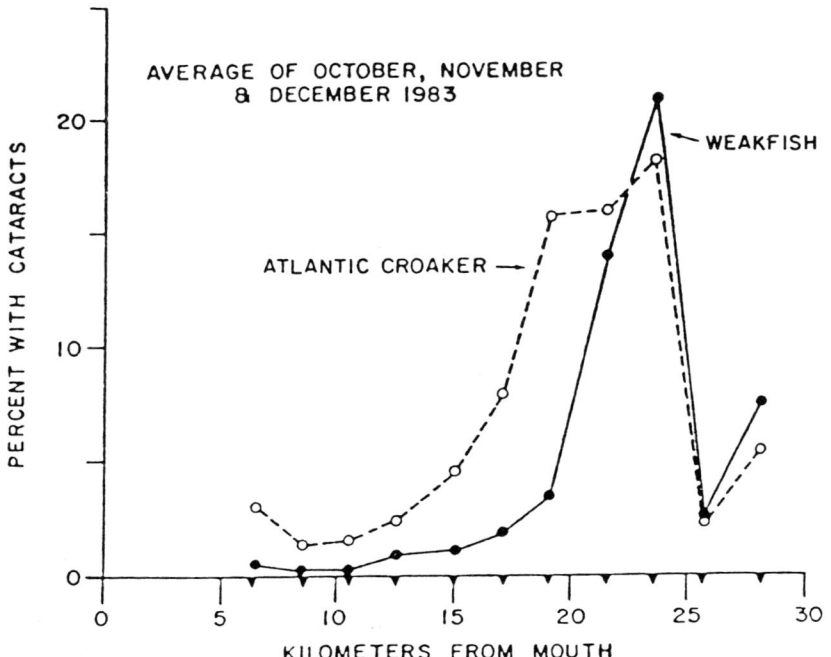

FIGURE 9. Average occurrence of cataracts in Atlantic croaker and weakfish from stations along the Elizabeth River, Virginia.[94]

In summary, PAHs are widespread contaminants in freshwater and estuarine systems and have also been implicated in adverse effects on finfish and shellfish in other areas such as the Niagara River,[97] Oregon Bay,[98] and Puget Sound.[99] Future studies focused on PAH contamination in Chesapeake Bay should include: (1) studies to define the levels of sediment contamination necessary to cause acute and chronic effects on fishes, (2) surveys of the mainstem Bay and tributaries to determine whether increases in the incidence of abnormalities in fishes are related to PAH presence and concentration, and (3) laboratory and field investigations to determine the specific PAH compounds responsible for observed abnormalities.

INORGANIC AND ORGANIC CONTAMINANT MIXTURES

Passage of the Chaffee Amendment to the Anadromous Fisheries Conservation Act in 1980 stimulated an ambitious research program aimed at determining why anadromous stocks of striped bass along the Atlantic coast had declined since the mid-1970's and how these stocks could be restored to former abundance levels.[100] The role that contaminants may have played in this decline was a major study objective led by the Columbia National Fisheries Research Laboratory (CNFRL) of the U.S. Fish and Wildlife Service in Columbia, Missouri.

During the first phase of the CNFRL contaminants program, a comprehensive survey was conducted in several major Atlantic coast spawning rivers (including Chesapeake Bay) and selected hatcheries. This survey analyzed the tissues of striped bass collected in these waters for residues of over 100 organic and inorganic contaminants.[101] The survey was complemented by other recent surveys of inorganic and organic contaminant residues in fish tissue samples collected in Chesapeake Bay[102,103,104]

The most prevalent organic contaminant residues found in the tissues of juvenile striped bass from the Hudson River, New York; the Nanticoke and Potomac Rivers in Chesapeake Bay; and the Edenton Fish Hatchery, North Carolina (control group) by Mehrle et al.[9] were polychlorinated biphenyls or PCBs (Aroclors 1248, 1254, 1260). Chlordane, DDT, DDD, and DDE were also detected, but at concentrations equal to or less than 0.06 ug/g (wet weight), and not considered to be significant residue levels.[9] Total organochlorine residues in Chesapeake Bay striped bass were higher in the Potomac River (0.21 - 0.40 ug/g wet weight) than in the Nanticoke River (0.06 - 0.09 ug/g wet weight).

The major inorganic constituents detected in juvenile striped bass tissues were cadmium, lead, zinc, arsenic, and selenium.[9] Selenium residues tended to be slightly higher in Potomac River juveniles (0.22 - 0.84 ug/g wet weight) compared to fish collected in the Nanticoke River (0.19—0.64 ug/g wet weight). Concentrations of other metal residues were similar in these two Bay populations.

Mehrle et al.[9] correlated tissue residues with vertebral development and suggested that contaminants such as cadmium, lead, and PCBs could decrease survival of striped bass larvae and early juveniles. The risk seemed highest in the Hudson River population where juveniles had the highest concentrations of contaminant residues and lowest structural-integrity indices for their vertebrae. Swimming stamina and condition factor indices were also low in Hudson River juveniles, suggesting that the poor condition symptoms were consistent with the uptake of environmental contaminants.[105] By comparison, vertebral integrity was intermediate for Nanticoke and Potomac river juveniles, and highest in fish of Hudson River origin that had been reared in the relatively uncontaminated waters of the Edenton Hatchery.[9]

Neither study[19,105] concluded that contaminants were not affecting the status of striped bass populations in the Potomac and Nanticoke rivers. Rather, the problem of contaminants appeared to be more critical in the Hudson River, where levels of PCBs and other chemicals constitute a major environmental issue[106,107,108] that resulted in the closure of the fisheries for striped bass and other finfishes in 1976 that continues to this day.[109,110] Interestingly, however, the Hudson River population of striped bass continued to produce average or above average year classes during the 1970's and early 1980's when reproductive success for populations in Chesapeake Bay and other Atlantic coast estuaries was dismal.[5] Although clear evidence is lacking, closure of the fishery for striped bass in the Hudson River may have contributed to the favorable trends in annual juvenile production.

The first phase of the CNFRL studies[19,101,105] detected relatively small quantities of several contaminants (PCBs, organochlorine pesticides, dioxins, dibenzofurans, petroleum hydrocarbons, cadmium, copper, lead, arsenic, selenium) in the tissues of juvenile striped bass collected in several Atlantic coast rivers. However, no single contaminant was found in sufficient concentration or frequency to explain the observed decline in coastal stocks.[101] This conclusion stimulated a series of laboratory toxicity studies which focused on the array of inorganic and organic contaminants measured in juvenile striped bass tissues.[9] These studies departed from traditional single contaminant experiments and evaluated acute and chronic effects of mixtures containing two or more compounds at environmentally-realistic concentrations.[16,111,112]

Several studies exposed the early life stages of striped bass to a complex mixture of contaminants (Table 4) in fresh and saline water.[24,101,113] Palawski et al.[113] also compared the relative acute toxicities of inorganic and organic components of this toxicant mixture and measured individual toxicities of cadmium chloride, copper sulfate, zinc chloride, nickle chloride, arsenic pentoxide, selenium selenite, and sodium chromate on 35 to 80-d old juveniles.

Growth of larvae and early juveniles was unaltered by exposure to the contaminant mixture at 25 to 400% of the environmental concentration (Table 5) in fresh water or 2 and 5 ppt saltwater.[101] Percent fertilization and hatching suc-

TABLE 4

Concentrations of organic and inorganic contaminants included in the contaminant mixture stock solution.[24,101,a] Organic compounds were dissolved in acetone; inorganics were dissolved in hydrochloric acid.

Aroclor 1248	10 ng/L
Aroclor 1254	10 ng/L
Aroclor 1260	10 ng/L
DDE	3 ng/L
Toxaphene	3 ng/L
Chlordane	5 ng/L
Kepone (chlordecone)	15 ng/L
Perylene	40 ng/L
Fluorene	40 ng/L
Phenanthrene	40 ng/L
Anthracene	40 ng/L
Fluoranthene	40 ng/L
Perene	40 ng/L
Benzoanthrene	40 ng/L
Chrysene	40 ng/L
Arsenic (as pentoxide)	1 ug/L
Selenium (as selenite)	2 ug/L
Lead (as nitrate)	1 ug/L
Cadmium (as chloride)	3 ug/L
Copper (as sulfate)	1 ug/L

[a] Stock solution of contaminant mixture[113] did not contain Anthracene or Fluoranthene.

cess were not diminished by various dilutions of the contaminant mixture; but a 100% concentration reduced survival of yolk-sac larvae after a 144-h continuous exposure from fertilization.[24]

Juvenile survival decreased during exposure to the contaminant mixture, which was most toxic in moderately soft fresh water (hardness of 40 mg/L as $CaCO_3$) compared to 1 or 5 ppt saltwater.[113] This increased toxicity in freshwater was attributed to differences in speciation of metals associated with water chemistry, especially for cadmium, copper, and zinc. Wright et al.[14] demonstrated that uptake and toxicity of cadmium to larval and juvenile striped bass are inversely related to calcium levels in the test medium. The organic chemical fraction of the contaminant mixture (Table 4) was not toxic to juvenile striped bass at concentrations 51 times greater than environmental levels.[113] This conclusion implies that inorganic rather than organic contaminants in the mixture pose a potentially greater risk to the survival of young striped bass during the late larval and early juvenile stages (35 to 80-d old). The relative toxicity of organic versus inorganic components in the contaminant mixture has not been determined for younger life stages of striped bass or other fish species.

The contaminant mixture tested by[24,101,113] did not contain all organic compounds that may pose a threat to Bay fishes. For example, the mixture did not contain the herbicide atrazine (2-chloro-4-ethylamino-6-isopropyl), that is widely used in the Chesapeake Bay watershed and present at low concentrations (up to 2 ug/L) in the water column.[115] The limited toxicity data for atrazine suggest that current levels in Bay habitats will not adversely affect finfishes.[16] The contaminant mixture also did not contain any organotin compounds, biocides considered to be a major contaminant problem in Chesapeake Bay (discussed above).

Sublethal effects of the CNFRL's contaminant mixture on swimming performance, feeding behavior, and predation avoidance for juvenile striped bass after 20 to 60-d exposures were inconclusive.[101] Whole body residues of inorganic and organic contaminants in juveniles exposed for up to 90 days were relatively low and in the range observed in wild juveniles collected from several Atlantic coast rivers.[9] These results support the premise that the CNFRL series of laboratory studies exposed test organisms to environmentally-realistic concentrations of contaminants.

Striped bass yolk-sac larvae were more sensitive to the contaminant mixture than embryos, older larvae, and juveniles.[24,101,113] This finding corroborates the general pattern of life stage sensitivities in finfishes reported by others.[53,114,117,118,119,120] Overall, striped bass were as sensitive as most salmonid fishes to seven metals and three organic pesticides, but much more sensitive than several cyprinids, ictalurids, and centrarchids (Table 5). Cadmium, copper, and zinc were extremely toxic to young striped bass. Wright et al.[114] observed that 7-d old larvae were very sensitive to cadmium (5 - 10 ug/L) when exposed in a low calcium (8 mg/L) medium. Pathological changes were induced in the visual system of 28-d old larvae after only 24-h exposures to 80-150 ug/L copper in a dose-dependent fashion.[121] By comparison, arsenic, selenium, nickel, and chromium were much less toxic to young striped bass.[113]

Klauda[122] also demonstrated that the acute toxicities of arsenic and selenium to striped bass eggs, larvae, and juveniles are relatively low, either as isolates

TABLE 5

Comparison of the relative sensitivity (96-h median lethal concentrations, ug/L) of four finfish species to seven metals tested in soft fresh water.[113]

Metal	Species			
	Striped Bass	Rainbow Trout	Fathead Minnow	Bluegill
Cadmium (as chloride)	4	1	630	1,940
Copper (as sulfate)	100	17	25	660
Zinc (as chloride)	120	93	780	5,370
Selenium (as selenite)	1,325	1,800	10,000	4,500
Nickel (as chloride)	3,900	15,000	4,580	5,180
Chromium (as chloride)	28,000	59,000	17,600	118,000
Arsenic (as pentoxide)	40,500	28,000	42,000	41,760

or mixtures in a 3-7 ppt salinity medium (Table 6). Klauda[122] tested arsenate (+5) and selenate (+6) because they are generally the dominant inorganic forms available to early life stages of fishes via waterborne pathways in estuarine waters.[123,124] Arsenite (+3) and selenite (+4) are more prevalent in freshwater and also more toxic to fishes than arsenate and selenate.[125] Various forms of arsenic (arsenate, arsenite, methylarsenic acid, dimethylarsenic acid) and selenium (selenate, selenite, elemental selenium, heavy metal selenides, methylated forms) can occur in aquatic environments.[126] About 30% of arsenic and selenium inputs to the environment come from coal combustion,[127] hence these contaminants can be expected to be present in aquatic habitats near coal-fired power plants operating in Chesapeake Bay.[125]

Klauda[122] showed that the joint toxicities of arsenate and selenate in mixtures were additive to striped bass yolk-sac larvae, but subadditive and suggestive of antagonism to post larvae and juveniles (Table 7). Selenium reduces the toxicity of mercury, cadmium, and copper in several aquatic organisms,[128,129,130,131] but antagonism with arsenic had been previously observed only in mammals.[132] Continuous exposure of young striped bass to sublethal levels of selenate (89 to 1,360 ug/L) for 60 days post-hatch was associated with an increased frequency (52%) of lower jaw deformities.[122] Cumulative toxicity during long-term exposures could decrease feeding ability in postlarvae and juveniles and alter survival probabilities.

In summary, these laboratory studies with early life stages of striped bass demonstrated that survival can be diminished by relatively brief encounters with environmentally-documented concentrations of inorganic and organic contaminant mixtures. Such findings suggest that contaminants cannot be ignored as a possible factor contributing to the decline of striped bass stocks in Chesapeake Bay and other Atlantic coast estuaries. Toxic forms of inorganic contaminants (especially cadmium, copper, zinc) appear to pose a major threat to young striped

TABLE 6

LC50 values (ug/L) for early life stages of striped bass exposed to sodium arsenate or sodium selenate for 96 hours in estuarine water (3-7 ppt).[122]

Life Stage	Age[a] (days after hatch)	LC50 (95% Confidence Interval)	
		Arsenate	Selenate
Yolk-sac larva	1	18,690 (16,780-20,590)	9,790 (8,260-11,310)
Post larva	17	7,280 (6,510-8,050)	13,020 (11,560-14,480)
Juvenile	72	18,960 (18,130-19,780)	85,840 (81,650-90,030)

[a]Age at start of 96-h test

bass in freshwater reaches of spawning and nursery areas.

The accuracy of this important conclusion will be influenced by the extent to which continuous exposure laboratory studies can accurately predict metal-induced effects on young striped bass survival in nature. These predictions may be reliable, based on results with other species,[133] unless metal concentrations in natural waters are temporally quite variable[134] or fish avoid potentially lethal levels. Avoidance responses to copper and zinc should be effective at reducing mortality, but fish appear to possess limited abilities to detect and avoid lethal concentrations of cadmium.[71]

As developing young striped bass migrate downstream and reach saline habitats, the toxicity of heavy metals should decrease and pose a less serious threat to their survival. Prior exposure of young striped bass to sublethal doses of metals as larvae or early juveniles could also enhance their chances for survival to maturity. Several studies have shown that pre-exposure of fishes to copper,[135] cadmium,[136] arsenic,[137] zinc,[138] and other toxicants can reduce their deterious affects. For some fish species, metal-binding proteins called metallothioneins, produced primarily in liver tissue, are presumably involved in the development of enhanced tolerance to some heavy metals.[139,140,141,142,143] Exposure of young fishes to sublethal concentrations of inorganic aluminum during acidic episodes may stimulate increased amounts of calmodulin, a calcium binding protein, in gill tissues and reduce the toxic effects of aluminum on ionic fluxes.[144]

ACIDIC DEPOSITION

Deposition of chemical pollutants from the atmosphere is a major environmental issue of international scope.[145] Concern that acidic deposition (often called acid rain) is a contaminant problem that can lead to aquatic habitat acidification and detrimental effects on finfish populations was first documented in southern Scandinavia in the 1950's[146] and about a decade later in eastern North America.[147] Acid precipitation in northeastern United States is 60 to 70% sulfuric acid and 30 to 40% nitric acid[148] and assumed to originate primarily from gaseous industrial emissions of oxides of sulfur and nitrogen produced during fossil fuel combustion and metals smelting. Changes in fish species composition and elimination of sensitive species, due to decreased recruitment of young individuals, have been well documented in Scandinavia, the Netherlands, Scotland, eastern Canada, and northeastern United States.[145,149,150,151,152] However, relationships among acid deposition, surface water acidity, and fish population status are much less definitive in other regions of the United States.[153]

Short-lived acidification events (also called episodes, pulses, spates) associated with snow-melt and intense rain storms can severely stress the early life stages of finfishes.[154] Acidic episodes may be more detrimental to fish populations than long-term, gradual habitat acidification processes. Adverse effects of habitat acidification on finfish have been attributed to increased hydrogen ion concentrations (i.e., more acidic pH), and elevated levels of metals. The metals prob-

TABLE 7
Cumulative percent mortality of striped bass early life stages exposed to sodium arsenate (As) and sodium selenate (Se) for 96 hours in estuarine water (3-6 ppt).[122]

Treatment	Yolk-Sac Larva[a]		Postlarva[b]		Juvenile[c]	
	Mean Concentration (ug/L)	Percent Mortality	Mean Concentration (ug/L)	Percent Mortality	Mean Concentration (ug/L)	Percent Mortality
As Only	0	22	0	6	0	17
	9,300	50	3,400	25	10,300	57
	17,200	60	7,300	60	18,400	37
	28,700	98	12,500	100	19,100	47
Se Only	0	22	0	6	0	17
	4,900	17	8,900	6	48,500	40
	9,800	39	14,300	7	90,100	97
	14,300	40	20,200	35	142,100	100
As/Se Mixture	10,300/ 4,700	60	3,500/7,600	15	9,800/ 40,600	87
	21,300/10,100	88	8,400/14,300	76	18,400/101,000	70
	29,800/14,200	100	11,900/21,000	96	25,200/146,900	80

[a] Age = 1-d old at start of 96-h test
[b] Age = 16-d old at start of 96-h test
[c] Age = 75-d old at start of 96-h test

lem is primarily due to pH-related mobilization or leaching of toxic metals (e.g., aluminum, cadmium, copper, zinc, lead, manganese) from watershed soils and aquatic sediments, and secondarily from acid precipitation itself which can contain several heavy metals, particularly near smelters.[155,156,157,158,159] Elevated hydrogen ion concentrations can also decrease fish tolerances to low dissolved oxygen levels[160] and enhance their sensitivities to an array of inorganic and organic contaminants via changes in compound toxicity or accumulation kinetics.[161,162,163] Other studies have shown that contaminant toxicities to fishes are either unrelated to pH[164,165] or ameliorated in low pH waters.[161,166,167] This problem is complicated by pH-related changes in chemical speciation of metals.

Acidic deposition is not limited to northern climates, but also occurs in the middle Atlantic and southeastern United States.[168,169,170,171,172,173] Recent investigations suggest that habitat acidification may be an important ecological

TABLE 8

Cumulative percent mortality and LT_{50} (time in hours to 50% mortality) for American shad yolk-sac larvae exposed to pH and aluminum during a 55-h continuous exposure experiment in the laboratory.

Nominal Treatment		Cumulative Mortality (%)		Time to 50% Mortality (h)
pH	Aluminum (ug/L)	After 24 hours	After 55 hours	
7.5	0[a]	0	6	—
	50	34	46	—
	100	30	50	—
	200	36	52	60.0
	400	40	80	30.0
6.7	0	6	14	—
	50	32	98	31.8
	100	24	92	37.5
	200	31	100	24.3
	300	100	100	b
6.2	0	29	100	29.7
	50	29	98	24.9
	100	52	100	b
	200	82	100	17.8
	400	100	100	b
5.7	0	23	100	28.0
	50	33	100	23.7
	100	46	100	24.7
	200	92	100	b
	400	100	100	b

[a] Control group
[b] All test organisms were dead within 16 hours.

problem in Chesapeake Bay,[174,175,176] especially in freshwater reaches of higher order streams which drain the Coastal Plain physiographic province.[177,178] This region is underlain by thick layers of unconsolidated sand and gravel, silty sand, clay, marl, and shell beds superimposed upon buried rocks of the Piedmont province.[179] The thickness of Coastal Plain sediments preclude interaction between acid deposition and bedrock, pH and base saturation characteristics of the soils are low, and alkalinity values in smaller tributaries are characteristic of acid-sensitive surface waters.[169,180]

Water chemistry data collected in 23 higher order streams draining inner and middle coastal plain areas of Maryland's eastern and western shores revealed acidic conditions during a relatively wet spring, March and April of 1983 (Figure 10). Several streams exhibited temporary, storm-associated depressions of pH (to 4.5) and alkalinity (to 0.3 mg/L as $CaCO_3$), accompanied by increases in dissolved aluminum levels to 4.0 mg/L.[177] Other recent studies in Maryland detected acidic pulses associated with rainstorms in freshwater sections of the Choptank River,[182] the Nanticoke River,[25] and Granny Finley Branch[183] on the eastern shore; and in Lyons Creek on the western shore.[26] pH depressions in these coastal plain streams are usually short-lived phenomena exhibiting rapid changes in hydrogen ion concentration (Figure 11) accompanied by equally rapid changes in stream stage, turbidity, and dissolved aluminum levels.[26,178,184]

Recent field and laboratory studies demonstrated that the early life stages of striped bass, blueback herring and American shad are very sensitive to

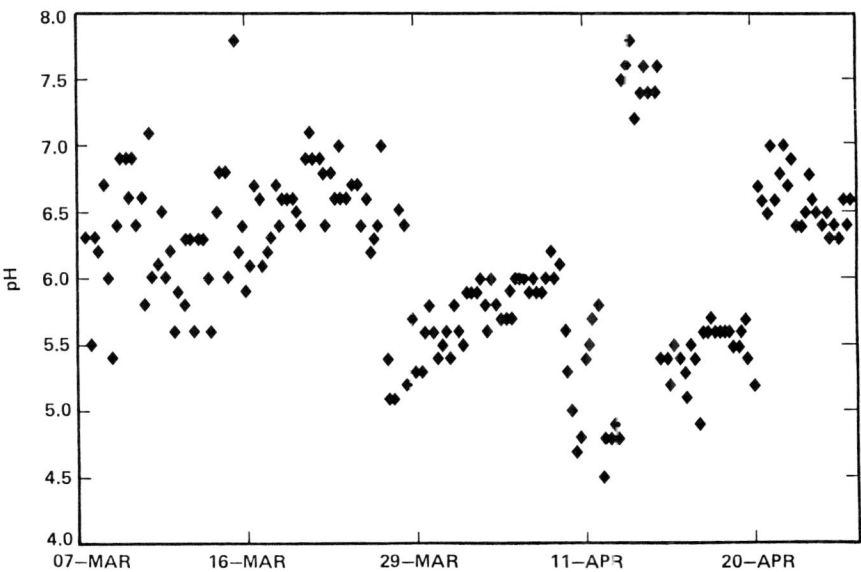

FIGURE 10. pH measurements in 23 higher order tributaries of Chesapeake Bay during spring 1983.[181]

moderate acidity. These three anadromous species are among the most sensitive finfishes in Maryland yet studied (Figure 12 and Table 8). Fertilized alosid eggs were generally more resistant than larvae to pH and dissolved aluminum in the laboratory. Blueback herring[26,117] and American shad (Table 8) larvae tolerated pH 6.5 in the laboratory, but succumbed to pH 5.7 or 6.2. The toxic effect of pH was intensified by simultaneous exposure to dissolved inorganic aluminum, especially for American shad. These laboratory-derived predictions of pH and aluminum toxicity to blueback herring and American shad must still be verified during in-situ field experiments.[26] Survival of striped bass larvae was diminished by exposure to pH and aluminum in the laboratory,[186] supporting field observations in the Nanticoke River.[25] Investigators are beginning to study the effects of acidic pulses on two semi-anadromous Bay species, yellow perch and white perch.[180]

In summary, research into the role of habitat acidification on the population dynamics of finfishes in Chesapeake Bay has just begun, and most studies have focused on Maryland waters.[138] The available data demonstrate that survival of striped bass larvae was diminished by storm-associated changes in pH, dissolved aluminum, and water hardness in one spawning-nursery area, the Nanticoke River, during spring 1984. Laboratory data also indicate that blueback herring and American shad eggs and larvae are very sensitive to pH and

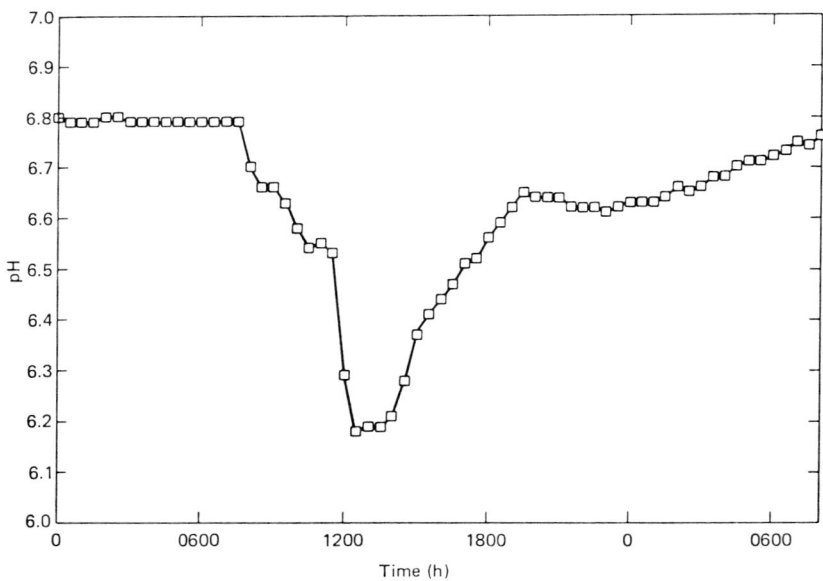

FIGURE 11. Moderately acidic pulse of approximately 4-h duration in Lyons Creek, Maryland, during June 1986 in response to a 2-d rainstorm measuring 0.94 cm. Stream pH was measured with an in-situ monitor, Hydrolab Datasonde model 2030-DS.[26]

aluminum conditions that have been measured in several coastal plain spawning sites. To date, however, no direct link between acidic deposition, habitat acidification, and fish mortality has been established for any Maryland watershed.[138]

RADIONUCLIDES

Nuclear power plants in the United States are licensed and regulated by the Nuclear Regulatory Commission (NRC). Conditions imposed in the operating licenses for each plant allow routine discharges of low levels of radioactivity to the environment. These releases must be within the guidelines of Federal regulations. Within the Chesapeake Bay system are five nuclear power plants: Peach Bottom and Three Mile Island on the Susquehanna River in Pennsylvania, Calvert Cliffs on the western shore of the Bay in Maryland, and the North Anna and Surrey plants on the James River in Virginia.

Radionuclide releases from the Calvert Cliffs Nuclear Power Plant to atmospheric, terrestrial, and aquatic environments are monitored by the utility company (Baltimore Gas and Electric Company) and two Maryland State agencies (Department of Health and Mental Hygiene, DHMH; Power Plant Research

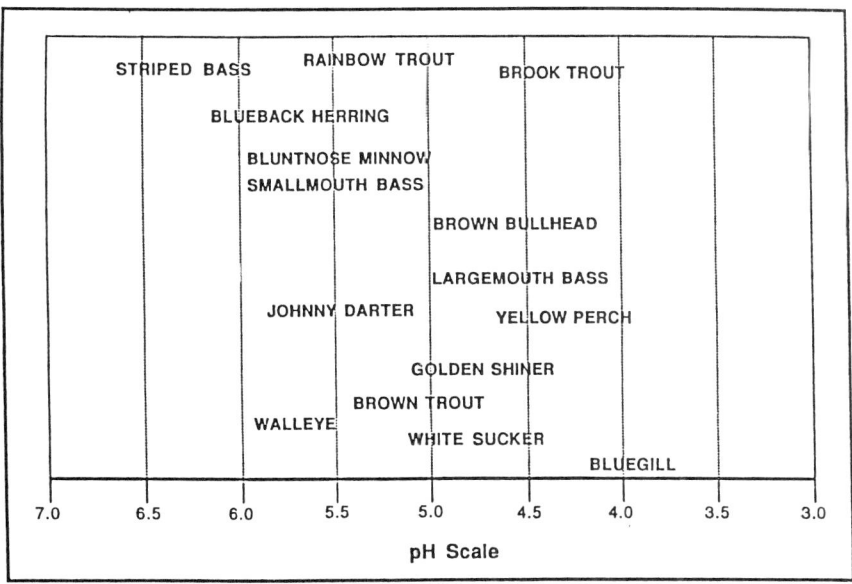

FIGURE 12. Critical pH values causing mortality in some segment of the life cycle for Maryland fishes.[185]

TABLE 9

Maximum concentrations of radionuclides from Calvert Cliffs Nuclear Power Plant in aquatic biota samples from Chesapeake Bay, 1983-1984.[187]

Radionuclide Concentration[a,b] (pCi/kg wet weight)

Sample	Co-58 1983	Co-58 1984	Co-60 1983	Co-60 1984	Zn-65 1983	Zn-65 1984	Ag-110m 1983	Ag-110m 1984
Seaduck (Flesh)	<13	<12	<14	<14	<28	<28	<17	<17
Edible Finfish (Flesh)	<12	<10	<10	<8	<20	<8	<12	<10
Forage Finfish (Whole)	<12	<8	<15	<8	<30	<12	<20	2±2
Oyster Meat	10±5	21±4	1±2	<8	67±13	21±8	420±16	250±10
Crab Meat	<15	<15	<12	<8	<12	<12	15±7	10±5 6±5
Crab Shell	<30	<20	<14	<10	<40	<16	<25	56±19 38±11
Grass Shrimp	<20	<20	±25	<15	<30	<20	8±9	46±18 6±4
Epifauna	2192±428	2344±153	661±244	351±76	±200	<100	339±250	307±89
Macroalgae	109±12[b]	16±6[c]	<10	<10	<15	<15	<15	31±9[c] 8±5[d]
Bay Sediment								
Clay	233±59	91±47	213±31	173±21	<30		39±17	58±25
Sand	87±21	70±18	50±10	52±8	±30		24±37	12±8

[a]Counting uncertainty at 95% confidence level
[b]Crab shell and sediment concentrations are in pCi/kg dry weight; epifauna concentrations are in pCi/kg ash weight.
[c]*Enteromorpha* sp.
[d]*Ulva* sp.

Program, PPRP). Radiological surveillance indicates that this plant is in compliance with operating license guidelines imposed and regulated by the NRC to assure no adverse human health or environmental effects.[187] Low levels of plant-related Co-58 and Co-60 were detected in Bay sediments in the Calvert Cliffs area. Zn-65 and Ag-110m were detected in some aquatic biota, but not

TABLE 10

Maximum concentrations of radionuclides in aquatic biota attributed to Peach Bottom Atomic Power Station, 1983-1984.[187]

Sample Type	Radionuclide Concentration (pCi/kg, wet weight)[a,b]			
	Co-60	Zn-65	Cs-134	Cs-137[c]
Edible Finfish (Flesh)				
Holtwood	<10	<20	<14	6 ± 6
Conowingo Pond	<10	82 ± 21	94 ± 15	288 ± 20
Conowingo Dam Tailrace	<10	21 ± 11	53 ± 17	81 ± 11
Susquehanna Flats	<10	<60	<27	22 ± 19
Forage Finfish (Whole)				
Holtwood	<10	<20	<10	14 ± 3
Conowingo Pond	19 ± 5	639 ± 200	51 ± 6	76 ± 15
Conowingo Dam Tailrace	<7	59 ± 11	49 ± 5	70 ± 7
Crayfish				
Holtwood Reservoir	<25	<50	<25	<30
Conowingo Pond	7 ± 15	106 ± 66	81 ± 44	94 ± 48
Mussel (*Elliptio complanata*)				
Holtwood Reservoir	<15	<30	<20	2 ± 8
Conowingo Pond	2 ± 2	269 ± 32	10 ± 7	11 ± 7
Susquehanna Flats	<15	13 ± 9	<7	1 ± 2
Submerged Aquatic Vegetation (*M. spicatum*)				
Susquehanna Flats	± 8	<8	<4	35 ± 7
Sediment				
Holtwood	0	0	0	334 ± 12
Conowingo Pond	988 ± 18	837 ± 67	308 ± 20	1163 ± 40
Susquehanna Flats	28 ± 12	45 ± 22	57 ± 11	383 ± 13

[a] Counting uncertainty at 95% confidence level.
[b] Sediment concentrations are pCi/kg wet weight.
[c] Primarily attributable to weapons testing fallout; however where Cs-134 was also present, a power plant produced Cs-137 increment is indicated.

in edible finfish (Table 9). Oysters in the vicinity of the plant discharge contained the highest levels of Zn-65 (67 ± 13 pCi/kg wet weight) and Ag-100m (420 ± 16 pCi/kg wet weight) of edible aquatic biota. These radionuclide concentrations fluctuated over time in response to variations in quantities of radioactivity released by the power plant and by oyster assimilation and depuration rates. If consumed by humans, the maximum concentrations of radionuclides detected in finfish or other aquatic biota could produce radiation doses that are orders of magnitude lower than doses resulting from naturally radioactive sources in the Bay.

Radiological surveillance of the Peach Bottom Atomic Power Station on the Susquehanna River in Pennsylvania, 4.8 km upstream from the Pennsylvania-Maryland border, was conducted by the utility company (Philadelphia Electric Company), DHMH, and PPRP.[87] The data indicate that the plant is in compliance with operating license guidelines. Low levels of plant-related Zn-65, Cs-134, and Cs-137 were detected in sediments and aquatic biota in the Conowingo Pond, the lower Susquehanna River, and the Susquehanna Flats portion of the upper Chesapeake Bay (Table 10). Edible finfish species with detectable concentrations of radionuclides included channel catfish, carp, hybrid (striped x white) bass, walleye, white perch, smallmouth bass, and largemouth bass.

The other nuclear power plant located on the Susquehanna River in Pennsylvania, about 67 km upstream from the Pennsylvania-Maryland border, is the Three Mile Island Nuclear Station. Owned jointly by Metropolitan Edison Company, Pennsylvania Electric Company, and Jersey Central Power and Light Company, the plant has not operated since the accident at Unit 2 in March 1979.[87]

In summary, routine discharges of radionuclides to Chesapeake Bay from nuclear power plants are not causing detectable adverse effects on finfishes, at least in Maryland waters. Due to NRC licensing requirements, these plants are closely monitored by the operating utilities and appropriate State agencies.

DISCUSSION

This review of current contaminant effects studies in Chesapeake Bay provides relatively convincing evidence that contaminants *may have been* or *may be* a factor responsible for the recent decline of several finfishes, or at least *may be* contributing to the continuing series of poor year classes. The growing body of tissue residue data show that many inorganic and organic contaminants present in Bay habitats are accumulated by finfishes. Laboratory studies reveal that exposure to mixtures of several contaminants can decrease survival of larvae and juveniles. Such observations are necessary to pursue the postulate that contaminants actually *have affected* or *are affecting* finfish populations in Chesapeake Bay. However, these observations are not, in themselves, sufficient to reach a definitive conclusion. It is also necessary to rigorously demonstrate that contaminants are adversely affecting finfish survival in nature, and compile evidence that the status of finfish populations has been altered by exposure to environmental contaminants.

In-situ cages, enclosures, or microcosms represent a useful first step toward verification of laboratory-derived predictions of contaminant effects on fish populations or in nature.[88] This approach has recently been used in Chesapeake Bay to assess the response of striped bass and blueback herring to acidic episodes and other habitat quality concerns.[25,26] In-situ studies with striped bass larvae in the Nanticoke River[25] successfully corroborrated laboratory study results on the toxicity of pH and aluminum.[86] In-situ studies with blueback herring are continuing. Although valuable in extrapolating laboratory study results to nature, the in-situ approach is logistically limited to a few species and a few spawning or nursery sites in any given year, unless a massive research effort is funded. Nevertheless, in-situ studies are feasible, underway in Chesapeake Bay, and yielding important results.

Taing the next step to a prediction of species population response to contaminant effects is exceedingly difficult.[89] Translation of a contaminant effect, via direct mortality, indirect mortality (mediated through growth, behavior, physiology), or changes in other community components that alter food availability or predator and competitor numbers, into a change in the size or productivity of a focal species population will necessarily become entangled in the array of inter- and intraspecific processes that regulate population size. The operation of these interacting process may eliminate or exaggerate the ultimate effect of contaminant-induced stress on fish populations, unless catastrophic levels of mortality result that are so obvious they cannot be easily masked. This level of ecological complexity suggests that efforts to evaluate fish population responses to contaminants will require concomittant study of other abiotic and biotic factors that could affect population dynamics.

The fisheries literature offers limited documentation that contaminants in Chesapeake Bay or other habitats have affected the status of finfish populations. Whenever serious decline or complete collapse of a commercial or sports fisheries has been documented, the proposed primary causes are usually an intensification of fishing pressure leading to overexploitation or severe changes in the aquatic environment (e.g., pollution) or both.[17] Identifying the relative importance of these two mortality sources, where both exist, is difficult. Examples of decimated fish species that were exposed to overfishing alone or overfishing and pollution together are numerous in the literature; but historical documentation of a declining fishery exposed to pollution in the absence of overfishing does not appear to exist.[17]

Mathematical models and multivariate statistics are two analytical techniques that have been applied to the study of contaminants and Chesapeake Bay finfish populations. Goodyear[20] addressed this topic for striped bass with a simulation modeling approach. He acknowledged evidence suggesting that environmental contaminants present in the Bay could impair survival of young fish in freshwater portions of spawning and nursery grounds. He also accurately stated that the level of excess mortality imposed on the striped bass population by toxic chemicals is unknown. Based on a Leslie matrix modeling approach, Goodyear concluded that an increase in population fecundity sufficient to offset even severe losses due to contaminant toxicities could be achieved by a reduction in fishing mortality. Such a reduction in fishing mortality could halt or even reverse the current stock decline.[20]

Use of various modeling approaches for assessing the effects of contaminants on fish populations was recently reviewed by Vaughan et al.[90] They compared five current approaches and concluded that a combination of bioenergetics and Leslie matrix approaches offers a powerful tool for estimating long-term impacts of toxic contaminants on fish populations, even though this modeling combination is very data intensive. A Leslie matrix model requires several age-class-specific parameters including: fecundity per mature female, proportion of females that are mature, sex ratios, mortality rates (natural and fishing), and first and last age classes having mature females. A Leslie matrix approach is advantageous because mortality (all age classes), individual growth (via growth rates), and reproduction (via condition factors, egg production, egg viability) can all be altered in the model to reflect specific contaminant effects on these processes.

Bioenergetics models examine the factors affecting growth of an individual fish, but these models can also be used to simulate the growth of representative individuals from each cohort over its lifetime.[90] When a bioenergetics approach is applied to an entire fish population, numbers of fish in each cohort must be obtained. Other data needs include a time series of ambient water temperature in the focal habitats and corresponding estimates of either body size or daily consumption. Additionally, estimates are needed for the physiological parameters describing rates of consumption, respiration, egestion, excretion, and reproductive loss as functions of body size, temperature, and other variables. One major advantage of the bioenergetics approach is that the predicted population response to contaminant stress can reflect the particular mode of action of that stress.

Vaughan et al.[90] concluded that a Leslie matrix-bioenergetics combination approach would allow detailed comparisons of stressed and unstressed fish populations. By comparison, surplus production and stock-recruitment models generally require long time series of data on population parameters that are difficult to estimate. Yield models are also less desirable for addressing contaminant effects on fish populations because any effect of toxicant stress on reproduction is confounded with mortality before the age of recruitment.

Vaughan et al.[90] cautioned that fisheries scientists should not expect too much from currently available fish population models. Many questions that we want to ask are beyond the current state of the art, whether the questions relate to the effects of contaminants or power plants.[191] Vaughan et al.[190] also stressed that modeling approaches should be used to compare stressed to unstressed fish populations rather than attempting the unrealistic goal of precisely predicting absolute population effects.[192]

Schaaf et al.[193] used a Leslie matrix model to simulate fish population changes through time and develop a technique for assessing the effects of acute and chronic pollution on several marine stocks via a comparison of stock vulnerability to pollution. Deterministic, stochastic, density-independent, and density-dependent versions of their simulations were achieved by modifying one element of the matrix, S_0, first year survival. They related various population responses among the fish stocks to V_x, the age distribution of expected egg production, and demonstrated that information on age-specific egg production of a stock can yield a prediction of that species' response to pollution perturbation.

Schaaf et al.[193] acknowledged the limitations of their modeling approach and the inherent difficulties in obtaining reliable estimates of S_o and compensation factors for even the most intensively studied fish stocks. They view their approach as useful for bounding the magnitude and time horizon of contaminant impacts, and thereby provide information useful to resource managers. For one of eight fish species examined, Atlantic menhaden, their modeling approach predicted that heavily exploited stocks are most susceptible to additional pollution stress.[193]

The other approach to assessing the effects of contaminants on finfish populations that has been applied to Bay species involves multivariate statistics. This approach is an extension of the method commonly used to relate commercial landings or juvenile abundance indices to environmental variables,[16,194] in an attempt to understand recruitment variability. Although more empirical than theoretical, a desirable advantage, use of regression statistics for assessing the role of contaminants has several limitations. Of major importance is that relationships identified by step-wise multivariate and time series regression statistics are correlative and not necessarily causal. However, if a large degree of annual variation in commercial landings or juvenile abundancce can be accounted for, statistically, by contaminant levels, one can be reasonably confident that contaminants play an important role in determining numbers of fish harvested or numbers of juveniles produced. Caution must be exercised, however, if contaminant levels are collinearly related to other key parameters that were either excluded or included in the regression analyses.

A multivariate statistical approach is being used by NOAA to study contaminant effects on fishes. The Ocean Assessments Division is funding studies to determine if populations of finfish and shellfish are threatened by contaminants present in estuarine and coastal waters.[195] A major objective of this program is to examine historical data and determine if past trends in stock abundance of important species can be correlated with contaminant inputs. One ambitious goal of this program is to "determine the necessary extent of control on chemical inputs to prevent their affecting fish populations."[195, p. 2).

As part of NOAA's program, Polgar et al.[96] investigated the relationships among pollutant loadings and stock levels in several northeastern United States estuaries, including the Potomac River. Using a reconstructed time series of long-term trends in commercial fisheries abundance, climate, and several indices of pollution loading,[97] Polgar et al.[96] evaluated hypotheses concerning effects of human population changes and dredging on stock trends for striped bass and American shad in the Potomac estuary. Human population history was one surrogate pollution variate. Dredging history, an indicator of habitat alteration, was the other pollution variate included in the analysis.

Climatic factors rather than the two surrogate pollution variates appeared to dominate striped bass dynamics in the Potomac estuary from 1929 through 1976.[96] The effect of human population change on the index of American shad stocks was significant compared to climatic factors. This result suggests that some aspects of anthropogenic pollution in the Potomac estuary watershed (e.g., industrialization, land use, municipal waste treatment, but not dredging) were somehow linked to variability in shad stock size from 1929 through 1976. Summers et al.[98] suggested that the primary pollutant variable was sewage loading in shad spawning habitats.

This review of current studies which have attempted to link variations in abundance of selected Chesapeake Bay finfish populations to anthropogenic factors (including contaminants) presents limited support for the hypothesis that contaminants have been or are playing an important role in the declining status of several species populations. Lack of strong quantitative evidence for adverse contaminant effects is, however, no cause for optimism that habitat quality in the Bay is good. Some fish populations may already be severely affected by contaminants, coupled with stress from other sources of mortality like fishing, but we may not be able to detect it.[99]

The current state of the art in our understanding and ability to model fish population dynamics may prevent us from detecting all but catastrophic effects of contaminant stress. Recognition of the limitations of our science is necessary but hardly reassuring to ecologists, resource managers, and administrators alike. However, as long as potentially toxic contaminants continue to enter Chesapeake Bay and alter habitat quality in important fish spawning and nursery areas, studies of contaminant effects should continue. To effectively manage finfish populations and knowledgeably harvest only surplus production, scientists must strive to understand the role of all mortality sources on these populations, including contaminants.

REFERENCES

1. Dahlberg, M.D. 1979. A review of survival rates of fish eggs and larvae

in relation to impact assessments. *Mar. Fish. Rev.* 41:1-12.
2. Setzler-Hamilton, E.M. 1987. Utilization of Chesapeake Bay by early life stages of fishes, pp. 63-93. *In:* S.K. Majumdar, L.W. Hall, Jr., and H.M. Austin (Eds.) *Contaminant problems and management of living Chesapeake Bay resources.* The Pennsylvania Academy of Science Publications, Easton, PA.
3. Murphy, G.I. 1968. Patterns in life history and the environment. *Amer. Natur.* 102:391-403.
4. Giesel, J.T. 1976. Reproductive strategies as adaptations to life in temporally heterogeneous environments. *Ann. Rev. Ecol. Sys.* 7:57-79.
5. Klauda, R.J., W.P. Dey, T.B. Hoff, J.B. McLaren and Q.E. Ross. 1980. Biology of Hudson River juvenile striped bass, pp. 101-123. *In:* H. Clepper (Ed.) *Marine recreational fisheries 5.* Sport Fishing Institute, Washington, D.C.
6. Rothschild, B.J., P.W. Jones and J.S. Wilson. 1981. Trends in Chesapeake Bay fisheries. *Trans. N. Amer. Wildl. Nat. Res. Conf.* 46:284-298.
7. Richkus, W.A. and G. DiNardo. 1984. Current status and biological characteristics of the anadromous alosid stocks of the eastern United States: American shad, hickory shad, alewife and blueback herring. Phase I. Interstate management plan for migratory alosids of the Atlantic Coast. NTIS No. PB84229483. Martin Marietta Environmental Systems, Columbia, MD.
8. Boreman, J. and H.M. Austin. 1985. Production and harvest of anadromous striped bass stocks along the Atlantic coast. *Trans. Amer. Fish. Soc.* 114:3-7.
9. Goodyear, C.P. 1985. Relationship between reported commercial landings and abundance of young striped bass in Chesapeake Bay, Maryland. *Trans. Amer. Fish. Soc.* 114:92-96.
10. Austin, H.M. 1987. Chesapeake Bay fisheries: An Overview. 33-53. *In:* S.K. Majumdar, L.W. Hall, Jr., and H.M. Austin (Eds.). *Contaminant problems and management of living Chesapeake Bay resources.* The Pennsylvania Academy of Sciences Publication, Easton, PA.
11. Galtsoff, P.S. 1947. Ecological and physiological studies of the effects of sulfate pulp mill wastes on oysters in the York River, Virginia. U.S. Fish. Wildl. Serv. 51:58-186.
12. Davis, C.C. 1948. Studies of the effects of industrial pollution in the lower Patapsco River area. 2. The effect of copper as pollution on plankton. Maryland Department of Research and Education, Chesapeake Biological Laboratory, Publ. No. 72, Solomons, MD.
13. Massman, W.H., E.C. Ladd and H.N. McCutcheon. 1952. A biological survey of the Rappahannock River, Virginia. Volumes 1 and 2. Virginia Fisheries Laboratory.

14. Mansueti, R.J. 1961. Effects of civilization on striped bass and other estuarine biota in Chesapeake Bay and tributaries. *Proc. Gulf Carib. Fish Inst.* 14:110-136.
15. Cronin, L.E. 1967. The condition of the Chesapeake Bay. *Trans. N. Amer. Wildl. Nat. Res. Conf.* 23:137-150.
16. Macalaster, E.G., D.A. Barker and M.E. Kasper (Eds.) 1983. Chesapeake Bay: a profile of environmental change. U.S. Environmental Protection Agency, Philadelphia, PA. 200 pp.
17. Wohlfarth, G.W. 1986. Decline in natural fisheries—a genetic analysis and suggestion for recovery. *Can. J. Fish. Aquat. Sci.* 43:1298-1306.
18. Egerton, F.N. 1985. Overfishing or pollution? Case history of a controversy on the Great Lakes. Tech. Rept. 41:1-28. Great Lakes Fisheries Commission, Ann Arbor, MI.
19. Mehrle, P.M., T.A. Haines, S. Hamilton, J.L. Ludke, F.L. Mayer and M.A. Ribick. 1982. Relationship between body contaminants and bone development of east-coast striped bass. *Trans. Amer. Fish. Soc.* 111:231-241.
20. Goodyear, C.P. 1985. Toxic materials, fishing, and environmental variation: simulated effects on striped bass population trends. *Trans. Amer. Fish. Soc.* 114:107-113.
21. Price, K.S., D.A. Flemer, J.L. Taft, G.B. Mackiernan, W. Nehlsen, R.B. Biggs, N.H. Burger and D.A. Blaylock. 1985. Nutrient enrichment of Chesapeake Bay and its impact on the habitat of striped bass: a speculative hypothesis. *Trans. Amer. Fish. Soc.* 114:97-106.
22. Coutant, C.C. 1985. Striped bass, temperature, and dissolved oxygen: a speculative hypothesis for environmental risk. *Trans. Amer. Fish. Soc.* 114:31-61.
23. Goodyear, C.P., J.E. Cohen and S.W. Christensen. 1985. Maryland striped bass: recruitment declining below replacement. *Trans. Amer. Fish. Soc.* 114:146-151.
24. Hall, L.W., Jr., L.O. Horseman and S. Zeger. 1984. Effects of multiple organic and inorganic contaminants on fertilization, hatching success and prolarval survival of striped bass, *Morone saxatilis*. *Arch. Environ. Contam. Toxicol.* 13:723-729.
25. Hall, L.W., Jr., A.E. Pinkney and L.O. Horseman. 1985. Mortality of striped bass larvae in relation to contaminants and water quality in a Chesapeake Bay tributary. *Trans. Amer. Fish. Soc.* 114:861-868 pp.
26. Klauda, R.J. and R.E. Palmer. 1986. Laboratory and field bioassay experiments on blueback herring from Maryland coastal plain streams. PPSP-AD-15. NTIS No. PB86/209541/AS. The Johns Hopkins University, Applied Physics Laboratory, Shady Side, MD. 228 pp.
27. Chesapeake Research Consortium, Inc. 1976. The effects of Tropical Storm Agnes on the Chesapeake Bay estuarine system. The Johns

Hopkins University Press, Baltimore, MD. 639 pp.
28. Boesch, D.F., R.J. Diaz and R.W. Virnstein. 1976. Effects of Tropical Storm Agnes on soft-bottom macrobenthic communities of the James and York estuarines and lower Chesapeake Bay. *Ches. Sci.* 17:246-259.
29. Bender, M.E. and R.J. Huggett. 1984. Fate and effects of Kepone in the James River estuary, pp. 5-50. In: E. Hodgson (Ed.) *Reviews in environmental toxicology.* Elsevier Science Publishers, New York, NY.
30. Loesch, J.G., R.J. Huggett and E.J. Foell. 1982. Kepone concentration in juvenile anadromous fishes. *Estuaries* 5:175-181.
31. Fisher, D.J., J.R. Clark, M.H. Roberts, Jr., J.P. Connolly and L.H. Mueller. 1986. Bioaccumulation of Kepone by spot (*Leiostomus xanthurus*): importance of dietary accumulation and ingestion rate. *Aquatic Toxicol.* 9:161-178.
32. Goodman, L.R., D.J. Hansen, C.S. Manning and L.F. Fars. 1982. Effects of Kepone on the sheepshead minnow (*Cyprinodon variegatus*) in an entire life-cycle test. *Arch. Environ. Contam. Toxicol.* 11:335-342.
33. Hall, L.W., Jr., G.R. Helz and D.T. Burton. 1981. Power plant chlorination perspective, pp. 1-5. *In:* L.W. Hall, Jr., G.R. Helz and D.T. Burton (Eds.) *Power plant chlorination. A biological and chemical assessment.* EA-1750. Electric Power Research Institute, Palo Alto, CA.
34. Morgan, R.P., II, D.A. Flemer, L.A. Noe, J.V. Rasin, Jr. and R.A. Murtagh. 1974. Biochemical studies of entrained phytoplankton at the Morgantown Maryland power plant, pp. 165-168. *In:* L.D. Jensen (Ed.) *Proceedings of second workshop on entrainment and intake screening.* The Johns Hopkins University, Baltimore, MD.
35. Burton, D.T. 1980. Biofouling control procedures for power plant cooling water systems. pp. 251-266. *In:* J.F. Garey, R.M. Jordan, A.H. Aitken, D.T. Burton and R.H. Gray (Eds.). *Condenser biofouling control symposium proceedings.* Ann Arbor Science Publications, Inc., Ann Arbor, MI.
36. Jolley, R.L., G. Jones, W.W. Pitt and J.E. Thompson. 1976. Chlorination of organics in cooling water and process effluents, pp. 115-152. *In:* R.L. Jolley (Ed.) *The environmental impact of water chlorination.* CONF-751096, Oak Ridge National Laboratory, Oak Ridge, TN.
37. Sugam, R. and G.R. Helz. 1977. Speciation of chlorine produced oxidants in marine waters: theoretical aspects. *Ches. Sci.* 18:113-115.
38. Jolley, R.L. 1973. Chlorination effects on organic constituents in effluents from domestic sanitary sewage treatment plants. CRNL-RM-4290. Oak Ridge National Laboratory, Oak Ridge, TN. 249 pp.
39. Glaze, W.H. and J.E. Henderson, IV. 1975. Formation of organochlorine compounds from the chlorination of a municipal secondary effluent. *J. Water Pollut. Control Fed.* 47:2511-2515.
40. Brungs, W.A. 1973. Effects of residual chlorine on aquatic life. *J. Water*

Pollut. Control Fed. 45:2180-2193.
41. Tsai, C.F. 1975. Effects of sewage treatment plant effluents on fish: a review of the literature. Ches. Res. Consortium Publ. No. 36. Chesapeake Research Consortium, Inc., College Park, MD. 229 pp.
42. Whitehouse, J.W. 1975. Chlorination of cooling water: a review of literature on the effects of chlorine on aquatic organisms. Central Electricity Research Laboratory, Rept. No. RD/L/M 496, Leatherhead, England. 22 pp.
43. Pike, D.J. 1971. Toxicity of chlorine to pike. *N.Z. Wildl.* 33:39.
44. Zillich, J.A. 1972. Toxicity of combined chlorine residuals to freshwater fish. *J. Water Pollut. Control Fed.* 2:212-220.
45. Tsai, C.F. 1973. Water quality and fish life below sewage outfalls. *Trans. Amer. Fish. Soc.* 102:281-292.
46. Bellanca, M.A. and D.S. Bailey. 1977. Effects of chlorinated effluents on the aquatic ecosystems of the lower James River. *J. Water Poll. Control Fed.* 49:639-645.
47. Coulter, J.B. 1984. Don't chlorinate sewage. Proceedings 2nd National Symposium on Wastewater Disinfection. U.S. Environmental Protection Agency, Orlando, FL. 14 pp.
48. Hall, L.W., Jr., G.R. Helz and D.T. Burton. 1981. Power plant chlorination: a biological and chemical assessment. Ann Arbor Science Publications, Inc., Ann Arbor, MI. 237 pp.
49. Hall, L.W. Jr., D.T. Burton and L.H. Liden. 1981. An interpretative literature analysis evaluating the effects of power plant chlorination on freshwater organisms. *CRC Crit. Rev. Toxicol.* 9:11-20.
50. Hall, L.W., Jr. and D.T. Burton. 1981. Freshwater toxicity studies, pp. 59-130. *In:* L.W. Hall, Jr., G.R. Helz and D.T. Burton (Eds.) *Power plant chlorination. A biological and chemical assessment.* EA-1750. Electric Power Research Institute, Palo Alto, CA.
51. Hall, L.W., Jr. and D.T. Burton. 1981. Estuarine and marine toxicity studies, pp. 131-217. *In:* L.W. Hall, Jr., G.R. Helz and D.T. Burton (Eds.) *Power plant chlorination. A biological and chemical assessment.* EA-1750. Electric Power Research Institute, Palo Alto, CA.
52. Hall, L.W., Jr., D.T. Burton and L.H. Liden. 1982. Power plant chlorination effects on estuarine and marine organisms. *CRC Crit. Rev. Toxicol.* 10:27-47.
53. Hall, L.W., Jr., W.C. Graves, D.T. Burton, S.L. Margrey, F.M. Hetrick and B.S. Roberson. 1982. A comparison of chlorine toxicity to three life stages of striped bass (*Morone saxatilis*). *Bull. Environ. Contam. Toxicol.* 29:631-636.
54. Hall, L.W., Jr., D.T. Burton, S.L. Margrey and W.C. Graves. 1983. Predicted mortality of Chesapeake Bay organisms exposed to simulated power plant chlorination conditions at various acclimation temperatures,

pp. 1005-1017. *In:* R.L. Jolley, W.A. Brungs, J.A. Cotruvo, R.B. Cumming, J.S. Mattice and V.A. Jacobs (Eds.). *Water chlorination environmental impact and health effects, vol. 4, book 2 of environment, health, and risk.* Ann Arbor Science Publications, Inc., Ann Arbor, MI.
55. Hall, L.W., Jr., D.T. Burton, S.L. Margrey and W.C. Graves. 1983. The effect of acclimation temperature on the interactions of chlorine, $\triangle$T and exposure duration to eggs, prolarvae and larvae of striped bass, *Morone saxatilis. Water Res.* 17:309-317.
56. Hall, L.W., Jr., D.T. Burton, S.T. Margrey and W.C. Graves. 1983. The influence of spring and fall temperatures on the avoidance response of juvenile Atlantic menhaden, *Brevoortia tyrannus,* exposed to simultaneous chlorine-$\triangle$T conditions. *Water Resources Bull.* 19:283-287.
57. Hall, L.W. Jr., S.L. Margrey, D.T. Burton and W.C. Graves. 1983. Avoidance behavior of juvenile striped bass, *Morone saxatilis,* subjected to simultaneous chlorine and elevated temperature conditions. *Arch. Environ. Contam. Toxicol.* 12:715-720.
58. Hetrick, F.M., L.W. Hall, Jr., S. Wolski, W.C. Graves, B.S. Roberson and D.T. Burton. 1984. Influence of chlorine on the susceptibility of striped bass (*Morone saxatilis)* to *Vibrio anguillarum. Can. J. Fish Aquat. Sci.* 41:1375-1380.
59. Bongers, L.H., T.P. O'Connor and D.T. Burton. 1977. Bromine chloride—an alternative to chlorine for fouling control in condenser cooling systems. EPA Rept. EPA-600/7-77-053.
60. Liden, L.H. and D.T. Burton. 1977. Survival of juvenile Atlantic menhaden (*Brevoortia tyrannus)* and spot (*Leiostomus xanthurus*) exposed to bromine chloride- and chlorine-treated estuarine waters. *J. Environ. Sci. Health* A12:375-388.
61. Burton, D.T. and S.L. Margrey. 1979. Control of fouling organisms in estuarine cooling water systems by chlorine and bromine chloride. *Environ. Sci. Tech.* 13:684-689.
62. Roberts, M.H., Jr. and R.A. Gleeson. 1978. Acute toxicity of bromochlorinated seawater to selected estuarine species with a comparison to chlorinated seawater toxicity. *Mar. Environ. Res.* 1:19-30.
63. Liden, L.H., D.T. Burton, L.H. Bongers and A.F. Holland. 1980. Effects of chlorobrominated and chlorinated cooling waters on estuarine organisms. *Jour. Water Pollut. Cont. Fed.* 52:173-182.
64. U.S. EPA (Environmental Protection Agency). 1983. Ambient aquatic life water quality criteria for chlorine. U.S. Environmental Protection Agency, Washington, D.C.
65. Gullens, S.R., R.M. Block, J.C. Rhoderick, D.T. Burton and L.H. Liden. 1977. Effects of continuous chlorination on white perch (*Morone americana)* and Atlantic menhaden *(Brevoortia tyrannus*) at two temperatures. *Assoc. Southeast. Biol. Bull.* 24:55.

66. Seegert, G.L. and A.S. Brooks. 1978. The effects of intermittent chlorination on coho salmon, alewife, spottail shiner, and rainbow trout. *Trans. Am. Fish. Soc.* 107:346-353.
67. Morgan, R.P. II and R.D. Prince. 1977. Chlorine toxicity to eggs and larvae of five Chesapeake Bay fishes. *Trans. Amer. Fish. Soc.* 106:380-395.
68. Fava, J.A. and J.W. Meldrim. 1985. Comparison of acute toxicity and avoidance responses of Atlantic silverside and white perch to chlorinated estuarine waters, pp. 493-508. *In:* R.L. Jolley, R.J. Bull, W.P. Davis, S. Katz, M.H. Roberts, Jr. and V.A. Jacobs (Eds.) *Water chlorination: chemistry, environmental impact and health effects.* Lewis Publishing Inc., Chelsea, MI.
69. Annual Environmental Operating Report (Non-Radiological) Salem Nuclear Generating Station—Union No. 1. 1978. Docket No. 50-272; Operating License No. DPR-70, Vol. 3 of 3. Public Service Electric and Gas Company. Newark, NJ.
70. Fava, J.A. and D.T. Burton. 1984. Toxicity of chlorine to aquatic life and its impact on the environment, pp. 1-20. *In:* Wastewater disinfection—the pros and cons. Proceedings, 1984 wastewater disinfection committee preconference workshop, New Orleans, LA.
71. Giattina, J.D. and R.R. Garton. 1983. A review of the preference—avoidance responses of fishes to aquatic contaminants. *Residue Reviews.* 87:44-90.
72. Morgan, R.P., II. 1980. Biocides and fish behavior, pp. 75-102. *In:* C.H. Hocutt, J.R. Stauffer, Jr., J.E. Edinger, L.W. Hall, Jr. and R.P. Morgan II (Eds.) *Power plants — effects on fish and shellfish behavior.* Academic Press, Inc., New York, NY.
73. Roberts, M.H., Jr., R.J. Diaz, M.E. Bender and R.J. Huggett. 1975. Acute toxicity of chlorine on selected estuarine species. *J. Fish. Res. Bd. Can.* 32:2525-2528.
74. Haven, D.S., W.J. Hargis, Jr. and P.C. Kendall. 1978. The oyster industry in Virginia: its status, problems and promise. *Special papers in marine science, No. 4,* Virginia Institute of Marine Science, Gloucester Point, VA. 1024 pp.
75. Hall, L.W., Jr. M.J. Lenkevich, W. Scott Hall, A.E. Pinkney and S.J. Bushong. 1986. Monitoring organotin concentrations in Maryland waters of Chesapeake Bay. Final report. The Johns Hopkins University, Applied Physics Laboratory, Shady Side, MD. 32 pp.
76. Department of the Environment. 1986. Organotin in antifouling paints: environmental considerations. Central Directorate of Environmental Protection. Pollution Paper No. 25 London, Her Majesty's Stationery Office.
77. Federal Register. 1985. Vol. 50, Issue 120, p. 25748 of June 21.
78. Stebbing, A.R.D. 1985. Organotins and water quality- some lessons to be learned. *Mar. Pollut. Bull.* 16:383-390.

79. Thain, J. 1983. The acute toxicity of bis (tributyltin) oxide to adults and larvae of some marine organisms. *ICES Paper CM1983:13.* International Council for the Exploration of the Sea, Copenhagen, Denmark.
80. Thain, J. and M.J. Waldock. 1983. The effect of suspended sediment and bis tributyltin oxide on the growth of *Crassostrea gigas* spat. *ICES Paper CM1983/EL19.* International Council for the Exploration of the Sea, Copenhagen, Denmark.
81. Evans, C.J. and P.J. Smith. 1975. Organotin-based antifouling systems. *J. Oil Col. Chem. Assoc.* 58:160-168.
82. Monaghan, C.P., V.I. Kulkarni, M. Ozcan and M.L. Good. 1980. Environmental fate of organotin antifoulants: chemical speciation of toxicants in aqueous solutions. Tech. Rep. No. 2, Project No. NR 356-709. Office of Naval Research, Arlington, VA.
83. Hall, L.W. Jr. and A.E. Pinkney. 1985. Acute and sublethal effects of organotin compounds on aquatic biota: an interpretative literature evaluation: *CRC Crit. Rev. Toxicol.* 14:159-209.
84. Seinen, W., T. Helder, H. Vernij, A. Penninks and P. Leewaugh. 1981. Short term toxicity of tri-n-butyltin in rainbow trout (*Salmo gairdneri* Richardson) yolk sac fry. *Sci. Total Environ.* 19:155-156.
85. Ward, G.S., G.C. Cramm, P.R. Parrish, H. Trachman and A. Slesinger. 1981. Bioaccumulation and chronic toxicity of bis (tributyltin) oxide (TBTO):tests with a saltwater fish, pp. 183-200. *In:* D.R. Branson and K.L. Dickson (Eds.) *Aquatic toxicology and hazard assessment.* ASTM STP 737, American Society of Testing and Materials, Philadelphia, PA.
86. M & T Chemicals Inc. 1976. Acute toxicity of tri-n-butyltin oxide to channel catfish (*Ictalurus punctatus*), the fresh water clam (*Elliptio complanatus*), the common mummichog (*Fundulus heteroclitus*), and the American oyster (*Crassostrea virginica*). Report submitted by EG and G Bionomics, Wareham, MA., to M & T Chemicals Inc., Rahway, NJ. 13 pp.
87. Pinkney, A.E., L.W. Hall, Jr., M.J. Lenkevich, D.T. Burton and S. Zeger. 1985. Comparison of avoidance responses of an estuarine fish, *Fundulus heteroclitus,* and crustacean, *Palaemonetes pugio,* to bis (tri-n-butyltin) oxide. *Water Air Soil Pollut.* 25:33-40.
88. Hall, L.W., Jr., A.E. Pinkney, S. Zeger, D.T. Burton and M.J. Lenkevich. 1984. Behavioral responses of two estuarine fish species subjected to bis (tri-n-butyltin) oxide. *Water Res. Bull.* 20:235-239.
89. M & T Chemicals, Inc. 1979. Acute toxicity of three samples of TBTO (tributyltin oxide) to juvenile sheepshead minnows (*Cyprinodon variegatus).* Report L14-500 submitted by EG and G Bionomics, Wareham, MA, to M & T Chemicals, Inc., Rahway, NJ.
90. M & T Chemicals, Inc. 1976. Acute toxicity of tri-n-butyltin oxide to bluegill (*Lepomis macrochirus*). Report submitted by EG and G

Bionomics, Wareham, MA, to M & T Chemicals Inc., Rahway, NJ. 8 pp.
91. Hall, L.W., Jr. 1987. Personal communication. The Johns Hopkins University, Applied Physics Laboratory, Shady Side, MD.
92. Zuckerman, J.J., R.P. Reisdorf, H.V. Ellis III and R.R. Wilkinson. 1978. Organotins in biology and the environment, pp. 388-424. *In:* F.E. Brinckman and J.M. Bellama (Eds.) *Organometals and organometalloids. Occurrence and fate in the environment.* ACS Symposium Series 82, American Chemical Society, Washington, DC.
93. Josephson, J. 1984. Polynuclear aromatic hydrocarbons. *Environ. Sci. Technol.* 18:93-95.
94. Huggett, R.J., M.E. Bender and M.A. Unger. (In Press). Polynuclear aromatic hydrocarbons in the Elizabeth River, Virginia. *In:* K.L. Dickson, A.W. Maki and W. Brungs (Eds.). *Fate and effects of sediment bound chemicals in aquatic systems.* SETAC Special Publication No. 2. Pergamon Press, New York, NY.
95. Hargis, W.J., Jr., M.H. Roberts, Jr. and D.E. Zwerner. 1984. Effects of contaminated sediments and sediment-exposed effluent water on an estuarine fish: acute toxicity. *Mar. Environ. Res.* 14:337-354.
96. Weeks, B.A. and J.E. Warinner. 1984. Effects of toxic chemicals on macrophage phagocytosis in two estuarine fishes. *Mar. Environ. Res.* 14:327-335.
97. Black, J.J. 1983. Field and laboratory studies of environmental carcinogenesis in Niagara River fish. *J. Great Lakes Res.* 9:326-334.
98. Mix, M.C. 1984. Polycyclic aromatic hydrocarbons in the aquatic environment: occurrence and biological monitoring, pp. 51-102. *In:* E. Hodgson (Ed.) *Reviews in environmental toxicology.* Elsevier Science Publications, New York, NY.
99. Malins, D.C., M.M. Krahn, M.S. Meyers, L.D. Rhodes, D.W. Brown, C.A. Krone, B.M. McCain and Sin-Lam Chan. 1985. Toxic chemicals in sediments and biota from a creosote-polluted harbor: relationships with hepatic neoplasms and other hapatic lesions in English sole (*Parophyrys vetulus*). *Carcinogenesis.* 6:1463-1469.
100. Anonymous. 1984. Emergency striped bass research study. Report for 1982-1983. U.S. Fish and Wildlife Service and National Oceanic and Atmospheric Administration, Washington, DC.
101. Mehrle, P.M. and L. Ludke. 1983. Impacts of contaminants on early life stages of striped bass. U.S. Fish and Wildlife Service, Columbia National Fishery Research Laboratory Progress Report 1980-1983, Columbia MD. 72 pp.
102. Schmitt, C.J., J.L. Zajicek and M.A. Ribick. 1985. National pesticide monitoring program: residues of organochlorine chemicals in freshwater fish, 1980-81. *Arch. Environ. Contam. Toxicol.* 14:225-260.
103. Lowe, T.P., T.W. May, W.G. Brumbaugh and D.A. Kane. 1985. National

contaminant biomonitoring program: concentrations of seven elements in freshwater fish, 1978-1981. *Arch. Environ. Contam. Toxicol.* 14:363-388.
104. Eisenberg, M. and J.J. Topping. 1986. Trace metal residues in finfish from Maryland waters, 1978-1979. *J. Environ. Sci. Health.* 821:87-102.
105. Buckley, L.J., T.A. Halavick, G.C. Laurence, S.J. Hamilton and P. Yevich. 1985. Comparative swimming stamina, biochemical composition, backbone mechanical properties, and histo-pathology of juvenile striped bass from rivers and hatcheries of the eastern United States. *Trans. Amer. Fish. Soc.* 114:114-124.
106. Horn, E.G., L.J. Hetling and T.J. Tofflemire. 1979. The problem of PCBs in the Hudson River system. *Ann. N.Y. Acad. Sci.* 320:591-609
107. Klauda, R.J., T.H. Peck and G.K. Rice. 1981. Accumulation of polychlorinated biphenyls in Atlantic tomcod *(Microgadus tomcod)* collected in the Hudson River estuary, New York. *Bull. Environm. Contam. Toxicol.* 27:829-835.
108. Smith, C.E., T.H. Peck, R.J. Klauda and J.B. McLaren. 1979. Hepatomas in Atlantic tomcod (*Microgadus tomcod*) collected in the Hudson River estuary, New York. *J. Fish Dis.* 2:313-319.
109. Brown, M.P., M.B. Werner, R.J. Sloan and K.W. Simpson. 1985. Polychlorinated biphenyls in the Hudson River. Recent trends in the distribution of PCBs in water, sediment, and fish. *Environ. Sci. Technol.* 19:656-661
110. Limburg, K.E., M.A. Moran and W.H. McDowell. 1986. The Hudson River ecosystem. Springer-Verlag New York, Inc., NY. 331 pp.
111. Haberman, D., G.B. Mackiernan and J. Macknis. 1983. Toxic compounds, pp. 51-59. *In:* E.G. Macalaster, D.A. Banker and M.E. Kasper (Eds.) *Chesapeake Bay: a profile of environmental change.* U.S. Environmental Protection Agency, Philadelphia, PA.
112. Bieri, R., O. Bricker, R. Byrne, R. Diaz, G. Helz, J. Hill, R. Huggett, R. Kerhin, M. Nichols, E. Reinharz, L. Schaffner, D. Wildikng and C. Strobel. 1982. Toxic substances, pp. 266-378. *In:* E.G. Macalaster, D.A. Barker and M.E. Kasper (Eds.) *Chesapeake Bay Program technical studies: a synthesis.* U.S. Environmental Protection Agency, Philadelphia, PA.
113. Palawski, D., J.B. Hunn and F.J. Dwyer. 1985. Sensitivity of young striped bass to organic and inorganic contaminants in fresh and saline waters. *Trans. Amer. Fish. Soc.* 114:748-753.
114. Wright, D.A., M.J. Meteyer and F.D. Martin. 1985. Effect of calcium on cadmium uptake and toxicity in larvae and juveniles of striped bass (*Morone saxatilis*). *Bull. Environ. Contam. Toxicol.* 34:196-204.
115. Wu, T.L. 1981. Atrazine residues in estuarine water and the aerial deposition of atrazine into Rhode River, Maryland. *Water Air Soil Pollut.* 15:173-184.

116. Ward, G.S. and L. Ballantine. 1985. Acute and chronic toxicity of atrazine to estuarine fauna. *Estuaries.* 8:22-27.
117. Klauda, R.J., R.E. Palmer and M.J. Lenkevich. 1987. Sensitivity of early life stages of blueback herring to moderate acidity and aluminum in fresh water. *Estuaries* 10:46-55.
118. Rosenthal, H. and D.F. Alderice. 1976. Sublethal effects of environmental stressors, natural and pollutional, on marine fish eggs and larvae. *J. Fish. Res. Bd. Can.* 33:2047-2065.
119. McKim, J.M. 1977. Evaluation of tests with early life stages for predicting long-term toxicity. *J. Fish. Res. Bd. Can.* 34:1148-1154.
120. Burton, D.T., L.W. Hall, Jr., S.L. Margrey and R.D. Small. 1979. Interactions of chlorine, temperature change ($\triangle$T) and exposure time on survival of striped bass (*Morone saxatilis*) eggs and prolarvae. *J. Fish. Res. Bd. Can.* 36:1108-1113.
121. Bodammer, J.E. 1985. Corneal damage in larvae of striped bass *Morone saxatilis* exposed to copper. *Trans. Amer. Fish. Soc.* 114:577-583.
122. Klauda, R.J. 1986. Acute and chronic effects of waterborne arsenic and selenium on the early life stages of striped bass (*Morone saxatilis*). PPRP-98. The Johns Hopkins University, Applied Physics Laboratory, Shady Side, MD.
123. Andreae, M.O. 1978. Distribution and speciation of arsenic in natural waters and some marine algae. *Deep-Sea Res.* 25:391-402.
124. Robberecht, H. and R. Van Grieken. 1982. Selenium in environmental waters: determination, speciation and concentration levels. *Talanta* 29:823-844.
125. Hall, L.W. Jr. and D.T. Burton. 1982. Effects of power plant coal waste runoff and leachate on aquatic biota: an overview with research recommendations. *CRC Crit. Rev. Toxicol.* 10:287-301.
126. Wood, J.M. 1974. Biological cycles for toxic elements in the environment. *Science.* 183:1049-1052.
127. Vouk, V.B. and W.T. Piver. 1983. Metabolic elements in fossil fuel combustion products: amounts and form of emissions and evaluation of carcinogenicity and mutagenicity. *Environ. Health Perspectives* 47:201-225.
128. Turner, M.A. and A.L. Swick. 1983. The English-Wabigoon River system: IV. Interaction between mercury and selenium accumulated from waterborne and dietary sources by northern pike (*Esox lucius*). *Can. J. Fish. Aquat. Sci.* 40:2241-2250.
129. Van Puymbroeck, S.L., W.J.J. Stips and O.L.J. Vanderboght. 1982. The antagonism between selenium and cadmium in a freshwater mollusc. *Arch. Environ. Contam. Toxicol.* 11:103-106.
130. Winner, R.W. 1984. Selenium effects on antennal integrity and chronic copper toxicity in *Daphnia pulex* (de Geer). *Bull. Environ. Contam. Tox-*

icol. 33:605-611.
131. Pelletier, E. 1985. Mercury-selenium interactions in aquatic organisms: a review. *Marine Environ. Res.* 18:111-132.
132. Levander, O.A. 1977. Metabolic interrelationships between arsenic and selenium. *Environ. Health Perspectives* 19:159-164.
133. Larsson, A., C. Haux and M.L. Sjobeck. 1985. Fish physiology and metal pollution: results and experiences from laboratory and field studies. *Ecotoxicol. Environ. Safety* 9:250-281.
134. Seim, W.K., L.R. Curtis, S.W. Glenn and G.A. Chapman. 1984. Growth and survival of developing steelhead trout (*Salmo gairdneri*) continuously or intermittently exposed to copper. *Can. J. Fish. Aquat. Sci.* 41:433-438.
135. McCarter, J.A. and M. Roch. 1983. Hepatic metallothionein and resistance to copper in juvenile coho salmon. *Comp. Biochem. Physiol.* 74C:133-137.
136. Duncan, D.A. and J.F. Klaverkamp. 1983. Tolerance and resistance to cadmium in white suckers (*Catostomus commersoni*) previously exposed to cadmium, mercury, zinc, or selenium. *Can. J. Fish. Aquat. Sci.* 40:128-138.
137. Dixon, D.G. and J.B. Sprague. 1981. Acclimation-induced changes in toxicity of arsenic and cyanide to rainbow trout. *J. Fish Biol.* 18:579-589.
138. Bradley, R.W., C. DuQuesnay and J.B. Sprague. 1985. Acclimation of rainbow trout, *Salmo gairdneri* Richardson, to zinc: kinetics and mechanism of enhanced tolerance induction. *J. Fish Biol.* 27:367-379.
139. Olsson, P.E. and C. Haux. 1985. Alterations in hepatic metallothionein content in perch, *Perca fluviatilis,* environmentally exposed to cadmium. *Marine Environ. Res.* 17:181-183.
140. Roch, M., P. Noonan and J.A. McCarter. 1986. Determination of no effect levels of heavy metals for rainbow trout using hepatic metallothionein. *Water. Res.* 20:771-774.
141. Kay, J., D.G. Thomas, M.W. Brown, A. Cryer, D. Shurben, J.F.G. Solbe and J.S. Garvey. 1986. Cadmium accumulation and protein binding patterns in tissues of the rainbow trout, *Salmo gairdneri. Environ. Hlth. Persp.* 65:133-139.
142. Harrison, F.L. and J.R. Lam. 1986. Copper-binding proteins in liver of bluegills exposed to increased soluble copper under field and laboratory conditions. *Environ. Hlth. Persp.* 65:125-132.
143. Weis, P., J.D. Bogelen and E.C. Enslee. 1986. Hg-and Cu- induced hepatocellular changes in the mummichog, *Fundulus heteroclitus. Environ. Hlth. Persp.* 65:167-173.
144. Siegel, N. and A. Haug. 1984. The involvement of calmodulin in aluminum toxicity. *Fed. Proceed.* 43:1521.
145. Haines, T.A. 1985. Acid rain: international aspects, pp. 65-71. *In:* P.M. Mehrle, Jr., R.H. Gray and R.L. Kendal (Eds.). *Toxic substances in the aquatic environment: an international aspect.* American Fisheries Society,

Bethesda, MD.
146. Barett, E. and C. Brodin. 1955. The acidity of Scandinavian precipitation. *Tellus* 7:251-257.
147. Cogbill, C.V. and G.E. Likens. 1974. Acid precipitation in the Northeastern U.S. *Water Resour. Res.* 10:1133-1137.
148. Stensland, G.J., D.M. Whelpdale and G. Oehlert. 1986. Precipitation chemistry, pp. 128-199. *In: Acid deposition, long-term trends.* National Academy Press, Washington, DC.
149. Watt, W.D., C.D. Scott and W.J. White. 1983. Evidence of acidification of some Nova Scotia rivers and its impact on Atlantic salmon, *Salmo salar. Can. J. Fish. Aquat. Sci.* 40:462-473.
150. Haines, T.A. 1981. Acidic precipitation and its consequences for aquatic ecosystems: a review. *Trans. Amer. Fish. Soc.* 110:669-707.
151. Harvey, H.H. 1982. Population responses of fishes in acidified water, pp. 227-242. *In:* T.A. Haines and R.E. Johnson (Eds.) *Acid rain/fisheries.* American Fisheries Society, Bethesda, MD.
152. Magnuson, J.J., J.P. Baker and E.J. Rahel. 1984. A critical assessment of effects of acidification on fisheries in North America. *Phil. Trans. R. Soc. Lond.* B 305:501-516.
153. Haines, T.A. 1986. Fish population trends in response to surface water acidification, pp. 300-334. *In: Acid deposition, long-term trends.* National Academy Press, Washington, DC.
154. Dillon, P.J., N.D. Yan and H.H. Harvey. 1984. Acidic deposition: effects on aquatic systems. *CRC Rev. Environ. Control.* 13:167-194.
155. Baker, J.P. 1982. Effects on fish of metals associated with acidification, pp. 165-176. *In:* T.A. Haines and R.E. Johnson (Eds.) *Acid rain/fisheries.* American Fisheries Society, Bethesda, MD.
156. Gorham, E. 1976. Acid precipitation and its influence upon aquatic ecosystems—an overview. *Water Air Soil Pollut.* 6:457-481.
157. Beamish, R.J. 1976. Acidification of lakes in Canada by acid precipitation and the resulting effects on fishes. *Water Air Soil Pollut.* 6:501-514.
158. Schindler, D.W. and M.A. Turner. 1982. Biological, chemical and physical responses of lakes to experimental acidification. *Water Air Soil Pollut.* 18:259-271.
159. Hutchinson, N.J. and J.B. Sprague. 1986. Toxicity of trace metal mixtures to American flagfish (*Jordanella floridae*) in soft, acidic water and implications for cultural acidification. *Can. J. Fish. Aquat. Sci.* 43:647-655.
160. Korwin-Kossakowski, M. and B. Jezierska. 1985. The effect of temperature on survival of carp fry, *Cyprinus carpio* L., in acidic water. *J. Fish Biol.* 26:43-47.
161. Broderius, S., R. Drummond, J. Frandt and C. Russom. 1985. Toxicity of ammonia to early life stages of the smallmouth bass at four pH values.

Environ. Toxicol. Chem. 4:87-96.
162. Spehar, R.L., H.P. Nelson, M.J. Swanson and J.W. Renoos. 1985. Pentachlorophenol toxicity to amphipods and fathead minnows at different pH values. *Environ. Toxicol. Chem.* 4:389-397.
163. Hodson, P.V., B.R. Blunt and D.J. Spry. 1978. pH-induced changes in the blood lead of lead-exposed rainbow trout (*Salmo gairdneri*). *J. Fish. Res. Board Can.* 35:437-445.
164. Lauren, D.J. and D.G. McDonald. 1986. Influence of water hardness, pH, and alkalinity on the mechanisms of copper toxicity in juvenile rainbow trout, *Salmo gairdneri. Can. J. Fish. Aquat. Sci.* 43:1488-1496.
165. Bendell-Young, L.I., H.H. Harvey and J.F. Young. 1986. Accumulation of cadmium by white suckers (*Catostomus commersoni*) in relation to fish growth and lake acidification. *Can. J. Fish. Aquat. Sci.* 43:806-811.
166. Bradley, R.W. and J.B. Sprague. 1985. The influence of pH, water hardness, and alkalinity on the acute lethality of zinc to rainbow trout (*Salmo gairdneri*). *Can. J. Fish. Aquat. Sci.* 42:731-736.
167. Cusimano, R.F., D.F. Brakke and G.A. Chapman. 1986. Effects of pH on the toxicities of cadmium, copper, and zinc to steelhead trout (*Salmo gairdneri*). *Can. J. Fish. Aquat. Sci.* 43:1497-1503.
168. Bowman, M.L. 1984. Acid deposition monitoring projects in Maryland, pp. 15-28. *In:* M.L. Bowman and S.S.G. Wierman (Eds.). *The potential effects of acid deposition in Maryland. Report of the Maryland interagency working group on acid deposition.* Maryland Department of Natural Resources, Power Plant Research Program, Annapolis, MD.
169. Correll, D.L., N.M. Golf and W.T. Peterjohn. 1984. Ion balances between precipitation inputs and the Rhode River watershed discharges, pp. 77-111. *In:* O.P. Bricker (Ed.). *Geological aspects of acid deposition.* Butterworth Publishers, Stoneham, MA.
170. Brezonik, P.L., E.S. Edgerton and C.D. Hendry. 1980. Acid precipitation and sulfate deposition in Florida. *Science.* 208:1027-1029.
171. Haines, B. 1980. Acid precipitation in southeastern United States: a brief review. *Georgia J. Sci.* 37:185-191.
172. Barrie, L.A. and J.M. Hales. 1984. The spatial distribution of precipitation acidity and major ion wet deposition in North America during 1980. *Tellus.* 36B:333-355.
173. Campbell, S., S. Kumar and R. Roig. 1987. Atmospheric transport, transformation, and deposition, pp. II-1 through II-42. *In:* R.A. Roig (Ed.) *Acid deposition in Maryland: a report to the Governor and General Assembly.* Maryland Power Plant Research Program, Annapolis, MD.
174. Harman, G.H. 1984. Assessment of Maryland's surface water sensitivity to atmospheric acid deposition, pp. 39-51. *In:* M.L. Bowman and S.S.G. Wierman (Eds.). *The potential effects of acid deposition in Maryland. Report of the Maryland interagency working group on acid deposition.*

Maryland Department of Natural Resources, Power Plant Research Program, Annapolis, MD.
175. Arnold, D.E., R.W. Light and E.A. Paul. 1985.Vulnerability of selected lakes and streams in the Middle Atlantic States to acidification: a regional survey. U.S. Fish and Wildlife Service, Eastern Energy and Land Use Team, Biol. Rep. 80 (40.19), 133 pp.
176. Janicki, A. 1987. Aquatic resources, pp. IV-1 through IV-14. *In:* R.A. Roig (Ed.) *Acid deposition in Maryland: a report to the Governor and General Assembly.* Maryland Power Plant Research Program, Annapolis, MD.
177. Bowman, M.L. 1984. Chesapeake Bay freshwater feeder stream chemistry, pp. 73-79. *In:* M.L. Bowman and S.S.G. Wierman (Eds.) *The potential effects of acid deposition in Maryland. Report of the Maryland interagency working group on acid deposition.* Maryland Department of Natural Resources, Power Plant Research Program, Annapolis, MD.
178. Petrimoulx, H., R. Keating, G. DeMuro and A. Janicki. 1987. Watersheds and their response to deposition, pp. III-1 through III-50. *In:* R.A. Roig (Ed.) *Acid deposition in Maryland: a report to the Governor and General Assembly.* Maryland Power Plant Research Program, Annapolis, MD.
179. Otton, E.G. 1970. Geologic and hydrologic factors bearing on subsurface storage of liquid wastes in Maryland. Maryland Geological Survey, Rept. 14:1-39.
180. Janicki, A. and H. Greening. 1986. A summary of Maryland stream chemistry data. Martin Marietta Environmental Systems, Columbia, MD.
181. Ecological Analysts. 1983. Survey of water quality and flow in Maryland coastal zone streams: a data report. PPSP-AD-6 (PPSP-MD-50). NTIS No. PB83-249375. Ecological Analysts, Inc., Sparks, MD.
182. Janicki, A.J., W.P. Saunders and E.A. Ross. 1986. A retrospective analysis of the frequency and magnitude of pH events in some Atlantic coast striped bass spawning grounds. Martin Marietta Environmental Systems, Columbia, MD.
183. Martin Marietta Environmental Systems. 1986. Coastal streams data analysis: phase I report. Tech. Mem. 3704-608. Martin Marietta Environmental Systems, Columbia, MD.
184. Bachman, L.J. and B.G. Katz. 1986. Relationship between precipitation quality, shallow groundwater geochemistry, and dissolved aluminum in eastern Maryland. PPSP-AD-14. U.S. Department of the Interior, Geological Survey, Washington, DC.
185. Roig, R.A. and M. Bowman. 1987. Introduction, summary, conclusions, and future actions, pp. I-1 through I-31. *In:* R.A. Roig (Ed.) *Acid deposition in Maryland: a report to the Governor and General Assembly.* Maryland Power Plant Research Program, Annapolis, MD.
186. Mehrle, P.M., D. Buckler, S. Finger and L. Ludke. 1984. Impact of contaminants on striped bass. Interim report. U.S. Fish and Wildlife Ser-

vice, Columbia, MD.
187. Miller, P.E. (Ed.). 1986. Power plant cumulative environmental impact report for Maryland. PPSP-CEIR-5. Department of Natural Resources, Power Plant Sitting Program, Annapolis, MD.
188. Taub, F.B. 1976. Demonstration of pollution effects in aquatic microcosms. *Int. J. Environ. Stud.* 10:23-33.
189. Wedemeyer, G.A., D.J. McLeay and C.P. Goodyear. 1984. Assessing the tolerance of fish and fish problems to environmental stress: the problems and methods of monitoring, pp. 163-165. *In:* V.W. Cairns, P.V. Hodson and J.O. Nriagu (Eds.) *Contaminant effects on fisheries.* John Wiley and Sons, Inc., New York, NY.
190. Vaughan, D.S., R.M. Yoshiyama, J.E. Breck and D.L. DeAngelis. 1984. Modeling approaches for assessing the effects of stress on fish populations, pp. 259-278. *In:* V.W. Cairns, P.V. Hodson and J.O. Nriagu (Eds.) *Contaminant effects on fisheries.* John Wiley & Sons, Inc., New York, NY.
191. Barnthouse, L.W., J. Boreman, S.W. Christensen, C.P. Goodyear, W. Van Winkle and D.S. Vaughan. 1984. Population biology in the courtroom: the Hudson River controversy. *BioScience* 34:14-19.
192. Van Winkle, W. 1977. Conclusions and recommendations for assessing the population-level effects of power plant exploitation: the optimist, the pessimist, and the realist, pp. 365-372. *In:* W. Van Winkle (Ed.) *Assessing the effects of power-plant-induced mortality on fish populations.* Pergamon Press, New York, NY.
193. Schaaf, W.E., D.S. Peters, D.S. Vaughan, L.C. Clements and C.W. Krouse. (In press). Fish population responses to chronic and acute pollution: the influence of life history strategies. *Proceedings, 8th Biennial International Estuarine Research Federation Conference.*
194. Summer, J.K., W.A. Richkus, R.N. Ross and R.E. Ulanowicz. 1982. A statistical model of white perch (*Morone americana*) harvest assessment in relation to environmental variation. *Fish. Res.* 1:289-298.
195. Anonymous. 1986. The consequences of contaminants to living marine resources and human health. FY 1986 programs description. Ocean Assessments Division, National Oceanic and Atmospheric Administration, Rockville, MD. 10 pp.
196. Polgar, T.T., J.K. Summers, R.A. Cummins, K.A. Rose and D.G. Heimbuch. 1985. Investigation of relationships among pollutant loadings and fish stock levels in northeastern estuaries. *Estuaries* 8:125-135.
197. Summers, J.K., T.T. Polgar, J.A. Tarr, K.A. Rose, D.G. Heimbuch, J. McCurley, R.A. Cummins, G.F. Johnson, K.T. Yetman, and G.T. DiNardo. 1985. Reconstruction of long-term time series for commercial fisheries abundance and estuarine pollution loadings. *Estuaries.* 8:114-124.

198. Summers, J.K., K.A. Rose, and T.T. Polgar. (In press). The role of interactions among environmental conditions in controlling historical fisheries variability. *Estuaries* 9:000-000.
199. Vaughan, D.S. and W. Van Winkle. 1982. Corrected analysis of the ability to detect reductions in year-class strength of the Hudson River white perch (*Morone americana*) population. *Can. J. Fish. Aquat. Sci.* 39:782-795.

Chapter Sixteen

CONTAMINANT EFFECTS ON CHESAPEAKE BAY SHELLFISH

MICHAEL E. BENDER and ROBERT J. HUGGETT

Virginia Institute of Marine Science
School of Marine Science
College of William and Mary
Gloucester Point, VA 23062

ABSTRACT

The paper reviews contaminant effects on Chesapeake Bay shellfish from two avenues (1) adverse biological effects on the organisms and (2) fisheries closures due to bacterial and chemical contamination. The use of shellfish to monitor anthropogenic inputs of chemical contaminants is also discussed. Fisheries closures due to bacterial contamination account for the greatest economic loss due to man's activities. Kepone contamination in the James River, Virginia caused fisheries closures but has not appeared to cause biological damage to the resources. Organotin compounds from antifouling paints appear to pose a threat to Chesapeake Bay shellfish.

INTRODUCTION

Shellfish resources can be damaged by contaminants (chemical and biological) in two ways. The first is most applicable to chemical contamination where concentrations of chemicals in water or sediments may reach levels that cause adverse biological effects on the organisms. This damage may be acute, causing death, or chronic, causing lowered rates of recruitment, growth, etc. The other avenue of impact is economic, an impact brought about by closures of fisheries because of chemical or microbial contamination. In the case of chemical contamination it must be pointed out that concentrations of toxic substances in animals which may cause fisheries closures are not necessarily the same as those which may cause biological effects on the animals. This point is illustrated in Figure 1 which shows the relationship between residues of Kepone in blue crabs, biological effects levels and closure levels. As can be seen in this figure the residue level at which the fishery is closed is 0.4 ppm while levels at which biological effects, such as carapace thinning do not occur until residues reach − 1.4 ppm. The reverse can be true also, i.e. effects levels expressed in terms of residue levels in the animals may be reached before residues climb to levels which are of public health concern.

Fisheries Closures — Bacteriological

Because certain species of shellfish, e.g. oysters and hard clams, are frequently consumed raw, growing areas are closed to direct marketing of the shellfish resource if they are contaminated by bacteria which can cause human diseases. The areal extent of these closures varies with (1) seasonal patterns of runoff which brings in potential disease causing organisms, (2) operational malfunctions at sewage treatment facilities, and (3) the proximity of the shellfish beds to marinas and other polluted areas. Bacterial monitoring programs to delineate areas of contamination are conducted routinely by the State Health Departments of Virginia and Maryland. Areas are opened and/or closed based on the results of these monitoring programs.

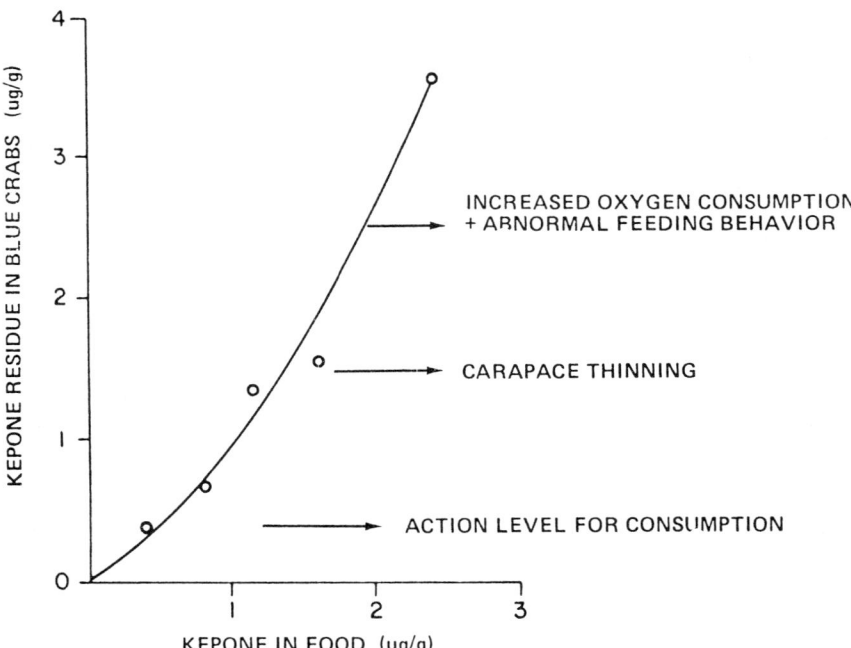

FIGURE 1. Kepone Residues (wet wt) in Blue Crabs vs. Effects Levels, data modified from Fisher, et al.[1]

Table 1 lists the acreages of oyster harvesting areas, public and leased, by basin; these oyster bars are shown in Figure 2. Areas closed to direct marketing because of bacterial contamination are listed in Table 2. The areas condemned during the summer of 1986 represent 35% of the total harvesting areas. Shellfish may be harvested from closed beds and transplanted to "clean" areas where they are allowed to depurate prior to reharvesting and sale. This process is expensive and often results in a loss of 20-30% of the animals which either die or cannot be recovered with normal harvesting techniques. New methods, e.g. the use of cages which are elevated off the bottom, are presently being evaluated for depuration of hard clams in Virginia.[3] This technique, although still in the experimental stage, appears to offer great promise in reducing mortality, increasing reharvesting efficiency and reducing time and expense.

Fisheries Closures — Chemical

Shellfish have the ability to concentrate hydrophobic chemicals orders of

TABLE 1

Acres of Public and Leased Oyster Grounds

Basin	Public Oyster Grounds	Leased Grounds	Total
Chesapeake Bay North	0	21	21
Chesapeake Bay Upper Central	19,038	0	19,038
Chester River	5,547	0	5,547
Eastern Bay	26,979	212	27,191
Choptank River	1,378	454	1,832
Chesapeake Bay Lower Central	29,173	778	29,951
Patuxent River	7,543	1,119	8,662
Honga River	15,475	1	15,476
Fishing Bay	11,811	333	12,144
Nanticoke River	577	190	767
Wicomico River	568	1,268	1,836
Chesapeake Bay South	32,315	0	32,315
Tangier Sound	31,043	889	31,932
Pocomoke Sound	4,899	4,303	9,202
Potomac River	28,523	9,389	37,912
Rappahannock River	44,254	19,022	63,276
Piankatank River	16,000	328	16,328
Chesapeake Bay General	35,566	20,170	55,736
Mobjack Bay	17,061	1,516	18,577
York River	2,381	26,729	29,110
Mattaponi River	0	0	0
Pamunkey River	0	0	0
Chickahominy River	0	0	0
James River	25,152	13,260	38,412
TOTAL	355,283	99,982	455,265

Source[2]

magnitude higher than those found in the aqueous phase. For those chemicals which pose a potential threat to human health, (e.g. chlorinated pesticides, PCBs, and certain dioxins), the Food and Drug Administration and/or the U.S.E.P.A. (United States Environmental Protection Agency) establishes limits above which

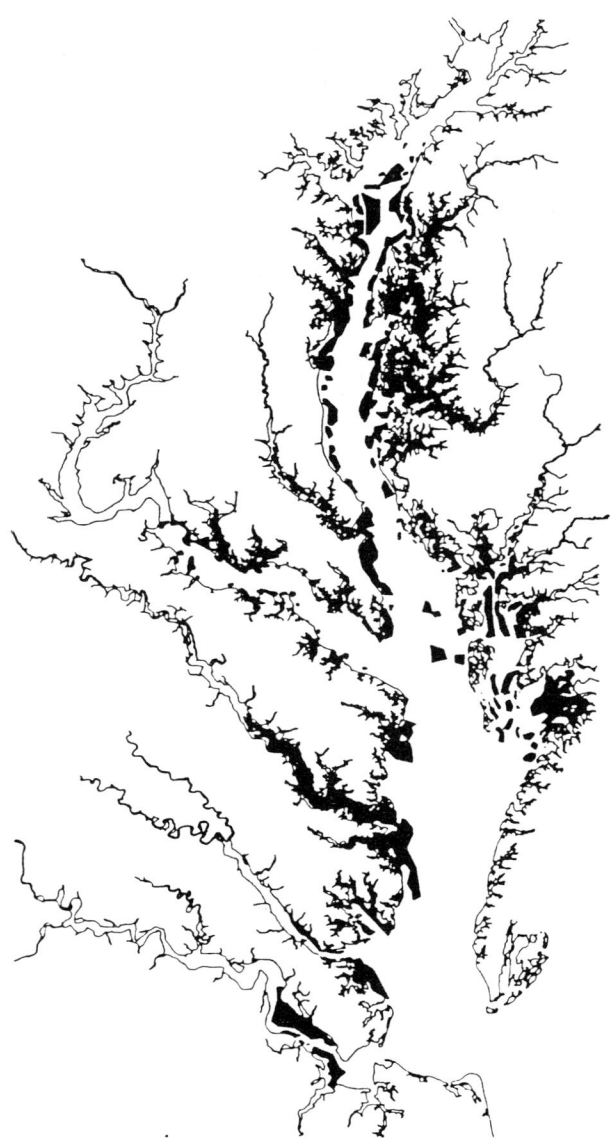

FIGURE 2. Chesapeake Bay Oyster Bars shown in black.[2]

interstate transport of the food item is restricted. State health departments frequently adopt these limits and impose closures for commercial and/or recreational harvest of species when necessary. Establishment of limits and subsequent terminology varies with, (1) the geographical extent of the contamination, (2) the type of food items contaminated, (3) the amount of the particular food item consumed by the average citizen and (4) the mammalian toxicity of the compound. Frequently additional warnings on limiting consumption are issued for groups of people considered special risks (e.g. pregnant women and children).

In the lower Chesapeake Bay, shellfish closures due to chemical contamination have been limited to those due to Kepone in the James River. Soon after the 1975 discovery of Kepone contamination in the James River, the oyster and crab fisheries were closed to commercial harvesting. The oyster fishery was reopened in 1976 when it was found that seed oysters, the major oyster resource

TABLE 2

Acres of Condemned Shellfish Areas
Maryland and Virginia

Basin	Acres Condemned[a]
Chesapeake Bay North	0
Chesapeake Bay Upper Central	17,600
Chester River	9,330
Eastern Bay	14,560
Choptank River	5,330
Chesapeake Bay Lower Central	1,000
Patuxent River	14,660
Honga River	350
Fishing Bay	0
Nanticoke River	300
Wicomico River	4,000
Chesapeake Bay South VA	915
Tangier Sound	1,098
Pocomoke Sound	1,485
Potomac River, MD	2,660
Potomac River, VA Tributaries	5,395
Rappahannock River	7,105
Piankatank River	700
Chesapeake Bay General, VA	7,040
Mobjack Bay + Tributaries	827
York River + Tributaries	9,105
Mattaponi River	0
Pamunkey River	0
Chickahominy River	0
James River + Tributaries	53,945[b]
Total	157,405

[a] productive areas only
[b] includes 35,509 acres in Hampton Roads, most of which are too deep to allow oyster harvest

in the river, rapidly depurated Kepone body burdens when transplanted to clean growing areas. The blue crab fishery was affected longer with closures remaining in effect for 4 years. The declining residues in crabs from 1976 through 1985 are presented in Figure 3. Residues in female and male crabs differed dramatically in the early years. Roberts and Leggett[4] concluded the loss of Kepone in the egg masses when female crabs spawn was in part responsible for the differing body burdens in males and females.

Figure 4 depicts the rate decline in Kepone residues for male crabs and oysters in the lower James River. These data are of interest because they show similar rates of decrease with time for these two species, yet it has been shown that crabs obtain most of their residues from food[5] while Kepone appears to be available to oysters from both solution and suspended particles.[6]

Economic losses due to shellfisheries closures in the James during 1976 were estimated at $50,000 for the oyster fishery and $67,000 for the blue crab fishery.[7] Losses due to limiting the harvest of crabs continued for another 3 years; however, the extent of economic loss is difficult to estimate because many fishermen moved their operations to other waters or obtained other employment.

Effects of Chemical Contamination — The potential effects of toxic chemicals

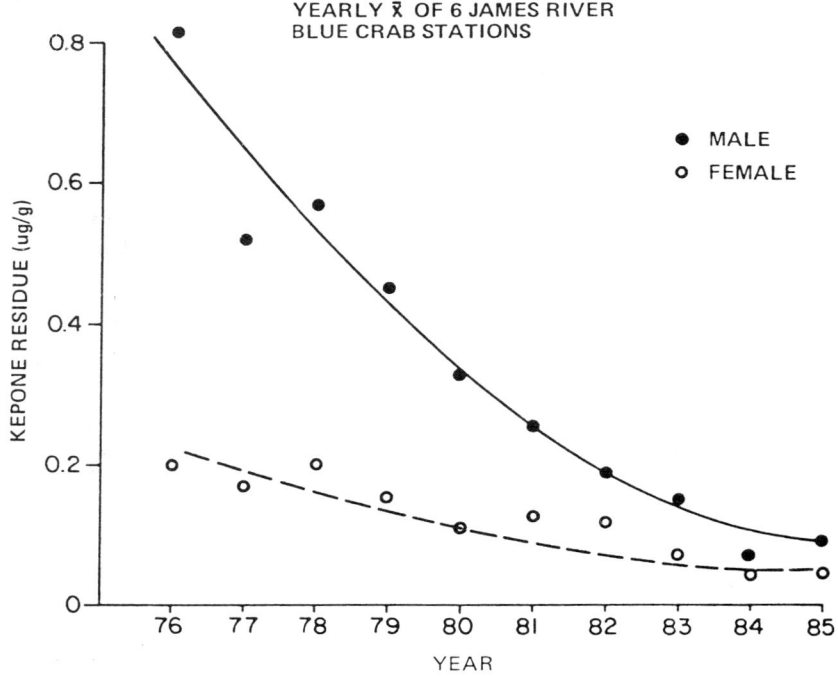

FIGURE 3. Kepone Residues (wet wt) in Male and Female Blue Crabs, vs. Time.

on aquatic animals are usually estimated by conducting laboratory bioassays, the results of which are then related to expected and/or measured environmental concentrations. Acute, partial chronic, and chronic toxicity tests have all been utilized to estimate effects. For shellfish, both molluscs and crustaceans, the larval stages are usually the most sensitive. In this volume, Roberts and Bradley review toxicity data for zooplankton, including larval stages of shellfish. We therefore will limit our discussions to effects on adults, except in the case of Kepone for which, to be complete, we have included the larval data.

KEPONE

After the discovery of Kepone in the James River, numerous studies were conducted to estimate its impact on the biota of the river. The objective of most tests was to estimate those concentrations which would have no deleterious effects on the animals. Once no-effect levels had been determined or estimated,

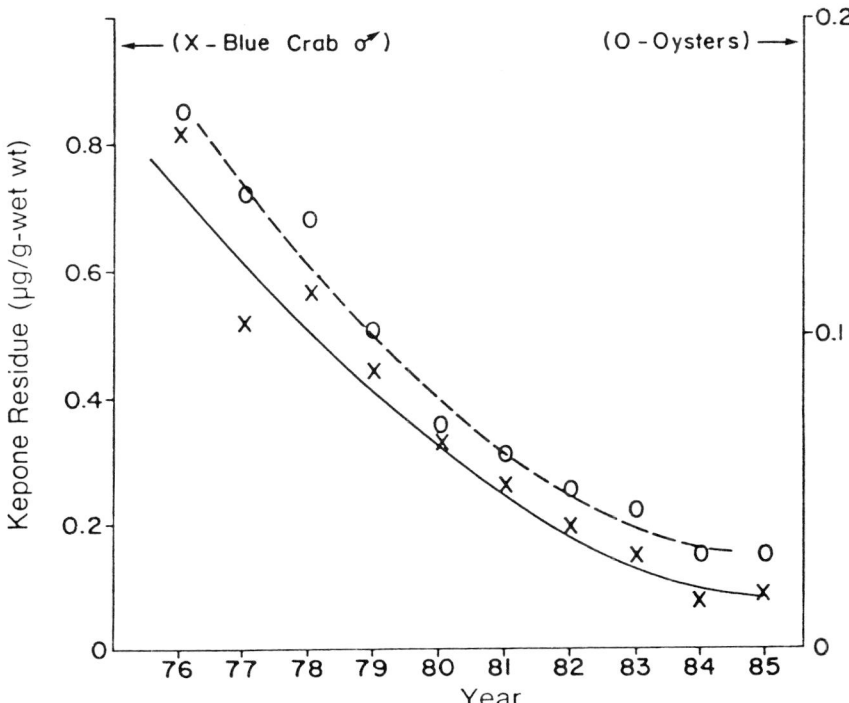

Figure 4. Yearly Mean Kepone Residues (wet wt) in Oysters and Male Blue Crabs vs. Time.

they could be compared with measured levels of Kepone in the river and estimates of the potential for effects could be made.

Acute, partial chronic and chronic toxicity tests have all been utilized to estimate Kepone effects. The ideal procedure to establish safe exposure concentrations involves chronic toxicity testing of the chemical on several different freshwater and/or marine species. Once these tests have been conducted, an application factor can be derived and used to estimate safe chronic exposure concentrations for other organisms from acute toxicity data. The application factor is defined as the ratio of the maximum acceptable toxicant concentration (MATC) to the 96 hour LC_{50}.

Results of acute Kepone toxicity tests conducted on some marine and estuarine animals are summarized in Table 3. Mysid shrimp had an LC_{50} of 10 μg/l while shell growth of oysters was inhibited by 12 μg/l. Dissolved Kepone levels measured in the saline portion of the James since 1979 have ranged between 0.8 and 7 ng/l or 3 orders of magnitude less than those concentrations acutely toxic to the most sensitive shellfish species.

Chronic toxicity studies have been conducted on a variety of marine, estuarine and freshwater animals by several techniques. In three cases natural populations of contaminated animals, obtained from the James River, were brought into the laboratory and evaluated for spawning ability and larval survival. Other studies involved exposure of animals in the laboratory to Kepone, either in food or water, while measuring various parameters. An attempt has been made to summarize the results of these studies in Table 4. Each will be discussed below.

TABLE 3

Acute Toxicity of Kepone to Some Marine and Estuarine Animals.

Species	96 hr $LC_{50}(\mu g/l)$	Reference
Eastern oyster		
Crassostrea virginica) larvae	66	Hansen et al.[8]
adults[a]	12	Butler[9]
Brown shrimp	28	Butler[9]
(*Penaeus aztecus*)		
Mysid shrimp	10	Nimmo et al.[10]
(*Mysidopsis bahia*)		
Grass shrimp	120	Schimmel & Wilson[5]
(*Palaemonetes pugio*)		
Sand shrimp	263	Hixon[11]
(*Crangon septemspinosa*)		
Blue crab	>210	Schimmel & Wilson[5]
(*Callinectes sapidus*)		
Mud crab	>35	Bookhout et al.[12]
(*Rhithropanopeus harrisii*)		

[a]Shell growth

In three experiments, oysters, blue crabs and grass shrimp, were collected from locations with different degrees of Kepone contamination and tested for spawning success. A strength of this type of study is that it evaluates animals which have obtained Kepone residues by their natural routes and have, in addition, been subjected to other stresses in the environment. However, since the experimenter has no control of exposure conditions, environmental levels responsible for effects may not be identified.

Oysters were obtained from the James River during July of 1976 and spawned in the laboratory within two days of collection.[13] As measures of the effects of Kepone contamination, egg production, i.e., numbers of eggs produced, larval abnormalities, and setting success were compared to those parameters for control animals from the York River. Kepone contaminated oysters (0.3 μg/g) produced 11 million viable eggs of which 9 million developed successfully to straight hinge larvae and set. Compared to control animals, no increase in larval abnormalities was detected and the development duration was normal. Residue

TABLE 4

Chronic Effects of Kepone on Some Marine Animals.

Species	Parameters Measured	Effect Level	No Effect Level	Reference
Eastern Oyster	Field Exposure			
	Egg production	ND[a]	0.3 μg/g residue	Bender & Huggett[13]
	Larval set	ND	0.3 μg/g residue	
Blue Crab	Field Exposure			
	Egg hatchability	ND	1.0 μg/g residue	Leggett[14]
	Larval survival	ND	1.0 μg/g residue	Leggett[14]
	Laboratory Exposure (food)			
	LC_{50} (65 day)	>2.5 μg/g	2.5 μg/g	Fisher[1]
	Growth	>2.5 μg/g	2.5 μg/g	Fisher[1]
	Carapace thickness	1.2 μg/g	0.8 μg/g	Fisher[1]
	Behavior	2.2 μg/g	1.2 μg/g	Fisher[1]
	Laboratory Exposure (water)			
	Larval survival	1.0 μg/g	0.1 μg/g	Bookhout et al.[12]
Mysid Shrimp	Laboratory Exposure (water)			
	LC_{50} (life cycle)	1.4 μg/l	0.4 μg/l	Nimmo et al.[10]
	Larval production	0.4 μg/l	ND	Nimmo et al.[10]
Grass shrimp	Field Exposure			
	Larval hatchability and survival	ND	0.6 μg/g residue	Provenzano et al.[15]
	Larval growth	ND	0.6 μg/g residue	

[a]ND = Not Determined
[b]Egg production greater, so effects not considered significant

levels in James River oysters collected from 10 locations have been monitored by the State Health Department monthly since November of 1975. Of the 120 samples collected in 1976 only 10% exceeded the 0.3 μg/g residue found acceptable in our experiments and in 1977 only 1 of 110 samples exceeded this level. As shown in Table 3, the acutely toxic concentration of Kepone to oysters was 12 μg/l for adults and 66 μg/l for larvae compared to dissolved Kepone levels in the lower saline portion of between 0.8 to 7 ng/l. Based on data from these two experiments it appears highly unlikely that Kepone residues either in the animals or in river water are detrimental to oysters in the James River.

Grass shrimp (*Palaemonetes pugio*), an important member of the food chain in the Chesapeake Bay, were tested in a similar experiment.[15] Shrimp were collected from 6 locations and egg hatchability, larval survival and larval growth were measured as a function of location (degree of Kepone contamination). Larvae hatched from females having Kepone residues of 0.6 μg/g with equal success to those from the control groups. In addition, no effects on development or survival were noted. Acute toxicity of Kepone to this species (Table 3) was estimated to be 120 μg/l.

To investigate the potential effects of Kepone on blue crabs, Leggett[14] studied blue crabs collected from 7 locations, 2 in the lower James and 5 in the lower Bay over a three month period during the summer of 1978. The hatchability and larval of several hundred eggs from each crab was determined and related to degree of Kepone contamination. Over the range of Kepone concentrations in contaminated eggs, i.e., from non-detectable to 1.45 μg/g, no effects of contamination could be demonstrated on embryogenesis, hatchability or larval survival.

Fisher, et al.[1] studied the long-term effects of Kepone exposure to juvenile blue crabs by feeding them with a series of concentrations in naturally contaminated striped bass, *Morone saxatilis,* flesh. Besides mortality, several sublethal effects were measured. Kepone uptake by crabs was linearly related to exposure concentration and reached a maximum of 4.6 μg/g in the first experiment (exposure to 2.5 μg/g). The average number of molt, percent increase in width, mid-body thickness and wet weight per molt did not differ significantly from controls at any Kepone concentration tested in either experiment. At the highest Kepone exposure in each experiment (2.5 and 2.3 μg/g) oxygen consumption was greater than at the other exposure levels and these crabs exhibited "excitable feeding behavior." Also, at high exposure levels (> 1.2 μg/g) in the first experiment, crabs which molted had low carapace thickness to width ratios. This effect was not observed in the second experiment. These sublethal effects occurred at tissue levels greater than the average tissue levels found in adult crabs from the James River (Figure 1).

Chronic toxicity of Kepone to mysid shrimp *Mysidopsis bahia,* an estuarine species native to the Gulf states, was studied by Nimmo *et al.*[10] They determined effects by measuring survival, egg production and larval growth after aqueous exposures. No mortalities were observed among shrimp exposed to Kepone concentrations of 0.4 µg/l, but egg production was reduced compared to that of control populations. Some reduction in growth of young was observed, but the results were erratic. They found growth reductions of 6% at exposure levels of 0.07 µg/l and only 3% at 0.23 µg/l. Although *Mysidopsis bahia* is not native to Chesapeake Bay, a related species, *Noemysis americana*, is resident in the Bay and its tributaries. Roberts et al.[16] compared the response of *M. bahia* and *N. americana* to three toxicants (cadmium, sodium lauryl sulfate and Lannate) and found very similar lethal concentrations. Similar sensitivities may hold for Kepone and therefore, we would not predict effects at environmental exposure levels.

ORGANOTINS

In recent years the potential impact of tributyltin (TBT) in Chesapeake Bay has surfaced as a major environmental issue. The following factors are responsible for this concern: (1) the increased use of TBT based paints as antifouling agents on pleasure craft; (2) the recent proposal by the U.S. Navy to utilize TBT on all Navy vessels;[17] and (3) laboratory and field studies in England, France and the U.S., which have implicated TBT in causing abnormalities and/or mortalities in a number of species of shellfish.[18,19,20,21]

Space limitations preclude a complete review of available literature on TBT; however, we have attempted to provide a brief summary of some relevant literature and our assessment of some of the more pressing research needs.

Recent studies in England and France, summarized by Stebbing,[18] have implicated tributyltin in causing decreased spatfall, decreased growth and shell malformations in oysters (*Crassostrea gigas)*). Thain and Waldick[22] showed that a low concentration of tributyltin oxide (TBTO), 0.15 µg/l, inhibited growth of young oysters (*C. gigas*). Thain and Waldock[23] found that the growth of European oyster spat (*Ostrea edulis*) was severely curtailed after 10 days exposure to 0.06 µg/l of TBT. Henderson[24] reported a mortality rate for the American oyster (*Crassostrea virginica*) of 50 percent after 30 days exposure to 2.5 µg/l of TBT. In the same experiment he determined that the oyster's condition index was reduced by exposure to 0.1 µg/l over a period of 57 days.

Stephenson, et al.[21] transplanted oysters (*Crassostrea gigas*), and two species of mussels (*Mytilus edulis*) and *(M. californianus)* in San Diego Bay along a gradient of known seawater TBT concentrations. Reduced shell growth in all three species was found at the stations with the highest levels of TBT. Beaumont and Budd[25] reported about 50% mortality of mussel larvae (*Mytilus edulis*) after 15 days exposure to TBTO concentrations of 0.1 µg/l. For adult mussels of the same species, 96 hr LC_{50} values of 20-60 µg/l have been reported.[26]

Smith[27] found strong evidence that exposure of American mud snails (*Nassarius obsoletus*) to TBT caused a phenomenon known as "imposex" (the superimposition of male sex characters onto the female). He concluded, however, that for the mud snail such effects produced no significant decrease in reproductive capacity. Gibbs and Bryan[19] described this phenomenon in the dog-whelk (*Nucella lapillus*) and also related its development to TBT exposure. In the case of the dog-whelk, however, they presented convincing evidence that exposures to TBT caused sterility and reproductive failure of the populations. The specific exposure concentrations necessary to induce full imposex development in *N. lapillus* remain to be determined. However, the authors found that exposure of dog-whelks to 0.02 μg/l of TBT for a period of 6 months induced the phenomenon to progress from early to late stages.

The above studies have shown that TBT is quite toxic to a variety of shellfish species. Huggett et al.[28] have shown that potentially toxic concentrations of TBT can exist in marinas in the southern Chesapeake Bay, and Hall et al.[29] found TBT concentrations in marina areas in the upper Chesapeake Bay, which would be toxic to sensitive aquatic animals. However, before the true magnitude of the problem can be determined, we must establish, through long-term exposures, the TBT concentrations which are non-toxic to oysters, clams, and other important shellfish. These studies at a minimum should include an evaluation of the effects of TBT on gametogenesis, larval survival, spat growth and the potential for imposition of "imposex" on certain species. At the present time some of these studies are being conducted.

SHELLFISH AS INDICATORS OF POLLUTION

Animals vary considerably in their ability to accumulate, depurate and metabolize both naturally occurring and xenobiotic chemicals. Chemicals may be taken up from the water across gill membranes, other exposed external body surfaces and/or from contaminated food. The relative importance of the three routes of uptake for aquatic species is often debated and is probably specific for each animal species and class of chemical substance, e.g. metallic ions, polar organics, etc.

Factors which make a given animal species well suited as an indicator of bioavailability of anthropogenic substances in the environment have been identified by various researchers.[30,31,32,33,34] In brief, these factors for oysters and clams are: (1) the pollutants are often accumulated without mortality; (2) the animals are sedentary in life habit; (3) they are often abundant; (4) they are relatively long-lived, (5) they are easily collected; (6) they are adaptable to laboratory studies, so that experimental work can be performed; (7) they usually have a high BCF (bioconcentration factor) for the pollutant of interest; (8) they usually attain a residue which is correlated with the concentration in the environment and (9) they have a limited ability to metabolize the substance.

Many bivalve species have most if not all of the above characteristics. However, monitoring of contamination by various pollutants in estuaries by the use of bivalves is complicated by the necessity to use different species as one progresses upstream along the salinity gradient. In the lower Chesapeake Bay, most tributary sub-estuaries contain three or four bivalve species, the oyster, (*Crassostrea virginica*), the mussel, (*Mytilus edulis*), the hard clam, (*Mercenaria mercenaria*) and the brackish water clam (*Rangia cuneata*). In this section of the chapter we describe the use of two of these species in detecting anthropogenic

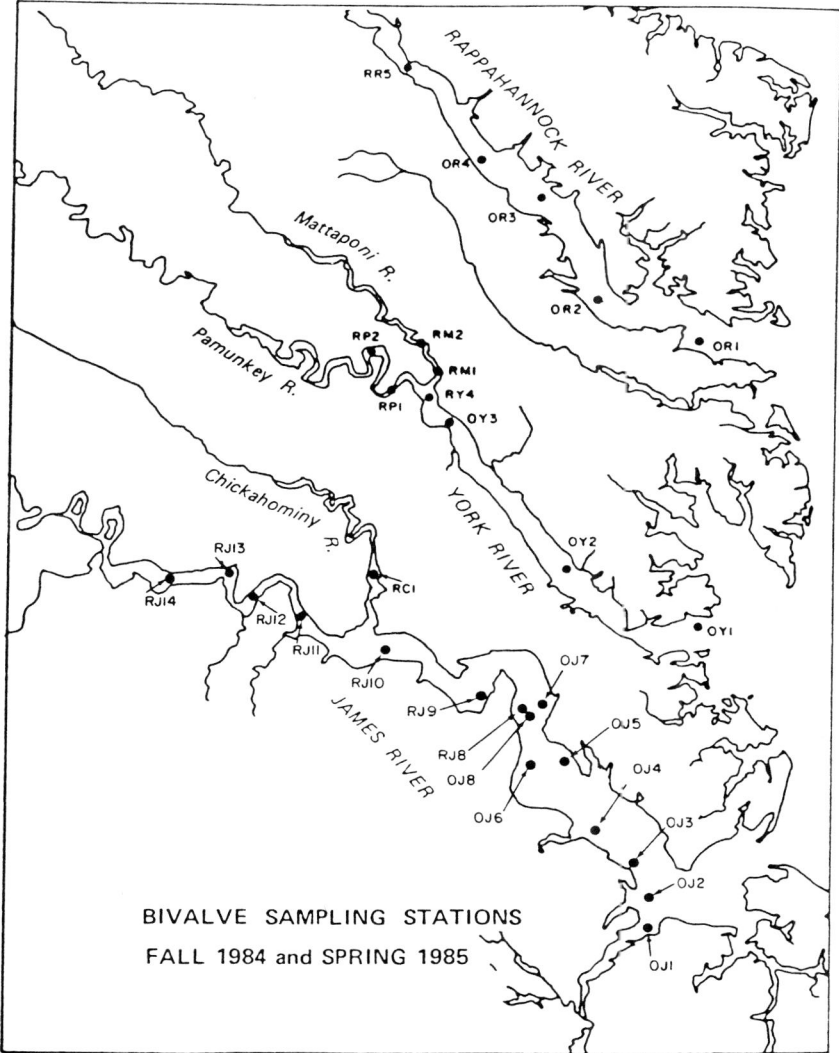

FIGURE 5. Bivalve Sampling Stations for Polynuclear Aromatic Hydrocarbons.

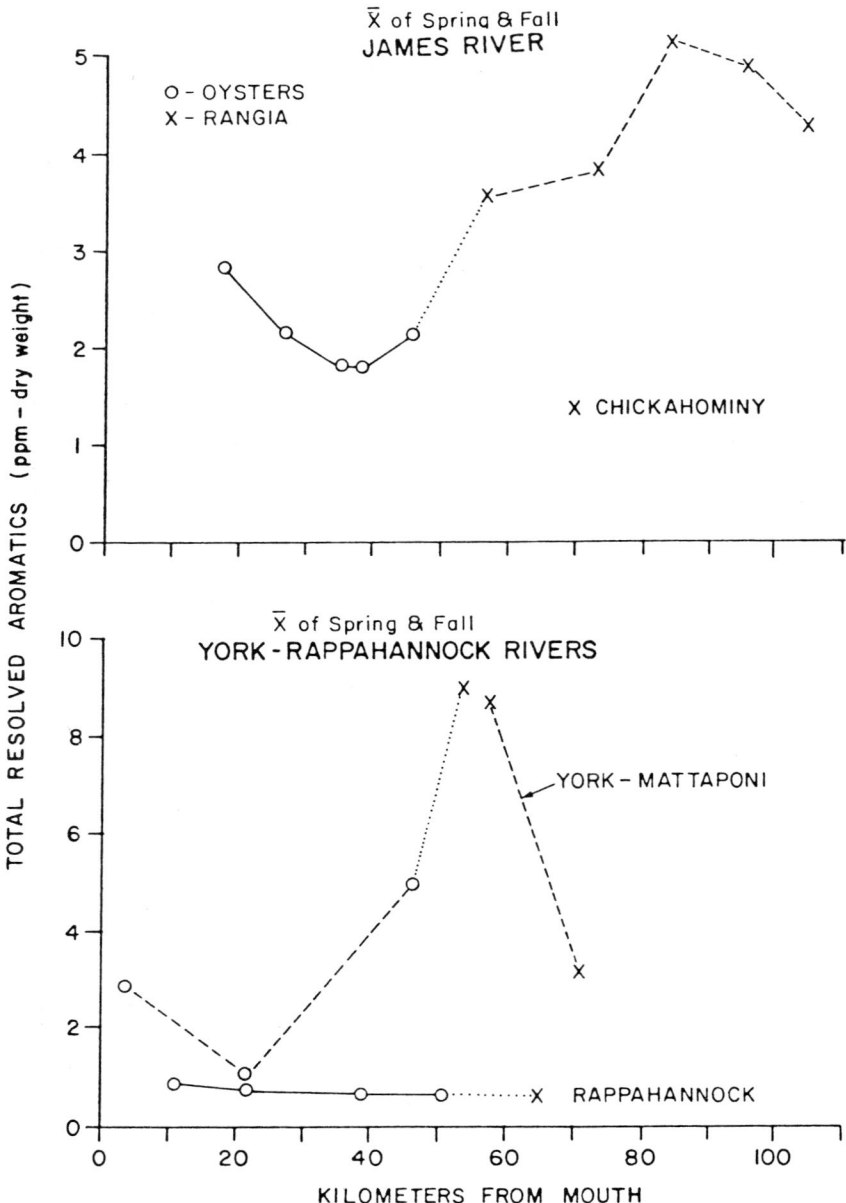

FIGURE 6. Total Resolved Aromatic Hydrocarbons (dry wt) in Oysters and *Rangia* vs. Distance in the James, York and Rappahannock Rivers.

inputs of polynuclear aromatic hydrocarbons (PAHs) in sub-estuaries of lower Chesapeake Bay.

PAHs are widespread contaminants of freshwater and estuarine systems and have been implicated in causing effects on fishes and shellfish in the Niagara River,[35] Oregon Bays[36] and Puget Sound.[36]

Recent surveys of PAH contamination in Virginia's major river systems (see Figure 5 for station locations) indicate high residues in shellfish collected from estuaries draining industrialized or highly populated basins. Figure 6 shows the mean residues of total resolved PAHs in oysters (*Crassostrea virginica*) and brackish water clams (*Rangia cuneata*) along the James, York, and Rappahannock rivers in the fall of 1984 and spring of 1985. In the James River, residues of total PAHs in oysters declined with increasing distance from the river mouth while residues in clams increased in an upstream direction. Residues in *Rangia* collected from the Chickahominy River (an undeveloped tributary of the James) were considerably lower than those in *Rangia* from the James River stations.

In the York River, concentrations of total aromatics were dramatically higher than elsewhere at the most upstream oyster rock sampled, and clams collected from just below West Point had the highest residues observed anywhere during

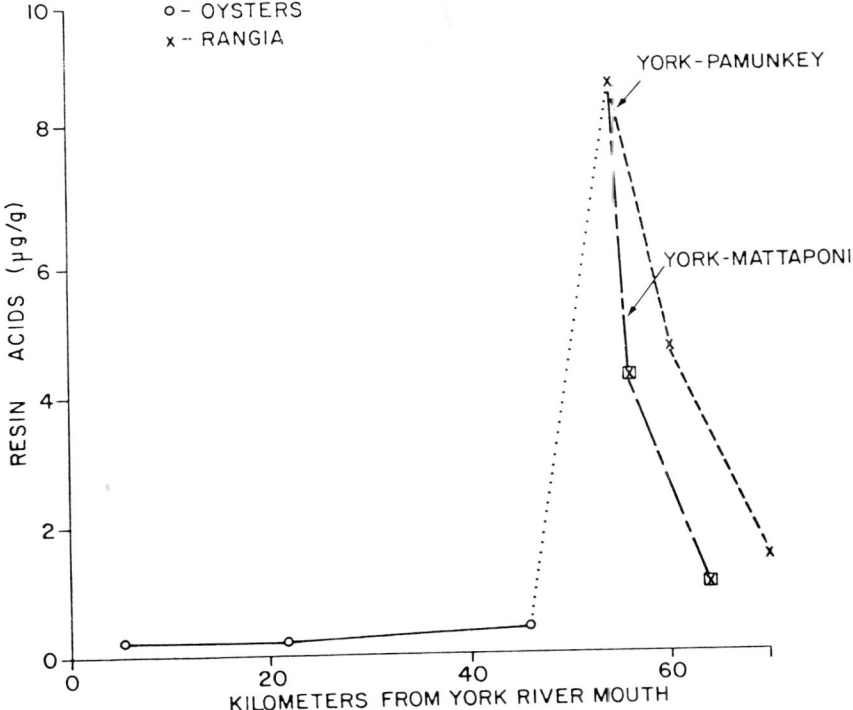

FIGURE 7. Resin Acids (dry wt) in Oysters and *Rangia*.

the survey (Figure 6). A detailed examination of clam samples from the York, Pamunkey and Mattaponi rivers indicated that compounds derived from resin acids of plants accounted for a significant proportion of the resolved aromatics in these samples. The concentrations of the "resin acid derived compounds"

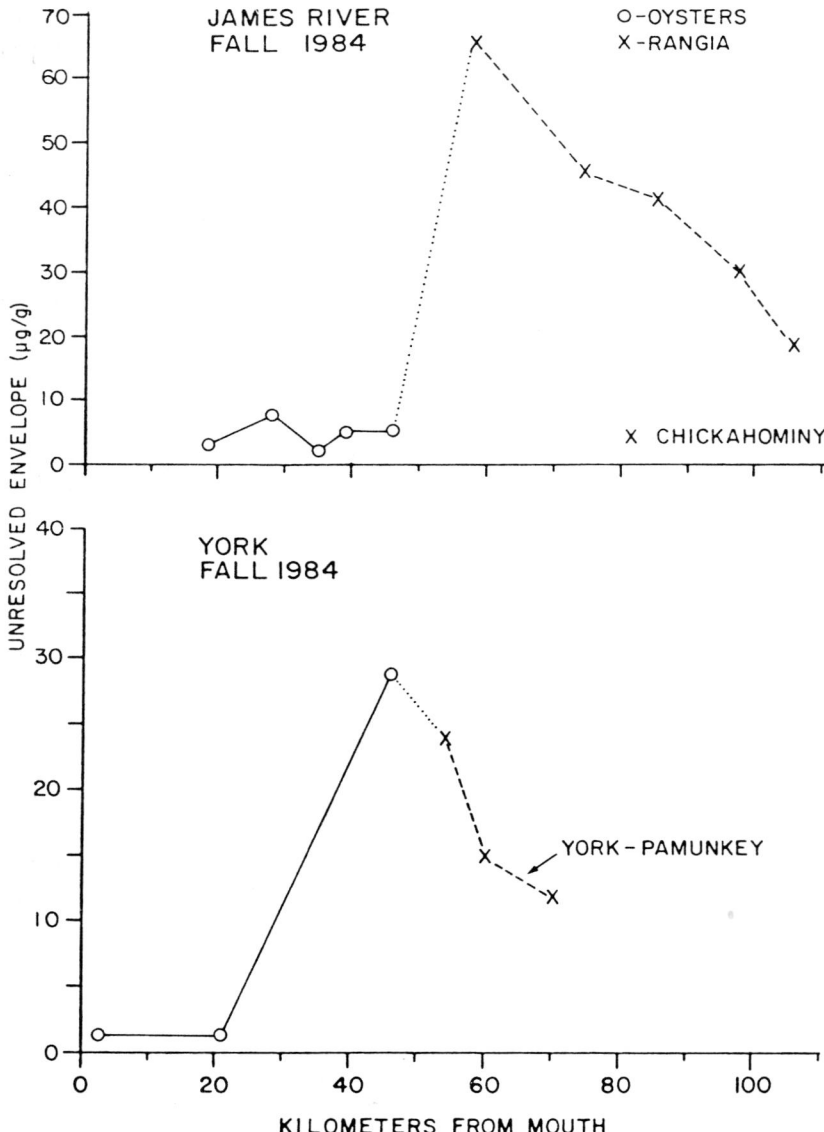

FIGURE 8. Unresolved Envelope (dry wt) in Oysters and *Rangia* in the James and York Rivers.

in the York, Pamunkey and Mattaponi rivers are shown in Figure 7.

Concentrations of hydrocarbons in the unresolved envelopes (mixtures of degraded and undegrated aromatic hydrocarbons) from the fall 1984 samples are shown in Figure 8. Oysters and clams collected from the Rappahannock showed no evidence of unresolved envelopes (UCMs). In both the York and James rivers, substantial increases in the UCM were observed in both oysters and clams collected near the turbidity maximum zone. The lack of a UCM in the Rappahannock samples and the relatively low concentration observed in the Chickahominy samples suggest anthropogenic origins for the envelopes.

At present we have no conclusive evidence to indicate that shellfish populations which show high PAH residues are adversely affected. It should be noted, however, that *Rangia* populations in the upper York and the lower Mattaponi and Pamunkey rivers are very small compared to those in the James and Rappahannock rivers. In addition, clams from these areas generally appear to be in poor condition, i.e. they have lower dry weight to wet weight ratios than clams from other river systems.

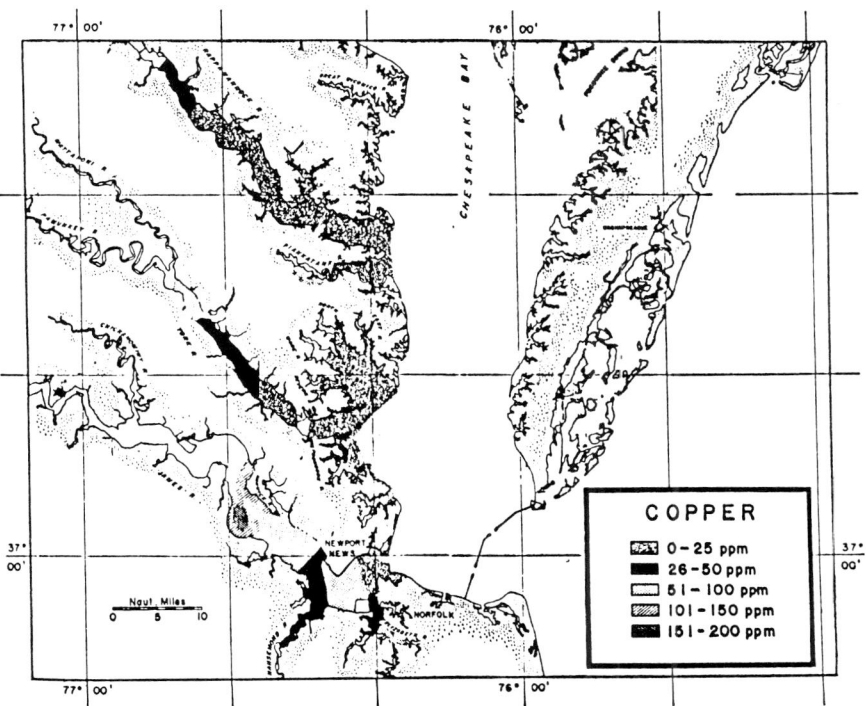

FIGURE 9. Copper Residues in Oysters from Southern Chesapeake Bay from Huggett, et al.[32]

TRACE METALS

Huggett, et al.[33] demonstrated that residues of certain heavy metals (Cd, Cu and Zn) in oysters (*Crassostrea virginica*) were a function of not only source but also the animal's position in the estuary. Figure 9 shows the distribution of copper in oysters from the James, York and Rappahannock rivers. In systems relatively less affected by anthropogenic inputs, e.g. the York and Rappahannock, concentrations are high in the upstream low salinity regimes of those estuaries. Similar distributions were observed for cadmium and zinc. The authors developed a method utilizing ratios of residues between Cu and Zn which allowed for determination of whether the body burdens were derived from natural or man-made sources.

The distribution of the metals Cr, As, Pb, Hg, Zn, Cu, and Cd in oyster tissues from the upper Chesapeake Bay and portions of the lower Bay were summarized from a number of studies in EPA's Chesapeake Bay Program report.[2] They concluded in part, (1) that certain metals, e.g. Cu, Cd, and Zn were high near urbanized areas and (2) that metal contamination levels in shellfish tissue did not violate FDA action levels.

DISCUSSION

In this paper we have limited our coverage to those subjects which we believe are of most importance to Chesapeake Bay shellfish. Chlorine, which has the potential for causing significant damage to oyster and clam resources, particularly through its effect on early life stages, was not included because Roberts and Bradley discuss this issue in Chapter 14.

There is no doubt that shellfish in parts of the Chesapeake Bay have been and are being affected by anthropogenic inputs both bacterial and chemical. The areas most effected are those near centers of urban and industrial development. Shellfishery closures due to bacterial contamination account for the greatest identifiable economic loss due to man's activities.

In general, the authors believe that the Bay is far from being overwhelmed by toxic substances. New standards and regulations and a new awareness of the Chesapeake Bay's environmental problems will hopefully result in an improved situation. There is no doubt that as scientists perform more monitoring and as new analytical and biological technologies emerge, new problems will be uncovered. This, however, should be viewed as a positive development because without continued vigilance, even a system as large as the Chesapeake Bay can be harmed.

ACKNOWLEDGEMENTS

The authors wish to thank Drs. W.J. Hargis and Morris H. Roberts, Jr. for reviewing the paper and Mrs. S. Sterling for preparing the manuscript. Financial assistance was provided in part by the Virginia State Water Control Board. Mr. Cloyde Wiley provided the data on shellfish closures in Virginia and Mr. Paul DiStefano supplied the information for Maryland.

Figure 9 is reprinted with permission from Water Research, No. 7. R. J. Huggett, M. E. Bender & H. D. Slone, Utilizing metal concentration relationships in the eastern oyster (*Crassostrea virginica*) to detect heavy metal pollution. 1973, Pergamon Press, Ltd.

REFERENCES

1. Fisher, D.J., M.E. Bender and M.H. Roberts, Jr. 1983. Effects of ingestion of Kepone-contaminated food by juvenile blue crabs (*Callinectes sapidus* Rathbun). Aquatic Toxicology 4:219-234.
2. U.S.E.P.A. 1983. Chesapeake Bay: A profile of environmental change. Region 3 Philadelphia, PA 19106. Appendix 1.
3. DuPaul, W. 1986. Personal Communication. Virginia Institute of Marine Science, Gloucester Point, VA 23062
4. Roberts, M.H., Jr. and A.T. Leggett, Jr. 1980. Egg extrusion as a Kepone-clearance route in the blue crab, *Callinectes sapidus*. Estuaries 3:192-199.
5. Schimmel, S.C. and A.S. Wilson. 1977. Acute toxicity of Kepone to four estuarine animals. Chesapeake Science 18:224-227.
6. Morales-Alamo, R. and D.S. Haven. 1983. Uptake of Kepone from sediment suspensions and subsequent loss by the oyster *Crassostrea virginica*. Marine Biology 74:187-201.
7. Stukenbroeker, G. 1985. Fishermen bore brunt of economic hardship. Kepone Enigma of the James. Daily Press, 22 July, Hampton, Va.
8. Hansen, D.J., D.W.R. Nimmo, S.C. Schimmel, G.E. Walsh and A.S. Wilson. 1977. Effects of Kepone on estuarine organisms, in R.A. Tubb (Ed.), Recent Advances in Fish Toxicology: A Symposium, EPA Ecological Research Series EPA-600/3-77-85.
9. Butler, P.A. 1963. Pesticide wildlife studies—a review of fish and wildlife investigations during 1961 and 1962. U.S. Fish and Wildlife Cir. 167:11-25.
10. Nimmo, D.W.R., L.H. Bahner, R.A. Rifby, J.M. Sheppard and A.J. Wilson. 1977. *Mysidopsis bahia:* an estuarine species suitable for life cycle toxicity tests to determine effects of a pollutant, in F.L. Mayer and J.J. Hamelenk (Eds.), Aquatic Toxicology and Hazard Evaluation, Amer. Soc. for Testing Materials STP 634. pp. 109-116.
11. Hixon, D.J. 1980. The determination of the acute toxicity, uptake and elimination rates for Kepone by *Crangon septemspinosa* Say. M.S. Thesis, College of William and Mary, Williamsburg, Virginia.
12. Bookhout, C.G., J.E. Costlow and R. Monroe. 1979. Kepone effects on development of *Callinectes sapidus* and *Rhithropanopeus*. U.S. EPA 600/3-79-104.

13. Bender, M.E. and R.J. Huggett. 1984. Fate and effects of Kepone in the James River estuary. In: *Reviews in Environmental Toxicology*. E. Hodgson (Ed.), Elsevier Science Publishers. New York, NY. pp. 5-50.
14. Leggett, A.T., Jr. 1979. The development of blue crabs, *Callinectes sapidus*, from Kepone contaminated eggs. M.S. Thesis, College of William and Mary, Williamsburg, Virginia.
15. Provenzano, A.J., K.B. Schmitz and M.A. Boston. 1978. Survival, duration of larval stages and size of postlarvae of grass shrimp, *Palaemonetes pugio*, reared from Kepone contaminated and uncontaminated populations in Chesapeake Bay. Estuaries 1:239-244.
16. Roberts, M.H., Jr., J.E. Warinner, Chu-Fa Tsai, D. Wright and L.E. Cronin. 1982. Comparison of estuarine species sensitivities to three toxicants. Arch. Environ. Contam. Toxicol. 11:681-692.
17. Federal Register. 1985. *Vol. 50*, Issue 120, p. 25748 of June 21.
18. Stebbing, A.R.D. 1985. organotins ans water quality-some lessons to be learned. *Mar. Poll. Bull.* 16:383-390.
19. Gibbs, P.E. and G.W. Bryan. 1986. Reproductive failure in populations of the dog-whelk, *Nucella lapillus*, caused by imposex induced by tributyltin from antifouling paints. J. Mar. Biol. Assoc. U.K. 66, 767-777.
20. Waldock, M.J. 1986. TBT in U.K. estuaries, 1982-86. Evaluation of the environmental problem. IEEE Oceans '86 Conf. Proceedings, Wash. D.C. Vol. 4 pp. 1324-1330.
21. Stephenson, M.D., D.R. Smith, J. Goetzl, G. Schikawa and M. Martin. 1986. Growth abnormalities in mussels and oysters from areas with high levels of tributyltin in San Diego Bay. IEEE Oceans '86 Conf. Proceedings, Wash. D.C. Vol. 4 pp. 1246-1251.
22. Thain, J. and M.J. Waldock. 1983. The effect of suspended sediment and bistributyltin oxide on the growth of *Crassostrea gigas* spat. *ICES Paper CM1983/E:10*. Inter. Counc. for the Explor. of the Sea, Copenhagen, Den.
23. Organotin in antifouling paints environmental considerations. 1986. Dept. of the Environment. Central Directorate of Environmental Protection. Pollution Paper No. 25. London, Her Majesty's Stationery Office. pp. 82.
24. Henderson, R.S. 1986. Effects of organotin antifouling paint leachates on Pearl Harbor organisms: a site specific flow-through bioassay. IEEE Oceans '86 Conf. Proceedings, Wash. D.C. Vol. 4 pp. 1226-1233.
25. Beaumont, A.R. and M.D. Budd. 1984. High mortality of larvae of the common mussel at low concentrations of tributyltin. Mar. Pollut. Bull. 15:402-405.
26. U.S. Navy. 1983. Information Paper. Navy proposed fleetwide use of organotin antifouling paints. *Memo*, Nov. 30.
27. Smith, B.S. 1981. Male characteristics on female mud snails caused by antifouling bottom paints. Jour. Applied Toxicology 1:22-25.
28. Huggett, R.J., M.A. Unger and D.J. Westbrook. 1986. Organotin concen-

trations in the southern Chesapeake Bay. IEEE Oceans '86 Conf. Proceedings, Wash. D.C. Vol. 4 pp, 1262-1265.
29. Hall, L.W. Jr., M.J. Lenkevich, W.S. Hall, A.E. Pinkney and S.J. Bushong. 1986. Monitoring organotin concentrations in Maryland waters of Chesapeake Bay. IEEE Oceans '86 Conf. Proceedings, Wash. D.C. Vol. 4 pp, 1275-1279.
30. Butler, P.A., L. Andren, G.J. Bonde, A. Jernelov, and D.J. Reisch. 1971. Monitoring organisms. In: *Food and Agricultural Organization Technical Conference on Marine Pollution and Its Effects on Living Resources and Fishing, Rome, 1970. Supplement 1: Methods of Detection, Measurement and Monitoring of Pollutants in the Marine Environment*, ed. by M. Ruivo, Fishing News (Books) Ltd., London. pp. 101-112.
31. Couch, J.A., L.A. Courtney, J.T. Winstead and S.S. Foss. 1979. The American oyster (*Crassostrea virginica*) as an indicator of carcinogens in the aquatic environment. In: *Animals as Monitors of Environmental Pollutants*, National Academy of Sciences, Washington, D.C., pp. 65-84.
32. Lee, R.F. 1984. Factors affecting bioaccumulation of organic pollutants by marine animals. In: *Concepts in Marine Pollution Measurements.* (Ed.) H.H. White, University of Maryland, Sea Grant College Publication No. Um-SG-TS-84-03, pp. 339-353.
33. Huggett, R.J., M.E. Bender, and H.D. Slone. 1973. Utilizing metal concentration relationships in the eastern oyster (*Crassostrea virginica*) to detect heavy metal pollution. Water Research 7:451-460.
34. Bender, M.E., P.O. deFur and R.J. Huggett. 1986. Polynuclear aromatic hydrocarbon monitoring in estuaries utilizing: oysters, brackish water clams and sediments. IEEE Oceans '86 Conf. Proceedings, Wash. D.C. Vol. 3 pp. 791-796.
35. Black, J.J. 1983. Field and laboratory studies of environmental carcinogenesis in Niagara River fish. *J. Great Lakes Res.* 9:326-334.
36. Mix, M.C. 1984. Polycyclic aromatic hydrocarbons in the aquatic environment: occurrence and biological monitoring. In: *Reviews in Environmental Toxicology.* E. Hodgson (Ed.) Elsevier Science Pub. New York, N.Y. pp. 51-102.
37. Malins, D.C., M.M. Krahn, M.S. Meyers, L.D. Rhodes, D.W. Brown, C.A. Krone, B.M. McCain and Sin-Lam Chan. 1985. Toxic chemicals in sediments and biota from a creosote-polluted harbor: relationships with epatic neoplasms and other hepatic lesions in English sole (*Parophyrys vetulus*). *Carcinogenesis* 6:1463-1469.

Chapter Seventeen
CONTAMINANT EFFECTS ON PRIMARY PRODUCERS IN CHESAPEAKE BAY

JAMES G. SANDERS
The Academy of Natural Sciences
Benedict Estuarine Research Laboratory
Benedict, MD 20612

ABSTRACT

Primary producers within the Chesapeake Bay (intertidal marsh communities, submerged aquatic vegetation, and phytoplankton) are subject to contaminants of many kinds, from a variety of sources. The magnitude of the threat posed, however, has not been determined. Localized impacts from industrial activity and inadvertant spills occur; the probability of continued occurrences in the future requires that adequate mechanisms for detection, cleanup, and prevention be implemented. In addition, some toxic compounds likely are present in concentrations high enough to exert sublethal pressure upon the structure of aquatic plant communities, but the significance of such alterations to the Chesapeake Bay is not known. Continued study of alterations of population dynamics of important plant species and of community and ecosystem structure is necessary for the formulation of appropriate management decisions for Chesapeake Bay resources.

INTRODUCTION

Many of the contaminants present within the Chesapeake Bay and its tributaries are reactive and biologically available. Many elements and compounds participate in the oxidative/reductive reactions that comprise cellular metabolism; some are necessary (generally in small quantities) for proper function. Others, however, are not, and can impair function, and, in some cases, cause organism failure. These substances and their impacts on autotrophic organisms are the subject of this chapter. Plant nutrients such as phosphorus and nitrogen also are ubiquitous contaminants within the Chesapeake Bay;

although generally not toxic (except at very high concentrations) the overstimulation of plant growth by such compounds can lead to disruption of an aquatic ecosystem. Therefore, such effects must also be considered in this review.

Primary producers of the Chesapeake Bay fall into three general categories: emergent grasses or marshes, submerged aquatic vegetation (SAV), and phytoplankton (Figure 1). The Chesapeake Bay contains some macroalgae as well, however its extent is minor and is largely omitted from this review. All three categories are subjected to contaminants from a variety of sources (Figure 1), depending upon their location within the estuary.

Marshes fringe much of the shoreline of Chesapeake Bay, and are composed of a wide variety of emergent grasses and other plants that are tolerant of water-soaked soil and have varying tolerance to saline water. The vast expanses of marshes within the bay system have decreased in size over the past decades in response to development. Large quantities still exist, however, and are vulnerable to impact from runoff from surrounding land, atmospheric deposition, and from pollutants carried onto the marsh surface during high tide.

Beds of submerged plants also are found throughout the Chesapeake Bay. Members of a variety of plant families adapted to living completely submersed in water, SAVs have exhibited dramatic declines over the past 20 years. SAV are subject to contaminants contained within the water column and those that diffuse out of the sediments.

Phytoplankton (microscopic, generally single-celled algae) are largely free-floating within the bay system and are responsible for the majority of the primary production of the entire Chesapeake Bay. Phytoplankton densities and species composition are highly variable on both a spatial and temporal scale. As free-floating cells, phytoplankton are vulnerable to contaminants contained within the water column.

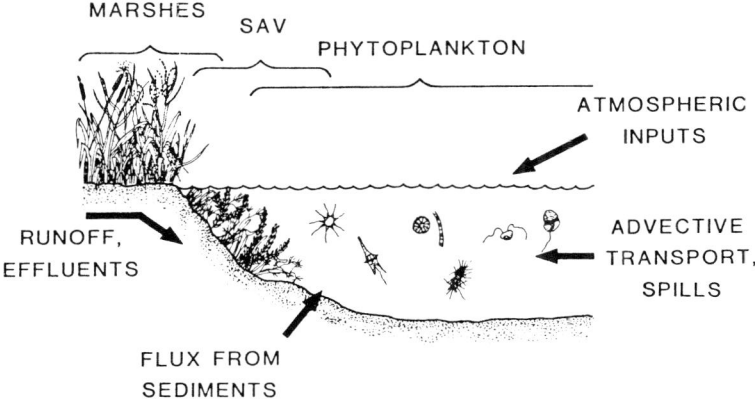

FIGURE 1. Categories and location of primary producers in Chesapeake Bay and potential sources of contaminants.

Each of the above primary producers react in a different manner to contaminants; thus, each will be addressed separately.

INTERTIDAL MARSHES

The soft-bottomed, intertidal marsh is a highly variable environment, subject to rapid changes in water level, temperature, and salinity. Accordingly, the plants that succeed within this zone are extremely hardy. Because they inhabit the shoreline, these plant communities are subject to attack both from the water and from the shore. Typical impacts include runoff from surrounding agricultural and urban land which may contain elevated nutrient concentrations, pesticides, herbicides, and a variety of organic and inorganic compounds. Often, the runoff is from agricultural land, because ~25% of the Chesapeake's shoreline is farmed. Agricultural runoff, depending upon season, can contain substantial quantities of nutrients, herbicides, and pesticides. Runoff from urban land, which constitutes 15% of the Chesapeake's shoreline, can contain a variety of inorganic and organic pollutants, such as trace metals and metalloids, polyaromatic hydrocarbons, and polychlorinated biphenyls. Marshes depend on frequent submergence (thus their location within the intertidal zone) and are also subject to whatever contaminants may be contained within the flooding waters.

Few studies have been designed to evaluate the effects of pollutants on Chesapeake Bay marshes; however, investigations in similar temperate marshes have shown that marsh plants are quite tolerant to chemical stress. These species have adapted to life in anaerobic, water-saturated sediments and have developed specific exclusion mechanisms within their subsurface rhizomes and the ability to actively pump out substances from their leaves[1,2,3] or to convert them to less harmful forms. For example, *Spartina alterniflora* is quite resistant to even high concentrations (100-1000 μg/L) of the herbicide atrazine; it has the ability to degrade the atrazine molecule to less toxic compounds.[4]

Marshes can accumulate substantial quantities of pollutants within plant biomass. Virginia marshes exposed to dredge spoil had significantly elevated concentrations of zinc and lead; *Phragmites communis* and *Spartina* spp. accumulated lead to an average of 7.6 times higher than control marshes.[5] Although little studied in the Chesapeake Bay region, metal enrichment of other U.S. marshes generally has led to increased plant metal concentrations and retention of metals within marsh sediments.[3,6,7,8]

Oil and other hydrocarbons can inhibit marsh productivity; however, the degree of inhibition is dependent upon season of application, toxicity of oil fraction released, and upon the physical energy of the affected marsh. Approximately 250,000 gallons of oil spilled in winter had only slight effects upon Virginia marsh productivity.[9] *Spartina alterniflora* was only slightly affected, exhibiting some modifications in its normal growth pattern; moreover, cleanup operations may have caused the greatest impact. In contrast, a smaller spill of oil in a protected portion of the Potomac estuary caused considerable damage to *Spartina*.[10] In areas where soil concentrations exceeded 2,000 μg/g, *Spartina* biomass was reduced; in areas of concentrations exceeding 10,000 μg/g, *Spartina* and underground rhizomes were killed and little regrowth occurred. Growth inhibition continued through at least two growing seasons in this marsh.[10] Experimental dosing of marshes to simulate chronic exposure to oil produced similar results: areas of *Spartina* death caused by high oil concentrations and larger areas of reduced growth and shortened growing season.[11]

SUBMERGED AQUATIC VEGETATION

The Chesapeake Bay system has experienced wide-spread losses of SAV since the mid 1960s.[12,13,14] Declines in all major SAV species have been noted, suggesting a general, system-wide impact is causing the loss of plant habitat or an inability to survive. Many possible mechanisms for the decline were postulated in the late 1970s; most were discarded because they were localized or transitory impacts. After extensive review,[3] three possible causes were identified for further study: herbicide runoff from agricultural land, increased sedimentation, and increased nutrient enrichment.

A major line of investigation focused on the effects of herbicides, primarily atrazine and linuron. Atrazine and other herbicides are rapidly taken up by SAV, predominately through plant shoots rather than by the root system.[15] Although differences in sensitivity to herbicides occur between different species,[16,17] all tested compounds were found to be quite toxic to SAV, with low doses (1-5 μg/L) causing short-term reductions in productivity and larger doses (10-100 μg/L) causing plant failure.[18,19,20]

Substantial damage to most species occurred only at concentrations of herbicides considerably above those measured in Chesapeake Bay. Concentrations of atrazine and linuron, major herbicides applied within the region, rarely exceed 1 μg/L,[21,22] although concentrations of 5 μg/L and higher have been measured during spring runoff.[19,22,23] Even assuming SAV communities can expect regular inputs of herbicides in the range of 5-10 μg/L, a maximum of 10-20% loss in photosynthesis would occur, and probably this loss would be short-term with full recovery.[18]

Suspended sediments can indirectly control SAV growth by scattering and reducing the amount of light available for growth. Measurements conducted to test whether increased sediment loading has led to SAV declines have shown that light may be limiting. Some species have the ability to adapt to lowered light levels by altering leaf shapes;[24] however, the light levels measured in SAV beds at times are below levels critical for plant survival.[18,24] Overall, reductions in available light have likely reduced the area of bottom that will support SAV growth. For example, the maximum depth at which SAV can live in the Patuxent River is postulated to have been reduced from 1.4 m to 0.6 m between 1960 and 1970.[18] Other tributaries and the upper Chesapeake Bay would be expected to show similar reductions.

Nutrient enrichment of the bay system can have both indirect and direct effects on SAV growth. Increased phytoplankton growth within the water column acts in a manner similar to suspended sediment as higher cell densities absorb and scatter light, reducing the amount of light reaching the submerged plants. In addition, an increase in nutrient loadings increases epiphyte growth upon the plants themselves,[25] reducing the quantity and quality of light that reaches the plant and the amount of plant surface that can participate in active transport of materials across the plant epidermis.[26,27] Experimental loadings of SAV communities indicated that nutrient enrichment (30-120 μM N) significantly increased phytoplankton and epiphytic biomass with a concurrent decrease in SAV biomass.[6] Along with biomass reductions, a decrease in vegetative reproductive ability was noted.[16]

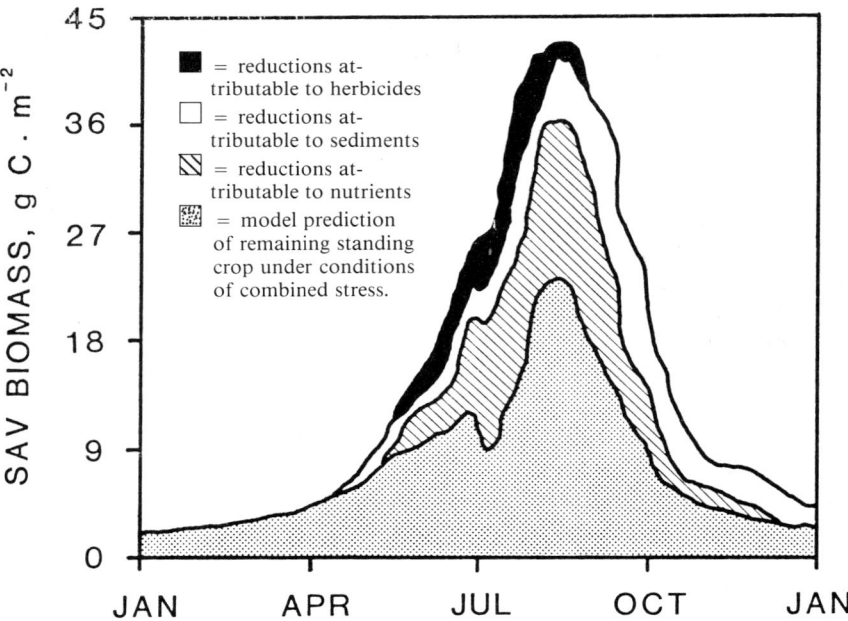

FIGURE 2. Model prediction of the separate and combined effects of stresses (herbicides, sediment loading, nutrient enrichment) on SAV standing crop over the year, relative to normal growth patterns. Taken from 18.

When all three potential impacts (herbicides, sedimentation, nutrient enrichment) are coupled, ecosystem models predict that combined effects will lead to a 35% decrease in biomass and a 56% decrease in net primary production, with nutrient enrichment being the primary cause in growth declines[18] (Figure 2). The models generated from experimental results indicate that such decreases in productivity and vegetative reproduction, when occurring over a period of several years, can lead to the disappearance of whole communities, consistent with field observations to date.[16,18]

On a local basis, other factors can affect SAV abundance and growth. For example, chlorine discharges were implicated in SAV declines around the Washington, D.C. area. Chlorine concentrations of 0.05-0.12 mg/L retard SAV growth and can lead to chlorophyll-a loss and plant death.[28] Other industrial effluents have the potential to inhibit SAV growth; however, no direct evidence of such impact exists[13] As with other plants, SAVs will take up and concentrate trace metals within their tissues. The range of tissue concentrations of four metals (copper, cadmium, lead, and zinc) in Chesapeake Bay SAV was very large, but, in general tissue concentrations were proportional to sediment concentrations, with highest concentrations found in areas of high sediment metal concentrations.[29] Although no evidence of direct effect by toxic metals has been shown, SAVs can act as a conduit for toxic metal transfer (and potential impact) to higher trophic levels[13]

PHYTOPLANKTON

Phytoplankton are vulnerable to many of the compounds present in Chesapeake Bay. Although research has determined the impact of a wide number of different types of contaminants, no general pattern has emerged from these studies. Phytoplankton communities are composed of many different species and algal classes which undergo rapid succession of dominants. Species composition varies widely, both spatially and temporally. Thus, studies of any given pollutant at any location in the Chesapeake Bay cannot be transferred easily to another location, or even to the same location at a later time.

Phytoplankton have shown widely varying sensitivity to pollutants. For example, sensitivity to toxic metals such as cadmium and arsenic can vary over orders of magnitude (Table 1). Even clones of the same algal species are variably sensitive to the same pollutant, including trace metals,[30,31,32] organic compounds,[33,34,35] and industrial wastes.[36] Thus, it is not surprising that little progress can be made toward generalizing the response of phytoplankton communities to toxic substances.

Studies of phytoplankton response to contaminants have included lab experiments and field studies. All major forms of contaminants have received study (to some extent) within the Chesapeake Bay. Again, the wide variety of

experimental designs, species studied, and investigators preclude generalization. However, studies have generally considered one or more of three general effects:
1. inhibition of growth, photosynthesis, or reduction of some cellular constituent,
2. accumulation of the pollutant within cellular tissue, and/or
3. alteration of species composition or species succession within the community, leading to a community with altered structure.

Each of these end results has a different effect upon the species studied and the community within which that species resides.

Chlorine and Power Plant Operations

Chlorine is an effective biocide that is quite toxic to phytoplankton. The toxicity of chlorine has been demonstrated in a number of laboratory and field experiments. Depending upon species, chlorine doses of 0.03-0.64 mg/L are sufficient to reduce cell growth by one-half.[37] In the laboratory, green algae were far more tolerant than were diatoms, dinoflagellates, and chrysophytes.[38,39]

Temperature increase can also inhibit phytoplankton growth. Increases of 10-20°C significantly inhibited growth of diatoms (*Phaeodactylum tricornutum*

TABLE 1

Sensitivity of phytoplankton to arsenic or cadmium. Concentrations (μg/L) listed are those necessary to cause a 50% reduction in growth rate (EC_{50}).

Algal Species	EC_{50}		Reference
	Cd^{2+}	Arsenate	
Chaetoceros debile	1.8		69
C. pseudocrinitum		20	67
Skeletonema costatum	0.8-4.7	5.2	70, 116, 117
Asterionella formosa	1.3		118
A. japonica	6.0		119
Rhizosolenia fragilissima		2.2	68
Cerataulina pelagica		6.0	120
Ditylum brightwelli	2.2		121
Thalassiosira pseudonana	7.5	>25	66, 67, 116
T. weissflogii	1		71
Cylindrotheca closterium		>100	67
Tabellaria flocculosa	11		122
			(33% growth reduction)
Phaeodactylum tricornutum	>150		116
Isochrysis galbana	2.4	2.2	60, 67
Amphidinium carterae		9.8	67
Peridinium trochoidium		90	123
Prorocentrum micans	0.5		73
Scrippsiella faeroense	0.4		72
Tetraselmis contracta		>100	67

and *Thalassiosira fluviatilus*) and flagellates (*Dunaliella tertiolecta* and *Pseudoisochrysis* sp.); temperature increases of less than 10°C did not affect growth.[40,41]

Field observations have been made at a number of power plants within the Chesapeake Bay region. In almost all cases, phytoplankton productivity was diminished by passage through a power plant's cooling system, where the cell is exposed to multiple stress of chlorine, increased temperature, and increased shear. This topic has been extensively reviewed in recent years;[42] relevant results are only summarized here. Primary productivity usually was decreased by passage, from 13-95%.[43,44,45,46,47,48] Species composition often was altered because of variable sensitivity. Mulford[49] reported that dinoflagellates and diatoms exhibited greatest sensitivity to passage through the power plant, with 30-60% of cells killed. Sellner et al.[50] also found that cryptophytes were extremely sensitive.

Natural phytoplankton assemblages exposed to chronic, low-level chlorination exhibit shifts in species dominance and species succession. In a series of experiments, Mackiernan et al.[38] demonstrated that dinoflagellates, particularly *Gymnodinium nelsoni*, were greatly inhibited by exposure to chlorine (0.01-0.05 mg/L total residual chlorine), and gradually were lost as dominants from a natural assemblage. Small flagellates and diatoms were much less affected; these flagellates quickly became predominant within the assemblage. At the highest chlorine concentrations, however, flagellate growth was inhibited and diatoms were able to flourish. Similar results were obtained in studies of natural phytoplankton assemblages from the coastal Atlantic Ocean: a shift from centric diatom dominance to pennate diatom/microflagellate dominance occurred after chronic exposure to low chlorine concentrations, 0.05-0.15 mg/L.[51] Both studies concluded that chronic exposure to chlorine could lead to shifts in phytoplankton species composition and succession within estuarine communities.

In spite of the large potential toxicity of chlorine, investigators have felt that actual harm to natural systems is not great and that the phytoplankton community can recover because of the rapid disappearance of chlorine-produced oxidants in seawater, the relatively short duration of exposure, the high degree of dilution that occurs after chlorine application, and the rapid growth rate of phytoplankton.[52,53] This philosophy may be in error, however. Dilution is often restricted. Chlorination of saline waters results in the formation of a complex group of compounds such as bromamines and chloramines, many of which are toxic and may persist for long periods.[54] In addition, chlorination of natural waters leads to a reduction in organic carbon concentrations,[55,56,57] increasing the activity of toxic metals.[55,58] In a series of experiments, chlorinated estuarine water was toxic to sensitive algal species for over 35 days, even though residual chlorine compounds were undetectable after 10 days. More tolerant species were able to grow after 35 days, but sensitivity to copper was greatly increased,[39] implying either an increase in copper ion activity or an interactive effect between copper and chlorine byproducts. Thus, the widespread release of chlorine into natural waters via sewage effluent and power plant discharges may have long-lasting effects.

Trace Metals and Other Inorganics

A large number of metals are present in trace quantities in estuarine systems; most have the potential for toxicity. Algal cells are able to discriminate between a wide variety of chemical compounds, incorporating the elements necessary for cell growth. However, many metal ions are quite similar chemically; thus, toxic ions can be taken up indiscriminately. Competition between chemical analogues (one essential, one toxic) for uptake or attachment sites has been described for a number of ion pairs.[59,60,61,62,63,64,65] Cellular incorporation of the toxic compound can be detrimental to the success of an algal species; thus, the ability to discriminate between competing ions becomes an important factor in the continuation of a particular species as a member of the phytoplankton community.[66]

Many laboratory determinations of toxic concentrations of a large number of trace metals have been made; the results are widely dispersed and difficult to compare because of the large number of different experimental designs utilized. A number of such studies have been performed within the Chesapeake Bay. Arsenic (as arsenate, the predominant inorganic form) is quite toxic to some species of estuarine phytoplankton, with concentrations as low as 2.2 μg/L inhibiting growth of several species.[66,67,68] Cadmium is also quite toxic; free ion concentrations of approximately 1 μg/L are sufficient to cause sublethal effects in diatoms from the Chesapeake Bay,[69] and in other species tested in other areas, as well.[70,71,72,73] Chromium, as the chromate ion, is an analogue of sulfate; thus its toxicity is greatest in low sulfate (i.e., less saline) waters. Typical EC_{50} concentrations (concentration at which growth was reduced by 50%) for sensitive species range from 10 to 20 μg/L.[74] The free copper ion has been shown to inhibit phytoplankton growth in sensitive species at concentrations just above ambient;[75,76,77,78] in the Chesapeake Bay, total copper concentrations of 5-10 μg/L caused significant reductions in growth of *Thalassiosira pseudonana* and *Isochrysis galbana*.[39,55] Silver appears to act somewhat differently from other cations because the monochlorinated complex, AgCl, appears to be available and toxic to cells along with the free ion; concentrations of 5 μg/L of total silver (~ 0.2 μg/L AgCl + Ag$^+$) caused significant reductions in growth of diatoms and dinoflagellates.[79]

The results for many different compounds from investigations worldwide have been reviewed; e.g., for arsenic,[67,80] cadmium,[69,81] chromium, [66,82] copper,[83,84,85] and lead.[86] When these reviews are examined and the necessary caveats applied (use of largely nonrepresentative laboratory clones, use of media containing high concentrations of nutrients relative to natural waters, toxic concentrations are usually based on *total* metal rather than on the toxic chemical form), the result is that concentrations of most metals in the surface waters of Chesapeake Bay, as in most coastal systems, do not approach levels that would be considered harmful.[78,87] However, it is difficult to assess accurately the potential for trace metal toxicity at the community level. Laboratory determination of the levels of toxic substances sufficient to kill a few species in bioassays under carefully controlled conditions do not provide adequate understanding of the consequences of release of the same substances into the natural environment. In addition to variable sensitivity, physical and geochemical processes within the ecosystem greatly affect the transport, reactivity, and potential impact that may occur. Cellular processes also can transform the original compound into a new suite of compounds that may have quite different geochemical and biological properties.

For example, chronic additions of contaminants such as arsenic, chromium, copper, and mercury at concentrations well below those generally considered to be detrimental to algal productivity cause significant shifts in species composition and species succession in natural phytoplankton communities from Chesapeake Bay and similar coastal systems.[68,88,89,90,91,92,93] Concentrations of copper and arsenic higher than those necessary to cause significant shifts in species dominance have been measured in a number of areas within tributaries and the mainstem of the Chesapeake Bay.[55,87,94,95,96,97] Thus it seems likely that at least these two elements have the potential for causing significant alterations in the community structure of Chesapeake Bay phytoplankton in some areas of the Bay and during some periods of the year. As our investigations of trace metal concentrations within the Bay increase, perhaps we will discover that other metals can cause similar effects.

Other metals probably have less potential for harm. Mercury levels reported (e.g., 98), for example, are unlikely to impact, except perhaps in isolated regions of Baltimore Harbor. Although cadmium concentrations of 5 μg/L and above caused reduction of resting spore formation by dominant diatom species, an important sublethal effect,[69] these concentrations are 1-2 orders of magnitude higher than measured cadmium concentrations.[69,87] Chromium concentrations reported for Chesapeake Bay, 0.1-0.9 μg/L,[69,99] are well below the concentrations shown to cause sublethal changes to phytoplankton community structure ($\sim$ 10 μg/L in water of very low salinity).[93] It is unlikely that concentrations of these several metals will increase enough in the future to cause significant change. However, isolated, industrialized areas receiving higher than average contamination could contain such concentrations. Further research is necessary to discover whether such instances occur, and to determine the significance of such impacts to the Chesapeake Bay system as a whole.

Aside from potential toxic effects, phytoplankton have the capability to alter the chemical form or partitioning of many elements, such as arsenic, silver, chromium, mercury, and copper.[32,66,100,101] Such shifts alter biological reactivity and toxicity of these compounds and can alter their rate of transport through the estuary as well as their eventual fate. For example, biological uptake of arsenic leads to the production of reduced and methylated arsenic compounds, some of which are more toxic to higher trophic levels than was the original arsenic compound.[100,102,103] The production of such compounds also alters the rate of transport of arsenic through the coastal zone by altering its biological and geochemical reactivity.[32] For any element, biological uptake may result in a shift of element partitioning between dissolved and particulate phases, altering transport rates and, indirectly, bioavailability and toxicity.[66,101] Potential metal/phytoplankton interactions are illustrated in Figure 3. Such alterations of metal biogeochemistry are extremely important to the estuarine ecosystem as a whole, and must be considered when impact predictions and management assessments are made. Too often, such biogeochemical concerns are ignored.

Organic and Industrial Effluents

Phytoplankton are sensitive to a wide range of organic compounds, at widely varying concentrations. As with toxic metals, a number of laboratory experiments have been performed with a variety of different species. For example, polychlorinated biphenyls (PCBs) were inhibitory to phytoplankton growth at concentrations ranging from 10 to 50 μg/L;[33,34,104] this wide-ranging variability in sensitivity can lead to changes in species composition of natural phytoplankton communities.[105,106] Phytoplankton showed similar variation in response to an industrial waste (DuPont Grasselli), with inhibition of various species occurring at waste concentrations of 0.01-0.1%.[36]

Little information is available concerning the susceptibility of Chesapeake Bay phytoplankton to toxic organics, despite the large number of potentially harmful compounds that are released to the Chesapeake Bay. Most studies have focused upon the effects of such compounds to higher trophic levels, perhaps because higher trophic levels are more susceptible, more visible, or hold a greater threat to man. However, phytoplankton form the first link in the food chain and may serve as the primary accumulator of dissolved compounds from water.

For example, phytoplankton in the James River, which was subjected to high loadings of Kepone during 1966-1975, accumulated cellular concentrations of 1-2 μg/g, a concentration factor of 10^4 over dissolved concentrations.[107,108] Although Kepone is toxic to phytoplankton in concentrations below 1 mg/L,[109] these concentrations were well above measured concentrations in the James River. Thus, the most important aspect of Kepone interactions was not its toxicity to phytoplankton but rather their ability to accumulate the substance and pass it along to higher trophic levels. Similar interactions likely occur with other organic compounds.

As with SAV, pesticides are toxic to phytoplankton at low concentrations.[110,111] Again, however, toxic concentrations are still well above usual concentrations within the Chesapeake Bay, and may impact cell growth significantly only in isolated circumstances.

During the EPA study of the Chesapeake Bay, a number of industrial effluents were tested for toxicity to *Skeletonema costatum*. Most of the effluents at low concentrations *stimulated Skeletonema* growth (presumably due to added nutrients), SC_{20} (the concentration at which growth was stimulated by 20%) values ranged from 0.02-20%; most were below 1%.[112] At higher concentrations, however, the same effluents were toxic. EC_{50} values ranged from 3.7% to 53%.

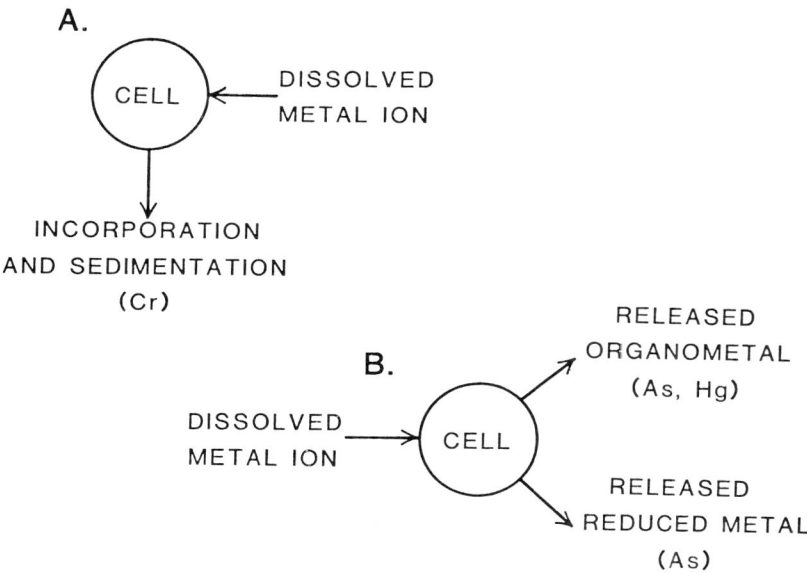

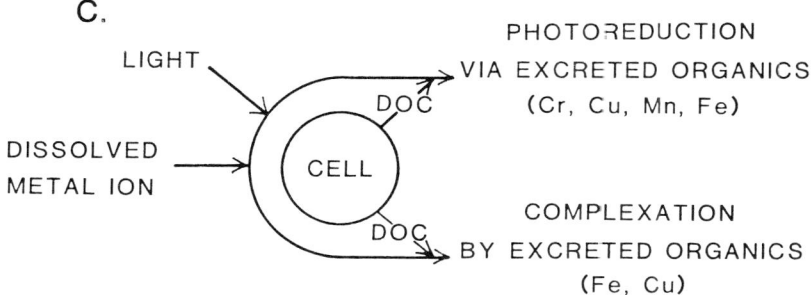

FIGURE 3. Possible interactions of dissolved metal ions with phytoplankton, including changes in partitioning (A), direct transformation of chemical speciation (B), and indirect facilitation of complexation and photo-reduction (C).

Generally industries producing products such as pulp and paper and pesticides, and municipal treatment plants were most toxic; steel plant and petroleum refinery effluents on average were least toxic.[12] Thus, potential for toxicity exists from the estimated 5000 effluent discharges, particularly in industrialized areas, where more than one effluent will be present, and in tributaries and harbors, where restricted circulation impedes dilution of effluents. The overall impact of these effluents, to my knowledge, has never been adequately considered.

One relationship has emerged from laboratory studies of phytoplankton sensitivity to stress. There seems to be a general increase in resistance to stress as one moves from oceanic communities through neritic communities to coastal communities and communities from polluted ecosystems. Such a general pattern has been shown for sensitivity to metals,[30] chlorinated hydrocarbons,[34] industrial effluents,[36] and even exotic compounds to which phytoplankton could not have had prior exposure.[35] Thus, estuarine phytoplankton may have an inherent protection from chemical stress, perhaps because these species are adapted to life in a highly variable system or because they have been exposed to a wide range of chemical compounds and have developed a general resistance.[36]

SUMMARY

Contaminants pose considerable threat to the primary producers of Chesapeake Bay; however, the magnitude of the threat is not known at present. Most attention has been focused upon the higher concentrations of toxic substances within the sediment and upon the effects of toxic substances on the more commercially-important and visible higher trophic levels. Although investigations have employed widely different methods and generalizations are difficult to make, concentrations of toxic substances within the water column generally are not high enough to cause toxicity to aquatic plants, when the results of laboratory bioassays are used as criteria. However, evidence exists that some compounds are present in concentrations high enough to exert sublethal pressures upon the structure of aquatic plant communities. Very little is known about the extent to which alteration is taking place or the significance of such alterations, which could be extremely important to the flow of carbon between trophic levels, the structure of estuarine food webs, and the ultimate yield of harvestable species (e.g., 113-115). These uncertainties are of concern, have been little studied, and must be given greater attention in future studies of estuarine impact. The significance of sublethal alteration of population dynamics of important plant species and of community and ecosystem structure is an important future area of research; management decisions for Chesapeake Bay resources cannot be made without such information.

In addition, localized impacts from industrial activity and inadvertant spills occur and will continue to occur as long as the Chesapeake Bay is a convenient source of water and transport for industry. Although these impacts are local in occurrence, the number of occurrences and the degree of severity of impact deserve attention. Adequate mechanisms for prevention must be implemented, as should procedures for spill containment, clean-up, and mitigation to follow inevitable accidents that will occur in the future.

ACKNOWLEDGEMENTS

This article was dependent upon the assistance of many area scientists, most notably, J.C. Stevenson, W.M. Kemp, and C. Tanner. I thank them for their many suggestions. My philosophy of research as presented in this review has benefited from interactions with S.J. Cibik and G.F. Riedel. I also thank L.W. Hall, Jr. and W.R. Boynton for their critical evaluation of the manuscript and numerous other scientists who took the time to respond to my requests and answer my questions. I am solely responsible, however, for errors of omission or judgement. The preparation of this review was supported by The Academy of Natural Sciences.

REFERENCES

1. Reimold, R.J. 1972. The movement of phosphorus through the salt marsh cord grass, *Spartina alterniflora* Loisel. *Limnol. Oceanogr.* 17:606-611.
2. McGovern, T.A., L.J. Laber, and B.C. Gram. 1979. Characteristics of the salt secreted by *Spartina alterniflora* Loisel and their relation to estuarine production. *Est. Coast. Mar. Sci.* 9:351-356.
3. Kraus, M.L., P. Weis, and J.H. Crow. 1986. The excretion of heavy metals by the salt marsh cord grass, *Spartina alterniflora,* and *Spartina's* role in mercury cycling. *Mar. Environ. Res.* 20:307-316.
4. Davis, D.E., J.D. Weete, C.G.P. Pillai, F.G. Plumley, J.T. McEnerney, J.W. Everest, B. Truelove, and A.M. Diner. 1979. Atrazine fate and effects in a salt marsh. EPA-600/3-79-111, 84 p.
5. Drifmeyer, J.E. and W.E. Odum. 1975. Lead, zinc, and manganese, in dredge-spoil pond ecosystems. *Environ. Conserv.* 2:39-45.
6. Banus, M., I. Valiela, and J.M. Teal. 1975. Lead, zinc, and cadmium budgets in experimentally enriched salt marsh ecosystems. *Est. Coast. Mar. Sci.* 3:421-430.
7. Windom, H.L. 1977. Ability of salt marshes to remove nutrients and heavy metals from dredged material disposal area effluents. Technical Report D-77-37. U.S. Army Engineer Waterways Experiment Station, Environmental Effects Laboratory, Vicksburg, MS.

8. Giblin, A.E., A. Bourg, I. Valiela, and J.M. Teal. 1980. Uptake and losses of heavy metals in sewage sludge by a New England salt marsh. *Am. J. Bot.* 67:1059-1068.
9. Hershner, C. and K. Moore. 1977. Effects of the Chesapeake Bay oil spill on salt marshes of the lower bay, p. 529-533. *In: Proceedings 1977 Oil Spill Conference,* American Petroleum Institute, Washington, D.C.
10. Krebs, C.T. and C.E. Tanner. 1981. Restoration of oiled marshes through sediment stripping and *Spartina* propagation, p. 375-385. *In: Proceedings 1981 Oil Spill Conference,* American Petroleum Institute, Washington, D.C.
11. Hershner, C. and J. Lake. 1980. Effects of chronic oil pollution on a saltmarsh grass community. *Mar. Biol.* 56:163-173.
12. Bayley, S., V.D. Stotts, P.F. Springer, and J. Steenis. 1978. Changes in submerged aquatic macrophyte populations at the head of the Chesapeake Bay, 1958-1975. *Estuaries* 1:171-182.
13. Stevenson, J.C. and N.M. Confer. 1978. Summary of available information on Chesapeake Bay submerged aquatic vegetation. Report # FWS/OBS-78/66. U.S. Dept. of the Interior, Fish and Wildlife Service, Washington, D.C.
14. Orth, R.J. and K.A. Moore. 1981. Submerged aquatic vegetation of the Chesapeake Bay: past, present and future. *Trans. N. Amer. Wildl. Nat. Res.* 46:271-283.
15. Jones, T.W., W.M. Kemp, P.S. Estes, and J.C. Stevenson. 1986. Atrazine uptake, photosynthetic inhibition, and short-term recovery for the submersed vascular plant, *Potamogeton perfoliatus* L. *Arch. Environ. Contam. Toxicol.* 15:277-283.
16. Kemp, W.M., J.C. Means, T.W. Jones, and J.C. Stevenson. 1982. Herbicides in Chesapeake Bay and their effects on submerged aquatic vegetation, p. 503-567. *In:* E.G. Macalaster, D.A. Barker, and M. Kasper (Eds.), *Chesapeake Bay Program Technical Studies: A Synthesis.* U.S. Environmental Protection Agency, Washington, D.C.
17. Correll, D.L. and T.L. Wu. 1982. Atrazine toxicity to submersed vascular plants in simulated estuarine microcosms. *Aquat. Bot.* 14:151-158.
18. Kemp, W.M., W.R. Boynton, R.R. Twilley, J.C. Stevenson, and J.C. Means. 1983. The decline of submerged vascular plants in upper Chesapeake Bay: summary of results concerning possible causes. *Mar. Technol. Soc. J.* 17:78-89.
19. Kemp, W.M., W.R. Boynton, J.J. Cunningham, J.C. Stevenson, T.W. Jones, and J.C. Means. 1985. Effects of atrazine and linuron on photosynthesis and growth of the macrophytes, *Potamogeton perfoliatus* L. and *Myriophyllum spicatum* L. in an estuarine environment. *Mar. Environ. Res.* 16:255-280.
20. Cunningham, J.J., W.M. Kemp, M.R. Lewis, and J.C. Stevenson. 1984.

Temporal responses of the macrophyte, *Potamogeton perfoliatus* L., and its associated autotrophic community to atrazine exposure in estuarine microcosms. *Estuaries* 7:519-530.
21. Zahnow, E.W. and J.D. Riggleman. 1980. Search for linuron residues in tributaries of the Chesapeake Bay. *J. Agric. Food Chem.* 28:974-978.
22. Means, J.C., R.D. Wijayaratne, and W.R. Boynton. 1983. Fate and transport of selected pesticides in estuarine environments. *Can. J. Fish. Aquat. Sci.* 40(Suppl. 2):337-345.
23. Wu, T.L. 1980. Dissipation of the herbicides atrazine and alachlor in a Maryland corn field. *J. Environ. Qual.* 9:459-465.
24. Goldsborough, W.J. 1983. Light and shade adaptation in the submerged vascular plant, *Potamogeton perfoliatus*. MS Thesis, Univ. of Maryland, College Park.
25. Phillips, G.L., D. Eminson, and B. Moss. 1978. A mechanism to account for macrophyte decline in progressively eutrophicated freshwaters. *Aquat. Bot.* 4:103-126.
26. Sand-Jensen, K. 1977. Effects of epiphytes on eelgrass photosynthesis. *Aquat. Bot.* 3:55-63.
27. Penhale, P. 1977. Macrophyte-epiphyte biomass and productivity in an eelgrass (*Zostera marina* L.) community. *J. Exp. Mar. Biol. Ecol.* 26:211-224.
28. Wester, H.V. and S.D. Rawles. 1976. Impact of chlorine pollution in the upper Potomac and Anacostia estuaries. U.S. Dept. of the Interior, Washington, D.C. (Abstract).
29. DiGiulio, R.T., and P.F. Scanlon. 1985. Heavy metals in aquatic plants, clams, and sediments from the Chesapeake Bay, U.S.A. Implications for waterfowl. *Sci. Tot. Environ.* 41:259-274.
30. Jensen, A., B. Rystad, and S. Melsom. 1974. Heavy metal tolerance of marine phytoplankton. I. The tolerance of three algal species to zinc in coastal seawater. *J. Exp. Mar. Biol. Ecol.* 15:145-157.
31. Murphy, L.S. and R.A. Belastock. 1980. The effect of environmental origin on the response of marine diatoms to chemical stress. *Limnol. Oceanogr.* 25:160-165.
32. Sanders, J.G. 1986. Alteration of arsenic transport and reactivity in coastal marine systems after biological transformation. *Rapp. P.-v. Réun. Cons. int. Explor. Mer* 186:185-192.
33. Mosser, J.L., N.S. Fisher, and C.F. Wurster. 1972. Polychlorinated biphenyls and DDT alter species composition in mixed cultures of algae. *Science* 176:533-535.
34. Fisher, N.S., L.B. Graham, E.J. Carpenter, and C.F. Wurster. 1973. Geographic differences in phytoplankton sensitivity to PCBs. *Nature* 241:548-549.
35. Fisher, N.S. 1977. On the differential sensitivity of estuarine and open-

ocean diatoms to exotic chemical stress. *Amer. Natur.* 111:871-895.
36. Murphy, L.S., P.R. Hoar, and R.A. Belastock. 1981. The effects of industrial wastes on marine phytoplankton, p. 399-410. *In:* B.H. Ketchum, D.R. Kester, and P.K. Park (Eds.) *Ocean Dumping of Industrial Wastes.* Plenum Press, New York.
37. Bender, M.E., M.H. Roberts, R. Diaz, and R.J. Huggett. 1977. Effects of residual chlorine on estuarine organisms, p. 101-108. *In:* L.D. Jensen (Ed.), *Biofouling Control Procedures Technology and Ecological Effects.* Marcel Dekker, Inc., New York.
38. Mackiernan, G.B., D.R. Heinle, and S.D. Van Valkenburg. 1978. The effects of chlorine-produced oxidants on the growth rates and survival of estuarine phytoplankton. University of Maryland technical report, UMCEES Ref. No. 78-55CBL, Solomons, MD, 55 p.
39. Sanders, J.G. 1984. The longevity of algal inhibition after chlorination of estuarine water. *Environ. Sci. Technol.* 18:383-385.
40. Domotor, S.L., K. Mountford, and C.F. D'Elia. 1982. Autoradiographic detection of species-specific thermal stress effects on natural phytoplankton assemblages. *Mar. Environ. Res.* 6:27-35.
41. Sellner, K.G., L. Lyons, E.S. Perry, and D.B. Heimark. 1982. Assessing physiological stress in *Thalassiosira fluviatilis* (Bacillariophyta) and *Dunaliella tertiolecta* (Chlorophyta) with DCMU-enhanced fluorescence. *J. Phycol.* 18:142-148.
42. Hall, L.W., Jr. and D.T. Burton. 1981. Estuarine and marine toxicity studies, p. 131-218. *In:* L.W. Hall, Jr., G.R. Helz, and D.T. Burton (Eds.) *Power Plant Chlorination—A Biological and Chemical Assessment.* Ann Arbor Science, Ann Arbor, MI.
43. Flemer, D.A. 1974. The effects of entrainment on phytoplankton at the Morgantown steam electric station Potomac River estuary, September 5-8, p. 163-164. *In:* L.D. Jensen (Ed.) *Proc. Second Workshop on Entrainment and Intake Screening.* The Johns Hopkins University, Dept. of Geography and Environmental Engineering, Baltimore, MD.
44. Morgan, R.P., II, D.A. Flemer, L.A. Noe, J.V. Rasin, Jr., and R.A. Murtagh. 1974. Biochemical studies of entrained phytoplankton at the Morgantown Maryland power plant, p. 165-168. *In:* L.D. Jensen (Ed.) *Proc. Second Workshop on Entrainment and Intake Screening.* The Johns Hopkins University, Dept. of Geography and Environmental Engineering, Baltimore, MD.
45. Flemer, D. and J.A. Sherk, Jr. 1977. The effects of steam electric station operation on entrained phytoplankton. *Hydrobiologia* 55:33-44.
46. Bongers, L.H., B. Bradley, D.T. Burton, A.F. Holland, and L.H. Liden. 1977. Biotoxicity of bromine chloride- and chlorine-treated power plant condenser cooling water effluent. Final Report, Martin Marietta Corporation, Baltimore, MD.

47. Marcy, B.C., Jr., A.D. Beck, and R.E. Ulanowicz. 1978. Effects and impacts of physical stress on entrained organisms, p. 135-188. *In:* J.R. Schubel, B.C. Marcy, Jr. (Eds.) *Power Plant Entrainment — A Biological Assessment.* Academic Press, Inc., NY.
48. Liden, L.H., D.T. Burton, L.H. Bongers, and A.F. Holland. 1980. Effects of chlorobrominated and chlorinated cooling waters on estuarine foodchain organisms. *J. Water Poll. Control Fed.* 52:173-182.
49. Mulford, R.A. 1974. Morgantown entrainment. Part IV, phytoplankton taxonomic studies, p. 169-175. *In:* L.D. Jensen (Ed.) *Proc. Second Workshop on Entrainment and Intake Screening.* The Johns Hopkins University, Dept. of Geography and Environmental Engineering, Baltimore, MD.
50. Sellner, K.G., M.E. Kachur, and L. Lyons. 1984. Alterations in carbon fixation during power plant entrainment of estuarine phytoplankton. *Wat. Air Soil Poll.* 21:359-374.
51. Sanders, J.G. and J.H. Ryther. 1980. The impact of chlorine on the species composition of marine phytoplankton, p. 631-639. *In:* R.L. Jolley, W.A. Brungs, and R.B. Cumming (Eds.) *Water Chlorination — Environmental Impact and Health Effects, Vol. 3.* Ann Arbor Sci. Ann Arbor, MI.
52. Goldman, J.C., J.M. Capuzzo, and G.T.F. Wong. 1978. Biological and chemical effects of chlorination at coastal power plants, p. 291-305. *In:* R.L. Jolley, H. Gorchev, and D.H. Hamilton, Jr. (Eds.) *Water Chlorination — Environmental Impact and Health Effects, Vol. 2.* Ann Arbor Science Publishers, Inc., Ann Arbor, MI.
53. Goldman, J.C. and H.L. Quinby. 1979. Recovery potential of phytoplankton after passage through power plant entrainments. *J. Water Poll. Control Fed.* 51:1816-1823.
54. Helz, G.R. 1981. Chlorine chemistry, p. 7-50. *In:* L.W. Hall, Jr., G.R. Helz, and D.T. Burton (Eds.) *Power Plant Chlorination — A Biological and Chemical Assessment.* Ann Arbor Science, Ann Arbor, MI.
55. Sanders, J.G. 1982a. Chlorination of estuarine water: the occurrence and magnitude of carbon oxidation and its impact on trace metal transport. *Environ. Sci. Technol.* 16:791-796.
56. Sigleo, A.C., G.R. Helz, and W.H. Zoller. 1980. Organic-rich colloidal material in estuaries and its alteration by chlorination. *Environ. Sci. Technol.* 14:673-679.
57. Helz, G.R., D.A. Dotson, and A.C. Sigleo. 1983. Chlorine demand: studies concerning its chemical basis, p. 181-189. *In:* R.L. Jolley, H. Gorchev, and D.H. Hamilton (Eds.) *Water Chlorination — Environmental Impact and Health Effects, Vol. 4.* Ann Arbor Science, Ann Arbor, MI.
58. Carpenter, J.H. and C.A. Smith. 1978. Reactions in chlorinated seawater, p. 195-207. *In:* R.L. Jolley, H. Gorchev, and D.H. Hamilton (Eds.) *Water Chlorination — Environmental Impact and Health Effects, Vol. 2.* Ann

Arbor Science, Ann Arbor, MI.
59. Blum, J.J. 1966. Phosphate uptake by phosphate-starved *Euglena*. *J. Gen. Physiol.* 49:1125-1136.
60. Li, W.K.W. 1978. Kinetic analysis of interactive effects of cadmium and nitrate on growth of *Thalassiosira fluviatilis* (Bacillariophyceae). *J. Phycol.* 14:454-460
61. Morel, N.M.L., J.G. Rueter, and F.M.M. Morel. 1978. Copper toxicity to *Skeletonema costatum* (Bacillariophyceae). *J. Phycol.* 14:43-48.
62. Sanders, J.G. and H.L. Windom. 1980. The uptake and reduction of arsenic species by marine algae. *Est. Coast. Mar. Sci.* 10:555-567.
63. Sunda, W.G., R.T. Barber, and S.A. Huntsman. 1981. Phytoplankton growth in nutrient rich seawater: importance of copper-manganese cellular interactions. *J. Mar. Res.* 39:567-586.
64. Rueter, J.G., Jr. 1983. Effect of copper on growth, silicic acid uptake and soluble pools of silicic acid in the marine diatom *Thalassiosira weissflogii* (Bacillariophyceae). *J. Phycol.* 19:101-104.
65. Riedel, G.F. 1985. The relationship between chromium (VI) uptake, sulfate uptake, and chromium (VI) toxicity in the estuarine diatom *Thalassiosira pseudonana*. *Aquat. Toxicol.* 7:191-204.
66. Sanders, J.G. and G.F. Riedel. 1987. Control of trace element toxicity by phytoplankton. *In:* E.E. Conn (Ed.) *Recent Advances in Phytochemistry, Vol. 21.* In press.
67. Sanders, J.G. and P.S. Vermersch. 1982. Response of marine phytoplankton to low levels of arsenate. *J. Plank. Res.* 4:881-893.
68. Sanders, J.G. and S.J. Cibik. 1985a. Adaptive behavior of euryhaline phytoplankton communities to arsenic stress. *Mar. Ecol. Prog. Ser.* 22:199-205.
69. Sanders, J.G. and S.J. Cibik. 1985b. Reduction of growth rate and resting spore formation in a marine diatom exposed to low levels of cadmium. *Mar. Environ. Res.* 16:165-180.
70. Berland, B.R., D.J. Bonin, O.J. Guérin-Ancey, V.I. Kapkov, and D.P. Arlhac. 1977. Action de métaux lourds à des doses sublétales sur les caractéristiques de la croissance chez la diatomée *Skeletonema costatum*. *Mar. Biol.* 42:17-30.
71. Foster, P.L. and F.M. Morel. 1982. Reversal of cadmium toxicity in a diatom: an interaction between cadmium activity and iron. *Limnol. Oceanogr.* 27:745-751.
72. Kayser, H. 1982. Cadmium effects in food chain experiments with marine plankton algae (Dinophyta) and benthic filter feeders (Tunicata). *Neth. J. Sea Res.* 16:444-454.
73. Kayser, H. and K. R. Sperling. 1980. Cadmium effects and accumulation in cultures of *Prorocentrum micans* (Dinophyta). *Helgol. Meeresunters.* 33:89-102.

74. Riedel, G.F. 1987. Interspecific and geographical variation of the chromium sensitivity of algae. Unpublished ms. submitted for publication.
75. Sunda, W.G. and R.R.L. Guillard. 1976. Relationship between cupric ion activity and the toxicity of copper to phytoplankton. *J. Mar. Res.* 34:511-529.
76. Anderson, D.M. and F.M. Morel. 1978. Copper sensitivity of *Gonyaulax tamarensis*. *Limnol. Oceanogr.* 23:283-295.
77. Knauer, G.A. and J.H. Martin. 1983. The cycle of living and dead particulate organic matter in the pelagic environment in relation to trace metals, p. 447-465. *In:* C.S. Wong, E. Boyle, K.W. Bruland, J.D. Burton, and E.D. Goldberg (Eds.) *Trace Metals in Sea Water.* Plenum Press, New York.
78. Morel, F.M.M. and N.M.L. Morel-Laurens. 1983. Trace metals and plankton in the oceans: facts and speculation, p. 841-869. *In:* C.S. Wong, E. Boyle, K.W. Bruland, J.D. Burton and E.D. Goldberg (Eds.) *Trace Metals in Sea Water.* Plenum Press, New York.
79. Sanders, J.G. 1987. Silver transport and impact in estuarine and marine systems. Unpublished ms. submitted for publication.
80. Environmental Protection Agency. 1985a. Ambient Water Quality Criteria for Arsenic—1984. EPA 440/5-84-033, National Technical Information Service, Springfield, VA, 66 p.
81. Environmental Protection Agency. 1985b. Ambient Water Quality Criteria for Cadmium—1984. EPA 440/5-84-032, National Technical Information Service, Springfield, VA, 127 p.
82. Environmental Protection Agency. 1985c. Ambient Water Quality Criteria for Chromium—1984. EPA 440/5-84-029, National Technical Information Service, Springfield, VA, 99 p.
83. Schmidt, R.L. 1978. Copper in the marine environment. Part II. *CRC Crit. Rev. Environ. Control* 8:247-291.
84. Eisler, R. 1979. Copper accumulations in coastal and marine biota, p. 383-449. *In:* J.O. Nriagu (Ed.) *Copper in the Environment, Part I: Ecological Cycling.* J. Wiley and Sons, New York.
85. Environmental Protection Agency. 1985d. Ambient Water Quality Criteria for Copper—1984. EPA 440/5-84-031, National Technical Information Service, Springfield, VA, 142 p.
86. Environmental Protection Agency. 1985e. Ambient Water Quality Criteria for Lead—1984. EPA 440/5-84-027, National Technical Information Service, Springfield, VA, 81 p.
87. Kingston, H.M., R.R. Greenberg, E.S. Beary, B.R. Hardas, J.R. Moody, T.C. Rains, and W.S. Liggett. 1982. The characterization of the Chesapeake Bay: a systematic analysis of toxic trace elements. EPA-600/S3-82-085, Environmental Protection Agency, Washington, D.C.
88. Thomas, W.H. and D.L.R. Seibert. 1977. Effects of copper on the

dominance and the diversity of algae: controlled ecosystem pollution experiment. *Bull. Mar. Sci.* 27:23-33.
89. Thomas, W.H., O. Holm-Hansen, D.L.R. Seibert, F. Azam, R. Hodson, and M. Takahashi. 1977. Effects of copper on phytoplankton standing crop and productivity: controlled ecosystem pollution experiment. *Bull. Mar. Sci.* 27:34-43.
90. Bishop, S.S. and W.M. Dunstan. 1980. Effects of ionic copper on natural populations of phytoplankton grown in continuous culture. Paper presented at the 43rd annual meeting, American Society of Limnology and Oceanography, Knoxville, TN.
91. Sanders, J.G., J.H. Ryther, and J.H. Batchelder. 1981. Effects of copper, chlorine, and thermal addition on the species composition of marine phytoplankton. *J. Exp. Mar. Biol. Ecol.* 49:81-102.
92. Kuiper, J. 1982. Ecotoxicological experiments with marine plankton communities in plastic bags, p. 181-194. *In:* G.D. Grice and M.R. Reeve (Eds.) *Marine Mesocosms.* Springer-Verlag, New York.
93. Frey, B.E., G.F. Riedel, A.E. Bass, and L.F. Small. 1983. Sensitivity of estuarine phytoplankton to hexavalent chromium. *Est. Coast. Shelf Sci.* 17:181-187.
94. Eaton, A. and C. Chamberlain. 1982. Copper cycling in the Patuxent estuary and condenser micro-fouling studies. Final report to MD Power Plant Siting Program #P42-78-04, 70 p.
95. Zamuda, C.D. 1984. The bioavailability of copper to the American oyster, *Crassostrea virginica.* Ph.D. Dissertation, Univ. of Maryland, College Park, MD, 213 p.
96. Abbe, G.R. and J.G. Sanders. 1986. Condenser replacement in a coastal power plant: copper uptake and incorporation in the American oyster, *Crassostrea virginica. Mar. Environ. Res.* 19:93-113.
97. Newell, A.D. and J.G. Sanders. 1986. Relative copper binding capacities of dissolved organic compounds in a coastal-plain estuary. *Environ. Sci. Technol.* 20:817-821.
98. Brinckman, F.E. and W.P. Iverson. 1975. Chemical and bacterial cycling of heavy metals in the estuarine system, p. 319-342. *In:* T.M. Church (Ed.), *Marine Chemistry in the Coastal Environment.* ACS Symposium Series 18, Washington, D.C.
99. Riedel, G.F. 1986. Personal communication. Academy of Natural Sciences.
100. Sanders, J.G. 1985. Arsenic geochemistry in Chesapeake Bay: dependence upon anthropogenic inputs and phytoplankton species composition. *Mar. Chem.* 17:329-340.
101. Sanders, J.G. and G.R. Abbe. 1987. The role of suspended sediments and natural phytoplankton in the partitioning and transport of silver in estuaries. *Cont. Shelf Res.* In press.
102. Sanders, J.G. 1980. Arsenic cycling in marine systems. *Mar. Environ. Res.*

3:257-266.
103. Nissen, P. and A.A. Benson. 1982. Arsenic metabolism in freshwater and terrestrial plants. *Physiol. Plant.* 54:446-450.
104. Cosper, E.M., C.F. Wurster, and R.G. Rowland. 1984. PCB resistance within phytoplankton populations in polluted and unpolluted marine environments. *Mar. Environ. Res.* 12:209-224.
105. Biggs, D.C., R.G. Rowland, H.B. O'Connors, C.D. Powers, and C.F. Wurster. 1978. A comparison of the effects of chlordane and PCB on the growth, photosynthesis, and cell size of estuarine phytoplankton. *Environ. Pollut.* 15:253-263.
106. O'Connors, H.B., Jr., C.F. Wurster, C.D. Powers, D.C. Biggs, and R.G. Rowland. 1978. Polychlorinated biphenyls may alter marine trophic pathways by reducing phytoplankton size and production. *Science* 201: 737-739.
107. Huggett, R.J., M.N. Nichols, and M.E. Bender. 1980. Kepone contamination of James River Estuary, p. 33-52. *In:* R.A. Baker (Ed.), *Contaminants and Sediments, Volume 1.* Ann Arbor Science, Ann Arbor, MI.
108. Lunsford, C.A. and C.R. Blem. 1982. Annual cycle of Kepone residue and lipid content of the estuarine clam, *Rangia cuneata. Estuaries* 5:121-130.
109. Walsh, G.E., K. Ainsworth, and A.J. Wilson. 1977. Toxicity and uptake of Kepone in marine unicellular algae. *Chesapeake Sci.* 18:222-223.
110. Plumley, F.G. and D.E. Davis. 1980. The effects of a photosynthesis inhibitor, atrazine, on salt marsh edaphic algae, in culture, micro-ecosystems, and in the field. *Estuaries* 3:271-277.
111. Karlander, E.P., J.M. Mayasich, and D.E. Terlizzi. 1983. Effects of the herbicide atrazine on an oyster-food organism. Maryland Water Resources Research Center, Univ. of Maryland, Tech. Rep. 73, 20 p.
112. Bieri, R., O. Bricker, R. Byrne, R. Diaz, G. Helz, J. Hill, R. Huggett, R. Kerhin, M. Nichols, E. Reinharz, L. Schaffner, D. Wilding, and C. Strobel. 1982. Toxic substances, p. 263-375. *In:* E.G. Macalaster, D.A. Barker, and M. Kasper (Eds.) *Chesapeake Bay Program Technical Studies: A Synthesis.* Environmental Protection Agency, Washington, D.C.
113. Ryther, J.H. 1969. Photosynthesis and fish production in the sea. *Science* 166:72-76.
114. Steele, J.H. and B.W. Frost. 1977. The structure of plankton communities. *Phil. Trans. R. Soc. (B)* 280:485-534.
115. Sanders, J.G. 1986. Direct and indirect effects of arsenic on the survival and fecundity of estuarine zooplankton. *Can. J. Fish. Aquat. Sci.* 43:694-699.
116. Braek, G.S., D. Malnes, and A. Jensen. 1980. Heavy metals tolerance of marine phytoplankton. IV. Combined effect of zinc and cadmium on growth and uptake in some marine diatoms. *J. Exp. Mar. Biol. Ecol.*

42:39-54.
117. Sanders, J.G. 1979. Effects of arsenic speciation and phosphate concentration on arsenic inhibition of *Skeletonema costatum* (Bacillariophyceae). *J. Phycol.* 15:424-428.
118. Conway, H.L. 1978. Sorption of arsenic and cadmium and their effects on growth, micronutrient utilization, and photosynthetic pigment composition of *Asterionella formosa. J. Fish. Res. Board Can.* 35:286-294.
119. Fisher, N.S. and G.J. Jones. 1981. Heavy metals and marine phytoplankton: correlation of toxicity and sulfhydryl-binding. *J. Phycol.* 17:108-111.
120. Sanders, J.G. 1982b. Adaptive behavior of euryhaline phytoplankton to stress: response to chronic, low-level additions of trace metals. Semiannual progress summary to NOAA under Grant #NOAA-NA81RAD00032, 44 p.
121. Canterford, G.S. and D.R. Canterford. 1980. Toxicity of heavy metals to the marine diatom *Ditylum brightwellii* (West) Grunow: correlation between toxicity and metal speciation. *J. Mar. Biol. Ass. U.K.* 60:227-242.
122. Adshead-Simonsen, P.C., G.E. Murray, and D.J. Kushner. 1981. Morphological changes in the diatom, *Tabellaria flocculosa*, induced by very low concentrations of cadmium. *Bull. Environ. Contam. Toxicol.* 26:745-748.
123. Sanders, J.G. 1978. Interactions between arsenic species and marine algae. Ph.D. Dissertation, Univ. of North Carolina, Chapel Hill, NC, 77 p.

Contaminant Problems and Management of Living Chesapeake Bay Resources. Edited by S. K. Majumdar, L. W. Hall, Jr. and H. M. Austin. © 1987, The Pennsylvania Academy of Science.

Chapter Eighteen
EFFECTS OF CONTAMINANTS ON ESTUARINE ZOOPLANKTON[1]

BRIAN P. BRADLEY[1] and MORRIS H. ROBERTS, JR.[2]

[1]Department of Biological Sciences
University of Maryland Baltimore County
Baltimore, Maryland 21228
and
[2]Virginia Institute of Marine Science
College of William and Mary
Gloucester Point, Virginia 23062

ABSTRACT

The objectives of the chapter are (1) to evaluate laboratory studies concerning effects of heavy metals, pesticides and oxidants on copepods, mysids, bivalve and decapod larvae (2) access field studies (mainly with copepods) on these and other contaminants which when coupled with laboratory data provide information on known and potential hazards of contaminants to zooplankton and (3) briefly review some bioassay methods used in these studies.

Mercury is the most toxic heavy metal by weight, followed by copper, silver and cadmium. Pesticides have been tested much less extensively than heavy metals. In general, bivalve larvae seem less sensitive than the crustacean taxa. Mysids, decapods and copepods seem comparable in sensitivity. Of the pesticides, tributyltin, an antifoulant, presents the greatest present or potential hazard. Chlorine, the most widely used oxidant in Chesapeake Bay, is highly toxic to all taxa reviewed, making zooplankton highly vulnerable. Lethal effects can be reduced or eliminated by dechlorination, but sublethal effects may persist.

Most field studies with copepods have dealt with uptake of heavy metals and pesticides, effects of oil residues and impacts of power plants. Acute lethal effects are rarely observed in the field. For this reason death is an inadequate basis

---

[1]Contribution Number *1390* from the Virginia Institute of Marine Science, College of William and Mary, Gloucester Point, VA.

from which to infer an ecotoxicological response. We believe that reliable sublethal endpoints for laboratory studies need to be developed to evaluate field observations on contaminant effects on zooplankton.

INTRODUCTION

The overall objective of this review is to evaluate the degree of potential or actual impact of contaminants on zooplankton within the Chesapeake Bay. To accomplish this, we made a comprehensive but not exhaustive review of both laboratory and field studies relating to contaminants and bay species. The review includes work done elsewhere on marine and estuarine species similar to those found in Chesapeake Bay.

To narrow the scope of the task, we focused on a limited array of contaminants, namely metals, pesticides, oxidants, and, to a lesser extent, other contaminants which are of widespread concern within Chesapeake Bay. We also narrowed the list of taxa to copepods, mysid shrimp, decapod larvae and bivalve larvae. This approach was used because of their dominance in the zooplankton community and the availability of extensive data.

The three most abundant species of holoplankters are the calanoid copepods, *Eurytemora affinis, Acartia tonsa* and *A. clausi*. These are also the most commonly used copepod species in toxicity studies, being readily cultured in the laboratory.[1,2,3,4] Copepods are the dominant holoplankton species in the Chesapeake Bay, and mysid shrimp represent a secondary dominant species. Several mysid species are known to occur in the Chesapeake Bay, notably *Neomysis americana* and *Mysidopsis bigelowi*. *N. americana* is not only numerically abundant, but also important as food for a variety of estuarine fishes.[5] While this species can be cultured, most toxicological research with mysids has centered on the subtropical species, *Mysidopsis bahia*. *M. bahia* seems comparable in sensitivity to *N. americana* for several toxicants;[6,7] therefore data for both species will be considered in this review.

Meroplankton species, a second important component of estuarine zooplankton communities, include larvae of a large proportion of benthic invertebrates as well as fish larvae. Two principal taxa in this group are larvae of molluscs and decapod crustaceans. Annelid larvae, while abundant, are less well studied within Chesapeake Bay. Data for fish larvae are discussed elsewhere in this volume.

Embryos and larvae of the oyster (*Crassostrea virginica*) and hardclam *(Mercenaria mercenaria)* are the most studied resident molluscan species in Chesapeake Bay. These species are of considerable interest because they are both commercially important and readily cultured in the laboratory.[8] Both species produce large numbers of gametes at each spawning, and hence are well represented in the plankton throughout the warm months of the year.

Larvae of decapod crustaceans are another major meroplanktonic component in Chesapeake Bay. A considerable diversity of species has been used in toxicity tests since many can be cultured using standard techniques.[9,10,11] Much of this work relates to species of no commercial interest. Relatively little work has been done with the commercially important blue crab (*Callinectes sapidus*) because the larvae are difficult to culture reliably. Many of the species most frequently used for toxicity tests are resident within Chesapeake Bay.

Since data for both Bay and non-Bay species are included, we have compared data for Bay and non-Bay species of a given genus or class (e.g. *C. virginica* and *C. gigas*) whenever both were available. These comparisons suggest that data for several non-Bay species of copepods, bivalve and decapod larvae are representative for Bay species.

EFFECTS OF HEAVY METALS

Considering mortality as the end point, zooplankters, regardless of taxon, exhibit a wide range in response to various metals with greatest sensitivity to mercury in the range 3-10 μg/L[4,12,13,14,15,16,17] and least sensitivity to lead (in the range 500-3000 μg/L for mysids,[13] chromium, (over 10,000 μg/L in bivalve larvae[18]) or selenium (over 10,000 μg/L in bivalve larvae and over 1,000 μg/L in decapod larvae[19]). The taxa differ in relative sensitivity to each metal only slightly (Table 1B).

There is no consistent trend in relative sensitivity to heavy metals among zooplankton. While bivalve larvae appear most sensitive to silver and mercury, mysids and copepods appear most sensitive to mercury and cadmium. These differences are also expressed to some degree in the different rank order of toxicities for the taxa; however, taxa seldom differ in sensitivity by more than an order of magnitude and generally by much less.

For some metals, notably cadmium, salinity has a dramatic effect on acute toxicity to various species. This has been interpreted to reflect complexation of metals by chloride ions, reducing their bioavailability; lethal concentrations calculated on the basis of free metal ion concentration seem unaffected by differences in exposure salinity.[20] Such an effect can be seen in data for *M. bahia* exposed to cadmium in different salinity regimes[6,13,21] and is substantiated by research in progress with some indication that the effect is not entirely the result of changes in free ion concentration.[22] A role of calcium in the observed uptake and acute toxicity of cadmium has been suggested.[23,24,25] An effect of salinity on cadmium toxicity has also been demonstrated in decapod larvae of several species.[26,27] Similar studies with copepods and bivalve larvae are lacking.

Studies of interactions between metals and both temperature and salinity are of special interest in estuaries such as Chesapeake Bay. Even in the absence of chemical interactions between a metal and chloride (i.e. salinity), there is still a basis for interaction of these variables, especially near the thermal and salinity tolerance limits for any test species. Most research relating temperature and salinity interactions with heavy metals has been done with decapod larvae. It seems safe to conclude that at least for these zooplankters there are significant interactions between metals and temperature or salinity. Often, however, the effect of temperature or salinity alone on survival or some other endpoint is much greater than that of the metal alone.[27,28]

Evaluation of metal toxicity over a range of salinities and temperatures is important regardless of whether or not there is a direct effect of salinity and temperature on speciation of the toxicant. Organisms tend to be less tolerant to any added stress, such as a toxicant, when they are already challenged.

Results of laboratory studies are difficult to apply in the field because they are usually conducted under constant conditions with single chemicals. In contrast, organisms in estuaries are faced with complex mixtures of metals and other contaminants. Only two experimental studies were identified which tested interaction effects of two or more metals on zooplankton; one on decapod larvae[29] and one on a marine copepod.[30] For decapod larvae, time from hatch to megalopa exposed to mixtures of lead and zinc was extended, primarily resulting from lead. There was a significant interaction between lead and zinc, but zinc

TABLE 1

Summary of the acute toxicity data with respect to heavy metals.
(All values are expressed as µg/L of the metal.)

A - LC50 data for Chesapeake Bay zooplankton species in four taxa.

Metal	Copepods	Mysids	Bivalve embryos		Decapod larvae	
Ag		249	5.8-21	(22)[a]		(55)
As		1740	750	(326)		(232)
Cd	90-130 (500-1,000)	11.3-113	3800	(611)	50-300	(247)
Cr	(5,000-8,000)	2030		(>10000)		(3440)
Cu	9-80 (80-1800)	181	16.4-103	(50-100)		(49)
Hg	10 (230)	3.5	5.6-4.8	(5.5-6.7)	9.9	(8.2)
Ni	(6000)	508	300 -1200	(349)		(1360)
Pb		3130	780 -2450	(758)		(575)
Se				(>10000)		(>1000)
Zn	(1450)	499	160 -3100	(119-310)	500-1000	(456)

[a]Values in parentheses were derived for non-Chesapeake Bay species; for bivalves, *Crassostrea gigas*, and for decapod larvae, *Cancer magister, Upogebia pugettensis* or *Callianassa californiensis*.

B - Rank order of metal toxicity to four taxa

Taxa	Rank Order
Copepods	Hg <· Cu <· Cd <· Cr <· Zn <· Ni
Mysids	Hg > Cd > Cu > Ag > Zn = Ni > As > Cr > Pb
Bivalve larvae	Hg > Ag > Cu > Zn > As = Ni > Cd ± Pb > Cr = Se
Decapod larvae	Hg > Ag = Cu > Cd ≥ As ≥ Zn > Pb > Ni > Cr ≥ Se

alone caused little or no increase in time to megalopa.[29] For copepods, copper, cadmium and chromium when presented individually or in mixtures caused mortality. There were clear interaction effects when presented in mixtures. Interestingly, the toxicity of the three metal mixture was higher than that of the metals tested separately but lower than that of any two metal mixtures.[30]

Death of individuals is only one criterion used to measure the response of zooplankters to heavy metals. Population extinction (an extension of the individual death concept) has also been used as an endpoint, especially for copepods for which tests often involve mixed age populations.[31] Various sublethal criteria are used including longevity, fecundity, respiration rate and feeding activity,[32] number of eggsacs and interval between eggsacs (broods),[33] swimming rate and development time.[34] Delay in development of larval stages is a recognized response to metals, especially for decapod larvae.[27,28]

Effective concentrations based on sublethal endpoints are sometimes lower than those based on mortality, but the significance of this is not always immediately clear. The physiological systems used to measure sublethal responses of larvae are, however, important in some way to the survival of the population of organisms. For example, an increase in duration of development affects the amount of predation pressure applied to a species during the vulnerable planktonic period. A prolonged larval period, even without other physiological consequences, could therefore reduce the numbers of larvae metamorphosing.

There is some evidence that copepod populations adjust genetically to chronic heavy metal stress. *Acartia tonsa* from polluted areas were determined to be resistant to sublethal copper stress,[32] with higher LC_{50} concentrations than observed for copepods from "clean" areas.[35] The genetic consequences may depend on the most sensitive stage and different developmental stages often differ markedly in sensitivity.[36]

PESTICIDES

Published information on the effects of pesticides on zooplankton species is relatively scarce given the many chemicals and formulations used for commercial pest control products. This becomes especially obvious when one considers the number of compounds in various classes of pesticides which have received attention.

Pesticides vary greatly in acute toxicity, ranging from 0.33 μg/L for *M. bahia* exposed for 96 h to phorate[37] to considerably over 10,000 μg/L for bivalve larvae exposed to several herbicides, bactericides, and monobutyltin.[38,39] In an attempt to simplify the comparison of available data for the several zooplankton taxa under consideration, the compounds were sorted into four classes: LC50s < 10 μg/L, 10-100 μg/L, 100-1000 μg/L, and > 1000 μg/L. Several points were immediately obvious from this summary of the data.

Many more compounds seem to have been tested with bivalve larvae than with copepods, mysids or decapod larvae; however, much of the information for bivalves is from a single study involving larvae of *C. virginica* and *M. mercenaria*.[38] In any case, copepods are clearly the least studied taxon with respect to mortality effects.

The crustacean taxa appear to be considerably more sensitive to pesticides than bivalve larvae. Data have been found for three or more taxonomic categories for only four compounds: atrazine, sevin, tributyltin (TBT) and toxaphene. Indeed, there are only two additional pesticides (malathion and dieldrin) for which data have been published for bivalve larvae and even one other zooplankton species. Thus the data used to compare relative sensitivities of species are limited. Atrazine is most toxic to copepods[40] and mysids, and least toxic to oyster larvae by several orders of magnitude. For sevin, mysids are most sensitive, and bivalve larvae least sensitive.[21,38,41] Data for toxaphene lead to the same conclusion.[21,38,42] One value for decapod larvae seems to belie this conclusion, but this value is questioned because of inconsistency in the reported data. The original data are no longer extant precluding a reevaluation. Only for TBT do bivalve larvae appear to be similar in sensitivity to crustaceans.[43,44,45]

Finally, mysids and decapod larvae seem quite similar in response to various pesticides (copepod data are too sparce to include in the comparison). Both taxa are very sensitive, with mysids slightly more sensitive than decapod larvae in most cases.

Relative toxicities based on laboratory tests must be interpreted in context with actual concentrations in the Bay. For example, tributyltin, used as an antimicrobial agent industrially or as an antifouling agent primarily in the marine and estuarine milieu, is highly toxic to the copepod *Acartia tonsa* (0.65 μg/L, 96 h LC50[45] and 1.1 μg/L, 48 h LC50),[46] to the copepod *Eurytemora affinis* (0.6 μg/L, 72 h LC50[46]) to mysids (0.61 μg/L)[43] and to bivalve larvae (1.1-1.6 μg/L)[44] and only slightly less toxic to decapod larvae (30-50 μg/L).[47] Field survey data in the Chesapeake Bay indicate that the ambient concentration in areas near marinas and shipbuilding/repair facilities average about 0.02 μg/L with periodic upward excursions to 0.9 μg/L; concentrations elsewhere are near the detection limit of < 0.002 μg/L.[48] Over the greatest areal extent of the Bay and its tributaries, TBT concentrations are usually over 100-fold below an acutely toxic concentration. The absence of any chronic data at present reduces the reliability of further hazard assessment since no acute/chronic ratio is available for TBT. Application factors derived from laboratory acute and chronic test data for several pesticides vary over at least an order of magnitude. An application factor of < 0.01 is not unknown for other compounds,[49] but there is no evidence that such a value is appropriate for TBT.

The effects of salinity and temperature on pesticide toxicity have not received attention comparable to that for metal toxicity. The limited data available[50,51] indicate that interactions do occur when testing decapod larvae. No data were found for copepods, mysids or bivalve larvae. This lack of data represents a potentially significant gap when one attempts to evaluate pesticide hazard to estuarine zooplankton.

One response of decapod larvae to pesticides is prolonged larval duration even at concentrations which do not produce a marked reduction in survival.[52] This seems to be a response to every pesticide tested (as well as most other toxicants). The implications of this observation were discussed previously.

Underlying all lethal and sublethal effects is bioaccumulation which includes accumulation both from water and food. In some cases, accumulation through the food chain may dominate,[53] though in general uptake from water predominates. In some cases, biota may play a role in degrading the pesticide.[54] The effect of some pesticides taken up may be mitigated by elimination, which itself is influenced by feeding, egglaying and excretion.[55]

There is good evidence that decapod larvae accumulate dieldrin faster from water than from food.[52,56] The body burdens resulting from ingesting dieldrin contaminated food were sufficient to produce both lethal and sublethal effects, at high dietary concentrations.[57] Different rates of accumulation based on source are not unusual. It has recently been shown for a fish that when a pesticide is present in both water and food, uptake from the two sources will be additive.[58] The implication is that, if the food is at equilibrium with the water, uptake from water will always dominate.

OXIDANTS

The major sources of oxidant residuals, primarily chlorine, are disinfection of treated sewage and antifouling activities in industrial cooling systems (electric generating plants). Chlorination of treated sewage results in discharge principally of monochloramine (plus low concentrations of organochlorines, many of which do not contribute to the measured residual). In contrast, direct chlorination of a cooling water can result in primarily free chlorine plus a variety of combined chlorine residuals. In marine and estuarine waters, bromine analogs are produced through reaction with bromide.

After residuals are introduced into estuarine water, additional reactions will occur to modify the mixture of compounds in the total oxidant residual. Ultimately the residual will decay or be diluted, but in the process small yet potentially significant concentrations of various toxic materials may be formed. These include various "combined" residuals such as mono- and dichloramine, mono- and dibromamine, and bromate. In all cases, many haloorganic compounds may be formed by reaction with dissolved organics in saline water. The principal haloorganics formed in chlorinated sewage or surface saline waters and reported to be in surface waters of the Bay are small concentrations of the trihalomethanes, chloroform and bromoform.[59] Other oxidants, primarily bromine chloride and ozone, have been proposed as alternatives to chlorine in the primary applications which affect receiving waters. In both cases, many of the same residual compounds may be formed.

Considerable research has been published regarding the toxicity of chlorine residuals to estuarine zooplankton including copepods, bivalve larvae and decapod larvae. Total residual chlorine is toxic to all of these taxa at concentrations of 20 to 100 μg/L in 48 or 96 h tests.[60,61,62,63,64,65,66,67] Toxicities of chlorine of around 400 μg/L were reported in two studies[68,69] although in 24 hour tests there was evidence of greater sensitivity in pre-adults. These estuarine zooplankters have been shown to be among the most sensitive organisms to oxidant residuals, rivaled in estuarine waters only by the eggs and larvae of some fishes.[70]

Oyster larvae seem to become more tolerant of oxidant residuals as they develop. The 48 h EC50 for oyster embryos is 26 μg/L,[60] whereas the 48 and 96 h EC50s are 300 and 60 μg/L, respectively.[65] There are somewhat conflicting data for pediveligers. In a study with chlorinated sewage added to estuarine water, settlement and subsequent metamorphosis were inhibited by total oxidant residual concentrations between 20 and 60 μg/L, quite similar to concentrations affecting embryos. In a contrasting study, more than 50% of recently attached oyster spat survived exposure to a total oxidant residual of 300 μg/L (no sewage effluent), perhaps reflecting the difference in methodology and the ability of the oyster spat, even newly set and metamorphosed, to close their valves under adverse conditions.[65]

Oxidant residuals of bromine chloride, expressed as molar equivalents of oxidant, are approximately as toxic to zooplankton and other estuarine organisms as chlorine produced oxidant residuals expressed on the same basis. However, these residuals appear less toxic on a mass concentration basis.[60,69] The similarity in toxicity based on molar equivalents can be interpreted to reflect a similarity in the mixture of oxidant residuals independent of the oxidant introduced.

Much less data have been published regarding ozone effects on estuarine zooplankton. Oyster larvae are extremely sensitive to ozone-produced oxidant residuals[72] which are presumably similar to those produced by chlorine or bromine chloride. Thus, though ozone decays rapidly, small barely measurable residuals could have adverse effects on estuarine zooplankton much as some think may now occur following chlorination.

The residuals of chlorination can be reduced to chloride by reaction with various reducing agents prior to release. When this is done, the acute lethal toxic effects associated with chlorination and oxidant residuals are reduced or eliminated.[63,71,73] Indeed, when secondary treated sewage is chlorinated and dechlorinated, the toxicity of the mixture when added to estuarine water may be reduced compared to the unchlorinated sewaged effluent.[71,73] Since some reducing agents are relatively inexpensive, it is realistic to consider dechlorination as one strategy for eliminating oxidant residuals in natural waters near treated sewage effluent discharges as has been proposed under the Bay Cleanup Program. It should be noted, however, that chlorination-dechlorination with sodium thiosulfate caused reproductive failure in copepods which appeared otherwise unaffected.[68]

FIELD STUDIES OF POLLUTANT EFFECTS

Laboratory studies with single species exposed to one or several toxic substances can yield information on the effects of toxicants on the zooplankton community. However, one can only understand the extent of impact through carefully designed in-situ field studies. In Chesapeake Bay, there have been relatively few studies to evaluate effects of pollutants on zooplankton under complex field conditions. In these few studies, copepod distributions and bivalve larval settlement in the field have been used to provide direct information regarding the effects of pollutants on the zooplankton community.

No field studies were identified involving mysids, early stage larvae of bivalves, or decapod larvae to assess the impact of toxicants in natural surface waters on zooplankton communities. Generally, surveys of these components of the zooplankton community have been broad scale, and therefore cannot be used to focus on effects of a particular point source discharge or center of a non-point source discharge. In contrast, the number of field studies on copepods, in the Bay and elsewhere, seem to exceed the number of laboratory studies.

FIELD STUDIES WITH COPEPODS

One group of field studies with copepods dealt with uptake, body burden and bioaccumulation of contaminants, particularly heavy metals and pesticides. Another, involving a wide range of contaminants including hydrocarbons, oxidants and waste heat as well as heavy metals and pesticides, focused on toxic effects.

Mercury uptake is greater in microzooplankton and algae than in macrozooplankton and fish larvae.[74] Higher levels of mercury have been found in phytoplankton and detritus than in zooplankton.[75,76] One should not infer from these observations that there is no bioaccumulation from phytoplankton by herbivorous zooplankters. Differences in body burden may simply reflect differences in equilibrium kinetics among species, perhaps due to differences in lipid quality or quantity.

Several uptake studies involve complex mixtures of metals.[77,78,79,80,81,82] Both seasonal and geographic variation and in some cases interspecies variation in the body burdens of various metals were reported. No consistent patterns emerged.

While arsenic, copper, iron and zinc concentrations were higher in fish than in copepod prey, the opposite was observed for cadmium.[81] In contrast, the mercury contents in fish and zooplankton in another study were nearly equal.[83] These differences may reflect general responses to specific metals and how they are handled in fish as opposed to zooplankton.

The degree of heavy metal toxicity may depend on various physiological factors in addition to uptake; for example, whether the metal occurs in the enzyme pool or bound to a metallothionein,[84] the pH or cationic strength[85] and nutrition.[86] In at least one case, the availability of trace metals depended on naturally occurring organic matter.[87]

In a study of DDT uptake, retention by feral copepods depended on phytoplankton densities, with 60-70% of that ingested at low density being retained, and only 10% under bloom conditions.[88] DDT is also taken up directly from the water. Pesticides are bioaccumulated,[53] depending partly on biodegradability. Both pesticides and metabolic by-products have been measured in copepods.[54,89] Using experimental enclosures, biodegradation rates may have been limited by lack of inorganic nutrients. Biodegradation of pesticides may not be sufficient to reduce toxicity if other stable and toxic compounds are released by hydrolysis of the pesticides.

Rates of elimination of PCBs from a copepod increased with feeding and egg production.[55] In a study with euphausiids, high concentrations of PCB's were reported in fecal pellets.[90]

In contrast to heavy metals and pesticides, studies of oil and related hydrocarbons have been focused mainly on toxicity. In some cases the studies followed major oil spills. In studies of sublethal effects[91,92,93,94] effects on reproduction have been observed at concentrations of 50 mg/L crude oil[91] and on ingestion rates, viability and swimming behavior at concentrations as low as 80 μg/L.[92] Sublethal oil concentrations may interfere with chemoreception and food perception in copepods.[93] These indirect effects, however, may be mitigated by the ability of copepods to metabolize petroleum hydrocarbons.[94]

The effects of major oil spills on zooplankton and other biota appear to be short-term,[95,96,97,98,99] although studies of oil spills often lack controls.[99] In simulated spills, no significant effects were observed on sediments, bacteria, or copepods exposed for several days to concentrations up to 200 mg/L Prudhoe Bay crude oil.[100] However, in another simulation, chronic exposures of 190 μg/L No. 2 fuel oil reduced zooplankton populations.[101] This large difference may reflect differences in proportions of water-soluble toxic components in the different hydrocarbon mixtures. It appears that, if the concentrations of oil as a result of a spill do not result in the disappearance of phytoplankton or zooplankton, recovery may occur within months. However, it is obvious that studies on zooplankton, particularly short-term studies, tell us little of the effects on the ecosystem in general.

Steam electric power plants are a source of both waste heat and biocides, (particularly chlorine) used to control bio-fouling. Some studies on heat and oxidants together have been done in Chesapeake Bay. Of the two contaminants, chlorine caused greater effects on copepod mortality than temperature.[102,103,104] A much greater impact has been suggested for larval zooplankton than phytoplankton,[104] so, in the environment, chlorine effects are most likely directly on the zooplankton.

In a short term assay for chlorine tolerance in individual copepods, genetic adaptation was observed in response to low concentrations of chlorine residuals (100 μg/L).[68] In addition, copepods collected from natural waters with measurable residual oxidant had significantly higher tolerances to chlorine than copepods from water with no detectable chlorine.[68] There is significant genetic variability in chlorine tolerance based on intra-class correlations within copepod families,[105] supporting the earlier observation of genetic changes in field and laboratory populations.

One possible alternative to chlorine as a biocide is bromine chloride.[106] There was little difference between toxicities of chlorinated and bromo-chlorinated water to copepods at various ages after inoculation.[69,107] Free chlorine appears more toxic than bromine chloride, but there is no difference when concentrations are expressed on a milliequivalent basis.[60]

Some studies on the effect of power plants on copepods have been concerned with entrainment effects due mainly to increased temperatures. The temperature increases which are tolerated depend on ambient temperatures.[108,109] Depending on the temperature differential, there may be genetic differences between progeny from intake and progeny from discharge samples of copepods.[110] These genetic differences may be delayed effects of entrainment[109,111] even if the immediate consequences measured by mortality differences are not significant.[112] Laboratory studies done in conjunction with some of these field studies have shown that genetic and physiological effects vary between the sexes and depend on rates and magnitudes of temperature changes.[113,114] The genetic expression of temperature tolerance, however, does not depend on ambient conditions, so laboratory and field trials can safely be compared.[115]

Power plants have an effect on non-entrained organisms in the surrounding receiving waters.[116,117,118] Mortalities of biota in receiving waters varied seasonally,[116] estuarine species were less sensitive than neritic forms to heated effluent[117] and, what may be a non-thermal effect, zooplankton seemed less able to avoid predators in the turbulent receiving waters.[118]

Dredging of contaminated sediments and sewage discharges are two additional human activities for which field studies of copepods have been made. Dredged material, which has some similarities in impact to drilling muds,[119] has been shown to affect biota least at a dredge site and most at a disposal site, with intermediate effects downstream from the dredge site.[120]

Impact assessment for dredging using zooplankton requires careful design and sampling.[121] Sediment toxicity testing is very difficult and perhaps more subjective than toxicity testing in water.[22] The methodology for testing sediments for toxicity is still evolving with new species and endpoints being proposed.[123,124,125,126]

The effects of sewage outfalls on zooplankton populations are seen most clearly in the immediate vicinity of the outfalls.[127,128] In the absence of toxins the organic loading may actually result in an enrichment of biota beyond the immediate area of discharge.

CHLORINATION EFFECTS ON BIVALVE LARVAL SETTLEMENT

Settlement of oyster spat, and secondarily, barnacles, polychaetes, and tunicates was monitored in the James River, VA at an array of stations surrounding the discharge from the James River Sewage Treatment Plant (JRSTP).[129] This study focused on effects of a toxicant (chlorinated sewage) on naturally occurring late meroplanktonic stages of sessile species in the macrofouling community as well as the metamorphic phase of development.

As noted earlier, oyster pediveliger larvae exposed to chlorinated sewage in laboratory tests exhibit depressed settlement and metamorphosis at low concentrations such as might generally occur in the vicinity of sewage treatment plants. Near the JRSTP, total chlorine residual concentrations at or above concentrations observed to have an effect in the laboratory have been measured only in the boil from the discharge.[130] In the recent field study of settlement, no further data regarding residuals in the field were obtained. Residuals in the sewage effluent prior to discharge closely approximated 2.0 mg/L, which would produce a maximum residual concentration in the boil of less than 0.1 mg/L.

At no time during the two-year study was the number of oyster spat observed at stations within 30-40 m of the boil reduced compared to more remote locations, up to 1.6 km away. The same can be said for settlement of *Balanus* sp., *Polydora* sp., *Hydroides dianthus,* and *Molgula manhattensis*.

In New Haven Harbor, CT, spatfall was reduced whenever total chlorine residuals were high (0.33 mg/L to 0.27 mg/L) in a surface water near sewage outfalls, compared to areas with low chlorine residuals (0.03 mg/L to 0.19 mg/L).[67] These results suggest that the failure to observe an effect in the James River reflects the absence of sufficiently high concentrations to have an impact. However, the highest residual in the boil may have exceeded the laboratory estimates of inhibitory residuals.[71]

Total chlorine residual concentrations at the point of discharge in the James River may slightly exceed a concentration shown to inhibit settlement in the laboratory. However, these concentrations do not exceed 0.1 mg/L.[29] The following question must therefore be addressed: "why is no impact detectable?"

One explanation is that the expected concentration was never in fact realized, and therefore no inhibition of settlement occurred. This could be the case if the dilution rate at JRSTP is greater than 95% or the decay rate for the chlorine residual is very high in the receiving water.

Alternatively, the effect of chlorine residuals may be so pervasive in the James River that the impact of chlorine residuals extends beyond the bounds of the present study. This is deemed unlikely since the results within this study area are comparable to those of a more extensive monitoring effort encompassing all major seed-rocks in the James River during the same period.[31]

A final alternative explanation rests on an important difference between the laboratory experiments and the field situation. Recruitment of pediveligers in the laboratory test is finite (5000 larvae added once at the start of the test) whereas in the field recruitment is continuous. Despite a low percent settlement and metamorphosis, the sheer number of pediveligers passing the setting substrate could lead to settlement indistinguishable from that elsewhere.

BIOLOGICAL WATER QUALITY TEST

As may be seen from the review, lethality is often used as a biological criterion for water quality. Sublethal criteria are also important and even essential if a range of water quality is to be tested. In this section we describe some lethal and sublethal tests not described earlier.

A biological test method with embryos of bivalves[132,133] in particular the Pacific oyster, was used in Puget Sound[133,134,135] to define areas of high, medium and low water quality. The test was based on larval survival and percentage abnormal larvae produced from embryos cultured in water collected from specific sites. This test was extremely sensitive; reliability of the data was enhanced by use of a reference toxicant control for condition of embryos used in different tests.[136]

The bivalve larval test[133] does not discriminate among a variety of anthropogenic toxicants nor between anthropogenic toxicants and naturally occurring toxicants such as phytotoxins. Excessive amounts of phytotoxins which would deteriorate overall water quality may reflect anthropogenic activities, and in that sense be anthropogenic. In any case, if the method were applied to locate the source of any toxicant, ancillary data regarding phytoplankton community structure and distribution should provide insight to identify the specific cause of decreased water quality.

This methodology has not been applied within the Chesapeake Bay although the technology exists to perform such evaluations using bivalve embryos. Such a test would provide a sensitive biological tool to help focus attention on those locations within the Bay and its tributaries at which there is decreased water quality. The oyster embryo test requires two days to complete and depends on availability of suitably conditioned broodstock which requires specialized equipment and considerable laboratory space. Oysters have the advantage for testing estuarine water over many zooplankters of broad euryhalinity.

There is clearly a need for several measures of water quality based on sublethal criteria. Suitable endpoints might include development time and reproduction, or various physiological responses to stress.[137] Several of these endpoints are specific to the organism or the stressor. Among them are osmoregulation, ion regulation, taurine: glycine and other amino acid ratios, enzyme activity, energy availability (as energy charge ratio), oxygen consumption, serum constituents and finally measures of metabolism which integrate several biochemical and physiological responses.[137] In another review,[138] criteria for useful indicators of biological effects of pollution are listed, together with criteria for indicator organisms. Additional responses are described, based on growth and reproduction.

There are two cellular endpoints currently under investigation in the laboratory of the first author, which have not been suggested elsewhere. These are being investigated with the copepod *Eurytemora affinis* but are usable in principle with any indicator organism and, quite likely, with a wide spectrum of contaminants. The first of these is plasma membrane fluidity. It is a well-documented fact that membrane fluidity or viscosity changes as an adaptive response to temperature and other stressors. Such phase changes can be observed spectroscopically in whole organisms.[139] The other endpoint is synthesis of novel proteins under stress conditions.[140] These proteins are known as "heat shock proteins", but have been shown to be induced by a variety of stimuli in addition to heat shock. These proteins are now referred to as "stress proteins". All five stress proteins identified to date are synthesized following chronic exposure to contaminants or to environmental changes within normal ecological limits. To have practical applications, both endpoints will have to be linked to a simpler response. In the case of the stress proteins, an immunological method of detection would allow large-scale yet inexpensive assays.

Whatever laboratory biological test of water quality is used, coupling it with chemical analyses of the same water samples will facilitate interpretation of field observations which are possibly the result of pollutant effects. If a water sample from a given area tests high in water quality (e.g. embryo survival is high, abnormal development is infrequent, or stress proteins are not synthesized), whereas the zooplankton community is depauperate, one might shift attention to food availability or predation and away from anthropogenic chemicals as the basis for the community stress.

IN SEARCH OF A BALANCED ASSESSMENT

In developing this review of contaminant effects on the zooplankton community, a major objective was to assess the impact of each class of contaminant on this community at the present time. To do this, we examined both laboratory and field data.

Laboratory studies, performed under controlled and reproducible conditions with single species and single compounds using death as the principal end-point, can identify certain compounds as extremely hazardous. One such compound is TBT since 1) it is highly toxic, 2) it is presently found within the Bay at concentrations which are possibly significant to the zooplankton communities in localized areas such as marinas, and 3) expanded use for macrofouling control on ship hulls or in cooling water systems is likely if no regulatory action is taken. Chlorine residuals are another pollutant identified since 1) they are highly toxic, 2) they are presently found at concentrations which may have an adverse effect, and 3) expanded or modified use for disinfection or macrofouling control could result in elevated concentrations. These types of laboratory data are generally considered sufficient to define a water quality criterion which is then used to regulate municipal and industrial discharges.

To demonstrate the existence of a real population effect in the field is a more complex task and is rarely attempted with zooplankton. It may, nevertheless, be an important step in the development and implementation of regulatory action. In the case of chlorine residuals, laboratory toxicity data were collected using valid methods. These methods could not include recruitment as would occur in the field. Using historical estimates of residuals at the point of discharge at one plant, it seemed that a real and present impact was highly likely at this and similar sites. Laboratory studies also demonstrated the benefit of dechlorination. On this basis, it would seem reasonable to propose dechlorination to eliminate the presumed present impact. However, a measurable impact of present chlorination practice could not be demonstrated in the field, either because of erroneous assumptions regarding chlorine residual concentrations near the discharge or because of real, albeit unavoidable, deficiencies in experimental design.

The important point is not to account for the precise reason(s) for the difference between conclusions based on laboratory tests and on actual field observations, but to recognize that any negative impact of chlorinated sewage discharge is difficult to demonstrate in the field. The field data thereby call into serious question the *necessity* for expensive dechlorination, although some general benefit of eliminating the release of a toxicant into receiving water must surely accrue.

Yet relevant field evaluations are difficult to design, time consuming and costly to implement, and results are often difficult to interpret. A balanced approach must be sought to prevent precipitous regulatory action based on laboratory data derived from standardized tests which were designed to define relative toxicity of substances rather than ecotoxicological effects. The answers may lie in two quite different directions. One is to make greater use of sublethal biological tests of water quality, particularly those based on the rapid responses discussed earlier. Parcels of ambient waters could be tested biologically and chemically and the critical chemicals identified in cases of positive biological responses, using further tests on single chemicals if necessary. The other approach is to use community indices of low water quality, using field data augmented by laboratory microcosm tests.[41] Whatever methods are used, extrapolation from laboratory to field is no more straightforward in environmental toxicology than it is in any other area of environmental biology.

LITERATURE CITED

1. Heinle, D.R. 1969. Culture of calanoid copepods in synthetic sea water. *J. Fish. Res. Bd. Canada.* 16:150-153.
2. Bradley, B.P. 1975. The amomalous influence of salinity on temperature tolerances of summer and winter populations of the copepod *Eurytemora affinis. Biol. Bull.* 148:26-34.
3. Moraitou-Apostolopoulou M. 1978. Acute toxicity of copper to a copepod. *Mar. Pollut. Bull.* 9:278-282.
4. Sosnowski, S.L. and J.H. Gentile. 1978. Toxicological comparison of natural and cultured populations of *Acartia tonsa* to cadmium, copper and mercury. *Fisheries Research Board of Canada J.* 35:1366-1379.
5. Mauchline, W. 1980. Biology of the mysids. *Adv. Mar. Biol.* 18:1-369.
6. Roberts, M.H. Jr., J.E. Warriner, C-F Tsai, D. Wright and L.E. Cronin. 1982. Comparison of estuarine species sensitivities to three toxicants, *Arch. Environ. Contam. Toxicol.* 11:681-692.
7. Gentile, S.M., J.H. Gentile, J. Walker and J.F. Heltshe. 1982. Chronic effects of cadmium on two species of mysid shrimp: *Mysidopsis bahia* and *Mysidopsis bigelowi. Hydrobiologia.* 93:195-204.
8. Loosanoff, V.L. and H.C. Davis. 1963. Rearing of bivalve mollusks. *Adv. Mar. Biol.* 1:1-136.
9. Costlow, J.D., Jr. and C.G. Bookhout. 1960. A method for developing brachyuran eggs in vitro. *Limnol. Oceanogr.* 5:212-215.
10. Provenzano, A.T., Jr. 1967. Recent advances in the laboratory culture of decapod larvae. pp. 940-945. in *Proc. Symp. Crustacea. Pt. II.*
11. Roberts, M.H., Jr. 1975. Culture techniques for decapod crustacean larvae, pp. 209-220. *In:* W.L. Smith and M.H. Chanley (Eds.). *Culture of Marine Invertebrate Animals* (Proceeding of Conference on Culture of Marine Invertebrate Animals), ed. Plenum Press, New York, N.Y.
12. Gentile, J.H. and S. Gentile. 1981. The effects of chronic mercury exposure on the survival, reproduction and population dynamics of *Mysidopsis bahia.* C.M. 1981/E:35 1,7.
13. Lussier, S.M., J.H. Gentile, and J. Walker. 1985. Acute and chronic effects of heavy metals and cyanide on *Mysidopsis bahia,* Crustacea, Mysidacea. *Aquat. Toxicol.* 7:25-36.

14. Calabrese, A., R.S. Collier, D.A. Nelson and J.R. MacInnes. 1973. The toxicity of heavy metals to embryos of the American oyster *Crassostrea virginica*. *Mar. Biol.* 18:162-166.
15. Calabrese, A. and D.A. Nelson. 1974. Inhibition of embryonic development of the hard clam, *Mercenaria mercenaria*, by heavy metals. *Bull. Environ. Contamin. Toxicol.* 11:92-97.
16. Calabrese, A., J.R. Macinnes, D.A. Nelson, and J.E. Miller. 1977. Survival and growth of bivalve larvae under heavy metal stress. *Mar. Biol.* 41:179-184.
17. Shealy, M.H., Jr. and P.A. Sandifer. 1975. Effects of mercury on survival and development of the larval grass shrimp *Palaemonetes vulgaris*. *Mar. Biol.* 33:7-16.
18. Martin, M., K.E. Osborn, P. Billig, and N. Glickstein. 1981. Toxicities of ten metals to *Crassostrea gigas* and *Mytilus edulis* embryos and *Cancer magister* larvae. *Mar. Pollut. Bull.* 12:305-308.
19. Glickstein, N. 1978. Acute toxicity of mercury and selenium to *Crassostrea gigas* embryos and *Carcer magister* larvae. *Mar. Biol.* 49:113-117.
20. Sunda, W.G., D.W. Engle, and R.M. Thuotte. 1978. Effects of chemical speciation on toxicity of cadmium to grass shrimp, *Palaemonetes pugio*: importance of free cadmium ion. *Environ. Sci. Technol.* 12:409-413.
21. Nimmo, D.R., R.A. Rigby, L.H. Bahner and J.M. Sheppard. 1978. The acute and chronic effects of cadmium on the estuarine mysid *Mysidopsis bahia*. *Bull. Envir. Contam. Tox.* 19:80-85.
22. DeLisle, P.S. 1986. Personal Communication. Virginia Institute of Marine Sciences.
23. Wright, D.A. 1977. The effect of calcium on cadmium uptake by the tissues of the shore crab *Carcinus meanus*. *J. Exp. Biol.* 67:163-173.
24. Wright, D.A. and J.W. Fain. 1981. Cadmium toxicity in *Marinogamarus obtusatus*: effect of external calcium. *Environ. Res.* 24:338-344.
25. Wright, D.A. and J.W. Fain. 1981. Effect of calcium on cadmium toxicity in the freshwater amphipod, *Gammarus pulex*. *Arch. Environ. Contam. Toxicol.* 10:321-328.
26. Middaugh, D.P. and G. Floyd. 1978. The effect of prehatch and posthatch exposure to cadmium on salinity tolerance of larval grass shrimp, *Palaemonetes pugio*. *Estuaries* 1:123-125.
27. Rosenberg, R. and J.D. Costlow, Jr. 1976. Synergistic effects of cadmium and salinity combined with constant and cycling temperatures on the larval development of two estuarine crab species. *Mar. Biol.* 38:291-303.
28. Vernberg, W.B., P.J. DeCoursey, and W.J. Paggett. 1973. Synergistic effects of environmental variables on larvae of *Uca pugilator*. *Mar. Biol.* 22:307-312.
29. Benijts-Claus, C. and F. Benijts. 1975. The effect of low lead and zinc concentrations on the larval development of the mud-crab

Rhithropanopeus harrisii Gould, pp. 43-52. *In:* J.H. Koeman and J.J.T.W.A. Strik (Eds.). *Sublethal Effects of Toxic Chemicals on Aquatic Organisms.* Elsevier Sci. Publ. Co., New York, NY.

30. Moraitou-Apostolopoulou, M. and G. Verriopoulos. 1982. Individual and combined toxicity of three heavy metals, copper, cadmium and chromium for the marine copepod *Tisbe holothuriae. Hydrobiologia* 87:83-88.
31. Hoppenheit, M. and K-R Sterling. 1977. On the dynamics of exploited populations of *Tisbe holothuriae* (Copepoda, Harpacticoida) V The Toxicity of cadmium. Response to lethal exposure. *Helolander wiss. Meeresunters.* 19:328-336.
32. Moraitou-Apostolopoulou, M. and G. Verriopoulos. 1979. Some effects of sublethal concentrations of copper on marine copepod. *Mar. Poll. Bull.* 10:88-96.
33. Verriopoulos, G. and M. Moraitou-Apostolopoulou. 1981. Impact of chromium to the population dynamics of *Tisbe holothuriae. Arch. Hydrobiol.* 93:59-67.
34. Sullivan, B.K., E. Buskey, D.C. Miller and P.J. Ritacco. 1983. Effects of copper and cadmium on growth, swimming and predator avoidance in *Eurytemora affinis* (Copepoda). *Mar. Biol.* 77:299-306.
35. Moraitou-Apostolopoulou, M. 1978. Acute toxicity of copper to a copepod. *Mar. Poll. Bull.* 9:278-281.
36. Verriopoulos, G. and M. Moraitou-Apostolopoulou. 1982. Differentiation of the sensitivity to copper and cadmium in different life stages of a copepod. *Mar. Poll. Bull.* 13:123-129.
37. Nimmo, D.R., T.L. Hamaker, E. Matthews and J.C. Moore. 1981. An overview of the acute and chronic effects of 1st and 2nd generation pesticides on an estuarine mysid, pp. 3-19. *In:* F.J. Vernberg, A. Calabrese, F.P. Thurberg and W.B. Vernberg, (Eds.). *Biological Monitoring of Marine Pollutants.* Academic Press, New York.
38. Davis, H.C. and H. Hidu, 1969. Effects of pesticides on embryonic development of clams and oysters and on survival and growth of the larvae. *Fish. Bull.* 67:393-404.
39. Becerra-Huerra, R.M. 1981. The effect of organotin and copper sulfate on the late development and presettlement behavior of the hard clam *Mercenaria mercenaria.* Masters Thesis, University of Maryland, 83 pp.
40. Ward, G.S. and L. Ballantine. 1985. Acute and chronic toxicity of atrazine to estuarine fauna. *Estuaries.* 8:22-27.
41. Stewart, N.E., R.E. Millemann, and W.P. Breese. 1967. Acute toxicity of the insecticide Sevin and its hydrolytic product 1-naphthol to some marine organisms. *Trans. Am. Fish. Soc.* 96:25-30.
42. Courtenay, W.R., Jr. and M.H. Roberts, Jr. 1973. Environmental effects on Toxaphene toxicity to selected fishes and crustaceans. U.S. EPA, EPA-R3-73-035. Ecological Research Series, Washington, D.C., 73 pp.

43. Valkirs, A., B. Davidson, and P. Seligman. 1985. Sublethal growth effects and mortality from long-term exposure to tributyltin with marine bivalves and fish. Naval Ocean Systems Center, Technical Report 1042, TR 1042, 36 pp + app.
44. Roberts, M.H., Jr. 1986. Acute toxicity of tributyltin chloride to embryos and larvae of two bivalve molluscs, *Crassostrea virginica* and *Mercenaria mercenaria*. (in preparation).
45. U'ren, S.C. 1983. Acute toxicity of bis(tributyltin) oxide to a marine copepod. *Mar. Pollut. Bull.* 14:303-306.
46. Hall, L.W., Jr. 1987. Personal Communication. Johns Hopkins Univ. Applied Physics Laboratory.
47. Laughlin, R., W. French and H.E. Guard. 1983. Acute and sublethal toxicity of tributyltin oxide (TBTO) and its putative environmental product, tributyltin sulfide (TBTS) to zoeal mud crabs. *Water. Air. Soil Pollut.* 20:69-79.
48. Huggett, R.J. 1986. Personal Communication. Virginia Institute Marine Science.
49. Nimmo, D.R., L.H. Bayner, R.A. Rigby, J.M. Sheppard, and A.J. Wilson, Jr. 1977. *Mysidopsis bahia:* an estuarine species suitable for life cycle toxicity tests to determine the effects of a pollutant, pp. 109-116. *In:* F.L. Mayer and J.L. Hamelin (Eds.). *Aquatic Toxicology and Hazard Evaluation.* Eds. ASTM STP No. 634, Philadelphia, Pa.
50. Christiansen, M.E., J.D. Costlow, Jr. and R.J. Monroe. 1977a. Effects of the juvenile hormone mimic ZR-515 (Altosid) on larval development of the mud-crab *Rhithropanopeus harrisii* in various salinities and cyclic temperatures. *Mar. Biol.* 39:269-279.
51. Christiansen, M.E., J.D. Costlow, Jr. and R.J. Monroe. 1977b. Effects of the juvenile hormone mimic ZR-512 (Altosid) on larval development of the mud-crab *Rhithropanopeus harrisii* at various cyclic temperatures. *Mar. Biol.* 39:281-288.
52. Epifanio, C.E. 1971. Effects of dieldrin in seawater on the development of two species of crab larvae, *Leptodius floridanus* and *Panopeus herbstii. Mar. Biol.* 11:356-362.
53. Tanabe, S., H. Tanaka, and R. Tatsukawa. 1984. Polychlorobipenyls, SDDT, and hexachlorocyclohexane isomers in the western north pacific ecosystems. *Arch. Environ. Contam. Toxicol.* 13:731-741.
54. Kuiper, J. and A. Hanstveit. 1984. Fate and effects of 4-chlorophenol and 2,4-dichlorophenol in marine plankton communities in experimental enclosures. *Ecotoxicol. Environ. Safety* 8:15-36.
55. McManus, G.B., K.D. Wyman, W.T. Peterson and C.F. Wurster. 1983. Factors affecting the elimination of PCBs in the marine copepod *Acartia Tonsa. Estuar. Coast. Shelf Sci.* 17:421-433.
56. Epifanio, C.E. 1973. Dieldrin uptake by larvae of the crab *Leptodius*

floridanus. Mar. Biol. 19:320-322.
57. Epifanio, C.E. 1972. Effects of dieldrin-contaminated food on the development of *Leptodius floridanus* larvae. *Mar. Biol.* 13:292-297.
58. Fisher, D.J., J.R. Clark, M.H. Roberts, Jr., J.P. Connolly and L.H. Mueller. 1986. Bioaccumulation of kepone by spot (*Leiostomus xanthurus*): Importance of dietary accumulation and ingestion rate. *Aquat. Toxicol.* 9:161-178.
59. Bieri, R.H., M.K. Cueman, R.J. Huggett, W. MacIntyre, P. Shou, C.L. Smith, C.W. Su, and G. Ho. 1980. Toxic organic compounds in the Chesapeake Bay. Virginia Institute of Marine Science, Report to Environmental Protection Agency for Grant No. R806012010, 23 pp.
60. Roberts, M.H., Jr. and R.A. Gleeson. 1978. Acute toxicity of bromochlorinated seawater to selected estuarine species with a comparison to chlorinated seawater toxicity. *Mar. Environ. Res.* 1:19-29.
61. Roberts, M.H., Jr. 1978. Effects of chlorinated sea water on decapod crustaceans. pp. 329-339. *In:* R.L. Jolley, H. Gorchev, and D.H. Hamilton, Jr. (Eds.). *Water Chlorination: Environmental Impact and Health Effects, Vol. 2.* Ann Arbor Science Publ. Inc., Ann Arbor, MI.
62. Roberts, M.H., Jr., C.E. Laird, and J.E. Illowsky. 1979. Effects of chlorinated seawater on decapod crustaceans and *Mulinia* larvae. EPA-600/3-79-031. 110 pp.
63. Roberts, M.H., Jr. 1980. Flow-through toxicity testing system for molluscan larvae as applied to halogen toxicity in estuarine water. pp. 131-139. *In:* A.L. Buikema, Jr. and J. Cairns, Jr., (Eds.). *Aquatic Invertebrate Bioassays, ASTM STP 715.* American Society of Testing and Materials.
64. Heinle, D.R. and M.S. Beaven. 1977. Effects of chlorine on the copepod *Acartia tonsa*. Ches. Sci. 18:140-145.
65. Roosenburg, W.H., J.C. Rhoderick, R.M. Block, V.S. Kennedy, S.R. Gullans, S.M. Breenegoor, A. Rosencranz and C. Collette. 1980. Effects of chlorine-produced oxidants on survival of larvae of the oyster *Crassostrea virginica. Mar. Ecol. Prog. Ser.* 3:93-96.
66. Stewart, M.E., W.J. Blogoslawski, R.Y. Hsu and G.R. Helz. 1979. Byproducts of oxidative biocides: Toxicity to oyster larvae. *Mar. Pollut. Bull.* 10:166-169.
67. Stewart, M.E. and W.J. Blogoslawski. 1985. Effect of selected chlorine-produced oxidants on oyster larvae. pp. 521-532. *In:* R.J. Jolley, R.J. Bull, W.P. Davis, S. Katz, M.H. Roberts, Jr. and V.A. Jacobs. (Eds.). *Water Chlorination, Chemistry, Environmental Impact and Health Effects.* Lewis Publishers, Inc. Chelsea, MI.
68. Bradley, B.P. 1977. Long term biotoxicity of chlorine species to copepod populations. Tech. Report No. 42, Water Resources Research Center, Univ. Md. College Park. 16 pp.

69. Bradley, B.P. 1978. Comparison of residual biotoxicity of chlorine and bromine chloride to copepods. Report No. 47, Water Resources Research Center, Univ. Md. College Park, 15 pp.
70. Burton, D.T., L.W. Hall, Jr., S.L. Margrey and R.D. Small. 1979. Interactions of chlorine, temperature change, and exposure time on survival of striped bass (*Morone saxatilis*) eggs and larvae. *J. Fish Res. Bd. Canada.* 36:1108-1113.
71. Roberts, M.H., Jr. and B.B. Casey. 1985. Depression of larval growth and metamorphosis of oysters exposed to chlorinated sewage. pp. 509-520. *In:* R.J. Jolley, R.J. Bull, W.P. Davis, S. Katz, M.H. Roberts, Jr. and V.A. Jacobs, (Eds.). *Water Chlorination, Chemistry, Environmental Impact and Health Effects.* Lewis Publishers, Inc. Chelsea, MI.
72. Demanche, J.M., P.L. Donaghay, W.P. Breese, and J.F. Small. 1975. Residual toxicity of ozonized seawater to oyster larvae. Oregon State Univ., Sea Grant College Program, Publ. No. ORESU-T-75-003, 7 pp.
73. Esvelt, L.A., W.J. Kaufman, and R.E. Selleck. 1973. Toxicity assessment of treated municipal wastewaters. *J. Water Pollut. Contr. Fed.* 45:1558-1572.
74. Zubarik, L.S. and J.M. O'Connor. 1977. A radioisotopic study of mercury uptake by Hudson River biota. *Energy and Environmental Stress in Aquatic Systems, Conference,* Georgia, Nov. 2-4, 1977, pp. 273-292.
75. Flegal, A.R. 1977. Mercury in the seston of the San Francisco Bay estuary. *Bull. Environ. Contam. Toxicol.* 17:733-741.
76. Skei, J.M., M. Saunders, and N.B. Price. 1976. Mercury in plankton from a polluted Norwegian Fjord. *Mar. Poll. Bull.* 7:34-39.
77. Polikarpov, G.G., B. Oregioni, D.S. Parchevskaya and G. Benayoun. 1979. Body burden of chromium, copper, cadmium and lead in the neustonic copepod *Anomalocera pattersoni* (Pontellidae) collected from the Mediterranean Sea. *Mar. Biol.* 53:79-82.
78. Zafiropoulos, D. and A.P. Grimanis. 1977. Trace elements in *Acartia clausi* from Elefsis bay of the upper Saronikos Gulf, Greece. *Mar. Poll. Bull.* 8:79-84.
79. Harstedt-Romeo, M. and F. Laumond. 1980. Zinc, copper, and cadmium in zooplankton from the N.W. Mediterranean. *Mar. Poll. Bull.* 11:133-141.
80. Greg, R.A., A. Adams and D.R. Wenzloff. 1977. Trace metal content of plankton and zooplankton collected from the New York Bight and Long Island Sound. *Bull. Environ. Contam. Toxicol.* 18:3-11.
81. Bohn, A. and R.O. McElroy. 1976. Trace metals (As, Cd, Cu, Fe, and Zn) in Artic cod *Boreogadus saida* and selected zooplankton from Strathcona. Sound, Northern Baffin Island. *J. Fish.Res. Bd. Canada.* 33:2836-2843.
82. Lear, D.W. 1974. Environmental survey of two interim dumpsites Middle Atlantic Bight: Supplement Report. Operation Fetch. Cruise Report 5-10 November 1973. NTIS Report PB-239 257 122 pp.

83. Nishimura, H. and M. Kumagai. 1983. Mercury pollution of fishes in Minamata Bay and surrounding water: analysis of pathway of mercury. *Water Air Soil Pollut.* 20:401-414.
84. Brown, D.A. and T.R. Parsons. 1978. Relationship between cytoplasmic distribution of mercury and toxic effects to zooplankton and chum salmon (*Oncorhynchus keta*) exposed to mercury in a controlled ecosystem. *J. Fish. Res. Bd. Canada.* 35:880-887.
85. Hutchinson, T.C. and F.W. Collins. 1978. Effect of H^+ ion activity and CA^{2+} on the toxicity of metals in the environment. *Environ. Health Persp.* 25:47-55.
86. Sosnowski, S.L., D.J. Germond and J.H. Gentile. 1979. The effect of nutrition on the response of field populations of the calanoid copepod *Acartia tonsa* to copper. *Water Res.* 13:449-452.
87. Whitfield, P.H. and A.G. Lewis. 1976. Control of the biological availability of trace metals to a calanoid copepod in a coastal fjord. *Estuar. Coast. Mar. Sci.* 4:255-269.
88. Harding, G.C., W.P. Vass and K.F. Drinkwater. 1981. Importance of feeding, direct uptake from seawater and transfer from generation to generation in the accumulation of an organochlorine (p, p' DDT.) by the marine planktonic copepod *Calanus finmarchicus*. *Canadian J. Fish. Aquat. Sci.* 38:101-119.
89. Kuiper, J. and A. Hanstveit. 1984. Fate and effects of 3,4-Dichloroaniline (DCA) in marine plankton communities in experimental enclosures. *Ecotoxicol. Environ. Safety.* 8:34-57.
90. Elder, D.L. and W.S. Fowler. 1977. Polychlorinated biphenyls: penetration into the deep ocean by zooplankton fecal pellet transport. *Science.* 197:459-463.
91. Ustach, J.F. 1979. Effects of sublethal oil concentrations on the copepod *Nitocra affinis*. *Estuaries* 2:273-276.
92. Cowles, T.J. and R.F. Remillard. 1983. Effects of exposure to sublethal concentrations of crude oil on the copepod *Centropages hamatus*. I. Feeding and egg production. *Mar. Biol.* 78:45-51.
93. Cowles, T.J. 1983. Effects of exposure to sublethal concentrations of crude oil on the copepod *Centropages hamatus*. II. Activity patterns. *Mar. Biol.* 78:53-57.
94. Berduggo, V., R.P. Harris, S.C. O'Hara. 1977. The effect of petroleum hydrocarbons on reproduction of the estuarine planktonic copepod *Eurytemora affinis* in laboratory cultures. *Mar. Pollut. Bull.* 8:138-146.
95. Lee, W.Y., A. Morris and D. Boatwright. 1980. Mexican oil spill: A toxicity study of oil accomodated in seawater on marine invertebrates. *Mar. Pollut. Bull.* 11:231-237.
96. Linden, O., R. Elmgren and P. Boehm. 1979. The Tsesis oil spill: Its impact on the coastal ecosystem of the Baltic Sea. *Ambio.* 8:244-256.

97. Johansson, S., V. Larsson and P. Boehm. 1980. The Tsesis oil spill. *Mar. Pollut. Bull.* 11:284-295.
98. Linden, O., J. Mattsson and M. Notini. 1983. A spill of light fuel oil in the Baltic Sea. API/EPA/USCG Oil Spill 8th Conf. San Antonio pp. 517-522.
99. Teal, J.M. and R.W. Howarth. 1984. Oil spill studies: A review of ecological effects. *Env. Management.* 8:27-46.
100. Naidu, A.S., H.M. Feder and S.A. Norrell. 1978. The effect of Prudhoe Bay crude oil on a tidal-flat ecosystem in Port Valdez, Alaska. Offshore Technology 10th Ann. Conf. Houston. 1:97-116.
101. Oviatt, C., J. Frithsen, J. Gearing and P. Gearing. 1982. Low chronic additions of no. 2 fuel oil: chemical behavior, biological impact and recovery in a lab estuarine environment. *Mar. Ecol.* 9:121-139.
102. Heinle, D.R. 1976. Effects of passage through power plant cooling systems on estuarine copepods. *Environ. Pollut.* 11:39-61.
103. Lanza, G.R., G.J. Lauer, T.C. Ginn, P.C. Storm and L. Zubarik. 1975. Biological effects of simulated discharge plume entrainment at Indian Point Nuclear Power Station, Hudson River Estuary, USA. IAEA/OECD NEA Symp., Stockholm, pp. 95-129.
104. Goldman, J.C., J.M. Capuzzo and G.F. Wong. 1978. Biological and chemical effects of chlorination at coastal power plants. pp. 291-305. *In:* R.L. Jolley, H. Gorchev, and D.H. Hamilton, Jr. (Eds.). *Water Chlorination: Environmental Impact and Health Effects.* Ann Arbor Sci. Publ., Inc., Ann Arbor, MI.
105. Davis, R.M. 1986. Responses of the Copepod *Eurytemora affinis* to three environmental stressors. PhD Diss., Univ. of Md. Baltimore County. 135 pp.
106. Mills, J.F. 1973. The chemistry of bromine chloride in waste water disinfection. *J. Am. Chem. Soc.* 13:65 (Abstr.).
107. Liden, L.H., D.T. Burton, L.H. Bongers and A.F. Holland. 1980. Effects of chlorobrominated and chlorinated cooling waters on estuarine organisms. *Water Pollut. Contr. Fed. J.* 52:173-186.
108. Kwik, J.K. and T.G. Dunstall. 1985. Mortality of zooplankton resulting from temperature regimes encountered in once-through cooling systems. *J. Great Lakes Res.* 11:26-36.
109. Bradley, B.P. 1980. Responses of copepod individuals and populations to increased temperature variability. Univ. of Maryland Water Resources Research Center Tech. Report No. 62. 25 pp.
110. LaBelle, R.P. and B.P. Bradley. 1982. Selection for temperature tolerance during power plant entrainment of copepods. *J. Thermal Biol.* 7:39-44.
111. Davies, R.M., C.H. Hanson, L.D. Jensen. 1975. Entrainment of estuarine zooplankton into a mid-atlantic power plant. Delayed effects. ERDA 2nd Thermal Ecology Sym., Augusta, Apr. 2-3, 1975, pp. 349-359.

112. Bradley, B.P. 1980. Calvert Cliffs zooplankton entrainment study. Report PPSP-CC-80-1 Md. Dept. Natural Resources, Annapolis, MD.
113. Ketzner, P.A. and B.P. Bradley. 1982. Rate of environmental change and adaptation in the copepod *Eurytemora affinis*. *Evolution*. 36:298-306.
114. Bradley, B.P. and P.D. Ketzner. 1982. Genetic and non-genetic variability in temperature tolerance of the copepod *Eurytemora affinis* in live temperature regimes. *Biol. Bull.* 162:233-245.
115. Bradley, B.P. 1986. Genetic expression of temperature tolerance in the copepod *Eurytemora affinis* in different salinity and temperature environments. *Mar. Biol.* 91:561-565.
116. Miller, M.C., G.R. Hater, T.W. Federle and J.P. Reed. 1975. Effects of power-plant operation on the biota of a thermal discharge channel. ERDA 2nd Thermal Ecology Symp., Augusta, 1975 pp. 251.
117. Alden, W. 1979. Effects of a thermal discharge on the mortality of copepods in a subtropical estuary. *Environ. Pollut.* 20:3-20.
118. Morris, P.A. 1984. Feeding habits of blacksmith, *Chromis punctipinnis*, associated with a thermal outfall. *Fish. Bull.* 82:199-206.
119. Jacobson, J.P. and M.P. Lynch. 1976. Mid-Atlantic Outer Continental Shelf Benchmark Studies. IEEE/Marine Tech. Soc. Oceans 76 Conf., Washington, D.C.
120. Decoursey, P.J. and W.B. Vernberg. 1975. The effect of dredging in a polluted estuary on the physiology of larval zooplankton. *Water Res.* 9:149-157.
121. Sullivan, B.K. and D. Hancock. 1977. Zooplankton and dredging: Research perspectives from a critical review. *Water Res.* 13:461-469.
122. Peddicord, R.K. (Ed.). 1977. Ecological evaluation of proposed discharge of dredged material into ocean waters. EPA/COE Tech. Comm. on criteria for dredged and fill material. ALMS No. 10101670.
123. Swartz, R.C., W.A. DeBen, J.R. Phillips, J.O. Lamberson and F.A. Cole. 1985. Phoxocephalid amphipod bioassay for marine sediment toxicity pp. 284-307. *In:* R.D. Cardwell, R. Purdy and R.C. Bahner (Eds.). *Aquatic Toxicology and Hazard Assessment: Seventh Symposium* ASTM STP 854. Philadelphia, Pa.
124. Williams, L.G., P.M. Chapman, and T.C. Ginn. 1986. A comparative evaluation of marine sediment toxicity using bacterial luminescence, oyster embryo and amphipod sediment bioassays. *Mar. Environ. Res.* 19:225-249.
125. Bengtsson, B.E. 1978. Use of a harpacticoid copepod in toxicity tests. *Mar. Poll. Bull.* 9:238-241.
126. Bengtsson, B.E. 1981. The harpacticoid copepod *Nitocra spinipes* (Crustacea) as a test organism in brackish water toxicological bioassays. Paper at Inserm Collogue International d'Eco toxicologie, Lille, France.
127. Arfi, R., A. Bianchi, M. Bianchi, G. Champalbert, F. Blanc, P. David and M.C. Bonin. 1979. Planktonic systems and urban pollution: study of

of plankton populations. *Oceanol. Acta.* 2:1-12.
128. Soule, D.F. and M. Oguri. 1976. Marine studies of San Pedro, California. Part 12: Bioenhancement studies of the receiving waters in outer Los Angeles Harbor. USC Inst. Marine and Coastal Studies Report USG-SG-5-76. 285 pp.
129. Roberts, M.H., B.B. Casey, C.S. Strobel and E. Van Montfrans. 1986. Field evaluation of the effect of chlorinated sewage on settlement of oysters and barnacles, pp. 106-117. *In: Chesapeake Bay Research Initiatives,* Virginia Institute of Marine Science, College of William and Mary, Gloucester Point, Va.
130. Bender, M.E., D.S. Haven and H.D. Slone. 1978. Report on the Effect of Chlorine on Oysters in the Warwick River, in Response to House Joint Resolution No. 162, Commonwealth of Virginia, 26 pp.
131. Whitcomb, J.P. 1986. Personal Communication. Virginia Institute of Marine Sciences.
132. Okubo, K. and T. Okubo. 1962. Study on the bio-assay method for the evaluation of water pollution. II. Use of the fertilized eggs of sea urchins and bivalves. *Bull. Tokai Reg. Fish. Res. Lab.* 32:131-140.
133. Woelke, C.E. 1972. Development of a receiving water quality bioassay criterion based on the 48-hour Pacific oyster embryo, *Crassostrea gigas. Tech. Rep. Wash. Dept. Fish.* 9:1-91.
134. Cardwell, R.D., C.E. Woelke, M.I. Carr, and E.W. Sanborn. 1977a. Evaluation of water quality of Puget Sound and Hood Canal in 1976. Tech. Memo No. 21, National Oceanic and Atmospheric Administration. 36 pp.
135. Cardwell, R.D., C.E. Woelke, M.I. Carr, and E.W. Sanborn, 1979. Toxic substances and water quality effects on larval marine organisms. Tech. Rep. Wash. Dept. Fish. 45:1-71.
136. Cardwell, R.D., C.E. Woelke, M.I. Carr, and E.W. Sanborn. 1977b. Appraisal of a reference toxicant for estimating the quality of oyster larvae. *Bull. Environ. Contam. Toxicol.* 18:719-725.
137. Dillon, T.M. and M.P. Lynch. 1981. Physiological responses as determinants of stress in marine and estuarine organisms. pp. 227-241. *In:* G.W. Barrett and R. Rosenberg (Eds.). *Stress Effects on Natural Ecosystems.* John Wiley, N.Y. 305 pp.
138. Widdows, J. 1985. Physiological responses to pollution. Marine Pollution Bull. 16:129-134.
139. Vincent, J.V. 1986. Personal Communication. Dept. of Chemistry, University of Maryland Baltimore County.
140. Schlesinger, M.J., M. Ashburner and A. Tissieres (Eds.). 1982. Heat Shock from bacteria to man. Cold Spring Harbor Laboratory, N.Y. 440 pp.
141. Levin, S.A. (Ed.). 1982. New perspectives in ecotoxicology ERC Report No. 14 Ecosystems Research Center, Cornell Univ., Ithaca, N.Y.

Contaminant Problems and Management of Living Chesapeake Bay Resources. Edited by S. K. Majumdar, L. W. Hall, Jr. and H. M. Austin. © 1987, The Pennsylvania Academy of Science.

Chapter Nineteen
ORIGIN, DEVELOPMENT AND SIGNIFICANCE OF CHESAPEAKE BAY ANOXIA

JON H. TUTTLE,[1] ROBERT B. JONAS,[2] and THOMAS C. MALONE[3]

[1]University of Maryland
Center for Environmental and Estuarine Studies
Chesapeake Biological Laboratory
Solomons, Maryland 20688-0038, USA

[2]Department of Biology
George Mason University
Fairfax, Virginia 22030, USA

[3]University of Maryland
Center for Environmental and Estuarine Studies
Horn Point Environmental Laboratories
Cambridge, Maryland 21613-0775, USA

ABSTRACT

Summer anoxia in deep waters of the mesohaline portion of Chesapeake Bay has been an annual event throughout the historical record. There is evidence that the areal extent and duration of oxygen depletion is increasing due to increasing nutrient loading and the consequent increase in phytoplankton production. Phytoplankton carbon fuels oxygen consumption, a significant portion of which is due to aerobic, heterotrophic bacterial metabolism and microbial sulfur cycling. The latter is comprised of two key processes: sulfate reduction catalyzed by obligately anaerobic bacteria in Bay sediments and sulfide oxidation occurring in surficial sediments or in subpycnoclinal waters. Anoxia results when rates of oxygen consumption exceed reaeration to the deep waters. Oxygen depletion begins in late winter, as temperature begins to increase, and is accelerated following the spring freshet. Aerobic metabolism is the most important process leading to the development of anoxia in the spring while sulfur cycling becomes increasingly more important to the maintenance of anoxia during the summer. Both water column and benthic oxygen consumption contribute to oxygen depletion. It appears that dramatic changes have occurred in the ecosystem of the mid-Bay such that a large portion of primary production is metabolized by bacteria rather than transferred to higher organisms in the food web. Widespread anoxia occurred in the summer of 1984, but was virtually absent from the mid-Bay in 1985. This may reflect above average freshwater flow in 1984 compared to 1985, a drought year. It appears likely that these differences in the magnitude of oxygen depletion are related to interannual climatic variability and associated changes in how organic matter is processed by pelagic and benthic consumers.

INTRODUCTION

In recent years, few environmental issues about Chesapeake Bay water quality have generated as much public concern as the problem of anoxia. Anoxia or hypoxia (less than 20% oxygen saturation) are features common to a variety of aquatic habitats including thermally stratified lakes, fjords, upwelling systems,[1] certain oceanic basins,[2,3] and several estuaries along the East Coast of the United States. Hypoxia and anoxia develop in deep waters of these aquatic systems when oxygen consumption exceeds reaeration. This condition usually occurs when the deep waters are isolated from the surface waters by a density discontinuity. A density discontinuity may be formed by thermal stratification or, in some marine and estuarine environments, by salinity stratification. In the latter case, the density discontinuity is termed the pycnocline.

Annual summer hypoxia or anoxia have been known in the mesohaline portion of Chesapeake Bay since at least 1936,[4] and have been reported repeatedly since that time.[5-12] Even as far back as 1912, dissolved oxygen concentrations less than 35% saturation were observed in the lower Potomac estuary.[13] Given the long-standing observations of oxygen depletion in the Bay, a key question for environmental management is whether the problem has grown worse over the decades. If it has not, then measures employed to control the extent and duration of anoxia and hypoxia will fail to show any measurable results. On the other hand, if the problem is becoming more severe and anthropogenic causes can be identified, then management protocols can be designed to control the situation.

Oxygen Depletion: Physical Factors

The Chesapeake Bay is a partially stratified estuary characterized by some degree of stratification throughout most of the year. This is a necessary condition for oxygen depletion both because it restricts vertical mixing between surface and bottom waters and because it permits vertical separation between photosynthetic production in surface water and heterotrophic respiration in the bottom water. The annual cycle of oxygen depletion in the Bay is initiated as bottom water temperatures begin to increase in the late winter.[14] Freshwater flow into the Bay influences both the degree of stratification and nutrient input. Both of these factors probably contribute to the rapid decline in deep water DO following the spring freshet which usually occurs in March or April.[9,10,11,14] For these reasons, the magnitude of the spring freshet may govern the rate at which anoxia is approached during May-June and the extent of oxygen depletion during the summer. Reaeration of the bottom water usually occurs during the fall as a consequence of surface cooling and wind-driven mixing.

Based on similar arguments, the spatial extent of oxygen depletion is most pronounced in the mesohaline reach of the Bay. This reflects the fact that the mid-Bay region is bounded by shallow zones at either end where wind-driven and tidal turbulence tend to mix surface with bottom water and to facilitate reaeration. The southern end consists of shallow flats having a depth of about 12m and the northern end is bounded by a sill having a depth of about 10m. This sill restricts the northerly penetration of salt and oxygen-depleted bottom water.

Oxygen Depletion: Biological Processes

While physical and hydrodynamic factors influence reaeration of mid-Bay deep waters, biological processes are directly responsible for oxygen consumption. The imbalance between consumption and reaeration results in hypoxia and, ultimately, anoxia. The set of ecological interactions which lead to oxygen depletion can be summarized as follows: 1) the input of plant nutrients from point and non-point sources stimulates photosynthetic production of organic matter by phytoplankton; 2) a portion of this production accumulates at the pycnocline and below; 3) this supply of organic matter stimulates metabolism by heterotrophic organisms, particularly bacteria, in deep water and the benthos; and 4) the resulting oxygen consumption reduces dissolved oxygen levels, depending on depth and the degree of vertical stratification of the water column.

The key end process causing oxygen depletion is the metabolism of plytoplankton-derived organic material,[15] much of which is probably due to bacterial metabolism. Two major types of bacterial metabolism are involved. In one of these, termed aerobic respiration and depicted by Equation 1,

$$[CH_2O] + O_2 \rightarrow CO_2 + H_2O \tag{1}$$

oxygen is consumed directly during the mineralization of organic matter, $[CH_2O]$, by aerobic heterotrophic bacteria located in the water column or in oxygen-containing surficial sediments. During episodes of anoxia this process is confined to the oxygen-bearing portions of the water column.

The second major bacterial process contributing to oxygen depletion is described by equations 2 and 3.

$$2[CH_2O] + SO_4^{2-} \rightarrow S^{2-} + 2CO_2 + 2H_2O \qquad (2)$$

$$S^{2-} + 2O_2 \rightarrow SO_4^{2-} \qquad (3)$$

Equation 2 depicts the mineralization of organic carbon by sulfate-reducing bacteria, obligately anaerobic prokaryotes which use sulfate, SO_4^{2-}, as an oxidant in much the same way that aerobic bacteria use oxygen (Eq. 1). From the standpoint of oxygen consumption, the most significant product of reaction 2 is sulfide, S^{2-}, which is found as HS^- in the anoxic portion of the mid-Bay water column during summer stratification. Its presence is easily detected by the typical "rotten egg" odor of volatile hydrogen sulfide, H_2S. Oxygen consumption by sulfide (Eq. 3) occurs abiologically, but can also be microbially catalyzed by sulfur-oxidizing bacteria.

This chapter describes the details of oxygen depletion in the Bay during the period 1984 to 1986. We consider physical, climatological and biological factors and compare these factors and resultant oxygen deficits with respect to seasonal cycles, vertical gradients in the water column, and geographic distribution along both north/south and east/west axes.

EXPERIMENTAL APPROACH

We established a series of multi-station transects which were oriented lateral (east/west) to the mainstem axis.[12,14,16] One station of each transect was located over the deepest part of the main channel while the others were in depths varying from about 8 to 15 meters. The stations were occupied at 3-4 day intervals

TABLE 1

Summary of Experimental Parameters.

HYDRODYNAMIC PARAMETERS: Depth, Temperature, Salinity, Dissolved Oxygen (electrode, confirmed by Winkler titration),[17] Sulfide,[18] Sulfate.[19]

NUTRIENTS: Nitrate, Nitrite, Ammonium, Phosphate (all by autoanalyzer).

OXYGEN DEMAND INDICES—ORGANIC SOURCES: POC and PON (Combustion followed by elemental analysis), Primary Production (^{14}C bicarbonate assimilation, 24 hour incubations), Chlorophyll a (fluorometry), < 20 µg Chlorophyll a (differential filtration and fluorometry), Phaeopigments (fluorometry after acidification), Biochemical Oxygen Demand (5-day BOD, Winkler titration), Soluble Biochemical Oxygen Demand (5-day BOD of glass fiber filtered water).

OXYGEN DEMAND INDICES—BIOLOGICAL/CHEMICAL OXYGEN CONSUMPTION: Bacterial Abundance acridine orange/direct microscopic count,[20] Bacterial Production incorporation of ^3H-thymidine,[21] Bacterial Amino Acid and Metabolism,[22] Oxygen Consumption Rate (6 hour, Winkler titration), Sulfate Reduction chromium reduction,[24] Sulfide Oxidation.[25]

during 1984 and weekly to biweekly during 1985. This orientation and sampling frequency permitted definition of the dynamics of the anoxic/hypoxic layers (e.g. east/west seiching) and more importantly, elucidation of the biological and chemical forces driving oxygen consumption. For the first time, a systematic east/west and north/south transect study was established to test the hypothesis that biological processes over the shallow flanks of the Bay are important in driving oxygen consumption.

We simultaneously measured hydrodynamic water quality parameters, nutrient concentrations, organic carbon pools, and rates of heterotrophic metabolism, oxygen consumption and sulfur cycling. Parameters measured are listed in Table 1. We emphasize that several of the techniques for measuring key parameters, especially those concerned with bacterial abundance, heterotrophic activity, and sulfur cycling have only become available during the last decade. For example, bacterial abundance is determined by direct counting of acridine orange-stained cells under epifluorescent illumination. This technique was refined for routine use in about 1977.[20] Similarly, bacterial production rates were estimated using the technique first described by Furhman and Azam,[26] and methods for estimating sulfide oxidation were even more recently developed by Tuttle.[25] Thus, our understanding of the biological processes controlling Chesapeake Bay oxygen dynamics has, until now, been limited by available experimental methods.

RESULTS AND DISCUSSION

Seasonal Oxygen Dynamics   Although it has been concluded that water quality in mesohaline Chesapeake Bay has deteriorated with respect to dissolved oxygen (DO) concentrations over the past few decades,[10,27,28] our data demonstrate that there are complex annual, seasonal, and even shorter time scale variations in oxygen concentrations. Some reports state that anoxic conditions develop by May of each year and extend uninterrupted into September.[10] During the summer of 1984, a year of widespread anoxia in Chesapeake Bay,[11,16] periods of anoxia were separated by periods of reaeration of bottom water, indicating the dynamic nature of the oxygen regime (Figure 1).

Anoxic periods were associated with increasing strength and decreasing depth of the pycnocline. Turbulent weather in late August and early September partially destratified the water column for over a month. However, in October the water column restratified and deep water DO declined in response to continued oxygen demand.

Seasonal profiles of salinity and DO during 1985 reveal a very different pattern than in 1984 (Figure 1). Surface water salinities were considerably greater in 1985 and the intensity of stratification was markedly reduced. Associated with this reduced stratification, bottom water DO of 1-2 mg/l was common

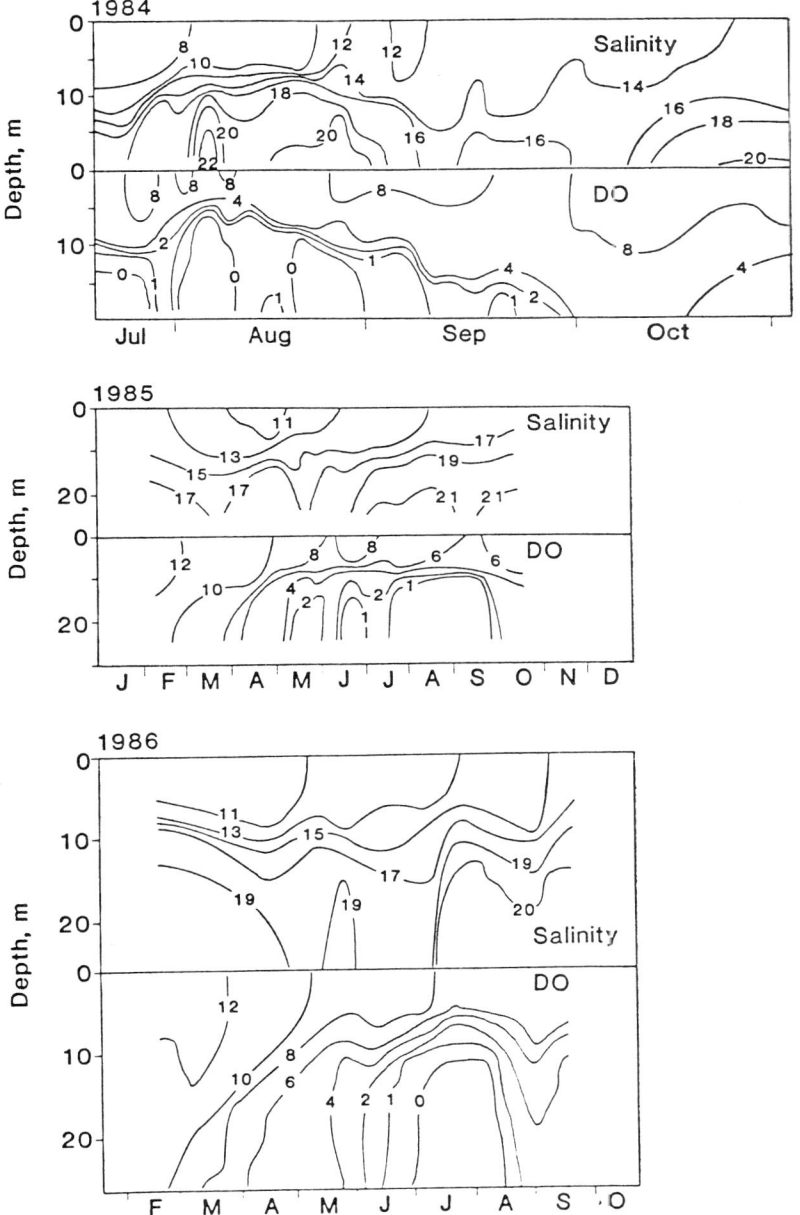

FIGURE 1. Salinity and oxygen isoclines at a deep channel station opposite the Choptank River, 1984-1986. Salinity is expressed as ppt and DO as mg/l. Data for 1984 are from Malone et al.[12] and 1985 data are from Tuttle et al.[14]

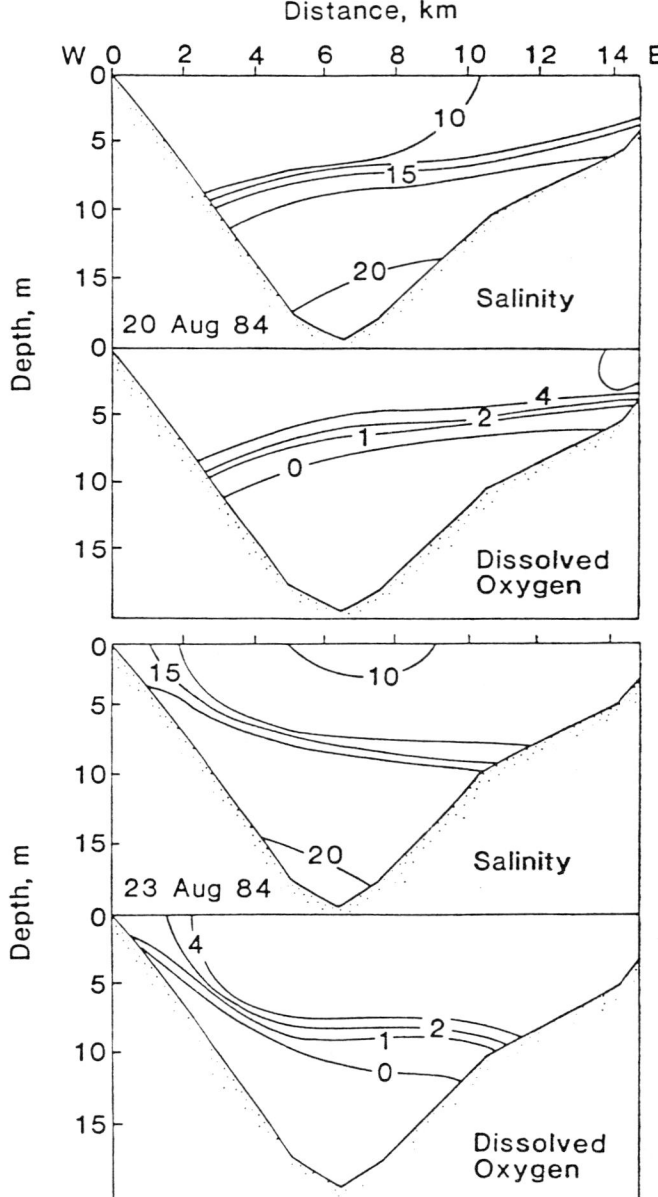

FIGURE 2. Salinity structure and DO along west/east transect of the Bay from Dares Beach to Choptank River on 20 August and 23 August 1984. Salinity is expressed as ppt and DO as mg/l. Data are from Malone et al.[12]

throughout the summer. A brief period of anoxia developed up-Bay in mid-August, but it lasted only about 2 weeks. The 1985 DO profile resembles that for 1937 and 1938[4] more closely than that for 1970 or 1980.[10] These results demonstrate the importance of annual climatic factors in the development of anoxia in the Bay.

Profiles for the same location in 1986 reveal surface water salinities comparable to 1985 but a more intense pycnocline at about 10-12 m depth (Figure 1). Deep water DO declined progressively throughout the spring of 1986, reaching 1 mg/l in mid-June and anoxia in late June to mid-August. Even though the water column remained moderately stratified throughout the summer, bottom water DO rose above 1 mg/l during mid-August when a series of strong storms passed over the area.

Freshwater streamflow, especially from the Susquehanna and Potomac Rivers, is a critical factor influencing the development and maintenance of anoxia in the Bay. During 1984 monthly streamflow entering Chesapeake Bay was above average during the period from February through August. Near-record high streamflows occurred in February (216,300 cfs) and April (251,000 cfs) of that year (U.S. Geological Survey, Towson, Md.). These high flows reduced surface salinities in the mid-Bay and provided the physical isolation of the bottom water which subsequently allowed biological processes to deplete deep water DO. In contrast, 1985 was an extremely dry year with streamflows below average from January until August. The low flow resulted in increased surface salinities, decreased intensity of stratification, and reduced nutrient input, all of which probably led to sustained aerobic conditions throughout the Bay during nearly the entire summer. Nevertheless, deep water DO was severely depleted (Figure 1). The somewhat higher summer bottom water DO in 1985 compared to 1984 is related to increased reaeration rates rather than to reduced rates of oxygen consumption.

Below average streamflows from April to August confirm that 1986 was another drought year. Nevertheless, deep water anoxia occurred from the Annapolis Bay Bridge to just south of the Great Wicomico River in Virginia for about 6 weeks during the summer. The difference between 1985 and 1986 oxygen regimes can be explained by slightly above average streamflows in February and March, 1986, and a lack of strong wind events during spring and summer of 1986. These two climatic factors resulted in moderately intense salinity stratification and, hence, development of widespread anoxia. Total streamflow and the timing of streamflow events by month appear to be important in controlling salinity stratification in the Bay.

Short-term Oxygen Dynamics

Wind events are a major factor controlling short-time scale oxygen dynamics in Chesapeake Bay. Two hydrodynamic processes, reaeration of water below

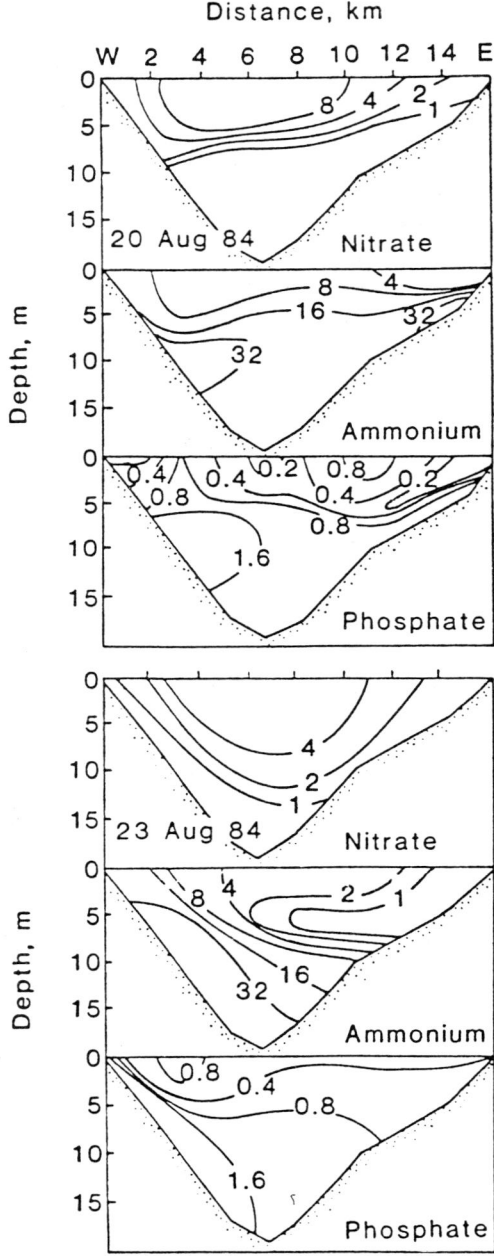

FIGURE 3. Nitrate, ammonium, and phosphate profiles along a west/east transect of the Bay from Dares Beach to Choptank River on 20 August and 23 August 1984. Nutrient concentrations are expressed as µg atoms/l. Data are from Tuttle et al.[14]

the pycnocline and tilting (or seiching) of the surface of the pycnocline itself[12] are of importance in this regard. Wind events can reoxygenate deep waters even in years when salinity stratification is very intense and without major disruption of the pycnocline as observed in mid-August 1984 (Figure 1). However, if the water column remains strongly stratified, as it did in the summer of 1984, DO is rapidly consumed and anoxic conditions return in the subpycnoclinal waters.

The phenomenon of pycnocline tilting and its influence is demonstrated by cross-Bay salinity, oxygen, and nutrient profiles for August 20 and August 23, 1984 (Figures 2 and 3). On August 20, salinity structure showed a marked tilt upward toward the eastern shore (Figure 2). The oxygen profile (Figure 2) followed that of salinity. A high pressure area passed over the Bay region on August 20 with accompanying winds of about 25 knots from the north/northwest. As a result, the pycnocline position was reversed so that on August 23 it tilted strongly upward toward the western shore (Figure 2). Along the western side of the

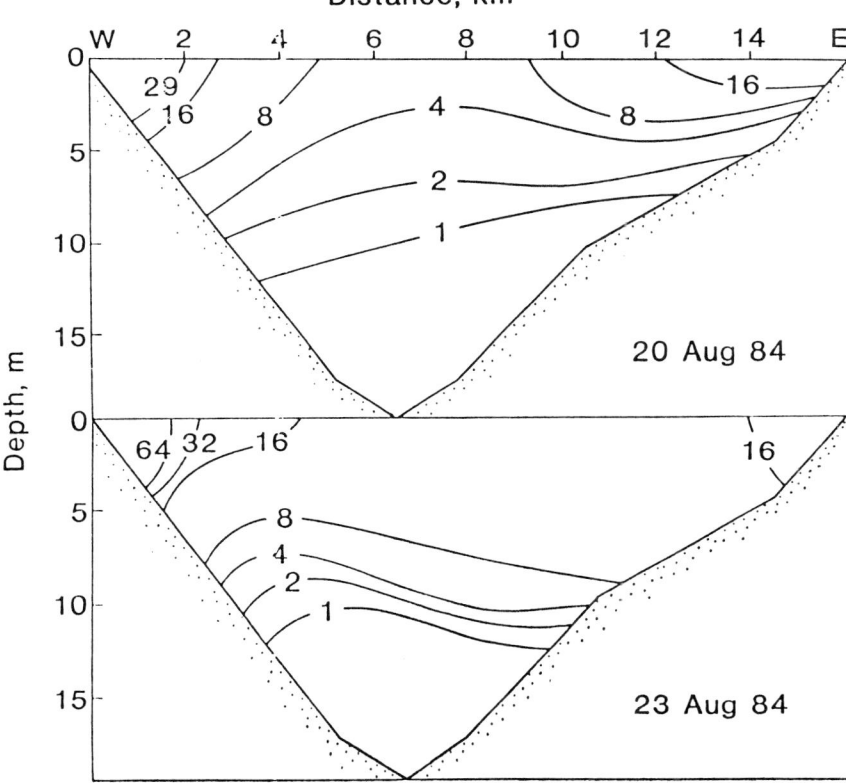

FIGURE 4. Chlorophyll a concentrations (μg/l) along a west/east transect of the Bay from Dares Beach to the Choptank River on 20 August and 23 August 1984. Data are from Malone et al.[12]

Bay the anoxic water mass, which had been several kilometers from shore and below 10 m depth, moved to within less than 1 kilometer of shore and rose to within 2 m of the surface. This change subjected vast areas of Bay bottom and a large volume of the Bay to sudden inundation by anoxic, sulfide-laden water which was accompanied by a major fish kill along the western shore.[12]

The frequency and temporal dynamics of the seiching phenomenon are still not well defined, nor has the effect on water column and benthic biota been completely described. However, surface water nutrient concentrations and phytoplankton biomass are clearly influenced by a seich. On August 20, 1984, nitrate concentrations near the surface were moderate and declined with depth (Figure 3). Due to the strong stratification and anoxia, ammonium and phosphate had accumulated in the deep water. By August 23 this nutrient-laden deep water had been advected upward toward the western shore and into the euphotic zone. A phytoplankton bloom subsequently developed along the western shore of the Bay (Figure 4).

The long-term consequences of these short-time scale events is still unclear. They may be responsible for advecting large masses of anoxic water into productive shellfish areas along the Bay flanks and up tributaries which would likely remain well oxygenated without seiching. Significant mortalities of benthic biota could then result.[11] From the management perspective, the rapid advection of water masses of significantly different quality could seriously influence the interpretation of water quality monitoring data, particularly when sampling frequency is biweekly, monthly or longer.

SPATIAL AND SEASONAL VARIATIONS—1985 ANNUAL CYCLE

Seasonal increases in temperature and freshwater flow combine to establish the physical conditions which set the stage for oxygen depletion. The 1985 seasonal cycle of conditions related to low DO are summarized in Figures 5-8. The three sample stations were located along the mainstem channel over an 80 km span of the mid-Bay in the low DO zone.

Temperature and Salinity

The temperature and salinity regimes (Figure 5) are typical of partially stratified estuaries located in temperate climates. Surface temperatures increased rapidly in April and May resulting in a marked thermocline at 10-12 m depth which persisted until about mid-July. By late July water temperature reached its seasonal maximum and was nearly constant with depth. There were no major differences in the temperature regime over the sample area.

Vertical salinity stratification occurred throughout the year at all deep water stations, but was particularly strong during April and again in July/August

Origin, Development and Significance of Chesapeake Bay Anoxia 453

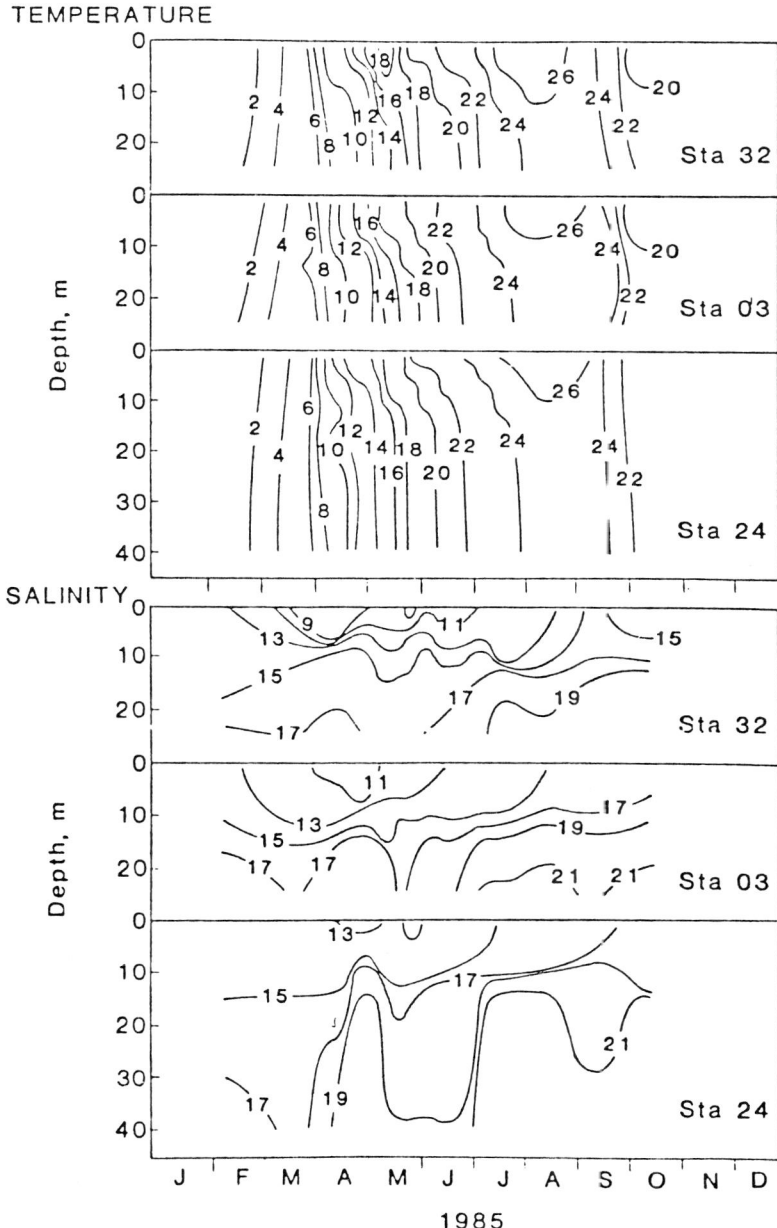

FIGURE 5. Seasonal variations in the vertical distributions of a) temperature (°C) and b) salinity (ppt) during 1985 along the deep channel at the Annapolis Chesapeake Bay Bridge (STA 32), the Choptank River (STA 3), and the Patuxent River (STA 24). Data are from Tuttle et al.[14]

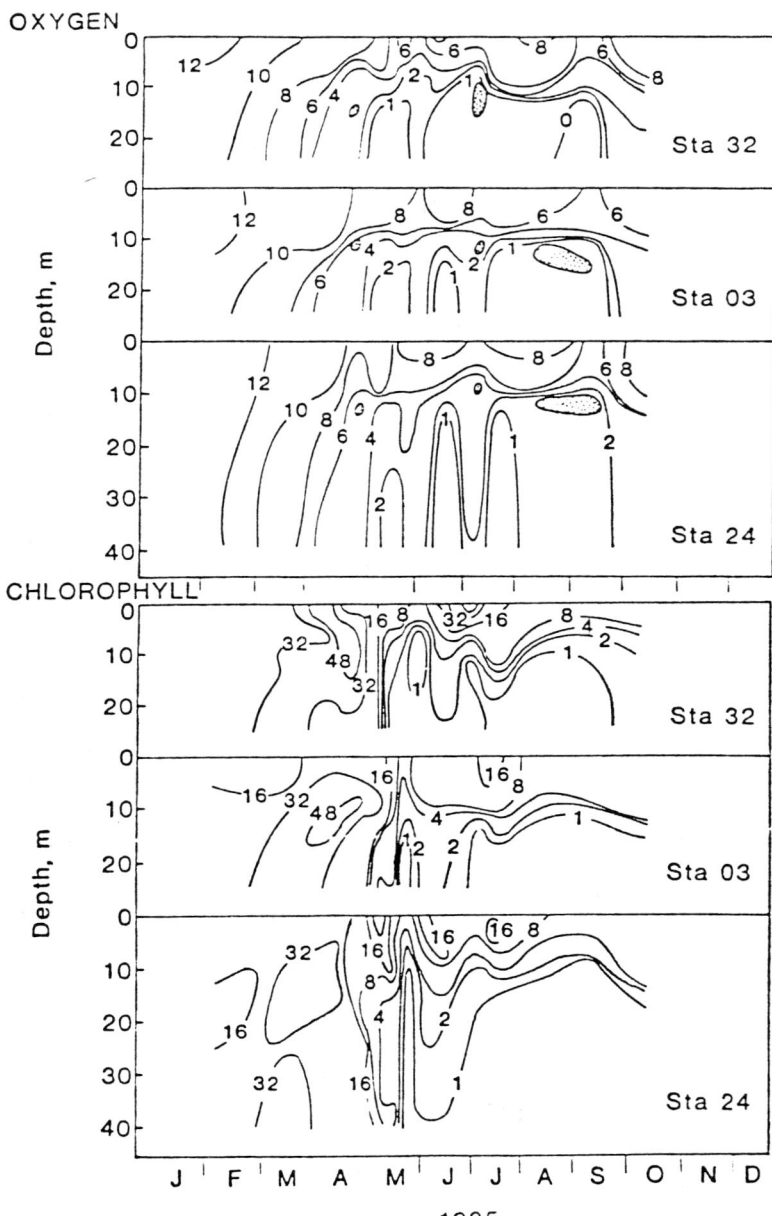

FIGURE 6. Seasonal variations in the vertical distributions of a) oxygen (mg/l) and b) chlorophyll a (μg/l) during 1985 along the deep channel at the Annapolis Bay Bridge (STA 32), the Choptank River (STA 3), and the Patuxent River (STA 24). Hatched areas represent mid-water oxygen minima. Data are from Tuttle et al.[14]

(Figure 5). The April salinity structure reflects the intrusion of fresh water from spring runoff. The downstream progression of this stratification peak indicates a lag time of about 2 weeks between the northern-most and the southern-most stations, indicating a mean drift rate of about 6 km/day. Reduced intensity of stratification during May and June reflects both reduced freshwater runoff and wind turbulence.

Dissolved Oxygen

Deep water DO declined rapidly during Spring, reaching 2 mg/l or less by mid-May (Figure 6). Oxygen concentrations were always lowest in the bottom waters of the northern-most station and higher to the south at any specific time during the spring and summer. However, the net rate of oxygen decline during February to mid-May was about 0.1 mg/l/d at all three stations. This rate agrees closely with the rates reported by Taft et al.[9] and calculated by Officer et al.[10] Hypoxic conditions persisted throughout spring/summer of 1985, but anoxic conditions were observed only at the northern-most station during a short period in August/September. Bottom water DO did not remain uniformly low but was interspersed with periods of partial reaeration. Reaeration events were more pronounced toward the south. The pattern of oxygen depletion from south to north and the stronger signal strength of reaeration events to the south indicate that oxygen is consumed as deep water moves northward up the Bay.

Vertical oxygen profiles show that even during cold weather DO was less than saturation in the bottom water, indicating that the processes which consume oxygen apparently continue, although at a reduced rate. Because the Bay is never completely mixed, rates of deep water oxygen consumption can exceed reaeration rates at any time of the year.

Mid-water oxygen minimum zones were occasionally observed in association with the pycnocline (Figure 6). This phenomenon may be caused by peaks of heterotrophic metabolism in the pycnocline layer, by advection of lower DO water from the north, or by advection of higher DO deep water from the south.

Chlorophyll a - Phytoplankton Biomass

The period of the spring maximum in chlorophyll *a* (Chl) was characterized by high Chl concentrations throughout the water column (Figure 6). A rapid decline in Chl occurred during the third week of May, resulting in a 5-fold decline in water column Chl content. During the remainder of the study, most of the Chl was confined to the layer above the pycnocline.

POC to PON ratios ranged from 4 to 20, much lower than for terrestrially derived organic matter or for material from the upper Bay where the ratios exceed 20.[29] These observations agree with the conclusions of Biggs and Flemer[15] that most suspended particulate organic matter in the mid-Bay is of local phytoplankton origin.

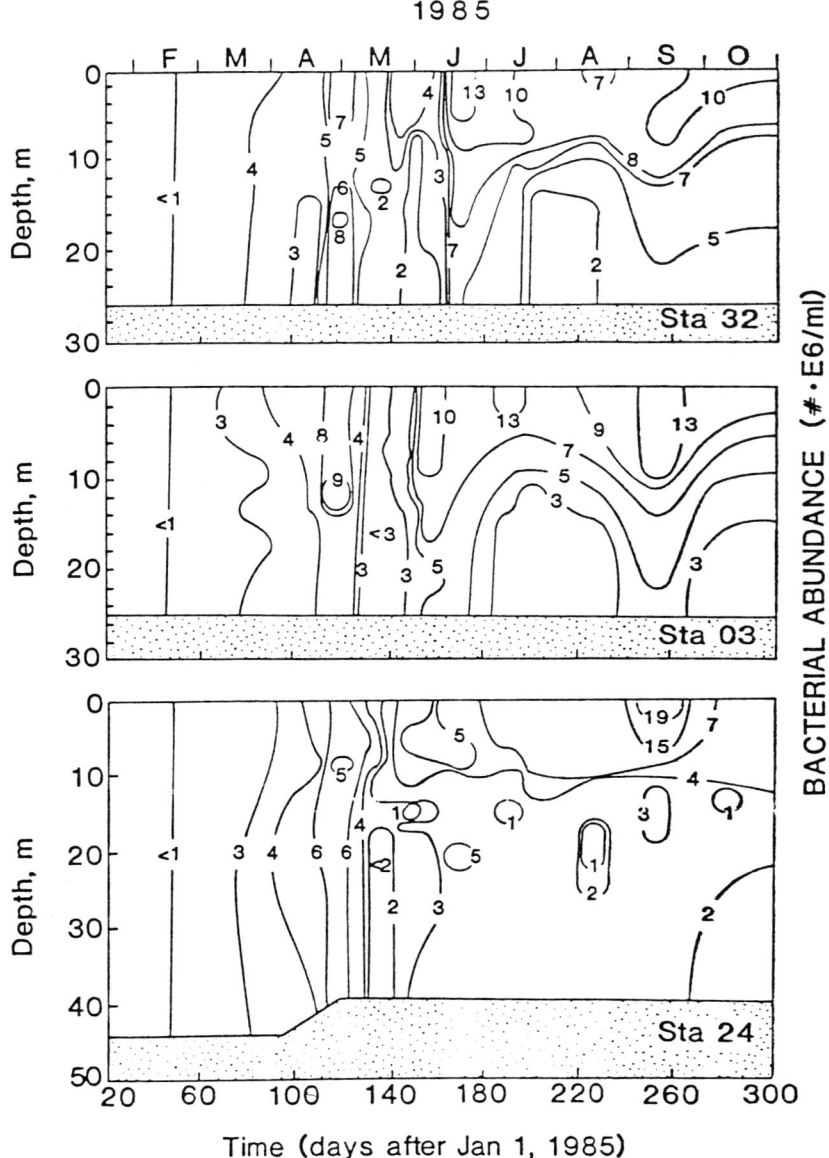

FIGURE 7. Seasonal variations in the vertical distribution of bacterial abundance during 1985 along the deep channel at the Annapolis Bay Bridge (STA 32), the Choptank River (STA 3), and the Patuxent River (STA 24). Data are from Tuttle et al.[14]

Bacterial Abundance

During late winter bacterial abundance throughout the mid-Bay (Figure 7) was relatively low. It increased markedly during the spring in association with both increasing phytoplankton biomass and water temperatures. Early in the year, bacteria were about evenly distributed throughout the water column, although at stations 3 and 32 there were April mid-water maxima which may be related to mid-water Chl maxima (Figure 6).

Following the April peak, bacterial abundance declined rapidly (Figure 7), preceding the phytoplankton decline by about one week (Figure 6). Bacterial abundance began to increase again in late May and early June, one to two weeks after the phytoplankton decline. By mid-June the summer pattern of high surface water bacterial abundance and a decline with depth had become established. Typically in the summer there were $7\text{-}10\times10^6$ bacteria/ml in surface water and $2\text{-}5\times10^6$/ml in deep water. Deep water abundances at stations 3 and 32 generally exceeded abundances at station 24. However, the highest bacterial abundances occurred in late summer in the surface waters at station 24.

Bacterial Production

During the late winter and spring, bacterial production (TdR, Figure 8) followed a pattern similar to bacterial abundance. Like bacterial abundance, increasing TdR was associated with increasing phytoplankton biomass and increasing water temperature. Mid-water and bottom water TdR maxima to the north during April were associated with bacterial abundance maxima at the same stations (Figure 7).

Bacterial production declined following the April peak (preceding the phytoplankton "crash") and then increased rapidly. Throughout the summer season highest TdR values occurred in the surface waters, while deep water values were lower. Episodes of very high bacterial production occurred in surface waters at all three stations, although these episodes were not always coincident. Very high production usually persisted longer at stations 3 and 32, in the north, than they did at station 24 to the south.

Geographic Initiation of Anoxia

Oxygen depletion appears to begin earliest in the northern portions of the mid-Bay (Figure 6), although the increase in deep water DO from north to south within this region is rather small. This may reflect progressive oxygen depletion of bottom water as it is transported to the north and an increase in bottom water BOD as the turbidity maximum is approached. With respect to the latter, there is evidence for gradients in Chl (Figure 6) and BOD (data not shown) increasing to the north. The patterns in both suggest somewhat elevated con-

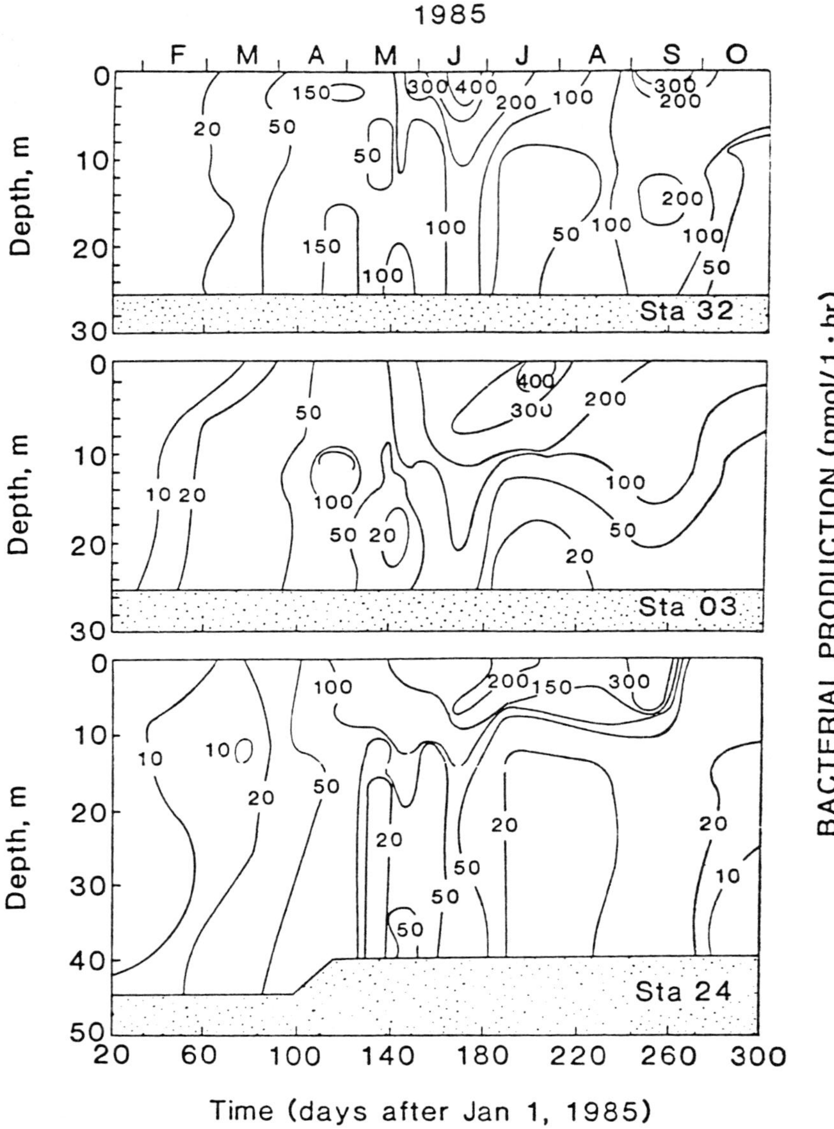

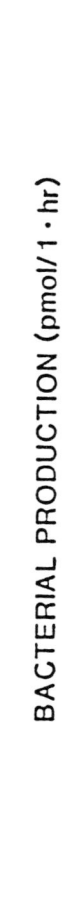

FIGURE 8. Seasonal variations in the vertical distribution of bacterial production during 1985 along the deep channel at the Annapolis Bay Bridge (STA 32), the Choptank River (STA 3), and the Patuxent River (STA 24). Data are from Tuttle et al.[14]

centrations of organic matter in the more northerly areas.

In terms of oxygen consumers and oxygen consumption itself, the northern deep water bacterial abundances were slightly higher, especially during the summer (Figure 7), and deep water bacterial production was somewhat greater than in areas farther to the south. Independently measured, gross mean oxygen consumption was also elevated in the more northern area (Figure 9).

All these data suggest that the processes to hypoxia and anoxia operate at somewhat greater rates in the more northern section of the mesohaline area of the Bay. However, the dominant feature of our data set is the widespread and pervasive nature of the factors which drive oxygen depletion. Throughout the mid-portion of Chesapeake Bay, high phytoplankton biomass and high

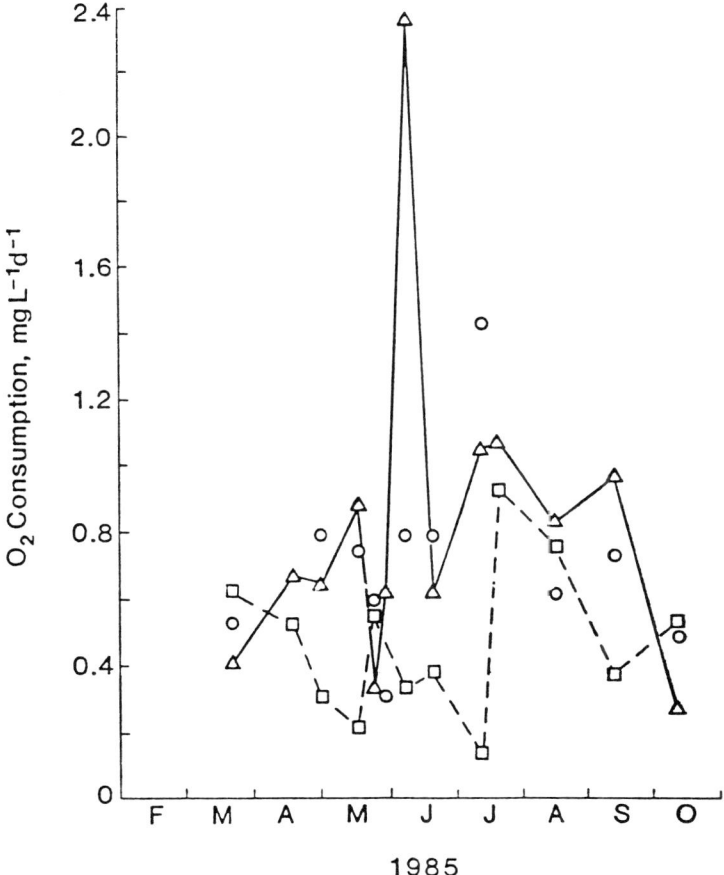

FIGURE 9. Mean oxygen concentration consumption beneath the euphotic zone along cross-Bay transects at the Annapolis Bay Bridge (△), the Choptank River (○), and the Patuxent River (□) during 1985. Data are from Tuttle et al.[14]

bacterial biomass combine with stratified conditions in the water column to drive oxygen to low levels.

Flanks versus the Main Channel

We have hypothesized that a significant proportion of phytoplankton biomass which acts as biochemical oxygen demand in subpycnoclinal waters is produced over the shallow Bay flanks. We reasoned that phytoplankton production will tend to be higher in shallow water because of a longer residence time within the euphotic zone and because of a closer coupling between phytoplankton production and nutrient regeneration. If phytoplankton production is the major source of organic material for bacteria, then bacterial biomass and production should also be elevated along the flanks of the Bay.

Data for Chl, BOD, and bacterial biomass and production are consistent with this hypothesis. Mean Chl concentrations along the eastern and western flanks were often greater than that over the main channel (Figure 10), especially between April and August when the water column was most strongly stratified. Similarly, bacterial abundance (Figure 11) and production (Figure 12) were considerably greater over the shallow flanks from May to October. Bacterial abundance tended to be highest along the eastern flank where, during the summer, it was often twice the mean abundance over the mainstem channel. In contrast, mean bacterial production was usually highest over the western flank. The concentration of readily metabolizable organic matter (BOD) was also greater along

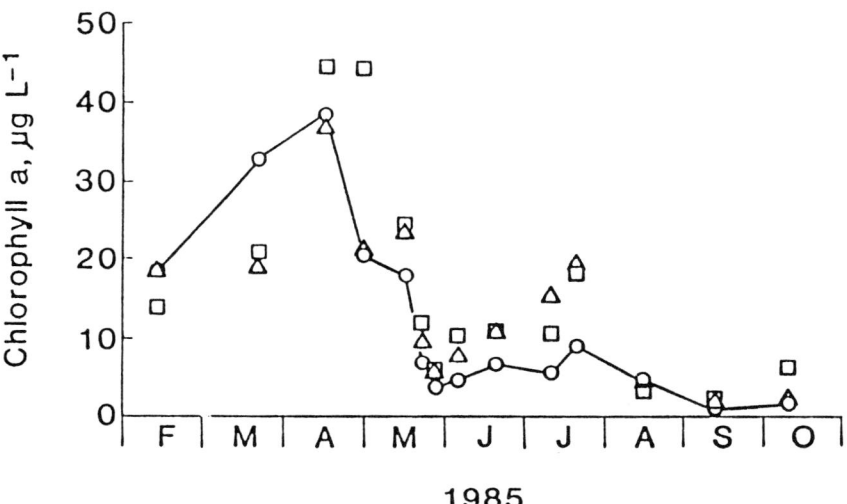

FIGURE 10. Mean chlorophyll *a* concentration along the west flank (△), main channel (O), and east flank (□) of Chesapeake Bay during 1985.

Origin, Development and Significance of Chesapeake Bay Anoxia 461

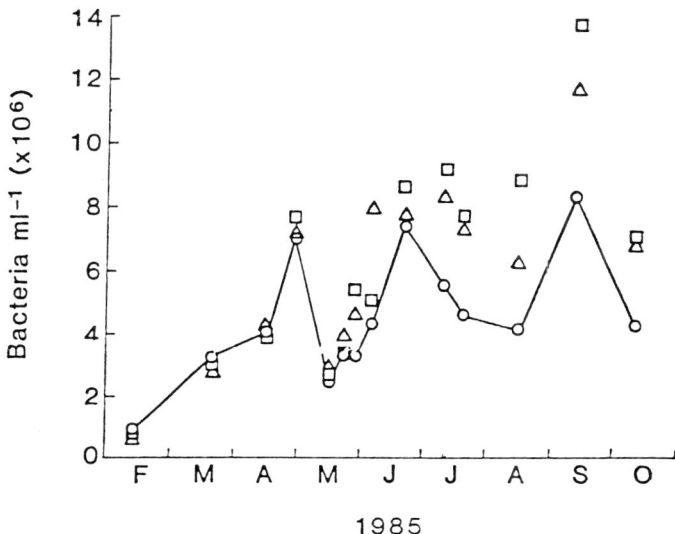

FIGURE 11. Mean bacterial abundances along the west flank (△), main channel (O), and east flank (□) of Chesapeake Bay during 1985. Data are from Tuttle et al.[14]

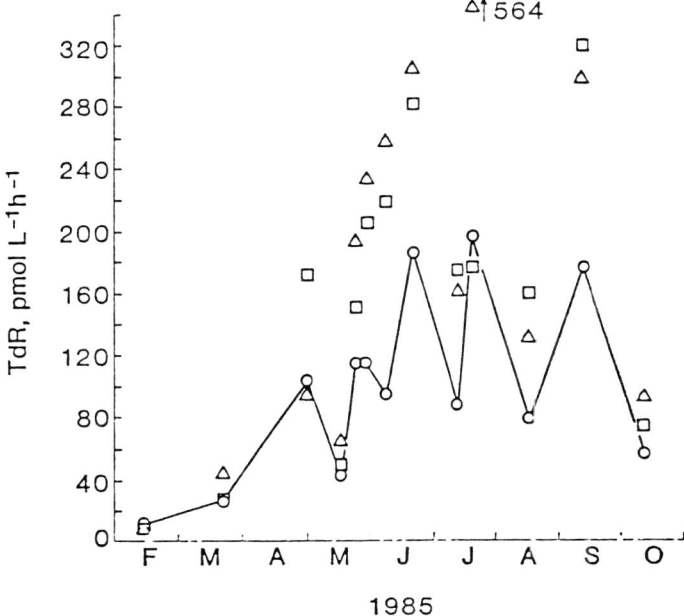

FIGURE 12. Mean bacterial production, expressed as TdR, along the west flank (△), main channel (O), and east flank (□) of Chesapeake Bay during 1985. Data are from Tuttle et al.[14]

the flanks than over the main channel during the period from mid-April to October (data not shown).

Despite the east/west variations in the parameters discussed above, there was a strong correspondence of changes across the entire Bay, both normal and parallel to the mainstem. Although individual exceptions occurred, changes in variables along one north/south or east/west transect were commonly mirrored by similar changes in the same variables along other transects. A striking example of this occurred during May when the phytoplankton decline occurred throughout the study area. During this period marked declines in BOD, bacterial abundance, and bacterial production were also observed throughout the mid-Bay.

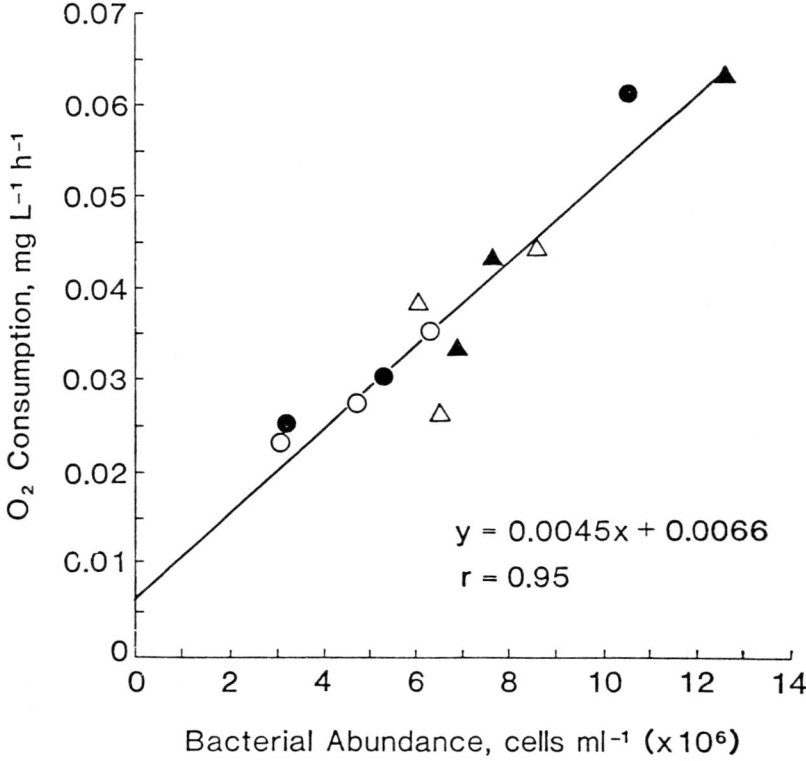

FIGURE 13. Linear regression of mean water column oxygen consumption on mean bacterial abundance. Symbols: (▲) euphotic zone means for the time periods 20 Aug.-11 Sep., 14 Sep.-3 Oct., and 5 Oct.-2 Nov. 1984; (•) beneath euphotic zone means for 1984 data collected in the time periods given above; (△) euphotic zone means for the time periods 12 Feb.-17 Apr., 29 Apr.-7 June, and 19 June-12 Sep. 1985; (○) beneath euphotic zone means for 1985 data collected in the time periods given above. The euphotic zone is defined as 2.7x Secchi disk depth. Data are from Tuttle et al.[14]

Bacterial Abundance and Oxygen Consumption

We have demonstrated that phytoplankton production, either as particulate biomass or as dissolved organic matter, is directly associated with microbial heterotrophic processes. It follows that if bacterial metabolic activity is responsible for oxygen depletion in the Bay, a strong correlation should exist between bacterial metabolism or abundance and oxygen consumption. Regression of oxygen consumption measurements made in 1984 and 1985 against corresponding bacterial abundances (Figure 13) gave a strong ($r = 0.95$), linear relationship. This confirms the contention that the water column heterotrophic bacterial community, responding to increases in labile organic carbon, is a dominant biological factor driving oxygen consumption in Chesapeake Bay. We suggest that inclusion of bacterial abundance measurements to monitoring efforts would be useful, both as an integrative index of the amount of organic carbon available to fuel oxygen depletion and as an indicator of the water column oxygen consumption rates.

Comparison of Selected Biological Parameters: 1984 and 1985

As discussed above, 1984 and 1985 were sharply contrasting years in terms of rainfall, water column stratification, surface salinity, and oxygen regime. It is of particular interest, therefore, to compare biological communities and their associated processes between these two years to discern whether the marked differences in the extent and duration of anoxia were related to contrasting

TABLE 2

Comparison of the means of key bacterial and phytoplankton parameters measured during summer conditions in 1984 and 1985 across an east/west transect at the Choptank River.

	EUPHOTIC ZONE		BELOW EUPHOTIC ZONE	
Phytoplankton Production	1.20	('84)	—	
(g C/m^2/d)	1.37	('85)	—	
Phytoplankton Biomass	56.10	('84)	—	
(mg Chl/m^2)	36.60	('85)	—	
Bacterial Production*	2.92×10^8	('84)	1.92×10^8	('84)
(cells/l/h)	5.32×10^8	('85)	3.08×10^8	('85)
Bacterial Abundance	1.37×10^{10}	('84)	8.44×10^9	('84)
(cells/l)	1.09×10^{10}	('85)	7.02×10^9	('85)
Bacterial Turnover	0.52	('84)	0.55	('84)
(d^{-1})	1.17	('85)	1.05	('85)
O$_2$ Consumption	1.48	('84)	1.04	('84)
(mg/l/d)	1.31	('85)	1.02	('85)
O$_2$ Consumption†	1.64	('84)	1.07	('84)
(mg/l/d)	1.34	('85)	0.92	('85)

*Based upon 2×10^8 cells produced/mol ^{3}H thymidine incorporated.

† Based upon bacterial abundance according to the relationship given in Figure 13.

climatological conditions, to changes in biological parameters associated with oxygen consumption, or to both acting in concert.

Phytoplankton biomass in the euphotic zone was higher in 1984 than in 1985, but phytoplankton production was slightly greater in the latter year (Table 2). These differences suggest that phytoplankton production and grazing were more tightly coupled in 1985 than in 1984 such that phytoplankton did not accumulate. As a consequence, the flux of organic matter of phytoplankton origin from the euphotic zone into bottom waters may have been lower in the summer of 1985. Poor coupling between zooplankton grazing and phytoplankton production during 1984 may have been caused by seiching and associated variations in vertical mixing[12] which permitted anoxic, sulfide-bearing deep waters to invade the shallows.

Bacterial abundances were remarkably similar in both years (Table 2). Therefore, despite higher summer phytoplankton biomass in 1984 than 1985, there appeared to be sufficient organic carbon input below the euphotic zone to support elevated levels of bacterial biomass similar to those found in 1984. The high bacterial cell densities ($> 10^{10}$ cells/l) maintained in the euphotic zone over successive summer seasons have not been previously documented and may indicate that Chesapeake Bay is one of the most eutrophic systems studied to date.[12]

Bacterial production increased in summer 1985 compared to 1984 (Table 2). The magnitude of this increase was 82% within the euphotic and 60% beneath it, resulting in doublings of summer bacterial turnover rates (biomass specific growth rates) in 1985 compared to 1984. The increase in bacterial production and turnover may have been caused by increased predation of bacteria in 1985 as appears to have occurred with phytoplankton. Thus, widespread anoxia with its attendant incursion of sulfide-bearing water into normally unaffected areas may exert a major inhibitory influence on grazing of both phytoplankton and bacterial communities.

Directly measured oxygen consumption rates were similar during the summers of both years, particularly below the euphotic zone (Table 2). Within the euphotic zone, directly measured oxygen consumption rates averaged only 11% lower in 1985 compared to 1984. Comparable oxygen consumption values were found by calculation from mean bacterial abundances (Table 2, Figure 13) suggesting that the majority of directly measured water column oxygen consumption can be attributed to heterotrophic bacteria.

We conclude that despite major differences in climatic conditions and in the DO regimes of 1984 and 1985, the bacterial parameters, especially biomass and oxygen consumption rates, were remarkably similar during the summers of both years. These similarities, in concert with nearly identical phytoplankton production in both years and increased bacterial production in 1985, suggest that a major ecological shift has occurred in the mesohaline portion of Chesapeake Bay during the summer (and perhaps during the spring as well), such that a large portion of phytoplankton production is processed directly by pelagic bacteria rather than by higher organisms. Indeed, carbon flow through bacteria in the euphotic zone alone during July - November 1984 has been estimated in excess of 50% of phytoplankton production.[12] These data imply that at the present state of water quality in the Bay, the extent and duration of anoxia in different years is related to both climatic variability and changes in the pathways and rates by which organic matter is metabolized. The existing biological conditions are such that widespread anoxia may occur whenever the appropriate physical conditions obtain.

Sulfur Cycling

Oxygen consumption in marine and estuarine environments is a complex process integrating biological consumers, differing nutrient sources, ambient oxygen concentrations, and the physical and chemical environment. An important example of the latter in the mid-Bay, particularly during summer anoxia, is the presence of sulfide in the water column. During episodes of anoxia, sulfide is normally found just beneath the pycnocline and can be detected in increasing concentrations to the bottom. Maximum water column sulfide concentrations in the mid-Bay usually range from about 8 to 15 μm.

Hydrogen sulfide is produced during the metabolism of organic matter by obligately anaerobic sulfate-reducing bacteria (see eq. 2). When the water column contains oxygen, as is the case for much of the year, sulfate reduction is confined to the sediments where oxidation of the sulfide produced (see eq. 3) contributes to sediment oxygen demand. However, during episodes of anoxia the potential exists for sulfate reduction to occur in the water column. Measurements made during anoxic events in the summers of 1984 and 1985 indicate that water column sulfate reduction accounts at most for only about 1% of the sulfide produced (Table 3). Sulfate reduction thus occurs chiefly in the sediments. It is especially intense in sediments of the deep channel where rates integrated over the top 15 cm of sediment averaged 13 and 11 times those measured in more shallow sediments during 1984 and 1985, respectively (Table 3). Little is known about sulfate reduction rates in estuaries other than Chesapeake Bay, but maximum rates of 335 and 205 mmol/m^2/day, measured in deep channel sediments in mid-September of 1984 and 1985, respectively, are among the highest reported and rival the maximum rates found by Howard and Teal[30] in a New England salt marsh. Bacterial sulfate reduction is clearly a dominant process in deep channel Bay sediments.

Rates of sediment sulfate reduction are strongly dependent upon temperature (Figure 14). However, the fact that much higher rates of sulfate reduction were found in October and November than in May (particularly in the deep channel) when the temperatures were comparable, suggests that sediment sulfate

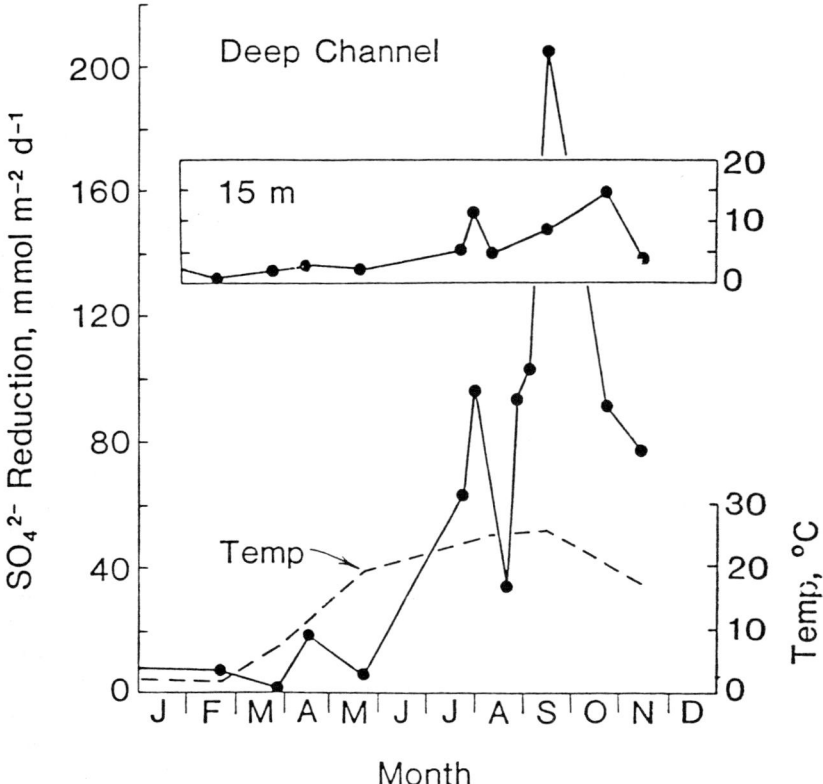

FIGURE 14. Rates of sulfate reduction in deep channel and 15m west flank Bay sediments during 1985. The rates represent integrated, discrete sulfate reduction measurements made at 2.5cm intervals over a total sediment depth of 15cm. Sulfate reduction below 15cm was negligible. Temperature values are for bottom water at 0.5-1m above the sediments.

TABLE 3

Daily Average Sulfate Reduction Rates, April to November.

Location	Year	Sulfate Reduced mmol/m^2/day
Deep Bay Sediment	1984	75.2
Deep Bay Sediment	1985	68.1
15m Bay Sediment	1984	5.7
15m Bay Sediment	1985	6.1
Bay Water Column*	1984	0.6
Bay Water Column*	1985	0.8

*Based on 10m deep anoxic zone.

reduction may also be responding to carbon inputs from high phytoplankton production during the later half of the summer and into early fall. This contention is supported by the fact that sulfate reduction in deep channel sediments peaked in mid-September in both 1984 and 1985. During 1985 the timing of the large increase in deep channel sulfate reduction from late May to July suggests that it may have been fueled by carbon from the annual spring phytoplankton bloom (Figures 6 and 10).

The timing of increased rates of sediment sulfate reduction (Figure 14) in relation to variations in heterotrophic bacterial production and biomass in the water column suggests that the two major oxygen-consuming processes operating in the mesohaline Bay (eq. 1-3) exhibit seasonal changes in their relative importance. Oxygen depletion of Bay deep waters begins early in the year when sediment sulfate reduction rates are low (Figure 14) but when aerobic, heterotrophic bacterial abundances and processes in the water column (and presumably in the surficial sediments) are rapidly increasing (Figures 6,7,8,11,12). Thus, direct oxygen consumption by heterotrophic processes appears to be instrumental in the establishment of anoxia whereas sulfur cycling, controlled primarily by sediment sulfate reduction, is a progressively more potent consumer of oxygen into the summer and early fall. If anoxic conditions fail to develop in the water column, sulfide oxidation in the sediments becomes a major component of sediment oxygen demand. When anoxia occurs in the water column, sulfide oxidation becomes the dominant process maintaining anoxia because direct heterotrophic oxygen consumption obviously cannot occur in subpycnoclinal waters in the absence of oxygen.

Mean sediment sulfate reduction rates varied little between 1984 and 1985 (Table 3) despite the large differences in physical factors and DO regimes. This supports our conclusion that climatological changes and not fundamental differences in biological processes were chiefly responsible for the failure of widespread and persistent anoxia to develop in 1985.

As Bay deep water becomes anoxic, sulfide rises from the sediments into the water column. The rate of sulfide ascent towards the pycnocline is controlled by the rate of sulfate reduction (eq. 2) in the sediments and by the net flux of oxygen into subpycnoclinal waters which, in turn, is a function of the difference between aerobic bacterial oxygen consumption (eq. 1) and reoxygenation of pycnoclinal and subpycnoclinal waters. Taft et al.[9] have estimated that only a 10% increase in oxygen consumption rates over reoxygenation rates is sufficient to cause deep water anoxia.

Unlike aerobic bacterial oxygen consumption, sulfide oxidation in the Bay appears to occur within a narrow depth band (ca. 1-2 m) in which oxygen and sulfide coexist. This band is usually found at or slightly below the pycnocline (Figure 15). Measurements of water column sulfide oxidation during anoxia episodes during the summers of 1984 and 1985 gave mean oxygen consumption rates of 9 mg O_2/l/day.[14] During summer periods when the deep water is not anoxic, mean aerobic bacterial oxygen consumption below the euphotic zone is about 1 mg O_2/l/day (Table 2). The latter, however, is not limited to a narrow depth band, but occurs throughout the deep water. At an average hypoxia thickness layer of 10 m, typical of the mid-Bay deep channel, the rate of oxygen consumption by aerobic bacteria in subpycnoclinal waters is comparable to that due to sulfide oxidation.

Benthic versus Water Columns Oxygen Consumption

Taft et al.[9] attribute oxygen depletion in the mid-Bay to heterotrophic metabolism in subpycnoclinal waters whereas Officer et al.[10] contend that oxygen depletion results primarily from benthic respiration. The latter authors considered deep water DO in the mid-Bay to be determined by the vertical exchange rate relation:

$$dC/dt \; k = - 1/d \; dR/dt + K_1(C_1 - C) - K_2C \quad \text{(Officer et al.[10])} \quad (4)$$

where C is the deep water DO (mg/l), C_1 is the DO of the water mass above the pycnocline, dR/dt is the benthic respiration rate (g O_2/m^2/day), d is the thickness of the subpycnoclinal water mass, K_1 is the nonadvective vertical exchange coefficient (reciprocal days), and K_2 is the advective vertical exchange coefficient. The first term of the equation describes oxygen consumption. The second and third terms represent reaeration to the deep water.

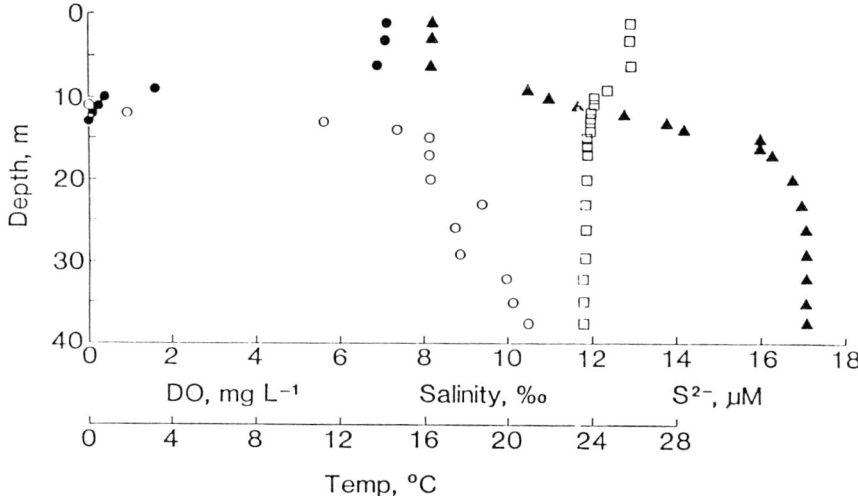

FIGURE 15. Water column profiles of oxygen (•), sulfide (O), temperature (■), and salinity (▲) at a deep channel station opposite the Patuxent River on 26 July 1984.

In order to arrive at the conclusion that benthic respiration dominates oxygen consumption, two key conditions have to be met. The first is that the advective term of equation 4 represents loss of oxygen from the deep layer and is insignificant. Officer et al.[10] appear to base this contention on their statement that "the anoxic water mass of Chesapeake Bay is bounded laterally by the Bay channel itself." Our observations clearly indicate that advective inputs of oxygen to the deep layer occur. During summer anoxia, we have noted intrusion of anoxic water far up the western flank to within 2m of the surface (Figure 2) in response to wind events. These events resulted in deep water reoxygenation on more than one occasion during the summer of 1984 when stratification was extreme (Figure 1). These observations indicate that advection cannot be ignored and may, at certain times, be a dominant process within the mid-Bay acting as a positive rather than a negative term in equation 4.

The second condition needed to buttress the conclusion that benthic processes control oxygen consumption and anoxia in the mid-Bay is that water column respiration is negligible compared to benthic respiration. Our data indicate that water column respiration is not only significant, but that it could easily exceed benthic oxygen consumption estimates of Officer et al.[10]

The measured rates of benthic respiration (1-2 g and 2-4 g O_2/m^2/day, spring and summer, respectively) used in the calculations of Officer et al.[10] were obtained by Boynton et al.[31] For purposes of comparison, mean spring oxygen consumption rates in deep waters from mid-March through May were 0.55 mg/l/day (Figure 9) and averaged 1.03 mg/l/day (Table 2) during the summer. Based upon a 20 m deep water column with a pycnocline at 10m, a typical mid-Bay condition identical to that assumed by Officer et al.,[10] the volume of water beneath the pycnocline is 10^4 1 per m^2 of bottom. Using the rates above, we calculate gross mean oxygen consumption rates of 5.5 and 10.3 g O_2/m^2/day for spring and summer, respectively. Thus, in a 20 m water column, oxygen consumption in water beneath the pycnocline would be 2.6 to 5.5 times greater than the sediment oxygen demand measured by Boynton et al.[31] and used in the calculations of Officer et al.[10] Using equation 4 with $K_1 = 0.012$/day,[10] $K_2 = 0.015$/day,[10] $C_1 = 9$ mg O_2/l, and C = 1 mg O_2/l, we calculate net water column oxygen consumption rates of -0.47 to -0.95 mg O_2/l/day for spring and summer, respectively. We conclude that the relative contributions of sediment oxygen demand and water column respiration to deep water oxygen consumption vary seasonally and with respect to the thickness of the subpycnoclinal layer over which water column respiration can occur. The latter changes with water depth (benthic processes are relatively more important in shallow zones) and as a function of the presence or absence of sulfide in the water column.

ACKNOWLEDGMENTS

Much of the research reported in this chapter was supported by grant

OCE-8208032 from the National Science Foundation, grants X-003311-01-0 and X-003310-02-0 from the U.S. Environmental Protection Agency, grant NA84AA-D-00014 from the National Oceanographic and Atmospheric Administration through the University of Maryland Sea Grant College and a grant from George Mason University, Office of Research and Advanced Studies. We are grateful to H.W. Ducklow and co:workers who furnished a portion of the bacterial production and abundance data. We thank D.G. Cargo, J.T. Bell, C. Divan, and D. Gluckman for excellent technical assistance. Computer assistance was contributed in part by the Computer Science Center of the University of Maryland, College Park. Contribution No. 1837, Center for Environmental and Estuarine Studies of the University of Maryland.

LITERATURE CITED

1. Walsh, J.J. 1981. A carbon budget for overfishing off Peru. Nature 20:300-304.
2. Richards, F.A. 1975. The Cariaco Basin (Trench). Oceanogr. Mar. Biol. Annu. Rev. 13:11-67.
3. Jannasch, H.W., H.W. Truper, and J.H. Tuttle. 1974. Microbial sulfur cycle in the Black Sea, pp. 419-425. In: E.T. Degens and D.A. Ross (Eds.) *The Black Sea: Its Geology, Chemistry and Biology.* Amer. Assoc. Petrol. Geol. Memoir No. 20.
4. Newcombe, C.L. and W.A. Horne. 1938. Oxygen-poor waters of the Chesapeake Bay. Science 88:80-81.
5. Carpenter, J.H. and D.G. Cargo. 1957. Oxygen requirement and mortality of the blue crab in the Chesapeake Bay. Ches. Bay Inst. Tech. Rept. No. 13 (Reference #57-2). Baltimore, MD.
6. Whaley, R.C., J.H. Carpenter, and R.L. Baker. 1966. Nutrient data summary 1964, 1965, 1966: Upper Chesapeake Bay (Smith Point to Turkey Point), Potomac, South, Severn, Magothy, Back, Chester, and Miles Rivers; and Eastern Bay. Ches. Bay Inst. Spec. Rept. 12, Baltimore, MD.
7. Taft, J.L. and W.R. Taylor. 1976. Phosphorus distribution in the Chesapeake Bay. Chesapeake Sci. 17:67-73.
8. Kemp, W.M. and W.R. Boynton. 1980. Influence of biological and physical processes on dissolved oxygen dynamics in an estuarine system: Implications for measurement of community metabolism. Estuarine Coastal Mar. Sci. 11:407-431.
9. Taft, J.L., W.R. Taylor, E.O. Hartwig and R. Loftus. 1980. Seasonal oxygen depletion in Chesapeake Bay. Estuaries 3:242-247.
10. Officer, C.B., R.B. Biggs, J.L. Taft, L.E. Cronin, M.A. Tyler and W.R. Boynton. 1984. Chesapeake Bay anoxia: origin, development, significance. Science 223:22-27.

11. Seliger, H.H., J.A. Boggs and W.H. Biggley. 1985. Catastrophic anoxia in the Chesapeake Bay in 1984. Science 228:70-73.
12. Malone, T.C., W.M. Kemp, H.W. Ducklow, W.R. Boynton, J.H. Tuttle and R.B. Jonas. 1986. Lateral variation in the production and fate of phytoplankton in a partially stratified estuary. Mar. Ecol. Prog. Ser. 32:149-160.
13. Sale, J.W. and W.W. Skinner. 1917. The vertical distribution of dissolved oxygen and the precipitation by salt water in certain tidal areas. J. Franklin Inst. 184 (Dec.): 837-848.
14. Tuttle, J., T. Malone, R. Jonas, H. Ducklow, and D. Cargo. 1986. Nutrient dissolved oxygen dynamics: roles of phytoplankton and microheterotrophs under summer conditions, 1985. Draft Final Report to U.S. Environmental Protection Agency, Chesapeake Bay Office, Anapolis, MD.
15. Biggs, R.B. and D.A. Flemer. 1972. The flux of particulate carbon in an estuary. Mar. Biol. 12:11-17.
16. Tuttle, J., T. Malone, R. Jonas, H. Ducklow and D. Cargo. 1985. Nutrient dissolved oxygen dynamics; roles of phytoplankton and microheterotrophs under summer conditions. Final Report to U.S. Environmental Protection Agency, Chesapeake Bay Office, Annapolis, MD.
17. Carpenter, J.H. 1965. The Chesapeake Bay Institute technique for the Winkler dissolved oxygen method. Limnol. Oceanogr. 10:141-143.
18. Truper, H.G. and H.G. Schlegel. 1964. Sulphur metabolism in Thiorhodaceae. Antonie Leeuwenhoek J. Microbiol. Serol. 30:225-238.
19. Howarth, R.W. 1978. A rapid and precise method for determining sulfate in seawater, estuarine water, and sediment pore water. Limnol. Oceanogr. 23:1066-1069.
20. Hobbie, J.E., R.J. Daley and S. Jasper. 1977. Use of Nucleopore filters for counting bacteria by fluorescence microscopy. Appl. Environ. Microbiol. 34:1225-1228.
21. Fuhrman, J.A. and F. Azam. 1982. Thymidine incorporation as a measure of heterotrophic bacterioplankton production in marine surface waters: evaluation and field results. Mar. Biol. 66:109-120.
22. Williams, P.J.LeB. and C. Askew. 1968. A method for measuring the mineralization by micro-organisms of organic compounds in seawater. Deep-Sea Res. 15:365-375.
23. Williams, P.J.LeB., T. Berman, and O. Holm-Hansen. 1976. Amino acid uptake and respiration by marine heteroptrophs. Mar. Biol. 35:41-47.
24. Howarth, R.W. and S. Merkel. 1984. Pyrite formation and the measurement of sulfate reduction in salt marsh sediments. Limnol. Oceanogr. 29:598-608.
25. Tuttle, J.H. 1985. The role of sulfur-oxidizing bacteria at deep-sea hydrothermal vents. Biol. Soc. Wash. Bull. 6:335-343.
26. Fuhrman, J.A. and F. Azam. 1980. Bacterioplankton secondary produc-

tion estimates for coastal waters of British Columbia, Antarctica, and California. Appl. Environ. Microbiol. 39:1085-1095.
27. Environmental Protection Agency (EPA). 1982. Chesapeake Bay Program Technical Studies: A Synthesis. U.S. Environmental Protection Agency, Washington, D.C. 634 pp.
28. Cargo, D.G., J.H. Tuttle, and R.B. Jonas. 1986. The low dissolved oxygen situation in Chesapeake Bay; then and now. Presented at the Spring 1986 meeting of the Atlantic Estuarine Research Society, Lewes, DE. 10-12 April 1986.
29. Flemer, D.A. and R.B. Biggs. 1971. Particulate carbon: nitrogen relations in northern Chesapeake Bay. J. Fish. Re. Bd. Canada 8:911-918.
30. Howarth, R.W. and J.M. Teal. 1979. Sulfate reduction in a New England salt marsh. Limnol. Oceanogr. 24:999-1013.
31. Boynton, W.R., W.M. Kemp and C.G. Osborne. 1980. Nutrient fluxes across the sediment-water interface in the turbid zone of a coastal plain estuary, pp. 93-109. *In:* V.S. Kennedy (Ed.) *Estuarine perspectives.* Academic Press, New York, NY.

Chapter Twenty

PENNSYLVANIA'S RESPONSE TO CHESAPEAKE BAY ECOLOGICAL PROBLEMS AND RESTORATION EFFORTS

PAUL O. SWARTZ,[1] C. VICTOR FUNK,[2] and LOUIS W. BERCHENI[3]

[1]Director, Bureau of Soil and Water Conservation
[2]Chief, Watershed Branch
[3]Assistant Director
Pennsylvania Department of Environmental Resources
Bureau of Water Quality Management
P.O. Box 2063
Harrisburg, PA 17120

ABSTRACT

In September 1983, the United States Environmental Protection Agency (EPA) published the results of its seven year Management Study[1] of the Chesapeake Bay. The study clearly indicated that the Bay is an ecosystem in decline.

On December 9, 1983, the Commonwealths of Pennsylvania and Virginia, the State of Maryland, the District of Columbia, the Chesapeake Bay Commission, and EPA pledged to restore the Bay. This commitment, known as the Chesapeake Bay Agreement, calls for the preparation and implementation of coordinated plans to improve and protect the water quality and living resources of the Bay.

In response to the study findings and recommendations, Pennsylvania is implementing various initiatives to meet this commitment.

INTRODUCTION

The major water pollutants of concern to the Bay from Pennsylvania are nutrients (phosphorus and nitrogen) in the Susquehanna River Basin. According to EPA's Management Study, the Susquehanna is dominated by nonpoint sources which account for 76% of the phosphorus and 90% of the nitrogen loads within the basin.

A high percentage of the nitrogen (85%) and phosphorus (60%) loadings delivered by the Susquehanna is attributable to runoff from cropland. The Study found that 41% of the Susquehanna's nonpoint source load comes from the intensively farmed area in the lower Susquehanna River Basin below Sunbury, Pennsylvania. Within the lower basin, soil loss from untreated cropland may be as high as 17.7 tons/acre/year. In addition, the large concentrations of livestock produce more manure than required for land application. Furthermore, the lower Susquehanna River Basin has a high percentage of conventional tillage cropland and a low percentage of forest land. This is significant because nutrient loadings from conventional tillage cropland are potentially the highest for all land uses, while those from forest lands are the smallest.

In accordance with the EPA Management Study recommendations, Pennsylvania is initially targeting the lower Susquehanna River Basin for a comprehensive implementation program to reduce agricultural nonpoint source pollution. The program focuses on commercial fertilizer management, animal waste application management, animal waste control, erosion and sediment control, and pesticide/herbicide application and control. The program also calls for continued implementation of: (1) Recommendations contained in the Pennsylvania Department of Environmental Resources (DER) Report, "An Assessment of Agricultural Nonpoint Source Pollution in Selected High Priority Watersheds in Pennsylvania"[2], (2) The Mason-Dixon Erosion Control Project, and (3) The Conestoga Headwaters Rural Clean Water Program.

The Management Study contains additional Baywide and Basin-specific recommendations relative to the signatory parties controlling nutrients and toxicants through (1) Water quality standards, (2) The National Pollutant Discharge Elimination System (NPDES) Program, (3) Pretreatment, (4) Innovative/alternative treatment technologies, (5) Construction grant funding, and (6) Regulating ground-water quality and hazardous wastes. Pennsylvania has developed and is implementing various programs and strategies to respond to these recommendations.

RESULTS AND DISCUSSIONS

NONPOINT SOURCE INITIATIVE

Background

In response to the data presented in the EPA Management Study, Pennsylvania accelerated its involvement on various committees and task forces that were formulating state level nonpoint source control programs. DER staff and agricultural organization representatives participated in interstate meetings such as the Resource Users Management Team and attempted to identify alternatives for the restoration and protection of the Bay.

The desire to better understand scientific information generated by the EPA study precipitated a landmark meeting in Hershey, Pennsylvania, in August 1982 bringing together scientists, government administrators, and agricultural leaders to discuss findings and plan an implementation program for Pennsylvania. Subsequent meetings spanning the balance of 1982 into most of 1983, hosted by the National Wildlife Federation at its facilities in Washington, D.C., were held by agricultural leaders from the Bay states to discuss mutual problems and share information.

Pennsylvania's major agricultural organizations—the Pennsylvania Farmers Association, the Pennsylvania State Grange, and the Pennsylvania Farmers Union—all made early commitments to assist in the development of an agricultural nutrient management program. This unique grass-roots support guaranteed an implementable program at the farm level. Despite the fact that the size or seriousness of the agricultural problem was not clearly understood, the opportunity existed to expand the state-of-the-art knowledge of nutrient movement, and farmers could benefit economically by reducing the application of expensive commercial fertilizers.

An ad hoc committee, expanded to include private, civic, and environmental organizations such as the League of Women Voters, the Sierra Club, the Audubon Society, and the National Wildlife Federation, remained in constant touch with government officials who were shaping the Pennsylvania control program.

As a result, a new initiative was included in Pennsylvania's FY 1984-85 budget providing $2,000,000 (including $1,000,000 of federal funds) to begin a comprehensive program to control excess nutrients from critical agricultural nonpoint source contribution areas.

Traditional approaches across the United States to address these types of problems were to develop programs patterned after the federal Agricultural Conservation Program. In Pennsylvania, a bolder concept was employed to require a more holistic approach, including a long-term commitment by farmers to install and maintain Best Management practices (BMPs) to address the major nutrient problems on their farms. The emphasis was on nutrient management with a goal to balance the application of nutrients (both from chemical fertilizers and animal wastes) with the crop production needs to prevent excess nutrients from leaving the farm in runoff to a stream or through infiltration to the ground water.

Explosion of animal numbers in south central Pennsylvania has created a vast surplus of animal manure. An example of this condition is the 173 percent increase in dairy cows in Lancaster County from 1960 to 1982. Hogs have increased 648 percent in the same time period. In the Lancaster area, to safely apply the waste from all livestock would require thousands more additional acres of cropland than are currently available. The intensification of agriculture in south central Pennsylvania has created additional pressure to develop new methods for animal waste disposal and utilization.

Realistically, all nutrient and erosion problems cannot be treated with the available technical and financial resources in the foreseeable future. Pennsylvania decided to implement an accelerated voluntary initiative in selected targeted high and medium priority watersheds. High and medium priority watersheds are those which show a high potential for agriculture related pollution based on an analysis of a number of factors such as the amount of cropland acres, the animal density, rainfall information, average slope conditions, acres in row crops, and erodibility of the soils within these watershed drainage basins. Drawing on earlier work to identify priority areas, the initial implementation area was selected from the study titled, "An Assessment of Agricultural Nonpoint Source Pollution in Selected High Priority Watersheds in Pennsylvania", conducted by DER under Section 208 of the Federal Clean Water Act during 1980 to 1982. Although this investigation was done statewide in ten large drainage basins, six were located in the lower Susquehanna River Basin in south central Pennsylvania. These six watersheds (within Adams, Chester, Dauphin, Lancaster, Lebanon and York Counties) were chosen as a foundation to begin implementing a control program through the refinement of the study data.

The Pennsylvania State Conservation Commission, an administrative policy body, was given the overall administrative responsibility to implement the new program. The State Conservation Commission plays an important role in the control of soil erosion from farmland. The Secretary of DER chairs the Commission, which is composed of individuals from the State Department of Agriculture, Federal government, Pennsylvania State University, and six members appointed by the Governor. Another major function of the Commission is to provide direction to the county conservation districts.

Guidance for Pennsylvania's Chesapeake Bay Nonpoint Source Abatement Program is provided in the Commission's statement of policy, effective April 12, 1985. The four programs established in this policy document are:

> *Planning Assistance Funding Program* — Assistance to conservation districts, agencies, and/or cooperating organizations to identify nonpoint source pollution sources, monitor water quality, measure soil and water runoff characteristics, and design treatment methods for nonpoint pollution sources.
>
> *Educational Assistance Funding Program* — Assistance to conservation districts, agencies, and/or cooperating organizations to sponsor educational programs for land owners and the public concerning the need for nutrient management, soil erosion control, and water quality management. The goal is to accelerate the adoption of soil and water conservation techniques through public awareness.
>
> *Technical Assistance Funding Program* — Assistance to conservation

districts, agencies and/or organizations to employ technical personnel and to purchase equipment and supplies to assist landowners in the planning and installation of BMPs to control nonpoint source pollution, and to foster the proper application and disposal of nutrients on land areas. *Financial Assistance Funding Program* — Assistance to Pennsylvania farmers through conservation districts to assist with the cost of installing BMPs to control soil and nutrient loss. Most of Pennsylvania's funds in the Chesapeake Bay program are directed towards this program.

Planning Assistance

All priority watersheds that become eligible for the Financial Assistance Funding Program must first have an assessment of the watershed completed that includes identifying agricultural land use, animal population densities, soil erodibility potential, subdrainage areas with high pollution potential, and a program implementation plan. These assessments provide the foundation for a detailed implementation program.

A monitoring plan has been developed for the Susquehanna River and its major tributaries to measure base flow and to sample storm events. Samples are analyzed for a variety of organic and inorganic constituents to determine nutrient and toxic levels.

Educational Assistance

Some of the more important groups that are involved in Bay restoration are the Pennsylvania Association of Conservation District Directors, Inc. (PACDD), the Cooperative Extension Service (CES), the Pennsylvania Farmers Association (PFA), Pennsylvania State Grange, Pennsylvania Farmers Union (PFU), the Soil Conservation Service (SCS), and the Agricultural Stabilization and Conservation Service (ASCS).

The Pennsylvania Association of Conservation District Directors, Inc. (PACDD) provides an overall educational program for farmers and the general public. The PACDD is responsible for the development of newsletters, fact sheets, brochures and public service announcements on both television and radio which inform the general public about Bay restoration efforts. Primarily through the PACDD, technical information is provided to farmers on how to install and maintain BMPs on the farm. The Association also coordinates activities for broader educational efforts such as the Pennsylvania Farm Show and has been primarily responsible for organizing major conferences and meetings to promote the Bay program.

The Cooperative Extension Service (CES) is part of the Pennsylvania State University and is also involved in the educational process, providing farmers, technicians, as well as the general public, with information on nutrient management and water quality through radio, television, and written materials. The nutrient management program includes computer software developed at Penn

State and distributed through the CES to provide technicians the expertise necessary to determine proper application rates of nutrients for maximum crop yields.

Through workshops, newsletters, and bulletins, agricultural organizations such as PFA, PFU, and the Grange provide information to their members concerning the best crop management practices and how to save soil. In addition, the Agricultural Stabilization and Conservation Service (ASCS) coordinates conservation programs locally with conservation districts to avoid duplication of effort and to maximize available funding.

Much of the nonpoint source nutrient problem to the Bay has been attributed to cropland erosion and over-fertilization, and little has been done in the past on technology for manure handling. Manure, as a resource, has not kept pace with other farm technology. Research projects at Penn State, with funding through the Chesapeake Bay program, are underway for such projects as manure spreader calibration, developing and implementing nutrient management systems for farms, predicting nitrogen fertilizer rates needed for economical optimum corn yields, soil chemical monitoring for nutrient management, and manure field application demonstration plots.

A special project has been initiated to help alleviate the problem of overabundance of animal manure in southeastern Pennsylvania, where animals produce more manure than farmers can safely spread on available farmland. As animal numbers continue to grow and land development continues to take land out of agriculture in southeastern Pennsylvania, a need to explore the use of poultry manure as an energy source is desirable. A study to analyze cogeneration using chicken manure in an environmentally safe manner is underway as a possible solution to the region's animal-related water quality problems.

Technical Assistance

The Soil Conservation Service (SCS) of the U.S. Department of Agriculture provides technical help to the conservation districts in developing needed conservation systems by developing nutrient management plans for the landowner. SCS personnel inspect the land, determine what BMPs are needed and then provide the technical system of design so that the BMPs will be implemented correctly. Once the plan has been developed, SCS personnel assist in the installation of practices according to technical specifications. They will certify to the district that practices are installed correctly, enabling the landowner to receive cost share reimbursement.

Nutrient management specialists employed by DER are operating a mobile nutrient testing demonstration lab and providing other consultation services to all conservation districts participating in the program.

Financial Assistance

The program, also known as the Cost Share Program, is open on a voluntary basis to landowners (farmers) in high priority watersheds that have nutrient problems. High priority watersheds are those with identified critical nonpoint source pollution. In our experience, such watersheds are those with high animal densities per unit of cropland. Farmers become cooperators with local conservation districts and sign an agreement (contract) to install and maintain BMPs in accordance with the nutrient management program (plan). BMPs eligible for cost sharing are nutrient management measures and associated erosion and water control measures necessary to provide a nutrient management program.

A nutrient management program is defined in the Commission's statement of policy as a system of BMPs to prevent the pollution of surface and ground waters by addressing the most critical farm nutrient problems through measures to manage fertilizer and animal waste and to reduce soil erosion. Included in the management plan are the proper timing, placement, and levels of nutrients required for an expected crop yield.

The agreement or contract between a landowner and a conservation district provides that reimbursement will be provided for up to 80 percent of the cost of installing BMPs to a maximum of $30,000 per landowner. The contract details when measures will be installed and guarantees funds are available to cover estimated costs. The contract stipulates, with few exceptions, that the landowner shall implement and maintain the nutrient management system as planned or, failing to do so, shall be required to refund to the district the monies provided under the agreement. Among other things, the landowner agrees not to apply nutrients in excess of the recommended rates prescribed in the nutrient management program.

The Cost Share Program is administered at the local level by the conservation districts and at the state level by the Bureau of Soil and Water Conservation within DER. Oversight responsibility rests with the State Conservation Commission and EPA.

Pennsylvania's Cost Share Program was officially begun June 17, 1985, in a contiguous six county area in southcentral Pennsylvania. The first contract was approved in Lancaster County, Pennsylvania, in November 1985. As of January 1, 1987, 325 landowners have made application for financial assistance in the six county area, and 105 contracts have been approved. Approximately $2,000,000 have been obligated to these contracts.

POINT SOURCE AND OTHER INITIATIVES

In addition to the agricultural nonpoint source pollution control recommendation, the EPA Management Study made other Baywide and Basin-specific recommendations that are relevant to Pennsylvania, including:

Pennsylvania should continue to implement its regulation requiring 80% phosphorus removal at all new or modified point source discharges in the lower Susquehanna River Basin.

The 80% removal is equivalent to a 2 mg/l discharge concentration limit and has been in effect since 1970. It specifically relates to all discharges to the main stem and all tributaries to the Susquehanna River in a zone extending from the mouth of the Juniata River to the Pennsylvania-Maryland line.

Pennsylvania completed Phase I of its triennial water quality standards review process following a public hearing on November 8, 1984. On December 18, 1984, the State's Environmental Quality Board approved various revisions to our standards—including our Statewide phosphorus control regulation for discharges to streams like the lower Susquehanna River and its tributaries. These revisions went into effect on February 16, 1985. Phosphorus controls will be based on a determination by DER of the specific level of control needed. Final implementation guidance for the lower Susquehanna River Basin was completed in December, 1985, and is being implemented. Existing dischargers with phosphorus controls in place will be required to continue to operate their facilities in accordance with their present permit requirements. New dischargers or existing dischargers (without phosphorus controls in place) will be required to meet a 2 mg/l effluent limitation if the discharge contributes 0.25% or more of the total point source phosphorus load in the lower Susquehanna pools.

The States, through the NPDES program, should control discharges of toxicants and nutrients.

Pennsylvania has been carrying out a federally-delegated NPDES program (permitting, compliance monitoring, and enforcement) since June 20, 1978. In Pennsylvania's review of NPDES permits, which must be renewed every 5 years, it applies the more stringent of EPA's effluent guidelines or the State water quality standards in setting effluent limitations. In developing effluent limits for toxic pollutants, DER uses its Point Source Toxics Management Strategy. The Strategy is a water quality-oriented approach which addresses EPA's priority pollutants and other toxics. It involves a step-wise permit application review process that takes into account: (1) Observed occurrence of priority pollutants as part of EPA's Best Available Technology (BAT) screening and verification surveys; (2) EPA's recommended water quality criteria; (3) Pennsylvania's general standards for water quality protection; (4) Current levels of detectability for these pollutants and available analytical techniques; (5) Pollutant treatability; and (6) Fate/transport of these pollutants in the aquatic environment.

There are 1,265 NPDES discharges within the Pennsylvania portion of the Chesapeake Bay Drainage Basin—115 are major discharges (i.e., sewage discharges of 1 mgd or greater, and industries that meet EPA's criteria for discharge volume, toxicity, and impact on receiving waters).

As a minimum, every major discharger and every Federal P.L. 92-500 funded minor municipal discharger is inspected annually. In accordance with the State/EPA Enforcement Agreement signed on October 15, 1984, Pennsylvania has also agreed to take enforcement actions involving significant non-compliance by major dischargers within 180 days.

Accelerate the development and implementation of Pretreatment programs.

Pennsylvania carries out a delegated NPDES program; therefore, in accordance with Federal requirements, we are also obligated to carry out the Federal Pretreatment program. DER has developed a draft proposal for program delegation. A package of regulations was approved by the Environmental Quality Board as proposed rulemaking in March, 1985. Pennsylvania anticipates applying for program delegation in Federal Fiscal Year 1987. In the meantime, DER is continuing to assist EPA Region III in implementing the program in Pennsylvania. Of the 90 municipal treatment plants in Pennsylvania that need a pretreatment program, 35 are in the Chesapeake Bay Drainage Basin—33 have approved programs, and the remaining two are expected to have their programs in place in Federal Fiscal year 1987.

Evaluate and utilize innovative and alternative nutrient removal processes.

In February 1986, Pennsylvania initiated the development of a comprehensive rural sewage management project in Millmont Village, Lewis Township, Union County. The project is designed to research and evaluate alternative and innovative technologies for sewage disposal. The project, which will be constructed in 1987, will be totally funded ($596,309) by the Commonwealth and designed by DER staff. It will be owned and operated by Lewis Township. The project will utilize a sequencing batch reactor and marsh/meadow treatment cells designed to allow for the research and monitoring of seven different treatment technologies.

The Department, in cooperation with the Pennsylvania State University, will conduct a five to ten year research and monitoring program of the project. This technology will provide the foundation for a comprehensive rural sewage management program designed specifically for the three million people living in Pennsylvania's rural areas (the largest rural population in the United States). Innovative technology is necessary as many of these rural communities (1,500) will never be able to afford existing conventional technology, and yet are in desparate need of adequate waste water treatment facilities to protect the surface and ground waters of the Commonwealth, as well as the Bay.

Consider the Management Study findings when funding construction grant projects.

Pennsylvania has been carrying out a federally-delegated Construction Management Assistance Grants program since June, 1979. During Federal Fiscal Years 1979-85, Pennsylvania approved $229 million in federal funds for the construction of 61 municipal sewerage projects within the Chesapeake Bay Drainage Basin. One project (York) was certified for grant funds totalling $19.8 million in Federal Fiscal Year 1986.

Increase Publicly Owned Treatment Works efficiency and improve operator training.

DER has developed and is implementing a Statewide Outreach Operator Training Program under Section 104(g) of the Federal Clean Water Act, using the training capabilities of the Pennsylvania Department of Community Affairs. Eight of sixteen municipal treatment plants selected for on-site training of Operation and Maintenance staff are within the Chesapeake Bay Drainage Basin. The pilot phase of this program showed that such training results in a noticeable operational and effluent quality improvement.

Evaluate the effects of contaminated ground water and hazardous wastes.

Development and implementation of a comprehensive Ground-Water Quality Management Program is continuing. As a pilot project, ambient ground-water quality monitoring systems were established within three high priority ground-water basins in Pennsylvania, including one in the Chesapeake Bay Drainage Basin. This enabled DER to assess the effectiveness of this method of monitoring before proceeding with its ground-water monitoring strategy. DER is now implementing the monitoring program, including the installation of seven more systems (including two more in the Chesapeake Bay Drainage Basin).

Pennsylvania received final program authorization (primacy) from EPA in January, 1986 to carry out a federally-delegated Resource Conservation and Recovery Act (RCRA) Program. DER is currently modifying its program to comply with new federal requirements. This will enable DER to develop and implement a comprehensive waste management program, including the control of hazardous wastes, under both State and Federal laws.

FUTURE DIRECTION

Funding for Chesapeake Bay Program implementation in Pennsylvania is dependent on continued EPA grants matched by state appropriations. The recently reauthorized Clean Water Act authorizes $13,000,000 per year to fund the Chesapeake Bay Program through Federal Fiscal Year 1990. These funds are distributed to Pennsylvania, Virginia, Maryland, and the District of Columbia on a proportionate formula. This is an increased level of funding and indicates a continuing commitment by the federal government to help fund Pennsylvania's program initiatives and those of the other signatory parties to the Agreement.

The Pennsylvania State Conservation Commission has established the Chesapeake Bay Advisory Committee to give direction and recommendations to the Commission on Bay issues. The advisory committee developed and submitted to the Commission a recommended long-range plan in 1986 for the Pennsylvania Bay Program. This plan addresses expected issues and requirements for maintaining a program over the next four years. The four major goals of this plan are:

1. Nutrient reduction (to reduce nutrient loadings through such measures as installation of BMPs, focusing on agricultural nonpoint sources of pollution following guidelines contained in DER's Manure Management Manual).[3]
2. Reduction or control of toxic materials.
3. Restoration and protection of living resources.
4. Development and management of related environmental programs.

In part, the report recommends that: (1) the Cost Share Program continue to focus primarily on agricultural nonpoint sources of pollution, and (2) upon the completion and verification of the Chesapeake Bay Watershed Model and the two-dimensional Chesapeake Bay Water Quality Model, Pennsylvania will consider the adoption of measurable nutrient reduction goals. These goals will be based on the available funding to Pennsylvania for the Cost Share Program to implement BMPs on farms and the technology that is available to assess the effectiveness of those BMPs in reducing nutrients and improving water quality.

DER's Manure Management Manual was revised in October, 1986 through the guidance of the Agricultural Advisory Committee to DER. This manual, first issued in 1977, provides a sound basis statewide for proper nutrient management. It has been updated by technical specialists of the U.S. Soil Conservation Service, the Cooperative Extension Service, DER personnel, farm organization representatives, and members of the legislature to respond to the changes in state-of-the-art for nutrient management. Primarily, the manual deals with animal waste storage facilities, with sections to recommend proper field application rates of manures as well as individual requirements for different types of animals. This manual will serve as the primary regulatory tool for animal waste management.

Presently, cost sharing for BMPs on farmland is only available in selected high priority watersheds in six contiguous southcentral Pennsylvania counties. Future program expansion will progress in the high and medium priority areas and also in areas more specifically identified using empirical water quality data. Other selected high priority watersheds in seven counties located in the Middle Susquehanna River Basin will become eligible for the Cost Share Program in the near future. Long range expansion of the Cost Share Program includes assessment of medium priority watersheds in the upper Susquehanna River Basin.

As a result of a study conducted by the Pennsylvania Department of Agriculture, an expanded educational program will be developed to respond to pesticide use for agricultural activities. The educational program will focus on a coordinated program involving Extension personnel, the Pennsylvania Department of Agriculture's field personnel, and pesticide manufacturers and is intended to provide information to pesticide applicators and farmers.

Baseline water quality and quantity monitoring on the mainstem Susquehanna River and its major tributaries will continue. The information will provide the basis for identifying changes in water quality resulting from the implementation of the Chesapeake Bay Nonpoint Source Pollution Abatement Program.

The restoration and protection of the Bay is an ongoing process in which both short-term and long-term goals need to be considered. Pennsylvania, as a good neighbor and as a signatory to the Chesapeake Bay Agreement, is committed to this process.

REFERENCES

1. Tippie, V.K., et. al. 1983. Chesapeake Bay: A Framework for Action. U.S. Environmental Protection Agency. Chesapeake Bay Program, Annapolis, MD.
2. Schueller, J.P. 1983. An Assessment of Agricultural Nonpoint Source Pollution in Selected High Priority Watersheds in Pennsylvania. Pennsylvania Department of Environmental Resources, Bureau of Soil and Water Conservation, Harrisburg, PA.
3. Agricultural Advisory Committee. 1986. Manure Management for Environmental Protection. Pennsylvania Department of Environmental Resources, Harrisburg, PA.

Chapter Twenty-One

VIRGINIA GOVERNMENT RESPONSE TO CHESAPEAKE BAY PROBLEMS

KEITH J. BUTTLEMAN and JANICE CARTER-LOVEJOY

Council of the Environment
903 Ninth Street Office Building
Richmond, VA 23219

ABSTRACT

The states which surround the Chesapeake Bay have acknowledged their stake in the resources of the Bay and their responsibilities for its current condition. More importantly, they have accepted the responsibility to do something about it. Virginia, Maryland, Pennsylvania and the District of Columbia have all embarked on ambitious programs to restore and protect their shared resource. And, while each state's programs are tailored to its own circumstances, they have many features in common. This chapter briefly describes the process that led up to these state commitments and outlines the basic content of each state's programs.

INTRODUCTION

The Chesapeake Bay is a national treasure and a resource of worldwide significance. Its ecological, economic and cultural importance are felt far beyond its waters and the communities that line its shores. And, while concerned citizens everywhere are rightfully dismayed at its decline, it is the people closest to it, and the state and local governments which represent them, that stand to lose the most if this decline is allowed to go unchecked. Naturally, therefore, it is they who must bear the greatest measure of responsibility for protecting the Bay from further deterioration and for restoring it to a more productive state.

Of course, the role of the federal government in dealing with the issues surrounding the Chesapeake Bay's decline has been important, from the leadership of Congress in authorizing EPA to institute a study of the Bay's problems to the ten-million-dollar annual funding provided by the Environmental Protection Agency. The real authority and resources to implement actual cleanup programs, however, are found at the state level. Similarly, the benefits of such programs will be felt most clearly at the local and state levels. Even more to the point, the political will necessary to sustain this effort (through the decade or more it will require) must come from a concerned and involved citizenry, best represented through state government. Simply put, the states have the most to gain from the success of a cleanup effort and the most to lose from inaction. The ultimate fate of the Bay depends in large part on how state government deals with the issues facing it.

An understanding of the nature of the state government response to Chesapeake Bay ecological problems, the subject of this chapter, is fundamental to an appreciation of the problems and prospects facing anyone who would seek to influence or contribute to this effort.

DISCUSSION

The Chesapeake's past, by geologic standards, is not a long one. It was created roughly 10,000 years ago with the rise in sea level at the end of the last ice age. At that time, the rising sea flooded the Susquehanna River Valley and the tributaries which flowed into it, The Potomac, Rappahannock, York, James, and others, to create what is now known as the Chesapeake Bay. The Chesapeake Bay drainage basin covers 64,000 square miles and extends into six states—Virginia, Maryland, West Virginia, Pennsylvania, New York, and Delaware—and the District of Columbia. The Bay proper is approximately 200 miles long and ranges in width from about 4 miles near Annapolis, Maryland to 30 miles at its widest point near the mouth of the Potomac. The water surface of the Bay encompasses more than 2,200 square miles. When the estuarine portions of the Bay's tributaries are included, this figure doubles.

Since near the time of its origin, the Chesapeake Bay has been important to the people living near it. By the time the English settlers first came to Virginia and Maryland, the Chesapeake Bay was already a major source of food and an avenue of commerce for the Indians living along its shores. The colonists recognized these values and made similar uses of the Bay and today our uses of the Bay and in many areas, our dependence on it are, if anything, more intense.

The Chesapeake Bay has grown in importance in a number of ways. In the area of seafood production, the Chesapeake Bay supplies the world's largest harvest of blue crabs, is a major source of oysters, and produces an annual catch of finfish of over $100 million.

In Virginia waters alone, over 800 million pounds of finfish and shellfish were taken in 1984. Over 12,000 fishermen and processors owe their livelihood to these resources with gross annual wages, sales and harvest for 1984 of $442 million. In the area of commerce, the Bay provides protected passage to two of the world's great ports, Hampton Roads and Baltimore. At Hampton Roads alone 50 million tons of foreign cargo (imports and exports) passed through in 1984 with a value of $11.1 billion. Of this tonnage, 36 million tons was coal. In 1984, employment in the Hampton Roads ports complex exceeded 51,000 with an indirect statewide employment of over 75,000.

The Bay is home to many industries that depend on a waterfront location. It is home to an even greater number of citizens who enjoy its beauty, like to live near its waters, and as a result, have a stake in its future. The Bay's very usefulness and attractiveness, however, have contributed to its decline. Its usefulness as an avenue of commerce and transportation has seen harbors and major cities grow up at strategic locations around the Bay and at the fall lines of the major tributaries that feed it. In its attractiveness as a place to live and a place for recreation, it has drawn many people to these cities and to the less developed areas around its shores.

In Virginia alone, over three quarters of the state's population lives within the Bay drainage area. Baywide the 1950 population was 8.5 million, the 1980 population was 12.7 million and the population in 2000 is projected to be 14.6 million. These figures translate into million of tons of municipal and industrial waste and sediment that enter the Bay and its tributaries each year as a result of man's daily activities. These activities have taken a toll on the Bay's resources, and its capacity for renewal is waning in the wake of decades of use and abuse.

By the mid 1970's, signs of stress on the Bay and its resources were noted by concerned citizens and state and federal authorities. Congress authorized the Environmental Protection agency to undertake an intensive study of the Bay to determine the factors causing its decline. After seven years of research and evaluation, the study results confirmed the hypothesis: the condition of the Bay was deteriorating, mainly due to point and nonpoint source of pollution. (Point sources refers to the discharge of wastewater from a specific location like a sewage treatment plant outfall pipe. Nonpoint source refers to runoff from non-discrete locations such as farms, lawns, and streets.)

The study report documented disturbing trends in three main areas:
- Excess Nutrients. Primarily phosphorus and nitrogen, these nutrients can foster the growth of aquatic plants such as algae when present in large quantities. When these blooms die off and decompose they reduce the dissolved oxygen which is critical to the survival of living resources in the Bay's waters. Excess nutrients are coming from a combination of agricultural, forestry, and urban runoff and municipal and industrial plant discharges. Since 1950, phosphorus and nitrogen entering Virginia's tributaries to the Chesapeake Bay have increased 44% and 87% respectively. If no additional nutrient controls are implemented, these loadings will increase by another 36% and 23% due to projected population increases by the year 2000.
- Decline of Submerged Aquatic Vegetation. Submerged aquatic vegetation

has all but disappeared in the Chesapeake Bay and its tributaries since the late 1960's. Submerged aquatic vegetation provides fish and crabs essential habitat and protection from predators, buffers wave energy, and produces oxygen needed by the living resources of the Bay. The decline of these grass beds is attributed to excess nutrients, turbidity, and sedimentation.
- Excess Toxics. Toxic substances have been found in specific areas of the Bay, primarily around urban and highly industrialized areas. Toxics contaminate waters, sediment, and living resources, and have the potential to affect humans as they accumulate in the food chain.

These are but the three most dramatic problems identified in the Chesapeake Bay study. Virginia and the other Bay area states, together with EPA and other federal agencies, continue to study the Bay to help determine sources of pollution problems and to assist in the formulation of new and alternative solutions to those problems.

Even before the full study report was released preliminary study findings began providing scientific validity to the casual observances of fishermen and recreationalists, and the individual Bay-area states began addressing the problem. Virginia and Maryland began informally discussing among themselves ways to get together to work on Chesapeake Bay restoration efforts.

The states soon began to take a more formal and comprehensive approach to shared problems and opportunities. In 1978, Virginia and Maryland established a Bi-State Working Committee, composed of cabinet and agency-head level representatives from the two states. In 1980 the two states extended their cooperative approach to Chesapeake Bay issues to include the legislative side by creating the Chesapeake Bay Commission. The Commission is made up mainly of legislators, augmented by a citizen member and the appropriate cabinet secretary from each state. The Commission's primary function is to coordinate legislative responses to Bay issues. In 1985 Pennsylvania joined the Commission.

While these mechanisms worked well for their intended purposes, the magnitude of the problems uncovered by the Chesapeake Bay study indicated that something more was necessary.

In December 1983 a region-wide conference was held to discuss ways to approach the pollution problems. Virginia, Maryland, and the Commission jointly conceived the conference, and the EPA and Pennsylvania later joined the process and contributed funds. The culmination of the conference was the signing of an agreement by the Commonwealths of Virginia and Pennsylvania, the State of Maryland, the District of Columbia, the EPA, and the Chesapeake Bay Commission. Known as "The Chesapeake Bay Agreement of 1983", the Agreement is a simple, concise commitment for a cooperative approach to restoring and protecting the Chesapeake Bay. The 250 word statement calls for the development of coordinated management plans and establishes the Chesapeake Executive Council to provide the leadership for their development, implementation, and evaluation.

The Council is comprised of cabinet-level officials from Virginia, Maryland, Pennsylvania, District of Columbia, and the Region III office of the EPA. Virginia is represented by the Secretary of Natural Resources and the Secretary of Human Resources. Similar top executives represent the other states. The chairmanship of the Council rotates annually between the EPA regional administrator and the governors. Virginia's Governor Gerald L. Baliles became the first state chairman in January 1987.

The Council is served by an Implementation Committee, made up of agency heads and program managers from the member jurisdictions. A wide variety of federal agencies regularly participate on the Implementation Committee. The Committee is designed to carry out the directives of the Executive Council and to coordinate the implementation of specific cleanup programs throughout the region.

The signing of the Chesapeake Bay Agreement at the 1983 conference also publicly committed the governors to return to their respective states and commence state programs to clean up their jurisdictions' share of the pollution problems contributing to the deteriorating condition of the Bay. The state programs are known as the Chesapeake Bay Initiatives.

Virginia's Chesapeake Bay program mustered good support from the beginning. As an initial biennial program budget for Virginia's Chesapeake Bay Initiatives former Governor Charles S. Robb recommended that the 1984 Virginia General Assembly appropriate $6 million. The General Assembly responded with over $13 million, and in mid-biennium increased the biennial budget by another $4.2 million. A second biennial budget for Virginia's Chesapeake Bay Initiatives totals $40.2 million, half of which goes into a revolving loan fund to help finance sewage treatment plant construction.

Spread among nearly a dozen state agencies within five cabinet secretariats, Virginia's Chesapeake Bay Initiatives are a compilation of state projects to restore and enhance the Chesapeake Bay. The projects include abating point and nonpoint source pollution, improving shell- and finfisheries for commercial and recreational use, restoring resource habitat, researching reasons for the decline of the Bay and seeking feasible solutions, and educating the citizenry about the Bay's problems and enlisting their continued support. The Initiatives also include a small amount of funding for administrative operations such as computerization and publications, but for the most part staffing costs are absorbed within normal state agency overhead budgets.

The initial grouping of state projects comprising Virginia's Chesapeake Bay Initiatives included both new and existing projects. New projects, in many cases, represent a different approach to an old problem, like using financial incentives to encourage farmers to conserve soil by installing best management practices. Soil conservation has been an education/economic issue for decades, but now the measures are being encouraged for the purpose of improving downstream water quality.

Existing projects were also deemed "initiatives" if they received an increase in funding over the previous years. More money meant expansion and an increased emphasis toward Bay improvement. For example, for years oyster rocks have been rebuilt to provide habitat for oyster larvae. The funds for this project were doubled so that twice as much repletion effort could take place to assist the foundering oyster industry.

Other state agencies which at first glance seem unlikely participants in the Bay program, have cooperated with new initiatives. For example, the Highway Department is improving access to the Bay through public landings, and the Department of Information Technology oversaw the production of a television documentary on the Chesapeake Bay.

Compiling new and existing State projects into a program called Virginia's Chesapeake Bay Initiatives was more than simply repackaging existing projects; but by including relevant ongoing state activities the Chesapeake Bay Initiatives could show some quick initial successes, and the political process in support of the Bay cleanup could quickly develop momentum.

Coordination, tracking, and reporting on Virginia's Chesapeake Bay program were assigned to one state agency, the Council on the Environment. The Council on the Environment is a non-regulatory, environmental policy-coordinating state agency whose role is to advise the governor and his cabinet on issues regarding the environmental interests of Virginia.

The Council on the Environment also serves as Virginia's chief representative to the Chesapeake Executive Council's Implementation Committee. In this central position, the COE not only can coordinate intrastate environmental policy positions on Chesapeake Bay matters, it can also represent coordinated environmental policy positions for Virginia to the other jurisdictions in the regional Bay restoration effort.

Many citizen and special interest groups have been involved in the planning, implementation, and monitoring of the cleanup effort from the beginning. As a result, public awareness and support for Virginia's program continues to grow. This, too, will serve Virginia's Bay cleanup effort well over the coming years by helping it to maintain the momentum it now enjoys.

It is important to keep in perspective the fact that it will be several decades before widespread improvements in water quality and living resource populations are evident. Nevertheless, some localized improvements have already been realized in a number of areas.

SUMMARY OF VIRGINIA'S CHESAPEAKE BAY INITIATIVES

Pollution Abatement

The greatest concentration of program effort in Virginia's Chesapeake Bay package includes a variety of individual programs designed to reduce the amount of pollutants entering the Bay and its tributary waters. Virginia is taking actions to reduce nutrient loadings on a large scale, and is dramatically increasing efforts to keep other pollutants out of Virginia's portion of the Bay.

Farms. Pollutant-carrying runoff from agricultural land is being reduced through a combination of education and cost-sharing grants designed to encourage farmers to use "Best Management Practices" (BMPs). During 1984 and 1985, 1,444 farmers installed BMPs on 58,595 acres as a direct result of state cost-sharing funds. From these actions, 333,930 tons of sediment which otherwise would have eroded off farm fields each year will not be retained in place. This also reduces the amount of sediment that would otherwise have actually reached a stream or river by approximately 31,260 tons. Besides reducing sedimentation, the BMPs reduce phosphorus from entering receiving streams. Phosphorus is carried by soil particles; 33,245 pounds of this "hitchhiking" nutrient are now being kept out of Bay and tributary waters. Another 48 farmers are installing facilities to manage 111,040 tons of animal waste each year from their livestock operations, thereby reducing the potential for additional nutrient pollution.

While these figures represent a promising beginning, it is really only a start. There are approximately 24,000 farmers in the Virginia portion of the Bay basin, operating on nearly 3.7 million acres of crop and pastureland. Based on Soil and Water Conservation District (SWCD) estimates, it would take about $170 million in state funds to bring all agricultural acreage and animal operations under BMPs if we rely on cost-sharing alone.

Consequently, the education component of our agricultural runoff control program is especially important, in order to demonstrate to farmers the value of using BMPs and also to convince them to install BMPs voluntarily. The Division of Soil and Water Conservation, along with Virginia Tech, has developed an educational program that illustrates to farmers the benefits of BMPs. There are no guarantees that a farmer will continue to use the BMP in subsequent years, or he may lease his land to another farmer who does not employ the BMP. For these reasons it is imperative that the education component of the BMP program continue each year.

One of the best means of encouragement is by demonstrating the value of BMPs in a clear, convincing way. Virginia Tech developed for the Commonwealth a rainfall simulator for providing such a demonstration. The rainfall simulator, a portable, modified spray irrigation system, creates, over a one-and-a-half acre area, the equivalent of a typical summer cloudburst. It is set up over a test area which contains two side-by-side farm plots, one of which has been conventionally tilled, the other using no-till. By "raining" on both plots equally, under controlled conditions the rainfall simulator provides a graphic demonstration of just how well no-till cropping reduces runoff compared to conventional practices. The runoff from both plots is channeled into two side-by-side flows, in which the difference in clarity (sediment) is clearly visible. One demonstration in Essex County showed the no-till plot to produce half the total runoff, and one-tenth of the sediment, and phosphorus loss and one-fourth the nitrogen loss, when compared with the conventional plot.

The local Soil and Water Conservation District offices and the federal Soil Conservation Service are also helping by providing technical assistance. Together with farmers, management plans are developed or conservation practices are recommended for circumstances particular to individual farms.

According to U.S. Department of Agriculture data, 1985 Virginia cropland under no-till practices had risen by about 7.6% from 1984, and from 58% to 61% of the total acreage planted. Through the education efforts targeted at farmers, including further use of rainfall demonstrations, as well as through the cost-sharing program, Virginia will continue to improve that ratio.

Urban Areas. In urban areas, runoff from streets, parking lots, and other impervious surfaces can carry contaminants into nearby waters. Just as in the case of farmland, certain best management practices can prevent or reduce this form of pollution. While not a major initiative area, the use of urban area BMPs is being encouraged through cost-sharing and technical assistance on selected demonstration sites. Eleven individual projects have been started in 7 localities. These include porous asphalt pavement, an infiltration trench and a grassed waterway, stormwater management, streambank stabilization, and an "urban marsh" and a "wet pond" (manmade rainwater detention basin). Monitoring at the "wet pond" site indicates that it is effective in removing up to 87% of the silt and 80% of the phosphorus from the runoff. It also removes up to 65% of the lead and zinc. While this project is relatively small, its efficiency at pollutant removal is significant. It and other projects serve to demonstrate the urban BMP concepts and promote voluntary use of similar practices in other urban areas.

Sewage Treatment Plants. Other significant sources of nutrients are the 476 municipal sewage treatment plants (STPs) in the Virginia portion of the Chesapeake Bay basin. The Virginia Water Control Board estimates a price tag of about $2 billion for the necessary construction, expansion, and improvement in levels of treatment at municipal STPs to carry Virginia to the year 2000; more than half of this need is in the Chesapeake Bay drainage area. In light of this need and the reductions in federal funds available for this purpose, Virginia has become directly involved in the financing of construction and repairs of municipal STPs. Prior to this, the majority of funding for such projects came from a federal construction grant program and some local sources.

Beginning in the 1986-88 biennium, the newly-created Virginia Water Facilities Revolving Fund makes available to localities construction loans at low interest rates. A limited amount of grant funds are also available and targeted for localities with a limited ability to pay. It is likely that a locality could design a financing package involving a loan from the Revolving Fund and/or bond financing through the Virginia Resources Authority.

The Virginia Resources Authority was created by the 1984 General Assembly in order to provide low-interest financing alternatives to localities to fund or refinance water, wastewater, and drainage facility projects. Three financings (bond issues) have taken place to date for a total of $63,620,000. Of the nine localities which have participated, six are within the Chesapeake Bay drainage area.

Other programs are underway to upgrade STPs, to reduce nutrients discharged by STPs as well as reduce sewerline infiltration and inflow problems (I&I). Fourteen municipal sewage treatment plants are scheduled to reduce or eliminate chlorine, ten with 1984-86 Chesapeake Bay Initiatives funds, the other four with 1986-88 funds. Eliminating the 40 infiltration and inflow problem areas existing in the Bay basin should result in significant reductions in the number of occasions that rainfall causes STP overflows and the discharges of untreated sewage into the Bay and its tributaries. Four I&I projects are currently underway with another six planned for fiscal year 1986-87 with the assistance of Initiative cost-share grants.

During the second year of the 1984-86 biennium, the Commonwealth instituted a pilot nutrient removal program. Grants were awarded to three localities to evaluate the costs and effectiveness of removing phosphorus and nitrogen at sewage treatment plants. Operations began in Fall 1986; preliminary results from the York River STP biological nutrient removal process indicate the level of phosphorus discharged has been reduced by 80%, from 8 mg/l to less than 1 mg/l.

During the 1984-86 biennium, the State has made significant strides in reissuing discharge permits so that the limitations imposed on treatment plants remain current. In addition, efforts are underway to develop nutrient standards and toxicity reduction strategies. The combined results of these regulatory programs and the financial assistance programs will contribute dramatically to the abatement of point source pollution in the years to come.

LIVING RESOURCES AND HABITAT IMPROVEMENT

A number of Virginia's Chesapeake Bay Initiatives have a continuing direct effect on marine habitat in the Bay and its tributary waters and complement on-going programs such as the management of tidal wetlands and subaqueous lands.

Chlorine removal or reduction. One effort is directed towards reducing the amount of chlorine used and discharged by sewage treatment plants (STPs). Chlorine, used as a disinfectant by STPs prior to discharging wastewater to the rivers, is acutely toxic to marine organisms, especially fish and oyster larvae. To address this problem, spawning areas of critical finfish populations and important shellfish areas have been identified and a program to reduce chlorine yet maintain a level of disinfection adequate to protect public health has been initiated. Localities with STPs adjacent to sensitive spawning and growing areas are being targeted for participation in the State's cost-share grant program to reduce the amount of chlorine being discharged. The cost-share program is really just an incentive to speed up chlorine control since the Virginia Water Control Board has adopted a water quality standard for chlorine.

For the 1984-86 biennium, ten localities were awarded cost-sharing grants totaling $1.8 million for either dechlorination or alternative disinfection at their sewage treatment plants. Another $1.7 million has been allocated for 1986-88 with four more projects approved to date. Other localities are reducing chlorine voluntarily, or under order in conjunction with state discharge permits. These actions will result in a 36% overall decrease in the amount of chlorine discharged to the Bay from Virginia tributaries. Prior to the Bay cleanup effort, 6670 lbs. of chlorine were being discharged each day; this amount will be reduced to 3905 lbs. per day. Because many of the localities reducing or eliminating chlorine in STP discharges are adjacent to spawning and nursery areas, an increase in fishery populations is anticipated as the young marine organisms reach maturity.

Finfish. Another effort to protect important commercial and recreational fisheries has been the development of fishery management plans. Plans set goals and objectives and include strategies for increasing available stock, improving habitat, managing harvest, and ensuring the proper collection of fisheries data. The first plans to be developed are for striped bass and oysters.

The State's agricultural cost-share program is also having direct impacts on marine habitat. 323,000 fewer tons of soil are eroding from farmland in the Bay basin as a result of new best management practices employed by farmers in 1985. In addition to keeping excessive nutrients out of the water, soil retention directly reduces the siltation of river bottoms, the burying of submerged grasses and bottom dwelling organisms, and decreases turbidity allowing better light penetration which is essential to good submerged aquatic vegetation growth.

Submerged Aquatic Vegetation. Beds of submerged aquatic vegetation (SAV) were once common features of the many shoal areas along the tributaries and Bay. An extensive experimental program was started in the first biennium to reestablish SAV beds and determine what causes their decline. Fifteen acres of eelgrass were transplanted to 10 locations in the Bay tributaries in the fall of 1984. Transplant survival ranged from moderate to poor. Losses are attributed to winter ice scour, turbidity, accidental dredging, cownose ray and crab uprootings, and other biological factors under investigation.

Another 15 acres were transplanted into 11 plots in four river systems in the fall of 1985, primarily in those areas where previous success had been demonstrated. As of June 1986 survival rates ranged from 10% to 75%. Growth in some areas has been phenomenal where at one site each transplanted plug has expanded an average 100-fold. Efforts to reestablish SAV beds, including using seeds in the planting process, monitoring of the key environmental parameters, and refinements of a conceptual model on eelgrass growth, are continuing in the 1986-88 biennium.

Artificial Reefs. Artificial fishing reefs continue to be constructed in order create habitat to attract and increase the production of recreationally important fish species. This program began in the mid-1970's but, with additional funding through the Chesapeake Bay Initiatives, the amount of reef material deployed increased by about 40% in each year of the biennium. Three reef sites continue to be added to each year: Parramore Reef—off the Wachapreague Inlet, Tower Reef - east of the Chesapeake Bay Light Tower, and Triangle Reef - east of Cape Charles. Other experimental reef sites are located in the Chesapeake Bay near Gwynn's Island and Cape Charles, and another in the Atlantic Ocean south of Wachapreague.

Shellfish Grounds. Virginia has led the nation in the production and export of shellfish in the past, but generally declining production threatens that prominent position. A variety of natural causes, such as predators and diseases, have been partly responsible. A substantial portion of Virginia's productive shellfish waters have been closed for public health reasons, however, due to their contamination with high bacterial levels usually associated with domestic sewage.

By 1982, over 91,000 acres of productive shellfish grounds had been permanently condemned due to contamination, out of the 450,000 total acres leased and public grounds available. As a result of the Chesapeake Bay Initiative program, however, Virginia has taken aggressive action to reverse this situation. Besides maintaining its firm commitment to protect the public health from contaminated seafood, the State is also working to reopen condemned shellfish grounds by correcting the causes of the contamination through the Shoreline Residential Sanitation Program and the Shellfish Enhancement Task Force.

This has been one of the most rewarding developments of the Chesapeake Bay Initiative program, and one that has shown dramatic results since its inception. Early in the process, the Virginia Marine Resources Commission reviewed all condemned shellfish areas and gave them priority ranking according to their value for shellfish production. Independently, the Health Department ranked areas in terms of their sources of contamination and the likelihood of their responsiveness to corrective actions. Sources of pollution include faulty septic tanks and pit privies, animal waste, industrial waste, sewage discharges, and marinas, among others. When the rankings of the two agencies were combined, the result was a priority ranking by both productivity and ease of cleanup. The State was then able to target its available funds to those areas where they would be most effective.

During the 1984-86 biennium, plus the first few months of 1986-88, 3,740 acres of productive shellfish grounds have been reopened, making available to commercial harvesting $1,288,288 in shellfish the first harvest year. If grounds are managed well, these areas should continue to produce shellfish valued at about half this amount in each year thereafter. The cost to the state has been $115,016, for an overall benefit-cost ratio of about 11 to 1. Another 756 acres with an estimated market value of $650,000 have tentatively been reopened under carefully monitored conditions.

A significant element of this program has been that once sources of contamination were identified, enforcement action was sufficient, in many cases, to correct the problem at no additional cost to the state. In addition to the 3,740 productive acres reopened, another 247 acres that are not now productive have been reopened, all through enforcement. Now that they are available, some of these acres could become productive in the future if developed by leaseholders.

Numerous areas remain condemned to shellfish harvest. In the first two years of the Chesapeake Bay Initiatives, those areas with easily identified problems were corrected first; those remaining will therefore be more difficult. In many of the remaining condemned areas the sources and causes of pollution are unknown. And in some cases reliable methods to identify and correct the problems range from poorly understood to non-existent.

Oyster Rock Repletion. Meanwhile, another initiative has expanded the existing program oyster repletion program to enhance the oyster industry in areas where production is on-going. Oysters depend on the availability of suitable bottom conditions in order for larvae to have a place to "set". One of the best substances for oyster larvae to set on is other oyster shells, but large scale harvesting, as well as siltation, has severely reduced the available oyster shell bottom in most areas. Therefore, the Virginia Marine Resources Commission (VMRC) has for many years planted oyster shells at appropriate locations. The Commission has also relocated seed oysters (very young oysters) to further encourage oyster development where natural set may not be sufficient. Since 1971 there has been a very strong correlation between VMRC repletion program shell planting and the subsequent harvest of marketable oysters three to five years later. This Chesapeake Bay Initiative added $1,000,000, or an increase of 50%, to the repletion program for the 1984-86 biennium enabling the Commission to plant approximately 3.8 million bushels of shell and 66,500 bushels of seed oysters by the end of 1986. There are plans to plant another 2 million bushels of shell in each year of the 1986-88 biennium as well as develop alternative methods of supplying shell for repletion.

Oyster Hatchery. The Virginia Institute of Marine Science (VIMS) has also been heavily involved in the restoration of the oyster industry with a major research project on seed oyster production and distribution. In the fall of 1985, VIMS began operation of an oyster hatchery which will in future years help ensure availability of seed oysters. The hatchery will produce eyed-larvae (those mature enough to attach to a substrate) for remote setting by oystermen as well as for scientific research. Ninety-eight million oyster larvae have been raised so far for in-house research and for industry use.

RESEARCH

Oysters. The Virginia Institute of Marine Science is studying the factors and processes influencing the productivity of the James River seed oyster beds. Water circulation studies suggest that these complex patterns play a vital role in the life cycle of the oysters. Beginning in 1987, these findings will be used in a three-dimensional model to help assess the impact of spoil island development and dredging on the oyster beds. Eventually, the model will be used to predict the movement of oyster larvae.

Finfish. Many factors affect the numbers of fish in Bay waters. Studies were undertaken to determine the trends and cyclic components of juvenile fish recruitment to the Bay, together with the climatological factors which may influence their populations. The viability of striped bass eggs in the Pamunkey River was also monitored. Egg mortality is a reliable indicator of spawning activity. Egg viability will be assessed again in 1987 as it indicates trends in the future size of fish populations.

Numerous other research projects are on-going including the analysis of water quality and living resource monitoring data. The findings of these studies are coordinated and shared throughout the state and Bay region.

CONCLUSION

Because the Bay area states, particularly Virginia and Maryland, have so much to gain from a healthy Chesapeake Bay, they have mobilized a tremendous amount of effort in a short period of time to try to ensure its restoration and protection. Once the main problems facing the Bay were documented, State response was quick and decisive. While not everyone agreed on all elements proposed for the cleanup programs, there was a consensus that the time had come for action. Those elements not based directly on strong scientific evidence were at least based on healthy doses of common sense. And, continuing monitoring and research has helped to refine those programs as they have proceeded. Meanwhile, progress is beginning to show and public support remains very high.

ACKNOWLEDGEMENT

Portions of this paper were previously published by the American Society of Civil Engineers in Coastal Zone '87 Proceedings (July 1987) in the paper entitled "Reversing the Decline of a National Resource: Coordination of Chesapeake Bay Cleanup Activities" (J. Carter-Lovejoy).

Contaminant Problems and Management of Living Chesapeake Bay Resources. Edited by S. K. Majumdar, L. W. Hall, Jr. and H. M. Austin. © 1987, The Pennsylvania Academy of Science.

Chapter Twenty-Two

THE STATE OF MARYLAND'S RESPONSE TO CHESAPEAKE BAY ECOLOGICAL PROBLEMS

MARY JO GARREIS

Office of Environmental Programs
Department of Health and Mental Hygiene
201 West Preston Street
Baltimore, Maryland 21201

ABSTRACT

In 1983, the U.S. Environmental Protection Agency published the results of its seven year study of the Chesapeake Bay. The study findings indicated that, if the resource was to be saved, immediate action was needed. The State of Maryland responded through seven initiatives which formed the focus of its Chesapeake Bay restoration effort. The initiatives built on the framework of existing Bay-related programs using increases in personnel and funding support and increased emphasis on program activities and results. In some instances, new programs which complimented or expanded the scope of existing programs were initiated. Two of the most significant programs to emerge from this effort were the creation of the Chesapeake Bay Critical Area Commission and the creation of the Chesapeake Bay Trust.

INTRODUCTION

The United States Environmental Protection Agency published the results of its seven year study in 1983. The results of the study indicated that the Chesapeake Bay had experienced water quality degradation and loss of aquatic resources and habitat. The study further recommended that the States and Washington, D.C. begin an intensive effort to restore and protect the Bay. The State of Maryland embraced the concept of the Bay restoration and clean up enthusiastically with an impressive array of commitments to improve water quality and protect aquatic habitat.

Central to the heart of the Maryland Bay restoration program is the recognition that improvement is dependent on changes occurring in the tributaries. In Maryland, all rivers with the exception of the Youghiogheny River in Garrett County drain to the Chesapeake Bay. Each major watershed is comprised of hundreds of streams and creeks, many with complex problems. Similarity exists in problems across these watersheds but the degree of the problem varies. Nutrient enrichment from agricultural activities, for example, is a problem in all watersheds but assumes major importance on the Eastern Shore. In more suburban drainage basins, toxics may be more important.

The Bay's tributaries serve as the foundation of the Bay ecosystem providing valuable spawning, nursery and adult habitats for aquatic species and waterfowl. Water quality in the tributaries determines water quality in the Bay. Recognizing the importance of the tributaries, Maryland has focused attention on these watersheds to manage man's activities which cause the abundance of nutrients, sediments and toxics which eventually find their way into the Bay. Existing water quality management plans are being revised and strengthened to establish management policy for coming years. Plans have been completed in the Office of Environmental Programs (OEP), for the Lower Susquehanna, Elk, Chester and Middle Potomac River systems.[1]

The Patuxent River is a watershed contained entirely within the State of Maryland. The watershed drains 7 counties and remains predominantly rural with 85% of its total land use committed to forest and agriculture.[2] The problems occurring with water quality and habitat loss in this watershed mirror the problems of the Chesapeake Bay on a smaller scale. Restoration efforts in the Patuxent basin are often viewed as a protype for the larger Bay effort.

In 1981, OEP developed a plan to focus on nutrient control in the Patuxent River. This plan was developed after negotiation with scientists, watermen, state and federal agency representatives, local elected officials and concerned citizens. Nutrient control was to address both phosphorus and nitrogen from point and nonpoint sources. Part of the strategy involved a new emphasis on best management practices for farmers and removal of nitrogen at selected waste treatment plants. The State agreed to provide financial assistance for this new approach.

In order to monitor the success of this strategy, the OEP agreed to develop the Patuxent River Estuarine model. The model will simulate the hydrodynamic regime and water quality characteristics of the river. The model will be used to demonstrate the effects on river water quality of non point and point source nutrient loading and the implementation of nutrient controls. The model is entering its third year of development and will cost approximately 3 million dollars.[3]

In 1980, the Maryland General Assembly enacted the Patuxent Watershed Act which created the Patuxent River Commission within the Department of State Planning. The Department was instructed to prepare a watershed policy plan to provide direction to local and state agencies conducting programs within this watershed. In 1984, the Patuxent River Policy Plan was ratified by the Maryland legislature and the seven counties within the watershed.

The primary thrust of the Patuxent River Policy Plan is control of non-point source pollution through a watershed land management strategy. Among the approaches used to protect and improve water quality are management of land along the river and its tributaries, provision of best management practices (including vegetative buffers), identification and correction of non point source pollution, retrofitting previously developed areas, provision of recreation and open space, accommodation of development, and preservation of aquaculture and forest land.[2] Central to the implementation of the plan was the designation of a primary management area (PMA). The PMA is the land from which pollution is most likely to be transported to the river. The PMA would be the focal area for implementation of the land management practices. In 1986, State Planning published the Primary Management Area Handbook to serve as a technical guide for local jurisdictions in the implementation of the PMA.[4] Implementation of the Policy Plan should result in improvement of aquatic and wildlife habitat as well as improvement in water quality through reduction in sedimentation and nutrient loading.

CHESAPEAKE BAY INITIATIVES

In planning its strategy to protect and restore the Chesapeake Bay, seven program areas were identified by Maryland as the focus of the Bay program. These areas were resource restoration, resource enhancement, protection of land resources, point source pollution control, nonpoint source pollution control, environmental education, and monitoring.[5] Changes, expansions or new focuses in these program areas were identified as initiatives. The initiatives built on the foundation of existing Bay related programs that preceded the Bay Program and represented an assemblage of the expertise and activities available throughout the State. Existing activities were strengthened by increases in personnel and funding support, or by increased emphasis on program activities and results. Thirty-six million dollars were appropriated for the start of these initiatives in FY 1985 with an additional $45 million appropriated in FY '86. In FY 1987, $41.5 million has been directed toward this effort.[6]

Resource Restoration

Resource restoration focuses directly on the flora, aquatic fauna and waterfowl historically common to the Chesapeake Bay. Through the Maryland Department of Natural Resources, the State has increased the funds expended for the spreading of oyster shell to encourage oyster reproduction. In 1984, the State planted 109,000 bushels of fresh shell and 4.4 million bushels of dredged shell in addition to 48,000 bushels of seed oysters. In 1986, this effort was expanded to the planting of 315,142 bushels of fresh shell and 5.64 million bushels of dredged shell on 1,035 acres of oyster bar. An additional 118 acres of oyster sanctuary and 900 acres of natural oyster bar were planted with 444,865 bushels of seed.[7]

Submerged aquatic vegetation, such as wild celery, is being replanted in critical locations as a first step toward reviving habitats where seagrasses once flourished. Grasses planted on Susquehanna Flats in 1984 are now entering their third winter. Test seagrass plots have been established in the Choptank River to study the effects of changes in water quality and sediment levels. Four ponds have been established and planted with seagrass to provide future nursery stock for future plantings. Hatcheries to help restore diminished stocks of striped bass and oysters are being expanded or newly constructed.[8]

Strategies such as replanting seagrasses, replanting oyster bars, and building hatcheries are important short term options. They can help important species survive a period of depletion. These activities also offer short term results which inspire public confidence that the estuary can be restored. As long term solutions, however, replanting and restocking can never replace the natural fecundity of the Bay or support major fisheries. The final restoration of the Bay is dependent on correction of the underlying problems.

Resource Enhancement

The objective of the resource enhancement initiative is long term improvement in the habitats and stocks of important Bay species, and provision of an objective basis for management of sport and commercial fisheries. Eighteen species were identified for development of comprehensive fisheries management plans. The management plans for american and hickory shad, alewifes, blueback herring, and american eel are substantially complete and are undergoing intensive review. Management plans for white and yellow perch and striped bass will be completed by December 1987.[9]

As part of this management effort, a moratorium on all striped bass fishing was imposed in Maryland in January 1985. The moratorium was imposed to protect the unusually large 1982 year class until these fish reached maturity and had the opportunity to spawn. Hopefully the increase in 1987 in spawning females provided by the 1982 year class will contribute significantly to reversing the decline in striped bass populations.

Close monitoring has been carried out annually in selected striped bass spawning rivers. The monitoring includes surveys of fish health and survival of all life stages. The associated water monitoring includes intensive sampling of food sources for young larvae, changes in pH and temperature, and concentrations of organic and inorganic contaminants in the water column.[10, 11] Results of the monitoring may offer clues to explain the poor survival of striped bass larvae. An assessment of the spawning, summer resident and winter resident striped bass stocks is being assembled for the Upper Bay, Chesapeake and Delaware Canal, Choptank River and Potomac River spawning grounds.[12, 13] The results of this assessment should provide insights for better stock management. The results of all these efforts should be available by January 1987. To help compensate watermen for loss of income associated with the striped bass moratorium, the Department of Natural Resources contracted some of the netting work for these projects to the watermen.

To help defray the costs of the studies of important recreational fish stocks, a salt water fishing license was established. The funds from the license purchase will also be used to acquire and maintain fishing piers, construct artificial reefs, and increase production from sportfish hatcheries.

Complimenting the studies and monitoring efforts is a program designed to rehabilitate or create natural resources habitat. A Youth Conservation Core has been established to provide summer employment for disadvantaged youth through projects designed to restore natural habitats by planting trees and grasses. In 1986, 614 youth participated in the restoration of over 40 miles of shoreline.[14]

Protection of Land Resources

Land use along the shoreline has immediate impact on adjacent water quality. The cumulative effects of human activity have resulted in deteriorating water quality and loss of productivity in the Bay and its tributaries. Wholesale conversion of shoreline areas to commercial development, housing, piers, and subdivisions with the associated impacts from humans introduced nutrients, toxics and other pollutants into the complex ecosystem of the Chesapeake Bay. Development along the shoreline has resulted in loss of more protective land uses such as forestland and agriculture. Recognizing that restoration of the Bay and its tributaries is dependent upon minimizing further impacts to water quality and natural habitats along the shoreline and adjacent lands, Maryland created the Critical Areas Commission. This controversial piece of legislation established a Commission comprised of state and local government officials and public members. The Commission was charged with developing standards and criteria which set forth requirements to be followed by affected local jurisdictions in developing protection programs for all land use planning and development within 1000 feet of the shoreline. These requirements were to embody the three protective goals of the Chesapeake Bay Critical Area Act of 1984:

1) Minimize adverse impacts on water quality that result from pollutants that are discharged from structures or conveyances or that have runoff from surrounding lands;
2) Conserve fish, wildlife and plant habitat; and
3) Establish land use policies for development in the Chesapeake Bay Critical Area which accommodate growth and also address the fact, that even if pollution is controlled, the number, movement and activities of persons in the area can create environmental impacts.

The standard and criteria were approved by the Legislation in 1986 and affect 16 counties and 44 municipalities. Local jurisdictions have approximately 18 months (until August 7, 1987) to develop their protection programs. Local programs, including periodic amendments are subject to Commission approval. Where local programs are not developed, the Commission can impose the program for local jurisdictions. Where local protection programs are not enforced, the Commission has authority to intervene and to seek other enforcement remedies.[15]

Two particularly controversial aspects of the critical areas legislation were the delineation of the critical area zone and the identification of the types of activities (particularly development) allowed to occur. The critical area zone was identified as a minimum, all land and water areas within 1,000 feet of the landward boundaries of State or private wetlands and the heads of tide. Certain types of closely managed agricultural and forestry activities with appropriate safeguards are permitted. Development is strictly controlled. Housing density is held to one house per 20 acres and all activities must maintain a 100 ft. setback (buffer) from the shoreline. The 100 ft setback was derived from the scientific literature and after consultation with States with similar programs.[16] Many benefits to water quality and living resource protection are anticipated as the critical areas of legislation become effective.

Conservation of tidal and non-tidal wetlands to maintain their nursery value, their natural infiltration capacity, and their ability to absorb erosive and flood impacts should be enhanced. Adverse impact on shellfish, and finfish productivity as well as submerged aquatic vegetation should be reduced. Land use that is compatible with protection of habitat, water quality and the shoreline will be encouraged. Some developmental growth will be permitted but with minimum impact to the aquatic environment.[16]

To compliment the critical area effort, an initiative was created to protect and maintain existing forest buffers. Currently forested critical land areas were inventoried and mapped, and assistance provided to landowners to prepare management plans to protect the forests. To encourage private landowners to preserve and protect undeveloped or low density areas along the shoreline of the Chesapeake Bay and its tributaries, the existing Maryland Environmental Trust easement program was intensified. Substantial increases in the land placed under perpetual conservation easement each year assure long term protection. Easements offer landowners an opportunity to make an individual contribution to protecting the Bay through the use of conservation management practices required as part of the easement. This type of land protection is believed to afford an excellent safeguard against point and non-point source pollution.

Point Source Pollution Control

Traditionally, efforts at controlling water pollution and improving water quality have been directed at controlling discharges from specific outlets such as industrial or sewage outfalls. Elimination of, or improvement in the quality of, these effluents remains a cornerstone in pollution abatement. In Maryland, two primary tools exist to control points sources: 1) the construction grants program which helps local jurisdictions finance sewage plant improvements, and 2) the National Pollution Discharge Elimination System (NPDES) that requires every discharger to apply for and adhere to a permit limiting the amount of waste discharged to a water body.

By regulation in 1974, Maryland established secondary treatment as the minimum standard for all sewage treatment plants. The State began a vigorous program to upgrade all existing facilities to this level of treatment. As part of this program, the State actively sought the 75% level of federal funding available for sewage treatment plant upgrading. To make the proposal more attractive to local jurisdictions, the State paid half the local share of costs. In 1985, faced with the cutback from 75% to 55% in the federal share of construction grant funds, Maryland agreed to assume, as part of the Bay Program, an additional 20% funding for sewage plant construction and upgrading. The local share remained 12.5%. At this time only one primary sewage treatment plant remains substandard with its upgrading to be completed by 1987. Substantial commitment remains to improve treatment at all sewage plants through continual upgrading.

In addition to the construction grant program, Maryland uses the state funded Water Quality Loan Fund to offer incentives for pollution abatement. Loans are available to pay all or part of the cost for installing dechlorination equipment at publicly owned waste treatment plants. Low interest loans are available to industries to help defray the costs of pretreating their wastes prior to discharge to sewage treatment plants. Funding is also available to assist with the cost of nitrogen removal at municipal sewage plants, and to treat urban storm water.

Maryland has enhanced its NPDES program by increasing personnel, providing more thorough surveillance of effluent discharges, and aggressively prosecuting violators. Increased enforcement against industrial violators has resulted in substantial fines and improved effluent discharges. The OEP working with the Attorney General's Hazardous Waste Strike Force combined inspectors, attorneys and police to bring civil and criminal actions against violators. In 1985, $421,000 in fines were collected.[5]

Phosphorus has been identified as an important Bay pollutant. Recognizing that the Upper Bay was experiencing increasing phosphorus loading, Maryland adopted a phosphorus control strategy for the Chesapeake Bay above the Bay Bridge in 1979.[17] An effluent limitation of 2 milligrams per liter (mg/l) total phosphorous was established for all sewage treatment facilities discharging more than 0.5 million gallons per day (mgd) to the Chesapeake Bay and its tributaries above the Baltimore Harbor. Waste water plants discharging 10 mgd in the area between the Baltimore Harbor and the Bay Bridge were also required to meet the 2 mg/l effluent limitation. Plants designed to comply with this limitation began to come into operation in 1982.

In 1985, the Maryland General Assembly carried the battle against phosphorus discharges one step further. A law was enacted banning the sale of laundry detergents that contain phosphates in excess of 0.5 percent by weight and dishwashing detergents exceeding 8.7 percent. Immediate short term benefits of the ban were projected as a 12% reduction in phosphorus loading to the Bay. Long term benefits include savings in operational costs at sewage treatment plants designed to remove phosphorus. Cost savings would come from a reduction in the treatment chemicals used and reduction in the amount of sludge produced. Sludge disposal is an increasing problem and any decrease in the amount of sludge generated lessens the pressure on available disposal sites.[18] Anticipated benefits to the Bay from all phosphorus controls include a reduction in algae blooms which retard growth of submerged aquatic vegetation. Excessive algae blooms also cause zones of low dissolved oxygen which impair aquatic habitat, disrupt aquatic communities and may kill fish.

The proper operation and maintenance of the waste water treatment facilities is important to good point source discharge control. Recognizing the facilities' performance is highly dependent on the operator, Maryland has directed efforts toward improved training and has strengthened certification requirements. Workshops were held to assist operators from major sewage plants in identifying and diagnosing operational problems.[5] The OEP has also employed a laboratory scientist to inspect all laboratories maintained by sewage treatment plants for testing their compliance with plant NPDES limitations. The laboratory scientist reviews laboratory practices and procedures to assure employment of proper laboratory protocols and works with plant operators to upgrade their laboratory capabilities.[19]

In 1981, The Maryland Legislature enacted a law prohibiting the use of chlorine or any chlorinated compound to disinfect wastes discharged into trout streams. In 1984, as part of the Bay Program, this law was amended to require any sewage plant using chlorine to install dechlorination equipment.[20] Funding was made available to assist municipalities in providing dechlorination facilities. By October 1986, only 2 municipalities had applied for this funding. To date, two hundred and ninety sewage treatment plants (60%) have installed dechlorination facilities. In 1986, this law was again amended to require all industries discharging chlorinated effluents to install dechlorination facilities by 1988. Currently, Maryland has the most stringent chlorine controls in the country. Anticipated benefits from this legislation are reduction in amounts and levels of chlorine discharged and increased protection for sensitive life stages.

In 1982, the State began an ambitious program to force industries to pretreat their wastes prior to discharge to sewage treatment plants. The first phase required all publicly owned waste water treatment plants (POTWs) with a capacity of 5 million gallons per day or all POTWs receiving wastes from a major industry and experiencing interference, pass through, or sludge contamination associated with industrial wastes to develop and operate a pretreatment program. In 1982, 12 major POTWs received grant monies to develop pretreatment programs. By 1984 an additional 15 POTWs were required to develop and implement similar programs.

In June 1984, a Maryland law was enacted authorizing the creation of a Stage pretreatment program. Final State regulations were adapted in August 1985 with full delegation of pretreatment authority from EPA in September 1985. Complete implementation of pretreatment requirements at the industrial level is anticipated to occur in 1988.[21] An additional incentive was offered in the form of loans available to industries to help pay the costs of pretreatment. In 1985 and 1986, three metal plating companies were awarded loans totaling 1.5 million dollars. By November 1986 a drum cleaning industry and an industrial laundry had applied for loans totaling $200,000 to assist with the costs of pretreating their wastes prior to discharge to municipal sewers.[20]

Anticipated benefits of this program are improved sewage treatment plant operation, and generation of sludges suitable for use in soil conditioning and fertilizing farm land. The Bay and its tributaries should experience improved water quality and aquatic habitat occasioned by a reduction in the amount of toxics discharged. Reduction in bioaccumulation of heavy metals and persistant organic chemicals by aquatic organisms is also projected.

NONPOINT SOURCE POLLUTION

Non point source pollution has long been recognized as a significant source of pollutants to the Chesapeake Bay system. Within this category lies agricultural runoff, storm water from urban and suburban areas, failing septic systems, animal wastes and a myriad of other diffuse sources. These sources are among the most difficult to control because the sources are diffuse and often small and because in many cases, the technology is limited, untried and expensive. As part of the Bay Program, several areas of nonpoint source pollution were targeted for improvement as Maryland launched an all-out effort to control runoff from agricultural and developed lands.

Agricultural lands were identified as major contributors of nutrients and potentially toxic chemicals. Seven million dollars in state funds were allocated in 1985 for agricultural cost sharing to install best management practices to control runoff. Best management practices include use of grassed or herbacious buffer strips, no till farming sediment and erosion control, and proper application of fertilizers. By December 1985, over 3000 applications for cost sharing assistance to install best management practices had been received from farmers. Almost 1300 projects were completed in 1985. By September 30, 1986, an additional 654 projects were approved at a cost of $3.5 million.[22] In addition, regulations were adopted in August 1986 requiring design, construction, operation and maintenance of agricultural drainage projects in accordance with best management practices.[23] Public drainage association which manage agricultural drainage projects have one year to submit a report to the Secretary of Agriculture describing the condition of their drainage systems. Within 5 years, the Association shall submit an operation and maintenance plan and an implementation schedule for their drainage system. Each drainage system shall be inspected by the Association every 2 years. Through the Maryland Department of Agriculture, new outreach personnel were hired to work with farmers in adopting necessary management practices.

Working together, the State Soil Conservation Committee, the Maryland Department of Agriculture and the OEP have identified priority creeks to be targeted for better runoff control in all major river systems. Double Pipe Creek in the Monocacy subbasin of the Potomac River has been identified as a Maryland watershed generating a high amount of nutrient runoff. Farms in this watershed have been encouraged to install best management practices. In order to determine the effectiveness of these best management practices, the Office of Environmental Programs has initiated a before and after study along the Creek.[24] Results will be used for designing and evaluating similar programs for priority areas in other heavily farmed watersheds.

Storm water runoff from developed areas was also identified as an important source of nonpoint source pollution. Stronger laws and regulations were adopted. The most significant of these was the law requiring that runoff rates and characteristics after development be similar to predevelopment runoff characteristics so that stream erosion, pollution, and local flooding would be curtailed. Another significant piece of legislation required the State to assume responsibility for enforcing sediment and erosion control laws in counties which failed to have effective enforcement. To encourage local governments to experiment with innovative storm water management, $1.2 million in grants were made available in 1985.[5]

Maryland moved in other ways to control nonpoint source pollution. The State has accelerated its efforts to maintain forest buffers along shorelines and farms to control erosion, filter nutrients, and provide wildlife corridors. State construction funds available for shoreline erosion control have been expanded. In 1985, 3,500 feet of shoreline was protected in 9 projects and an additional 55 projects were started.[5]

Anticipated benefits of improved water quality from nonpoint source pollution control are many. Reduced inflow of nutrients should result in a reduction in algae blooms in the Bay. Algae blooms cause zones with reduced dissolved oxygen which disrupt aquatic communities and kill or impair fish. Reduction in algae blooms and turbidity should benefit SAVs and fish habitat by improving sunlight penetration. Control of erosion and sediment transport protects important SAV beds, spawning habitats and oyster beds from burial by siltation. Retention of herbicides, pesticides, and heavy metals on land through improved runoff management protects aquatic plants, enhances the survival of aquatic larvae and juveniles, reduces bioaccumulation in adult organisms throughout the food chain and protects critical habitats.

Environmental Education

Essential to sustaining the effort to protect and restore the Chesapeake Bay is the citizen's appreciation and understanding of the value of the Bay as a unique and important natural habitat as well as a natural resource with aesthetic, commercial and recreational value. To enhance this appreciation, Maryland through its Department of Education has prepared and distributed materials which address the scientific and political issues of the Bay region. Through the school system, a summer program for sharpening teachers' environmental awareness has been instituted and 12,000 students have participated in the Chesapeake Bay Foundation estuarine studies program.[5] Through these and similar efforts, Maryland hopes to assure educated citizens aware of the importance of protecting the Bay.

Monitoring

The Chesapeake Bay is constantly changing in response to both man's activities and natural events throughout its watershed. To track these changes, to document the success of current pollution control and restoration programs, and to investigate important relationships and processes, an active research and monitoring program is necessary. To fully encompass the changes occurring within the system, the monitoring must address water quality, the living flora and fauna, and their habitat. Like the other initiatives, this effort is integrated and built on existing programs.

In its water monitoring, the Office of Environmental Programs samples both the Bay mainstem and its tributaries using key indicators of chemical, physical and biological quality. Twenty 3-day cruises are made on the mainstream each year and involve 22 stations. An additional 65 stations are sampled in the tributaries. Besides the usual physical and chemical properties of the water column, plankton are measured as indicators of short term water quality. Sediments are tested at selected stations as long term accumulators of pollutants. Seasonal processes such as sedimentation, nutrient cycling and oxygen creation and consumption are measured. In addition, benthics are monitored as indicators of long term change. A report summarizing the findings of the monitoring efforts will be available in January 1987.[3]

Three rivers on the western shore, the Susquehanna, Potomac, Patuxent, and the Choptank River on the eastern shore were selected by Office of Environmental Programs for fall-line monitoring. The U.S. Geologic Survey collects monthly base flow samples on 3 rivers while the Occoquan Watershed Monitoring Laboratory samples the Potomac. Hopefully, this fall line monitoring effort will provide information concerning the levels of nutrients and toxics entering the Bay from upland sources. The fall line monitoring stations are also part of the National Stream Quality Accounting Network administered by U.S. Geologic Survey and the data generated can be integrated into valuable historical background.[3]

The information from the Baywide monitoring program has many uses. Over a five to ten year period, the data will create a baseline picture of the estuary and its tributaries as an ecosystem. This holistic overview could provide insights into many hidden connections between changes in water quality and alterations in life forms depending on that water quality. Monitoring will indicate how the Bay system responded to specific management strategies and help fine tune the on-going clean up work.

Chesapeake Bay Trust

The Chesapeake Bay Trust is a special fund established to help support Bay restoration and education efforts. The basic concept of the Trust is to serve as a broker for Bay involvement by matching financial support with worthy, necessary projects. These may be projects conducted by groups participating in Bay cleanup activities or projects by others seeking to become involved for the first time. The Chesapeake Bay Trust does not develop its own program; instead it provides financial support for public awareness, education and citizen involvement. By law, the Bay Trust cannot take an advocacy position on specific issues but supports projects that will help inform the public or improve the quality of the Bay. The Trust is financially supported by private donations from corporations, individuals and small businesses. Between July 1, 1985 and June 30, 1986, $107,000 was donated to the fund by individuals and corporations.[25]

SUMMARY

While all the programs described above are important to protect and restore the Bay, real pollution control can only be achieved by the thousands of people who work, live and play in the Chesapeake watershed. The Bay restoration requires that citizens make conscious decisions about the way they conduct their businesses and the way they live. Without the continuing commitment, support and cooperation of the residents in the Bay's watershed, the programs as described can have only minimal effect.

REFERENCES

1. Halka, M.C. 1986. Personal communication. Office of Environmental Programs, Maryland Department of Health and Mental Hygiene.
2. Maryland Department of State Planning. 1984. Patuxent river policy plan. Baltimore, Maryland.
3. Haire, M.C. 1986. Personal communication. Office of Environmental Programs, Maryland Department of Health and Mental Hygiene.
4. Maryland Department of State Planning. 1986. Primary management area handbook. Baltimore, Maryland.
5. State of Maryland. 1984. Proposed Chesapeake Bay initiatives: summary of programmatic, operating and capital budgets for fiscal year 1985. Annapolis, Maryland.
6. McElroy, K.E. 1986. Personal communication. Office of Environmental Programs, Maryland Department of Health and Mental Hygiene.
7. Outten, W.A. 1986. Personal communication. Maryland Department of Natural Resources.
8. University of Maryland Sea Grant College. 1986. Maryland's Chesapeake Bay Program. NOAA. NA 86AA-D-SG006. UMD Sea Grant Prog., College Park, MD.
9. Rugolo, L.J. 1986. Personal communication. Maryland Department of Natural Resources.
10. Maryland Department of Natural Resources. 1986. Estuarine fish management. U.S. Fish and Wildlife Serv. 27-R, Washington, D.C.
11. Jordan, S. 1986. Personal communication. Maryland Department of Natural Resources.
12. Maryland Department of Natural Resources. 1986. Characterization of Potomac River spawning stocks and fecundities of Maryland striped bass. U.S. Fish and Wildlife Serv. F39-R. Washington, D.C.
13. Maryland Department of Natural Resources. 1986. Characteristics of striped bass populations in Maryland's Chesapeake Bay. National Marine Fish. Serv. AFC-15. Washington, D.C.
14. Underwood, J. 1986. Personal communication. Maryland Department of Natural Resources.
15. Subtitle 15. Chesapeake Bay critical area commission criteria for local critical area program development. COMAR 14.15.01 through COMAR 14.15.11. 1986.
16. Taylor, S. 1986. Personal communication. Chesapeake Bay Critical Areas Commission.
17. Maryland Water Resources Administration. 1979. Upper Chesapeake Bay phosphorus limitation policy. Maryland Department of Natural Resources, Annapolis, MD.

18. Maryland Department of Health and Mental Hygiene. 1985. Legislative Testimony.
19. Rein, J. 1986. Personal communication. Office of Environmental Programs, Maryland Department of Health and Mental Hygiene.
20. Quance, E.S. 1986. Personal communication, Office of Environmental Programs, Maryland Department of Health and Mental Hygiene.
21. Zaw-Mon, Merrylin. 1986. Personal communication. Office of Environmental Programs, Department of Health and Mental Hygiene.
22. Kay, C.S. 1986. Personal communication. Office of Environmental Programs, Maryland Department of Health and Mental Hygiene.
23. Code of Maryland Regulations, Subtitle 20, Soil and Conservation. 15.20.01 Agricultural Drainage Projects, 1986.
24. Versar Inc. 1986. Nonpoint Source Assessment of the Monocacy Review with Special emphasis in Double Pipe Creek Watershed. Versar Inc. Springfield, VA.
25. Burden, T. 1986. Personal communication. Maryland Department of Natural Resources.

Contaminant Problems and Management of Living Chesapeake Bay Resources. Edited by S. K. Majumdar, L. W. Hall, Jr. and H. M. Austin. © 1987, The Pennsylvania Academy of Science.

Chapter Twenty-Three

PUBLIC INVOLVEMENT IN THE CHESAPEAKE BAY PROGRAM

FRANCES H. FLANIGAN

Program Director
Citizens Program for The Chesapeake Bay, Inc.
6600 York Road
Baltimore, Maryland 21212

ABSTRACT

Science can't save the Chesapeake Bay.

This may sound like heresy in a book devoted to the role of science in the preservation of Chesapeake Bay. Unfortunately, it is true. The Chesapeake has deteriorated in terms of water quality and habitat availability, not because of lack of science but rather because people have made political decisions that did not take the Bay's well-being into account.

This chapter will briefly describe the public involvement aspect of the Chesapeake Bay Program, attempting to give some perspective on how a government research program that had a rather inauspicious beginning evolved into a catalyst for significant political commitments. The theme of this chapter is that estuarine management is an essentially political activity, and that if the Bay is to be saved, it will happen because of people.

BACKGROUND

The story of the renaissance of the Chesapeake began long ago, and has many elements familiar to all who have been involved in resource management issues. There were the periodic newspaper articles with banner headlines trumpeting a dying Bay; there were studies to look at this and look at that; there were laws to prohibit and regulations to control, and there were organizations, meetings, newsletters, hearings, boat trips and speeches all dedicated to getting government to do something about the Bay. Everyone was certain that something was wrong, although no one seemed to know quite what, and solutions were elusive.

The Chesapeake Bay Program was launched in 1976 after Congress directed EPA to conduct an in-depth study of water quality conditions in the Bay and make recommendations for improved management. The Bay Program succeeded in doing several things. First, it represented a successful effort to assemble a synoptic, bay-wide data set that would provide a baseline for comparison with historical data and for measuring future change. Second, it attempted to assess the sources of pollution and to understand the relative contributions of each source. Third, the study tried to link water quality and pollution (traditional EPA concerns) with changes in living resources. And finally, the study recognized the importance of the governmental decision-making network in place on the Bay, and made a good faith and fairly successful effort to involve representatives of affected entities in the program.

The study concluded that most of the Bay system is nutrient-enriched; localized sediments are contaminated with metallic and organic toxics; the volume of water experiencing very low or no dissolved oxygen in summertime has increased 15-fold in thirty years; and living resources, especially those that depend on freshwater in their juvenile stages, are at historic lows. The EPA study made a significant contribution to our technical understanding of how the Bay works and how it is being affected by contaminants generated on the land. The study also provided a springboard for response and action, and four years after its conclusion, the action continues to accelerate.

SPECIAL PROBLEMS WITH ESTUARIES

The evolution of the Bay Program's technical findings into a set of management plans and an implementation strategy is a unique study. Before looking at the specifics of the Chesapeake's story, some general observations can be made that may help set the stage for understanding the phenomenon of the Chesapeake's restoration. The management of estuaries seems to present special difficulties similar to other natural resource management situations, and it may be useful to explore for a moment why that is so.

First, the inherent technical and resource issues are complex and difficult for the layman to understand. Managing estuaries assumes that one knows about circulation patterns, the salt wedge, nutrients and phytoplankton productivity, dissolved oxygen, toxic chemicals in the sediment, loading rates from rivers and the watershed, the relation of rivers and the ocean to the estuary. Most people know little of this, and yet an elementary grasp of the concepts is important to being able to make sound judgements on management actions. The poor science background of most of today's adults hampers their ability to automatically comprehend these ideas.

Second, it is often true that the layman isn't the only one who doesn't understand estuarine issues. Estuaries appear to be under-studied, with rather poor long term data sets and until recently, relatively little research directed specifically at them. For example, until quite recently, the common belief among laymen on the Chesapeake was that pollutants coming into the Bay eventually found their way to the ocean. The circulation studies conducted by the Chesapeake Bay Program taught us otherwise—the Bay is, in fact, a big sink, and its current problems are partly the result of years of pollutant build-up in bottom sediments. On specific issues, there are often conflicting expert opinions, which further undermine the ability of the lay public to understand issues and arrive at an opinion on solutions. Right now, on the Chesapeake, such a situation exists with regard to the role of nitrogen in fueling the growth of algae. A fundamental lack of knowledge has set the stage for a major policy confrontation over nitrogen removal at sewage treatment plants.

Third, we are not always sure of the effectiveness of proposed strategies. Agricultural BMP's are a good example. In the Chesapeake area, a lot of money is being spent to encourage farmers to use filter strips, no-till cropping, contour plowing, terraces and other measures to retard the surface movement of water and sediment off the fields and into nearby streams. Belatedly, monitoring data is suggesting that filter strips may be only minimally effective and that other practices, such as terracing, may actually increase the nitrate contamination of groundwater. Another current example is the biological process being employed to remove nutrients at sewage treatment plants. While appearing promising, this treatment technology needs further demonstration before one would adopt it on a large scale as public policy.

Fourth, almost always the solutions designed to fix a natural resource problem impose an additional burden on one or more segments of the population. No solution is cost-free, and questions of equity inevitably come up. Who benefits and who pays need to be identified, and the equity issue dealt with openly.

PUBLIC INVOLVEMENT ON THE CHESAPEAKE

In the Chesapeake Bay Program, an early and substantial commitment was made to public involvement. The EPA awarded a grant to the Citizens Program for the Chesapeake Bay, Inc., a non-profit coalition of organizations, to provide non-technical information to the public and to create opportunities for citizen involvement in the study.

What is the role of the public in technical programs? Two points seem relevant. First, no program is purely technical. Especially in the case of estuarine management, the programs and issues have important technical components, but also economic and political ones. Science in these programs should be focused on illuminating public policy questions. Public input on these questions is equally important, because no solution gets implemented without public and political support. To achieve that support at the end of the program, managers have to start at the beginning to build confidence in the problem solving process.

By the time the study's final reports were submitted to Congress in September 1983, large constituencies had been informed about what the new scientific information meant in terms of their relationship to the Bay. A variety of techniques were used—newsletters, public meetings, and advisory committees—and relationships were developed with groups not previously involved in Bay discussions, such as the farm community.

A hallmark of the public involvement program conducted by the Citizens Program was (and continues to be) personal contact. CPCB has developed a network of interested individuals and organizations numbering into the thousands. Leaders of many of the key organizations participated on an advisory committee called the Resource Users Management Team, a group brought together to review the recommendations as they were being developed by Bay Program staff. This novel approach to involvement and citizen advice had two benefits: it provided EPA with an instant reality check on the feasibility of its recommendations, and it got the leaders of interest groups intimately involved in the study and with each other. These citizen advisers were among the first to go to bat on behalf of the management recommendations upon their release in 1983, and were instrumental in converting recommendations into actual legislative proposals and implementation strategies at the state level. Developing a substantive role for these "bay users," as distinct from the expected involvement of environmental activists, laid a sound foundation for broad-based political support.

The emphasis on bay users—sailors, shippers, utilities, fishermen, marinas, shorefront communities, and industry—was matched by efforts to couch Bay problems in terms of people and land. People in Richmond and Roanoke, Baltimore and Chestertown, Lancaster and Loudon County have begun to understand the connection of corn grown in the hills of the Piedmont with fish spawned in the headwaters of the Potomac. The pathways taken by water on its way to the Bay—across the endless parking lots and highways, through millions of dishwashers and washing machines, over thousands of acres of farm fields and suburban lawns, are beginning to be understood not just by scientists, but by people. Awareness of the inter-connectedness of the web of users with the land and water has disarmed those who would blame the Bay's decline on someone else. The growing sense that it is not "them," it's "us," has created an atmosphere where creative problem-solving can take place.

The Citizens Program's principal role in the Chesapeake Bay Program was to facilitate dialogue between the private sector and government, and within the private sector, among the wide array of organizations having an interest in the Bay. As a result, an extensive network of people who know each other and who know the issues now exists. Because the Bay restoration program was launched from a broad base of support, its chances of being sustained over a long period of time seem fairly good.

The necessary long term support must come from the same soundly conceived problem solving process that led to the crafting of the initial set of solutions. The problem solving process must include federal, state and local governments; legislators; the scientific community; and representatives of affected segments of the population. The problem solving process forces research to be focused on relevant management questions. The problem solving process pushes agencies, states and interest groups to work together, emphasizing the common ground rather than the divisions. The problem solving process is a search for solutions. A problem solving process that is open and founded on the notion of consensus has the potential for success, as the Chesapeake clearly demonstrates.

In the Chesapeake basin today there exists consensus on problems, a feasible range of solutions, and the first phase of an implementation strategy backed by significant federal and state dollars.

LESSONS LEARNED

What have we learned from the Chesapeake Bay Program and its remarkable political success? A number of lessons have emerged. First, the public is generally more interested in policy than in scientific and technical issues. To equip the public to participate in policy discussions, scientific material has to be translated into language that is relevant to the audience. It then has to be transmitted to that audience, which means for example, taking information about the effects of fertilizer over-application directly to the farmer. This is essentially a communication task which requires thoughtfulness and skill to be effectively carried out.

To build support for an action plan, such as the restoration program on the Chesapeake, the public has to concur in the definition of the problem. To accomplish this, public interaction has to begin at the beginning of the study process. The objective is to develop concurrence among public interest and user groups, politicians and scientists. Too often, the public is presented, after the fact, with a statement of problems or issues and a range of solutions, and asked to choose among the solutions. If they do not agree that there is a problem, they won't agree on solutions. If all the solutions are unpalatable, the easiest way to avoid acting is to disagree on the problem. The consensus building process is the critical backbone of complex management programs, and it needs to start at the beginning and be carried through to the end.

It is also a truism that the public is less interested in water quality numbers such as those expressed in permits or standards, than in measurable benefits such as recreational opportunities, fish to catch, and increased property values. Therefore, the goals of the program need to be expressed in these kinds of terms. Additionally, costs and impacts need to be translated into terms that are relevant and measurable. People are always concerned about how a proposed program will affect them directly—what will it mean to me if the Bay continues to decline, and what cost will I have to bear if I support a decision to clean it up? A Maryland official responded to this need for a tangible measuring stick when he said that the $15 million Maryland was at that time proposing to spend to restore the Bay was the equivalent of a six pack of beer per citizen. In Pennsylvania, concern about agricultural pollution was converted into dollars out of the individual farmer's pocket, by calculating the value of nutrients being lost from farm fields. Efforts to describe problems and solutions in a concrete and relevant way continue to be an important part of the clean-up strategy.

Finally, we've learned what wise men always knew, that management is a process that is highly political. This is not a bad thing; in fact, in a democracy, it can be a very healthy and constructive process. But to be effective, great care needs to be given to the process itself and to the people who are part of it. Personal and professional relationships need to be nurtured, consensus needs to be gradually developed, and opportunities for creative, collaborative problem solving carefully designed. All of this is as important, if not more so, than scientific experiments, data gathering and analyses, and report writing that frequently constitute the beginning and end of a study.

WHAT DOES THE FUTURE HOLD?

On the Chesapeake, we're not done yet; in fact we've just begun. We have learned that managing estuaries is a long term proposition, and that if we are to succeed, we had best be prepared for the long haul. Scientists cautioned that there are no "quick fixes," and fortunately, people listened. We talk about the future of the Chesapeake in terms of decades, not years.

As we look to the future, a number of issues, both technical and political, clearly remain to be dealt with. The first is the sustainability of political support. Politicians have short horizons. Elections bring constantly rotating groups of people to power. Sustaining the political support that is necessary for budgets and for tough pollution control regulations will be an ongoing challenge that will have to be met by concerned citizens and user and public interest groups. Keeping the Bay on the political agenda is the pre-eminent task for the future.

Another important issue yet to be decided is "how clean is clean?" Setting specific numeric goals for the reduction of nutrients and toxics in the tributary rivers and in the Bay itself is an important step that hasn't yet been taken. The increasing sophistication of mathematical models and improved ability to monitor what actually gets into the stream may make this task easier, but the goal setting process is as much a political process as a scientific one. It will need to be based on excellent science as well as on consensus about the relative merits of a clean Bay.

Funding the clean-up looms as another large issue on the Chesapeake. As federal dollars become scarcer, especially for expensive capital projects such as sewage treatment plants, the challenge to states and localities grows. The public's willingness to pay is likely to be tested in the near future, while both bureaucrats and bankers look for creative ways to finance clean-up programs.

New knowledge will undoubtedly point us in altered directions in the years ahead. Staying the course while at the same time maintaining enough flexibility to modify programs based on new research and the results of monitoring programs will require skill and wisdom.

Making the clean-up program a success will also depend on the ability of governments to enforce the regulations they write and to maintain the technological solutions they pay for. Recent history has shown how difficult it can be, in a litigious society, to write and enforce permits. And the technology of sewage treatment plants has to work day in and day out, something rarely achieved. Improvements in both these areas are essential in the decade ahead.

It is abundantly clear that the Bay clean-up program will succeed only as long as people support it. This implies a personal willingness on the part of the thirteen million people who now live in the drainage basin and the millions yet to come, to modify their life styles to be more sensitive to the environment. Conserving water, eliminating toxic home products, and being good stewards of the land are ideas that will carry us into the twenty-first century. The growing awareness that wise management of our land and water resources is an essential and ongoing responsibility bodes well for the future of Chesapeake Bay.

Contaminant Problems and Management of Living Chesapeake Bay Resources. Edited by S. K. Majumdar, L. W. Hall, Jr. and H. M. Austin. © 1987, The Pennsylvania Academy of Science.

Chapter Twenty-Four

USE OF AQUATIC BIOLOGICAL TESTING UNDER THE NPDES PERMIT SYSTEM TO REDUCE TOXIC POLLUTION OF THE CHESAPEAKE BAY

RICHARD L. WILLIAMSON, JR.[1] and DENNIS T. BURTON[2]

[1]Cleary, Gottlieb, Steen & Hamilton
1752 N. St., N.W.
Washington, DC 20036
and
[2] Aquatic Ecology Section
Environmental Sciences Group
The Applied Physics Laboratory
Johns Hopkins University
Shady Side, MD 20764

NOTE: This chapter provides an overview of the legal and scientific issues arising from the use of biological testing to control toxicity under the NPDES system. Of necessity, the discussion is a general one, and should not be taken as a definitive statement of either the relevant laws and regulations, or of the current state of scientific knowledge.

ABSTRACT

A vital tool in national efforts, as well as in the Chesapeake Bay, to control water pollution is the requirement that any person wishing to discharge any substance from a fixed location must first obtain a National Pollution Discharge Elimination System (NPDES) permit. This chapter provides an overview of the legal and scientific issues arising from the use of biological testing to control toxicity in the Chesapeake Bay under the NPDES permit system. A brief history of the control of effluents from point sources under the Clean Water Act is discussed. The use of toxicity testing in NPDES permits as well as the problems associated with the current use of toxicity tests to deal with the specific problems of the Chesapeake Bay is also presented. Several regulatory and scientific recommendations are made to improve the effectiveness of the NPDES permit system in controlling pollutants in the Chesapeake Bay.

INTRODUCTION

A vital tool in national efforts to control water pollution is the requirement that any person wishing to discharge any pollutant from a fixed location must first obtain a National Pollution Discharge Elimination System (NPDES) permit. NPDES permits regulate many, but not all, sources of pollution reaching the Chesapeake Bay.

It is possible to assess the toxicity of a discharge through the use of aquatic bioassay techniques, even in cases where the chemical composition of the discharge is not fully known. In the most commonly used whole effluent test, small sensitive fish or macroinvertebrates are subjected to various concentrations of the effluent and dilution water to determine the concentration which is lethal to 50% of the organisms in a specific period of time.

For the purposes of this chapter, we assume (and believe) that the Chesapeake Bay requires special measures for its protection from pollution, above and beyond those applicable to water bodies elsewhere in the country.[1] We further assume—though the data are sparse, and we have some doubts on the issue— that toxicity is a substantial contributor to the problems of the Bay. If these assumptions are correct, the combination of these two tools, (i) the requirement for dischargers to possess a valid NPDES permit, and (ii) the availability of bioassay techniques to measure the toxicity of whole effluents, could play a significant role in dealing with certain toxic contaminant problems facing the Chesapeake Bay. Both environmental and economic interests could benefit greatly from a well-designed and applied program. However, doing so poses numerous scientific and regulatory issues.

CONTROL OF EFFLUENTS FROM POINT SOURCES UNDER THE CLEAN WATER ACT

Brief History of Efforts to Control Water Pollution

Prior to 1972, there were relatively few legal tools available to control the pollution of the nation's surface waters. To be sure, the Refuse Act of 1899 was still on the books, and some states had their own controls, but these had proven to be ineffective. To deal with this problem, the Congress passed the Federal Water Pollution Control Act Amendments of 1972, now more commonly known as the Clean Water Act.[2] The CWA, amended several times, still serves as the primary legal framework for the control of surface water pollution in the United States.[3]

In the early days of the administration of the Act, NPDES permits focused primarily on the control of conventional and known non-conventional pollutants. Conventional pollutants served as general indicators of pollution and their control was desirable because these properties in water can lead to undesirable consequences. For example, high biochemical oxygen demand depletes surface waters of essential oxygen. By mid-1977, all dischargers were required to have permits which controlled conventional pollutants by the application of "best practical control technology," commonly called BPT.[4]

While controlling toxicity was not an initial focus of NPDES permits, the scientific basis for the assessment of toxicity to aquatic organisms had already been laid through the pioneering work of Hart, Carins, Sprague, Anderson, Burdick and others. This work concentrated primarily on individual toxicants, with most of the early attention devoted to pesticides and heavy metals. This focus on individual toxicants was understandable because it was widely assumed by regulatory agencies that the toxic constituents of a discharge were known, and that it would be sufficient to control each such constituent.[5]

Important amendments were adopted by the Congress in 1977.[5] In light of a growing concern over toxicants, the new provisions required the extension of "best available technology economically achievable," or BAT, to a list of "priority pollutants" (now numbering 126).[6] These changes reflected a growing recognition that controls on conventional and non-conventional pollutants did not always assure that discharges would not be toxic. These provisions control the individual chemical substances in a discharge, not the toxicity of the discharge taken as a whole. However, in 1984, EPA announced a policy which strongly encourages measures to control the toxicity of the whole effluent.[7]

Controls Under the NPDES Permit System

The Clean Water Act allows EPA's authority to issue NPDES permits to be delegated to the states under certain conditions.[8] All of the relevant jurisdictions with drainage into the Chesapeake Bay, except the District of Columbia, have authority to issue NPDES permits. Except pursuant to an NPDES permit or state equivalent, the discharge of any pollutant is illegal, and the CWA authorizes the imposition of heavy civil penalties for violations,[9] and for a willful violation, stiff fines and jail sentences.[10]

A typical NPDES permit operates primarily by placing concentration or mass limits on the amount of a pollutant or category of pollutants which can be contained in a discharged effluent. The legal authority to impose such limitations can derive from either federal or state law. Currently, limitations on toxic substances can be technology-based, water-quality based, or federal or state effluent standards. In addition, biologically-based toxicity limitations are beginning to be used.

The key question is: assuming that the Chesapeake Bay requires special measures to reduce toxicity, beyond those applicable to water-bodies in the rest of the country, which of these four approaches can best serve that end?

Technology-Based Limits

Technology-based limits have been included in thousands of NPDES permits. They are typically based primarily on an estimate of the discharge concentrations which can be achieved by treatment technologies in well-run plants in a given industrial category. There are, accordingly, wide discrepancies between industries as to the concentration of a particular pollutant which may legally be discharged. Technology-based limits for a given industry are theoretically national in scope, and should not vary from state to state. Indeed, states generally do not have authority to impose technology-based limits which vary from national effluent limitations guidelines. There is no requirement that the EPA or state permit writer show that BPT and BAT technology-based limits are either necessary or sufficient for the protection of the environment. In our view, technology-based controls are not a promising means of establishing special measures to deal with the unique problems of the Chesapeake Bay.

Water Quality Standards

The CWA requires each state to establish standards of water quality for the water bodies in the state.[11] These can and do differ considerably from water body, and from state to state. A state is allowed to decide, for example, that a particular water body is of critical importance to the propagation of wildlife and that accordingly, extraordinarily stringent standards must be met. In theory, it is not relevant whether the standards can be met at a reasonable cost; unless a variance or other exception is granted, the only choice available to a discharger which cannot meet the standard is to cease the discharge.

The states within the Chesapeake Bay watershed can impose more stringent effluent limits in NPDES permits for discharges in their state—if necessary to protect their own state waters, including the Bay—simply by promulgating more stringent water quality standards. It is less clear that states have the legal authority to promulgate more stringent water quality standards if the only purpose is to protect water bodies in some other state. If they attempted to do so, an affected discharger might successfully argue that the state had exceeded its authority.

Fortunately, this is not an insurmountable problem where there is sufficient political will, because the Constitution authorizes the states to conclude interstate compacts which, if approved by Congress, expand the legal authority of the states. There is precedent in the Atlantic watershed for compacts to deal with interstate water pollution, such as the Delaware River Basin Compact and the Susquehanna River Basin Compact.[12] A substantial degree of interjurisdictional cooperation on the Chesapeake Bay already exists, as exemplified by the Chesapeake Bay Agreement of December 1983, to which Maryland, Virginia, Pennsylvania, the District of Colombia and EPA are parties. (This agreement is not an interstate compact in the Constitutional sense). The Susquehanna River Basin Commission is monitoring non-point source nutrient and sediment pollution in the Susquehanna and selected tributaries which could reach the Bay.[13]

A further step should be taken: all the jurisdictions in the watershed should negotiate a Chesapeake Bay Basin Compact, which would increase the authority of the states over discharges which reach the Bay and where appropriate, allow Basin-wide standards. Given the existing cooperation on the Bay, prospects appear to be good for successful negotiations and for subsequent Congressional approval.[14]

Federal and State Effluent Standards

Section 307 of the CWA allows EPA to impose effluent standards on certain toxic pollutants. Where applicable, these standards set maximum concentration limits for a substance for dischargers in all states, irrespective of the water quality impact. EPA has found setting such limits to be difficult, and very few have been promulgated.[15] Some states have maximum concentration standards for other particular pollutants.

The use of toxicity standards to solve the problems of the Chesapeake Bay would be highly problematical. With a handful of exceptions (e.g., creosote in the Elizabeth River), no one currently knows which specific substances pose special risks to the biota of the Bay. Identifying substances of particular concern would be an investigative task of monumental proportions. There are nearly 60,000 chemicals on the EPA inventory of substances currently used in commerce; over 7,000,000 chemical substances are listed by the Chemical Abstracts Service.[16] Yet fewer than 5,000 chemical substances have undergone at least one aquatic toxicity test.[17] Only a trivial fraction of that testing has been done on species of special importance to the Chesapeake, such as the striped bass (*Morone saxatilis*) or the American oyster (*Crassostrea virginica*).

Determining which chemical substances could have an impact on the aquatic biota of the Chesapeake Bay would only be the beginning. It would then be necessary to determine which of these were being inadequately treated by the approximately 6,000 NPDES permitted dischargers in the Bay's watershed. Given the possibility of additive, synergistic or antagonistic effects among the many chemical substances which could be contained in a single discharge, the task becomes a practical impossibility.

A modest level of chemical-specific effort should be continued, in the hope of developing scientific data which would provide a basis for special regulation of particular chemicals. Such work could be especially valuable in setting limits on simple effluents containing only a few pollutants. However, this research is far less useful where highly complex and variable effluents are at issue, as is often the case for sewage treatment facilities, organic chemical plants, paper mills, etc. These can have hundreds or even thousands of substances in their discharge. Because a better approach is available, we believe that attempting to establish a special program for the Chesapeake Bay which focusses exclusively or primarily on specific chemicals would be a serious misuse of scarce resources.

Biologically-Based Whole Effluent Toxicity Limits

In recent years, increasing attention has been given to an alternative approach which does not depend on a knowledge of the toxicity of the constituent chemicals in a discharge, or even identification of the chemicals. The technique involves the use of aquatic toxicity bioassays, in which the test organisms are subjected to various concentrations of the effluent being tested. Most commonly, the test is a short-term acute test run to determine the concentration which is lethal to 50% of the test organisms in a 48- or 96-hour period, commonly expressed as the "LC50" value. Increasingly, a longer test to determine the concentration which brings about some other adverse chronic consequence, such as a reduction in reproductive success or growth, is also used.

In EPA's view, biologically-based toxicity limits derive their legal status primarily from state water quality standards. All relevant jurisdictions have narrative requirements—variously worded—which bar the discharge of toxic substances in concentrations which are harmful to aquatic organisms. Bioassay tests, it is argued, can establish that the effluent is sufficiently toxic to aquatic organisms to violate the state's narrative standard.

This authority was used sparingly prior to 1984. Washington state and California issued permits in the mid-1970s containing complex effluent toxicity limits,[18] and several more states issued permits requiring whole effluent toxicity monitoring. However, in response to the 1984 EPA national policy statement,[7] EPA and several states greatly expanded requirements for dischargers to undertake routine self-monitoring of their effluents by use of bioassays. As this is being written, nearly all states have some kind of biological testing program for whole effluents in place or in the planning stage. An even more recent development is the requirement in some states (including Virginia) that dischargers found to have toxic discharges must undertake a toxicity reduction evaluation, which may lead to specific measures to reduce toxicity.

Some states outside the Bay region have placed effluent limits in NPDES permits, compliance with which depends on the results of bioassay tests. In such a case, the permit contains a clause stating that the effluent may not have an LC50 value lower than some stipulated percentage of effluent. That percentage is generally determined by reference to the dilution and the other characteristics of the receiving stream. Because of the variability and other problems inherent in the use of bioassays discussed below, the use of such permit limits raises serious scientific and legal issues. Moreover, this approach is not necessary to the goal of limiting toxicity, which is better served by toxicity reduction programs. A few states instead set a state-wide numerical limit without reference to in-stream impact. For example, in New Jersey, no discharge can be more toxic than that shown by an acute LC50 of 50% effluent by volume. This approach lacks a scientific basis, and has led to numerous legal and practical difficulties.

While biologically-based toxicity controls are said to derive their legal authority from state water quality standards, they are better thought of as a fourth, separate way of regulating a discharge, which differs significantly from the technology-based and water quality-based limits, and toxicity standards discussed above.

A key conclusion of this chapter is that the careful use of biologically-based toxicity testing in the NPDES permit program offers the potential for cost-effective improvements in the ecological well being of the Chesapeake Bay. This is partially because of certain inherent benefits, and in part because the other three approaches discussed above appear to be approaching their limits of marginal utility.

Limitations on the Use of NPDES Permits as a Means to Deal with Pollution Problems of the Chesapeake Bay

Although the use of biologically-based toxicity tests to control toxicity in effluents reaching the Chesapeake Bay has substantial promise, there are important limitations. Some of these derive from the legal nature of the NPDES permit system; others are peculiar to bioassay techniques; some are political.

Statutory Limitations

Inherent limitations which derive from the statutory provisions of the CWA sharply restrict the use of the NPDES system as a means of dealing with all the contaminant problems of the Chesapeake Bay. First, the NPDES system regulates discharges from point sources. The term has been very broadly construed by the courts, and includes not only industrial plants and municipal sewage treatment works, but also a number of miscellaneous sources such as coal mines and aquaculture facilities. However, several important sources of water pollution in the Bay are considered non-point sources. Among these are agricultural run-off of pesticides and nutrients; mobile sources such as shipping, fishing and pleasure boats; and many sources of siltation. Indeed, if every industrial and municipal discharger in the Chesapeake Bay watershed were to stop discharging tomorrow, we have no data suggesting (and substantial reason to doubt) that the pollution problems facing the Bay would disappear. Fortunately, there are other legal authorities to deal with non-NPDES regulated pollutants. The 1987 amendments to the CWA require the establishment of state non-point source control programs.[19] Other federal statutes provide means to control particular toxic substances or sources.[20] State law can provide the authority to reduce growth-induced siltation, as exemplified by the newly passed Maryland legislation.

Second, NPDES permits provide a legal basis for regulating certain problems, but deal with them badly in practice. For example, the requirement to obtain and abide by such permits applies equally to publicly owned sewage treatment facilities. Yet the compliance record of some sewage treatment plants is appalling. In some states, they have been determined to be not only the prime cause of conventional pollution, but also a major source of toxicity.[21] Finally, the NPDES permit system is not always the most cost-effective means to deal with pollution problems. For example, incentives for a reduction of the amount of waste generated may be far more effective and less costly in some cases.

Scientific Limits

EPA's Chesapeake Bay Program (CBP) has concluded that the toxic materials presently in the estuarine environment of the Bay, as well as continued inputs, represent potentially serious threats to the integrity of the Bay.[22] The toxic materials of concern include heavy metals, synthetic organic compounds (including pesticides and herbicides), petroleum hydrocarbons, and other chemical substances such as chlorine. Many of these are non-persistent (i.e., degrade rapidly), while others have been shown to accumulate in the sediments and/or the tissues of aquatic organisms.

It is reasonable to assume that virtually all toxic substances in the Bay arise from human activities located within the Chesapeake Bay basin. In addition to many point source discharges, non-point sources of toxic materials which may enter the Bay include: (i) urban runoff — primarily heavy metals and hydrocarbons; (ii) agricultural runoff—pesticides and herbicides; (iii) acid mine drainage—heavy metals; (iv) atmospheric deposition — heavy metals; (v) marine activities—inorganic and organic pollutants; (vi) dredging and disposal of contaminated sediments—inorganic and organic pollutants; and (vii) weathering and erosion of land forms—minor amount of heavy metals.

While these source categories have been identified, the precise sources and levels of contamination are not well understood. Indeed, we do not even have good estimates of either the absolute or the relative amounts of persistent toxic materials which originate from point vs non-point sources. We certainly do not have enough data on quantities of toxicants being released, their degradation, and the Bay's natural flushing rate. Without that information, it is impossible to say whether the concentrations of toxicants in either water or sediment of the Bay are increasing or decreasing. Fortunately, information on contaminants in more limited locations provides an indication of the magnitude of the problem. For example, the CBP identified several areas in the Bay where high levels of toxic pollutants occur in the sediment.[23] Most of the high contamination areas occur or originate within urbanized areas. In some cases, certain toxic substances have been identified as originating from specific point source discharges. CBP also found that in such metropolitan areas as Baltimore, Washington, D.C., and Norfolk, urban non-point source runoff contributes significant loadings of toxic substances.[24]

Since the relative contributions of toxic materials entering the Bay from point and non-point sources cannot be adequately estimated, it is not possible to determine whether strengthening the NPDES permit system or imposing a strong non-point source program would bring about larger overall reductions of toxic materials in the Chesapeake Bay. A related question is how large a reduction in toxic substances from each major source type would be necessary to eliminate any significant adverse impacts. Answers to these questions would be invaluable and would provide the necessary framework to ensure that appropriate, cost-effective actions are taken to reduce the flow of toxic pollutants in the Bay and to restore and maintain the Bay's ecological integrity. Unfortunately, early answers to these questions are unlikely, given the magnitude of the problem and the depths of our current ignorance. This is not a rationale for inaction, but does mean that policy choices will have to be made on an inadequate scientific basis. On the other hand, the lack of adequate data is not a rationale for capricious regulation.

Political Limits

Even assuming that all of the states in the watershed have the legal authority to strengthen NPDES permits enough to meet fully the needs of the Chesapeake Bay, it is unclear that they have the political will to do so. The problem is potentially the greatest with respect to the non-littoral states, which may resist undertaking strict new limitations on their own agriculture, industry, and sewage treatment plants, in order to protect a body of water from which they derive little direct benefit. This suggests that any special restrictions on up-stream dischargers should also benefit the environment in the non-littoral states. Fortunately, carefully considered biologically-based toxicity testing and toxicity reduction measures should have that effect in nearly all cases.

GROWING USE OF TOXICITY TESTING IN NPDES PERMITS

The placement of biological testing requirements in NPDES permits has been growing rapidly since EPA issued its 1984 national policy statement.[7] According to Wall and Hanmer,[21] a July 1986 survey of the states and EPA regional offices that write permits showed that more than 1400 major industrial facilities (approximately 38% of major industrial permits) and almost 400 major municipal treatment plants (approximately 10% of major municipal sewage permits) in the United States have some form of requirement for whole effluent toxicity testing in their NPDES permits.

Description of the Technique

A bioassay is a toxicity test in which separate groups of test organisms of the same species are subjected to various dilutions of a chemical substance or mixture (such as an effluent) for a specific period of time. A control group is subjected to dilution water with no test material added, to provide a measure of the acceptability of the test. Toxic effects may be divided into a number of overlapping categories;[24] however, in existing NPDES permits, the two most widely used are acute lethal toxicity, and to a much lesser degree, sublethal chronic toxicity.

The acute toxicity of chemical substances to aquatic organisms is assessed by one of several standard test procedures,[25-30] all of which give approximately the same results. Most methods calculate the concentration of a substance which kills (or in some cases immobilizes) half the test organisms in a specific period of time, often 48 or 96 hours. Experimentally, this median lethal concentration (the LC50) is the most reproducible and easily determined measure of toxicity.

Chronic toxicity is a longer-term measure of the effects a chemical substance may have on survival, growth and reproduction. The exposure period for a chronic toxicity bioassay may be a portion of an organism's life cycle, or a complete life cycle (days to months depending on the species). Chronic toxicity bioassays have been used to estimate the long-term safe level (the maximum acceptable toxic concentration or, less precisely, the no-effect concentration) for an organism exposed to a chemical substance.

A number of important factors must be considered when conducting toxicity tests with aquatic organisms (see references 24-29 for details). Failure to pay strict attention to any of these factors can yield poor results. Indeed, poor test conditions in some laboratories are a serious problem in the use of biological testing for regulatory purposes. Among the critical factors are: (i) proper equipment; (ii) good quality dilution water; (iii) proper test organisms; (iv) the use of good test procedures including good quality assurance practices; and (v) analysis of data. For example, construction materials and equipment that contact water in which test organisms are placed should not contain any substances that can be leached or dissolved even in trace quantities.

The precision of a toxicity test is also inherently limited by the normal biological variation among individuals of a species. When properly done, the results of acute toxicity tests which use the same test species, dilution water, toxicant, and test procedures should ideally not differ by more than a factor of two.[27] Yet for regulatory purposes, a factor of two is a very large variation, and even that goal is often not achieved. Fortunately, where multiple tests are run, an accurate picture of the toxicity of an effluent is possible, since the average of many tests typically varies far less from the "true" value than will a given single test. Where permit compliance depends on the results of a single test, such analytical variability can pose severe legal difficulties, and in our view should be avoided.

Uses and Purposes of Aquatic Bioassays

The acute toxicity test has been in use since the 1930s when it was used to study the toxicity of individual substances. The pioneering work on whole effluent toxicity began in 1945, with the work by Hart et al.[31] The first "standard method" for the evaluation of the acute toxicity of industrial wastes to fish was published in 1951.[32] Acute toxicity evaluations of effluents have achieved widespread acceptance among aquatic toxicologists because they are reasonably quick and easy to conduct, inexpensive, and provide a great deal of information in a short time period.

Early investigators recognized that the effects of long-term exposures to toxicants could be as important as the acute effects. Such chronic effects would, of course, occur at lower concentrations of a toxicant or mixture than would cause lethality in acute tests. However, tests over the whole life cycle of an organism were far more costly and were wholly impractical for routine use. To circumvent those difficulties, several techniques (e.g., "application factors") were devised to provide a crude extrapolation from the acute test data to an estimate of the "safe" concentration of the chemical substance in the environment.

Short-term methods (under eight days) for estimating chronic toxicity have recently been developed for effluent toxicity testing, which offer substantial promise of reducing the cost and time of chronic tests, while still providing (it is hoped) good estimates of the safe concentration.[30] Other types of biological testing can also be important because organisms can be adversely affected by the accumulation of toxic residues, or by flavor impairment in recreational and commercial species.

Current Uses of Whole Effluent Bioassays by States in the Chesapeake Bay Watershed

Each of the jurisdictions in the Chesapeake Bay watershed has taken at least some steps toward the use of whole effluent toxicity testing of point source discharges. However, no jurisdiction has completed implementation of a full-fledged program, by which we mean (i) a comprehensive screening of all major dischargers for possible toxicity, (ii) the imposition of both acute and chronic monitoring requirements in all permits where justified by the screening tests, and (iii) the imposition of toxicity evaluation and reduction programs in those cases where significant toxicity is demonstrated. Generally speaking, acute toxicity testing has been used for monitoring purposes in the relevant jurisdictions, though chronic testing is beginning to take place. Imposition of toxicity reduction programs has been relatively rare. The fact that no relevant jurisdiction currently has a fully developed program is ironically something of a blessing for the Chesapeake Bay. Problems which have arisen in other jurisdictions can be avoided, and the special circumstances of the Bay can be taken into account.

The following is a brief description of the current use and plans of the relevant jurisdictions for bioassay testing and compliance actions in conjunction with NPDES permits. Much of this information was supplied directly by involved personnel, and in some cases differs from published plans and policy statements.

Delaware: Most of the population of the State of Delaware and virtually all of its major industry are located in the Delaware Bay watershed. However, a portion of the southern and western areas of the state, including some NPDES-permitted facilities, are located in the Chesapeake Bay watershed. For a number of years, the state has conducted an active toxicity testing monitoring program, and more recently has been moving toward the imposition of toxicity reduction programs. Toxicity screening will be undertaken for all dischargers. Where testing has indicated a toxicity problem, specific requirements have been placed in NPDES permits.

District of Columbia: The government of the District of Columbia is the only relevant jurisdiction which does not have the authority to issue NPDES permits. Instead, the permitting authority for dischargers in the District is EPA Region III. This makes some sense, since the only NPDES permitted facility of consequence is the vast Blue Plains Sewage Treatment Facility, which serves the District and a substantial portion of the Maryland suburbs. The NPDES permit for Blue Plains was renewed in September, 1986. The new permit requires chronic testing, initially using the water flea (*Ceriodaphnia dubia*) and the fathead minnow (*Pimephales promelas*).

Maryland: For a number of years, Maryland has required toxicity monitoring of a limited number of dischargers, particularly those in industries where toxicity of the effluent was suspected. However, no routine program of toxicity testing was required, in part because state officials saw little evidence of acute toxicity. More recently, following consultations with EPA, Maryland has announced a program of biomonitoring. Beginning in 1987, the state has committed to require toxicity screening by industrial and municipal discharges as a condition of permit renewal. In the first month of a new permit, a permittee is required to submit a plan of study to evaluate wastewater toxicity, which must provide for quarterly short-term chronic testing, utilizing fathead minnow and *Ceriodaphnia dubia*. Dischargers to saline waters may substitute Sheepshead minnow (*Cyprinodon variegatus*) and mysid shrimp. At least once during the first year, the discharger must carry out tests on a locally important fish and invertebrate species.

There is no single standard for toxicity under the state's regulatory regime. Declaring an effluent to be excessively toxic, and therefore requiring a toxicity reduction program or other remedial measures, would depend on such factors as the tested toxicity; the location, volume and chemical parameters of the discharge; and the dilution in the receiving stream.

The State has launched a testing and research program under contract to the University of Maryland with a subcontract to The Johns Hopkins University. In seeking to deal with the special problems of the Bay, one key weakness mentioned by state officials is the lack of estuarine test organisms. They hope the research aspects of the program will develop test protocols for short-term chronic toxicity testing using sensitive organisms which play a vital role in the ecosystems of the Bay.

New York: The Susquehanna River is the single largest source of fresh water entering the Bay, and a substantial portion of the upper reaches of the River's basin is located in the state, including numerous industrial and municipal NPDES-permitted dischargers.

Whether the toxicity of discharges from New York or other non-littoral states contributes to toxicity problems in the Bay is uncertain. Many substances degrade over periods ranging from minutes to months by chemical and/or biological processes. Of those which do not, many of the metals and complex organic substances tend to sorb to sediment in the general vicinity of the discharge, (where they can pose long-term problems, PCBs being the best known example). As a result, toxicity problems studied to date tend to be localized. The Susquehanna River, for example, shows substantial signs of recovery of water quality and of biological communities downstream of industrial and municipal dischargers.[13] However, there are soluble metal compounds and semi-volatile organic substances which are highly persistent and likely to remain in solution. It is possible that a continued buildup of these could pose toxicity problems as far away as the Bay. However, we are unaware of any body of scientific research which has addressed the precise question. We consider this a useful topic for future research.

New York has conducted little whole effluent bioassay testing. However, it has had the most ambitious chemical-specific research and control program of any state. There are indications that state officials are beginning to recognize the limits of a chemical-specific approach, and have expressed interest in supplementing it with some whole effluent testing. Indeed, a New York state manual for toxicity testing of industrial and municipal effluents has been developed for use in the state.[33]

Pennsylvania: Most of the watershed of the Susquehanna River is in Pennsylvania. Several cities and a substantial amount of industry are located on the river and its tributaries. As discussed under New York, we do not know the significance of toxic discharges to this fresh water system to the conditions in the Bay.

The Commonwealth has no biomonitoring program, though it has had an aggressive chemical-specific effluent control program similar to New York's. However, key officials believe their chemical-specific approach should be supplemented by whole effluent toxicity testing. Funds have been requested for qualified personnel and a laboratory, but the prospects for approval are uncertain as this is being written. If the program is approved, the Commonwealth would incorporate testing requirements into permits for both municipalities and industrial discharges as necessary. There are a few scattered requirements in such permits now.

There is currently no precise regulatory standard as to what constitutes an excessively toxic effluent, and cases are decided on the basis of professional judgment. Dilution in the receiving waters is taken into account. Commonwealth technical personnel believe that short-term chronic tests are more appropriate to their needs than acute tests. They are considering EPA's recently announced short-term tests on *Ceriodaphnia* and fathead minnows.[30] These may be modified to hold down testing costs.

Virginia: In most respects, the whole effluent testing program of the Commonwealth is the most advanced of any of the relevant jurisdictions. It has a state compliance laboratory which also does some research work. Its mobile lab can do freshwater acute and chronic testing, and is developing the ability to do saltwater testing. The Commonwealth has a number of qualified personnel, and gets additional technical and scientific support from the Virginia Institute of Marine Science.

Virginia began its whole effluent toxicity program in 1980, and has been using bioassay techniques to monitor discharges for nearly four years. It has imposed acute testing requirements as a routine part of its NPDES permit-issuing program. Primarily freshwater test species have been used, though there is some experience with marine organisms, and some research work on Chesapeake Bay species has been done in conjunction with EPA laboratories.

Chronic testing has been done on selected dischargers, primarily those located in the lower James River, and more recently in the Elizabeth River. This testing has been coupled with efforts to determine the ecological well-being of the water bodies downstream of the discharge. Commonwealth officials plan to extend this successful program to the Rappahannock and York Rivers and to the Bay. These programs would not be permit-oriented per se, but the results of the testing would include recommendations to permit writers, who could require the permittee to undertake toxicity reduction measures. Virginia has no fixed criteria for what constitutes unacceptable toxicity, and much depends on the level of severity, the in-stream impact, the pollutants involved, and the degree of dilution in the receiving waters.

Although confident of its legal authority to impose toxicity monitoring, the Commonwealth's Water Control Board also wanted a firm legal basis for imposing toxicity reduction evaluations on dischargers as a regular part of its permit process. The Board recently set in motion the steps to obtain additional regulatory authority under the Commonwealth's Administrative Process Act. The future shape of that program is unclear. It will probably involve both industrial and municipal dischargers, as the latter had been shown to have unacceptable levels of toxicity in some cases. More widespread use may be made of chronic toxicity testing and site specific evaluations, with special emphasis placed on the Chesapeake Bay. Virginia technical personnel would like to see whole effluent chronic testing protocols developed for Chesapeake Bay organisms, especially those which live in low salinity waters.

West Virginia: Although most of the state's chemical and other heavy industries are located on the Ohio River and its tributaries, a substantial fraction of the eastern portion of the state drains into the Potomac River. However, as noted, the significance of such discharges to the conditions in the Bay is unclear. The State has a small bioassay monitoring program. For compliance monitoring, acute tests on freshwater organisms have been used. However, the state has a cooperative program with EPA laboratories, and is switching to short-term chronic tests on *Ceriodaphnia.* Like the other relevant jurisdictions, West Virginia has no numerical standard for toxicity. Three problems with the West Virginia program are limited resources, the small percent of permits covered, and the fact that the state has so far limited its whole effluent toxicity program to industrial dischargers.

EPA: Region III covers all the relevant jurisdictions except New York, which is in EPA Region II. Region III has a laboratory in Wheeling, and a mobile facility which can do both fresh and saltwater testing. Region III headquarters in Philadelphia has been encouraging the states to adopt aquatic toxicity testing programs. The Chesapeake Bay Program is located in Annapolis, Maryland. Except for this program, EPA's national headquarters has devoted few resources to low salinity estuaries, and the EPA environmental research laboratories have worked almost exclusively on freshwater and high salinity test organisms. The Congressional blessing of the Bay Program, and establishment of a separate National Estuary Program in the 1987 Amendments to the CWA,[34] if followed with adequate funding, may finally herald the beginning of greater EPA research on and attention to toxicity matters of unique concern to estuaries.

Problems Inherent in the Use of Toxicity Tests in NPDES Permits

The purpose and objectives of an effluent toxicity test should be clearly defined as a first step to help reduce inherent problems in effluent testing. This should be obvious, but there are numerous examples where the nature of the testing poorly matches the stated purpose. When that happens, resources can be wasted and both economic and environmental interests harmed.

Effluent toxicity testing is often considered imprecise, and as Mount has pointed out, it gives the appearance of being a "sloppy science or maybe no science at all."[35] Two critical reasons previously discussed are poor conduct of the test and high natural (biological) variability. A third reason is the fact that effluents can be highly complex, and vary considerably in chemical composition over time. A highly complex effluent can contain hundreds of chemicals, some of which may not even be identified. Some compounds may be persistent, while others may break down quickly because of biodegradation and other processes. In general, perhaps as a consequence of their longer average treatment times, industrial wastewater treatment systems have smaller variation in effluent toxicity than municipal systems, although notable exceptions can be found.[36] Changing chemical complexity and high variability can make it difficult to assess the toxicity of effluents. If the effluent is highly variable, far more effluent samples must be tested, arguing for short-term rather than long-term tests.

An important consideration in the use of bioassays in the NPDES permit system is testing costs. These can vary considerably depending on the length and type of the test, whether it is conducted on or off-site, etc. Contrary to popular view, the vast majority of NPDES-permitted dischargers are small businesses and small sewage treatment plants. Testing costs which may be reasonable for a large industrial plant or major urban wastewater treatment plant may be prohibitive for smaller dischargers. In cautioning against imposing extensive testing requirements on small dischargers, we are not suggesting that they be exempt from controls on toxicity. Instead, we believe inexpensive screening methods should first be used to determine which smaller dischargers in the watershed merit a more thorough evaluation. An innovative approach has been undertaken in Maryland, where an initial screening for toxicity will be conducted by the State through funds provided by a modest permit fee levied on all dischargers.

While there are inherent problems in testing effluents for toxicity, we believe a properly designed program can shed light on the nature and extent of the problems which may be caused by toxic releases from point sources in the Chesapeake Bay watershed. Remediation efforts should then be required where significant toxicity is demonstrated. Despite the problems discussed below, toxicologists and regulators should apply their best professional judgement and currently-available methods to bring about near term improvements, while at the same time continuing research to devise better testing methods for complex effluents.

PROBLEMS WITH THE USE OF TOXICITY TESTS TO DEAL WITH THE SPECIFIC PROBLEMS OF THE CHESAPEAKE BAY

The Selection of Appropriate Test Species

There is a wide range of sensitivity among various species exposed to individual toxicants. The pesticide and heavy metal data base shows that the differences in acute LC50s between the most resistant and most sensitive species can vary up to 1,000 fold and in a few cases 10,000 fold.[35,37] There are virtually no data to tell if such ranges are true for complex effluents.[35] In theory, the differences in the sensitivity of test organisms to complex effluents should become apparent as toxicity testing requirements are written into more NPDES permits. In practice, it will be impossible to make statistical comparisons unless all states routinely report all whole effluent toxicity test results to the Complex Effluent Testing Information Service, EPA's computerized data base on aquatic toxicity testing of effluents.

Most of the existing data on inter-species sensitivity has been obtained on freshwater species; more limited LC50 data for salt-water species exposed to pure compounds suggest they have the same general ranges in sensitivity.[38] However, a substantial data base on estuarine species is lacking, using either pure compounds or complex effluents.

This paucity of data on complex effluents and on estuarine species raises a question about the appropriate test species to be used in NPDES toxicity testing for discharges to the Chesapeake Bay watershed. Most industrial and municipal plants discharge an effluent which contains chemical substances dissolved and/or suspended in freshwater. Freshwater species should be appropriate for discharges located in the Chesapeake Bay basin above the fall-line unless the effluent itself is very saline. However, it is difficult at best to extrapolate toxicity data from freshwater organisms (or marine organisms) to predict impact on an estuarine ecosystem. Thus, while freshwater species could be used to evaluate the toxicity of the effluent itself, it would be preferable to use estuarine organisms to predict the ecological impact of dischargers below the fall line. Unfortunately, although reasonably good guidance is available for selecting both freshwater and marine species,[27,29] a void exists for low salinity (less than 10 ppt.) Chesapeake Bay organisms. There is accordingly an urgent need to identify, culture, prepare standard methods for, and determine the inherent variability of low salinity estuarine species.

Some individuals concerned about the protection of the aquatic environment feel that the most sensitive species to a given discharge should be used for NPDES permit toxicity tests, no matter how many different species that might involve. While the idea has some superficial appeal, it is impractical and of dubious scientific validity. First, there is no reasonable way, in a system as dynamic as the Chesapeake Bay, to determine the organism that would be the most sensitive to all effluents. Secondly, as Tebo,[18] points out, it is likely that the difference between species is not as great as other errors involved in extrapolating from the bioassay to the environment. Third, it is essential that a few "standard" indigenous Bay species be used which can easily be reared and maintained in the laboratory. A species or life stage which is highly sensitive to a certain type of effluent is not useful as a test organism if it is exceedingly difficult to culture, or if it causes excessively high variability in test results.

Relevance of Toxicity in Tests to Predict the Impact in the Chesapeake Bay

The relevance of laboratory toxicity test data for predicting environmental impact is frequently raised by critics of bioassays. Indeed, there are scientific and regulatory issues involved in using bioassay data to predict the impact of a given discharge on the Chesapeake Bay or any other body of water. One fundamental problem is defining the dynamics of the populations, communities and/or ecosystem to be protected. The amount of change in a biologcal system which is detrimental must be defined to determine impact. Natural fluctuations in a biological system can mask changes resulting from the presence of toxic contaminants.[39]

Evaluating the potential environmental impact of a toxic effluent is a difficult but generally not impossible task. Acute and chronic toxicity tests and chemical fate data can be used to determine the biological response to a complex effluent, integrate the biological effects, provide an estimate of the persistence of toxicity, and define the bioavailability of materials. Predictive modeling techniques can then be used to extrapolate the toxicity data from the laboratory to the water column of the receiving stream. Exposure models in common use are steady state, although time-variable and more recently, probabilistic water quality models are gaining acceptance.[40]

Of particular importance in the Chesapeake Bay are the effects that contaminated sediments may have on benthic communities (e.g., oysters and clams). The concentrations of particular toxicants from effluents may never be sufficient to produce significant toxicity to organisms in the water column. Yet, there may be a gradual build-up of the toxic components in the benthic system. Predictive models to assess sediment-contaminant-water interactions are largely unvalidated,[41] as are quantitative predictions of the transport and distribution of sediment-associated contaminants.[42]

SUMMARY OF MAJOR RECOMMENDATIONS

We believe that despite the scientific uncertainties, a careful program of biologically-based toxicity testing under the NPDES system can provide a better understanding of the sources of toxicants in the Bay, at least from point sources, and suggest needed remediation efforts. Nevertheless, some caution is needed in doing so, and further research is required.

We recommend the following concrete legal and regulatory steps be taken:

(1) A Chesapeake Bay Basin Compact should be negotiated and approved by the Congress, covering all the states of the watershed, in order to provide a firmer legal basis for common water quality standards for the Bay where warranted, to fund research and testing of common interest, and to further the aims of the existing Chesapeake Bay Agreement.

(2) A comprehensive screening program should be undertaken for all of the

370 "major" point source dischargers throughout the watershed and selected "minor" ones, using short duration screening toxicity tests. Two goals would be served: (a) it would allow a better understanding of whether, and to what extent, point sources are contributing significant levels of toxicants; and (b) screening results can indicate which discharges merit further testing. Because the purpose is in part research, and because many dischargers are small entities, it would be preferable if the screening program were financed from public funds, such as shared funding from the states and the Bay Program.[43]

(3) Both acute and short-term chronic bioassays using fresh water test organisms should be used now to evaluate freshwater dischargers where the screening test results indicates there may be significant toxicity. Acute tests should be used now on appropriate salt-water organisms for dischargers to saline waters. Short-term chronic tests for salt-water organisms could be used once test protocols become available, and after sufficient experience with them has been gained. Full life cycle chronic tests should not be imposed except in extraordinary circumstances.

(4) Dischargers clearly demonstrated to have sufficient toxicity to have a significant impact on biological communities in the receiving waters should be required to undertake a toxicity evaluation and reduction program. However, in light of the variability problem, the placement of toxicity limits in NPDES permits—whereby the failure to achieve a specified LC50 value constitutes a violation which subjects the discharger to civil or criminal penalties— should be avoided except in highly unusual circumstances.

(5) All test results should be forwarded to EPA's computerized CETIS data base, so that comparisons can be made. Industry, regulatory agencies, and the public will then have a far better basis to evaluate whether additional efforts should be undertaken.

We recommend that additional research be undertaken, including the following:

(1) Short-term chronic bioassays should be developed on an urgent basis for low salinity estuarine organisms generally, and especially on Chesapeake Bay species. Early attention should be devoted to modifying and then validating the existing protocols under the Toxic Substances Control Act to test the toxicity of complex effluents on commercially and historically important Bay species, such as oysters.

(2) Because of its short spawning season, the rockfish (striped bass) is not a promising bioassay test organism. However, given its commercial and historical importance, testing should be done to determine a surrogate fin fish species with a toxicologically similar response.

(3) Research should be conducted to establish whether a relationship exists between toxic discharges to fresh water in the watershed and toxic impacts on the Bay. One possible approach might be to simulate the conditions under which

a complex effluent may degrade on route to the Bay, and then run short-term chronic tests to see if the residual effluent is toxic to Bay organisms.

(4) Further research should be devoted to the potential for bioaccumulation of toxic substances in the biota of the Bay from both point and non-point sources. A more active participation in the international Mussel Watch program should allow a better understanding of the bioaccumulation of metals by benthic organisms. Obtaining comparable data on the bioaccumulation of organic substances will be far more difficult. The approach of Veith and Morris,[44] whereby high pressure liquid chromatography techniques are used to calculate the n-octanol/water partition coefficient, can be extended to allow an estimate of the propensity of a treated whole effluent to biaccumulate. Useful work has already been done in Virginia and in a few other locations on this approach.

REFERENCES AND NOTES

1. Congressional affirmation of the view that special efforts are needed to protect the Bay can be seen in the newly enacted Water Quality Act of 1987, Pub. L. No. 100-4, which adds a new Section 117 to the Federal Water Pollution Control Act. In addition to providing a Congressional blessing for the Chesapeake Bay Program, the Act authorized (but did not appropriate) $3,000,000 per fiscal year for the Program office and $10,000,000 per fiscal year for grants to the states over a four-year period.
2. Pub. L. No. 92-500, 86 Stat. 816, 33 U.S.C. §§ 1251 *et seq.*
3. The most recent amendment was by Pub. L. No. 100-4, 1987 ("the 1987 Act").
4. 33 U.S.C. § 1311(b)(2)(A).
5. Pub. L. No. 95-217, 1977.
6. 33 U.S.C. § 1311(b)(1)(A).
7. U.S. EPA. Development of water quality-based permit limitations for toxic pollutants; national policy. 49 Fed. Reg. 9016-9019 (March 9, 1984).
8. 33 U.S.C. §§ 1314(b)(2), 1342(c) & (d).
9. Section 313 of the 1987 Act raised the maximum civil penalty to $25,000 per day of each violation.
10. Section 312 of the 1987 Act greatly increased the criminal penalties. For example, a knowing violation by an organization which results in imminent danger of death or serious bodily injury could result in a fine as high as $1,000,000.
11. 33 U.S.C. § 1313(a)(3)(A).
12. Following enactment by the Congress and the respective state legislatures, the Delaware River Basin Compact was signed Nov. 2, 1961. The Susquehanna River Basin Compact was signed Dec. 24, 1970. See also Pub. L. No. 87-328, 75 Stat. 688 (1961) and Pub. L. No. 91-575, 84 Stat. 1509 (1970). While these compacts contain provisions dealing with pollution control,

to date their activities have been more heavily focused on flood and drought control and drinking water supplies.
13. Susquehanna River Basin Commission Annual Report, 1985, pp. 12-13.
14. Use of a compact to deal with nutrients in the Bay is controversial; we express no opinion on the desirability of such use.
15. 40 C.F.R. § 129.4.
16. Toxic Substances Control Act Chemical Substances Inventory, EPA-560/7-85-002a, Vol. I, Introduction, (unnumbered page).
17. The AQUIRE data base (originally developed by EPA and now available through CIS) lists 4,179 substances reported in the literature as having undergone at least one aquatic bioassay. Several hundred other substances have been tested and reported to EPA under TSCA or FIFRA (see reference 20).
18. Tebo, L.B., Jr. 1986. Effluent monitoring: Historical perspective, pp. 13-31. *In:* H.L. Bergman, R.A. Kimberle and A.W. Maki (Eds.) *Environmental Hazard Assessment of Effluents.* Pergammon Press, New York, N.Y., pp. 366.
19. Current EPA regulations explicitly exclude many non-point sources from the requirement for an NPDES permit. 40 C.F.R. § 122.3. However, section 316 of the 1987 Act requires states to establish management programs to control non-point-source pollution, and authorizes substantial funds for grants to the states. It remains to be seen whether sufficient funds will be appropriated, and whether such a management plan approach can actually bring about a significant reduction in non-point-source pollution.
20. The Federal Insecticide, Fungicide and Rodenticide Act (FIFRA), 7 U.S.C. § 136 *et seq.*, provides a basis for regulating the toxicity of pesticides generally, though not on a water-body-specific basis. The Comprehensive Environmental Response, Compensation and Liability Act (CERCLA), 42 U.S.C. §§ 9601-9675, more commonly called Superfund, provides legal authority for dealing with toxic waste sites and cleaning up oil and other spills from vessels. The Toxic Substances Control Act (TSCA), 15 U.S.C. § 2601 *et seq.*, provides a (cumbersome) basis for the nationwide regulation of individual toxicants.
21. Wall, T.M. and R.W. Hanmer. 1987. Biological testing to control toxic water pollutants. *J. Water Pollut. Control Fed.* 59:7-12.
22. Gillelan, M.E., D. Haberman, G.B. Mackiernan, J. Macknis and H.W. Wells, Jr. 1983. Chesapeake Bay: A framework for action. U.S. Environmental Protection Agency Report (No Publ. No.) U.S. EPA, Region 3, Philadelphia, Pa.
23. U.S. EPA. 1983. Chesapeake Bay Program: findings and recommendations. U.S. Environmental Protection Agency Report (No Publ. No.) U.S. EPA, Region 3, Philadelphia, Pa.
24. Burton, D.T. 1977. General test conditions and procedures for chlorine tox-

icity tests with estuarine and marine macroinvertebrates and fish. *Chesapeake Sci.* 18:130-136.
25. Sprague, J.B. 1973. The ABC's of pollutant bioassay using fish, pp. 6-30. *In:* J. Cairns, Jr. and K.L. Dickson (Eds.) *Biological Methods for the Assessment of Water Quality.* ASTM Spec. Tech. Publ. 528, Amer. Soc. Testing Materials, Philadelphia, Pa., 256 pp.
26. Peltier, W. 1978. Methods for measuring the acute toxicity of effluents to aquatic organisms. U.S. Environmental Protection Agency Rep. No. EPA-600/4-78-012.
27. American Society for Testing and Materials. 1980. Standard practice for conducting acute toxicity tests with fishes, macroinvertebrates, and amphibians, pp. 272-296. ASTM Designation E 729-80. *In: Annual Book of ASTM Standards.* Amer. Soc. Testing Materials, Philadelphia, Pa., 1404 pp.
28. American Public Health Association. 1985. Standard methods for the examination of water and wastewater. 16th ed. APHA, Washington, D.C., 1268 pp.
29. Peltier, W.H. and C.I. Weber (Eds.). 1985. Methods for measuring the acute toxicity of effluents to freshwater and marine organisms. U.S. Environmental Protection Agency Rep. No. EPA/600-4-85/013.
30. Horning, W.B., II. and C.I. Weber (Eds.) 1985. Short-term methods for estimating the chronic toxicity of effluents and receiving waters to freshwater organisms. U.S. Environmental Protection Agency Rep. No. EPA/600-4-85/014.
31. Hart, W.B., P. Doudoroff and J. Greenbank. 1945. The evaluation of the toxicity of industrial wastes, chemicals and other substances to fresh-water fishes. Atlantic Refining Co., Philadelphia, Pa.
32. Doudoroff, P., B.G. Anderson, G.E. Burdick, P.S. Galtsoff, W.B. Hart, R. Patrick, E.R. Strong, E.W. Surber and W.M. VanHorn. 1951. Bioassay methods for the evaluation of acute toxicity of industrial wastes to fish. *Sewage Industrial Wastes* 23:1380-1397.
33. Jones, P.A. (Ed.). 1985. New York State manual for toxicity testing of industrial and municipal effluents. New York State Department of Environmental Conservation. Division of Water, Albany, N.Y.
34. Pub. L. No. 100-4, § 320.
35. Mount, D.I. 1986. Principles and concepts of effluent testing, pp. 61-65. *In:* H.L. Bergman, R.A. Kimerle and A.W. Maki (Eds.). *Environmental Hazard Assessment of Effluents.* Pergammon Press, New York, N.Y. 366 pp.
36. Bender, E.S. 1986. Effluent sampling for biological effects testing, pp. 81-91. *In:* H.L. Bergman, R.A. Kimerle and A.W. Maki (Eds.). *Environmental Hazard Assessment of Effluents.* Pergammon Press, New York, N.Y., 366 pp.
37. Mayer, F.L., Jr. and M.R. Ellersick. 1986. Manual of acute toxicity: Interpretation and data base for 410 chemicals and 66 species of freshwater animals. U.S. Fish and Wildlife Service Resour. Publ. 160. U.S. FWS,

Washington, D.C.
38. Mayer, F.L., Jr., 1987. Personal communication. U.S. EPA Gulf Breeze Environmental Research Laboratory, Gulf Breeze, Fla.
39. National Research Council. 1981. Testing for effects of chemicals on ecosystems. National Academy Press, Washington, D.C.
40. Dickson, K.L., A.W. Maki and J. Cairns, Jr. (Eds.). 1982. *Modeling the fate of chemicals in the aquatic environment.* Ann Arbor Sciences Publ., Ann Arbor, Mich., 413 pp.
41. Herbes, S.E. 1986. Predictive models and field studies of the fate of complex mixtures, pp. 172-190. *In:* H.L. Bergman, R.A. Kimberle and A.W. Maki (Eds.). *Environmental Hazard Assessment of Effluents.* Pergammon Press, New York, N.Y., 366 pp.
42. Uchin, C.G. and W.J. Weber, Jr. 1980. Modeling of transport processes for suspended solids and associated pollutants in river-harbor-lake systems, pp. 407-425. *In:* R.A. Baker (Ed.). *Contaminants and Sediments, Vol. 1, Fate and Transport, Case Studies, Modeling, and Toxicity.* Ann Arbor Sci. Publ., Ann Arbor, Mich., 558 pp.
43. If spread over a five period, we estimate that a comprehensive screening program for NPDES permittees in the watershed could be carried out for less than 1.0% of the amount Congress has authorized for the Chesapeake Bay Program.
44. Veith, G.T. and R.T. Morris. 1978. A rapid method for estimating Log P for organic chemicals. U.S. Environmental Protection Agency Rep. No. EPA-600/3-78-049.

Contaminant Problems and Management of Living Chesapeake Bay Resources. Edited by S. K. Majumdar, L. W. Hall, Jr. and H. M. Austin. © 1987, The Pennsylvania Academy of Science.

Chapter Twenty-Five

THE RESTORATION OF LIVING CHESAPEAKE BAY RESOURCES

CHARLES S. SPOONER

U.S. Environmental Protection Agency
Chesapeake Bay Liaison Office
410 Severn Avenue
Annapolis, Maryland 21403

ABSTRACT

Changes in the productivity of the Chesapeake Bay and the intrinsic value of its resources have generated a wide-ranging, innovative, and coordinated restoration and protection program. The program was initiated in the mid-1970's with the identification of ten areas of greatest concern in which both effort and monies were to be concentrated. Of these, declining submerged aquatic vegetation, high concentrations of toxicants, and nutrient enrichment of the Bay were designated highest priority. An intensive seven year research phase followed.

In 1983, based on the findings of the research program, a landmark conference was held which formed the foundation for the first phase of a plan of action to restore and protect the Chesapeake Bay: the Chesapeake Bay Agreement. The specific programs of the states and the Chesapeake Executive Council were assembled in the 1985 "Chesapeake Bay Restoration and Protection Plan." The Plan established goals and defined the means to attain them in the areas of nutrients, toxics, living resources, related environmental programs, and institutional/management strategies.

In conjunction with the research phase, the Chesapeake Bay Program was created. The Program is a composite of many state, federal and regional efforts united by the common purpose and commitment of restoring and protecting Bay waters and living resources. Its Executive Council provides the leadership and focus that has enabled these efforts to become a coordinated, cooperative partnership that functions as a single program. The operational arm of the Council is the Implementation Committee which is assisted in technical concerns by the Scientific and Technical Advisory Committee.

The U.S. Environmental Protection Agency's Chesapeake Bay Liaison Office in Annapolis provides administrative, technical, and public information support to all the organizations formed under the Chesapeake Bay Agreement. EPA supplements state initiatives through grants which has enabled the jurisdictions to expand their programs. Additionally, six federal agencies have signed special agreements with EPA, pledging to coordinate their activities on the Bay with the Bay Program.

The Chesapeake Bay Program represents a working example of coordinated governmental action working to solve a broad base environmental problem. Both the interest and support for the Program are expected to continue and be strengthened by the provisions of the Clean Water Act of 1987.

INTRODUCTION

Public and scientific concerns over the changes in the productivity of the Chesapeake Bay and the intrinsic value of its resources have generated a wide-ranging, innovative, and coordinated restoration and protection program. The program was initiated in the mid-1970s when P.L. 94-116 provided funds and guidance to the Environmental Protection Agency (EPA) to coordinate a research program designed to identify changes in the Chesapeake Bay and to recommend management strategies to correct the problem areas.

The EPA, after consultation with the scientific/technical community and the public, identified ten program areas of concern. From these, three areas were concluded to be of greatest concern: declining submerged aquatic vegetation, high concentrations of toxic chemicals, and nutrient enrichment. After an intensive seven-year research program with an investment of $27 million, the EPA summarized the Bay's plight. Acres of submerged aquatic vegetation were at the lowest levels in recorded history, with the prime suspect being reduced light penetration. Toxic chemicals were found in the sediments of certain parts of the Bay in concentrations exceeding 100 times the natural background levels. These water column concentrations were in excess of EPA's recommended water quality criteria. Nutrient enrichment was found to be high in large portions of the Bay causing reduced levels of dissolved oxygen vital to all living organisms and contributing to the turbidity responsible for diminished light penetration.

In addition to these three primary problems, Bay scientists observed that over time, shifts in the types of living resources of the Bay were occurring. Oyster harvests had declined dramatically and there were changes in the species diversity and composition of finfish. Populations of marine spawning fish like menhaden and bluefish were increasing, while populations of fish that spawn in the freshwater sub-estuaries of the Chesapeake such a striped bass, herring, *Alosa aestivalis,* American shad, *Alosa sapidissima* and white perch, *Morone americana,* were significantly declining.

The findings of the Research Program substantiated the concerns that many public officials and private citizens had voiced concerning the ongoing degradation of the Bay. Goals and approaches were recommended in the publication, "A Framework for Action."[1] These recommendations formed the foundation for a landmark conference convened in late 1983 by the governors of Virginia, Maryland, Pennsylvania, the Mayor of the District of Columbia, the Administrator of the Environmental Protection Agency and the Chesapeake Bay Commission. The conference not only solidified political and public support for a restoration and protection effort, but also determined goals for the Chesapeake Bay and established a management mechanism, the Chesapeake Bay Executive Council, to ensure that those goals would be achieved.

The conference in December 1983 was the basis for the first phase of a plan of action to restore and protect the Chesapeake Bay. The problems had been identified, a political commitment existed, federal and state funds were available, and the public supported the action.

The specific programs of the states and the Chesapeake Executive Council were assembled in the 1985 "Chesapeake Bay Restoration and Protection Plan"[2] and most continue in effect today. The programs focus on efforts to reduce pollutant loadings through the use of known and affordable technology, including the institution of basinwide nonpoint source programs in priority watersheds, phosphorus treatment and phosphate detergent bans in critical portions of the Upper Chesapeake Bay, and reduced chlorine loadings to tributaries during critical stages in the development of young fish.

The Restoration and Protection Plan contains traditional state and federal programs for pollution control activities including the construction and maintenance of sewage treatment plants and the regulation of treatment plants through the NPDES permit system. It includes other efforts to improve the environment through such programs as Coastal Zone Management and wetlands protection.

The 1985 Chesapeake Bay Restoration and Protection Plan, for the first time, established goals and defined the means to attain them in the areas of nutrients, toxics, living resources, related environmental programs, and institutional/management strategies. The goals follow:

- Nutrient Reduction—
    by reducing point and nonpoint source loadings to attain nutrient and dissolved oxygen concentrations necessary to support the living resources of the Bay;
- Toxicity Reduction—
    by reducing or controlling point and nonpoint sources of toxic materials to attain or maintain levels of toxicants not harmful to humans or the living resources of the Bay;
- Living Resources Restoration and Protection—
    by providing for the restoration and protection of the living resources,

their habitat, and the renewal or sustenance of the constituent food chain organisms supporting the harvestable species;
- Development of Related Environmental Programs—
  by developing and managing related environmental programs with a concern for their impact on the Bay; and,
- Coordination and Management of Bay Restoration and Protection—
  by supporting and enhancing a cooperative approach toward Bay management at all levels of government.

The success of programs implemented under the Restoration and Protection Plan and information collected in the course of their implementation are providing managers with insights that are helping to refine both the goals of the program and plans for new Bay programs.

DISCUSSION

Progress in Implementing the Plan

The responsibility to implement the Restoration and Protection Plan falls on state and federal pollution control agencies. The Executive Council's Implementation Committee oversees programs with multi-state significance. The Implementation Committee is aided in this by the Chesapeake Bay Liaison Office staff, consisting largely of EPA personnel.

Nutrient Discharge Reductions

Since the Plan was first adopted, the Chesapeake Bay Program has funded cost/share programs for the control of agricultural nonpoint sources at the State level which have assisted 2,990 farms and treated over 89,000 acres. It is calculated that the cost/share programs are reducing discharges related to this land use by an estimated 523,500 tons of topsoil and 576,000 pounds of phosphorus each year. Management practices that are used most frequently in this program are cropland conservation till, critical area treatment and terracing. Animal waste control projects are removing additional nutrients through proper manure storage. A nutrient management program has been structured in Pennsylvania that shows promise for effectively controlling nitrogen loss from agricultural lands.

The importance of nutrient controls in the Bay watershed was highlighted by the submission of the report "Nutrient Controls in the Chesapeake Bay"[3] by the Implementation Committee's Scientific and Technical Advisory Committee (STAC). The report recommended an examination of biological nutrient removal, a point source control technology that has been largely ignored in this country. This technology controls nitrogen and may do so at reasonable costs. Municipal sewage treatment plants have not controlled nitrogen as Figure 1 shows. New information about biological nutrient removal has prompted plans

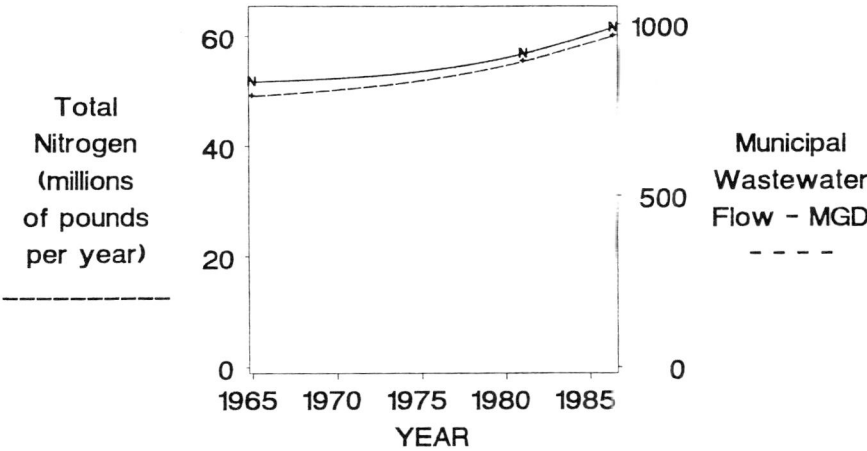

FIGURE 1. Annual nitrogen discharges and wastewater flows below the fall line and in the lower Susquehanna river resulting from municipal sewage treatment plants.

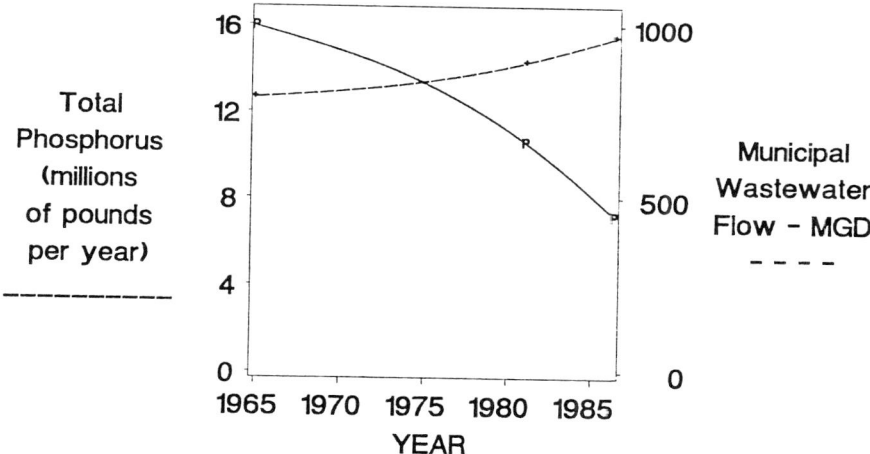

FIGURE 2. Annual phosphorus discharges and wastewater flows below the fall line and in the lower Susquehanna river resulting from municipal sewage treatment plants.

for further study and demonstration in the Bay watershed.

Plans to examine treatment options for nutrients have parallelled interest in the success of Maryland and the District of Columbia for the reduction of phosphorus concentration of generated wastewater. Phosphate detergent bans have reduced the phosphate content of wastewater coming in to the Bay's largest treatment plant (Blue Plains), lowering operating costs by an estimated 15 percent. The overall success in controlling phosphorus discharges is shown in Figure 2.

Toxicity Reduction Efforts

The magnitude and distribution of toxics in the Bay's sediments were documented in the research phase. Agricultural herbicides were extensively examined as a cause of the decline in submerged aquatic vegetation but were concluded not to be the principal problem. The "Framework for Action", and subsequent EPA policies focused on the control of toxicity from point sources. In the implementation phase, plans for reducing levels of toxicity in the Bay's waters have focused on both immediate action and on taking steps to better understand the extent of toxic threats.

Recent discovery of the adverse impact of chlorine on both human health and living resources has resulted in a series of tough state and federal control measures. In 1985, both Maryland and Virginia implemented stringent discharge requirements. Maryland required the reduction to nondetectable limits for all chlorinated waste waters discharged to state waters. Virginia established a chlorine water quality standard of .011 mg/l for freshwater and .0075 mg/l for saltwater.

Currently, 74 percent of Maryland's 160 publicly owned waste treatment works (POTWs) dechlorinate, and another 17 percent are scheduled to install dechlorination facilities in 1987. State money ($1.6 million) is available to assist the remaining 9 percent of the POTWs to dechlorinate. Virginia's Chlorination Discharge Control Initiative, established in fiscal year 1984 allocates $3.4 million in cost share money for dechlorination equipment or alternative disinfection to municipal treatment plants.

New Bay Program studies will develop additional information that should lead to better studies in the future and an array of additional control actions. The Chesapeake Bay Program has undertaken the coordination of two studies of the highly toxic pesticide tributyltin, used on boat hulls as an additive to anti-fouling paints, and has conducted monitoring in support of a third study. High concentrations of tributyltin found in the water column will prompt a thorough consideration of control options in the near future. Options include independent state actions as well as EPA regulatory efforts under the Federal Insecticide, Fungicide and Rodenticide Act (FIFRA).[4]

Controls of point sources of toxicity have come under increased scrutiny in the Bay using biomonitoring methods advocated for the Bay Program in the "Framework for Action."[1] Holders of major NPDES permits are now required to test the toxicity of their effluents. It is the goal of EPA and the states to complete the testing of all major dischargers in FY87 and to begin the reduction of toxicity using the general format called a Toxicity Reduction Evaluation (TRE)[5] as a next step. Beyond immediate concerns for point sources of toxicity, studies of both diffuse and sediment sources of toxics are needed.

Living Resource Restoration and Protection

There has been progress in living resource restoration and protection. While this progress has been achieved during a period when pollution control and habitat protection programs are also progressing, there is no strong evidence linking progress in living resource abundance with reduced pollution. This is expected to be possible in future years.

Increased abundance of striped bass is the direct result of catch restrictions enforced in the Bay. Striped bass have made a slight comeback in Maryland since their record low levels at the beginning of the decade. Distressed by the low population levels and virtual absence of females, Maryland Bay managers enacted a ban on striped bass fishing effective January 1985. Other states in the region, including Virginia, raised the size minimum to 24" and lowered daily catch allowances. (Bass reach the former 14" size limit in approximately two years even though reproduction ability is not achieved until 4-6 years of age.) The 1982 year class, freed from fishing pressures in Maryland, showed substantially greater increases in 1985 over the 1981 and 1982 year classes in their third year.

The two most important causes for striped bass decline are believed to be overfishing and elevated early life stage mortality. Maryland's striped bass moratorium and Virginia's increase in the size minimum address the first of the two problems. Baywide habitat restoration efforts address the second.

One of the best indicators of the quality of the Bay's habitats is Submerged Aquatic Vegetation (SAV). SAV has been extensively surveyed for three consecutive summers under the Bay program. The results of the 1985 SAV survey show a 26% increase from 1984 levels. A total of 19,390 hectares of SAV were mapped: 3,025 in the upper Bay, mostly on the Susquehanna flats; 4,986 hectares in the middle Bay, and 11,379 hectares in the lower Bay. The middle Bay zone showed the most dramatic increase with a 389 percent rise over last year's coverage. A multi-year assessment of trends in the distribution and abundance of Bay grasses is planned.

High Bay salinities have had a severe impact on other resources, particularly oysters which suffer increased incidence of MSX, a fatal viral infection, when reduced fresh water flows from the Bay's major tributaries allow the intrusion of salty marine waters. Such natural occurrences remind us that many aspects of the Bay's environment are essentially uncontrollable.

Recognizing that other important factors are controllable however, the Implementation Committee established a Living Resources Task Force that has begun to examine the criteria that should be explicitly considered to protect a range of species in the Bay. These criteria will influence both pollution control and habitat protection measures. The Chesapeake Bay Stock Assessment Committee is continuing to study ways to better manage the fisheries resources of the Bay within the limits of the Bay's ability to sustain them.

Related Environmental Programs

Several related environmental programs have been developed and refined within the context of the Chesapeake Bay Program. A national increase in problems associated with acid rain and groundwater contamination has led to specific monitoring efforts in the Chesapeake Bay region. Pennsylvania's Department of Environmental Programs has implemented a comprehensive groundwater management program to monitor and protect current groundwater resources. It has also developed an acid precipitation monitoring network. Virginia's Department of Health monitors the public drinking water to insure that it does not become contaminated. Shoreline sanitation corrections made in the Potomac basin have resulted in the reopening of 187 acres of productive shellfish grounds.

State recreational officials around the Bay continue to improve and maintain access to public parks, beaches and boat ramps in the Chesapeake. Virginia has spent over $5 million since FY 84 on this work. In addition, major efforts are focused on sewage sludge, dredge spoil and hazardous waste management under Superfund and the Restoration Conservation and Recovery Act.

Coordinated Bay Management

An increased interest in integrating Bay management activities at all levels of government has made local government more conscious of the influences of land use planning on Bay water quality. This heightened interest has led to the formation of a number of new local conservation groups and county or area programs in Maryland for water quality protection and enhancement. The Chester River Association was created this year to promote participation in educational and protection efforts. The Sassafras River Bi-County Advisory Committee was created in 1984 to advise Kent and Cecil County Commissioners of Sassafras River environmental concerns. The Zekiah Swamp Ad Hoc Committee recently recommended placing the swamp on the Charles County Chesapeake Bay Critical Areas Program and creating a Zekiah Swamp advisory board. The plan is being reviewed by the Tri-County Council and Charles County Planning Council. Montgomery County has initiated a rural sanitation program to review issues and prevent problems associated with on-site wastewater disposal and water supply collection in rural areas.

In Pennsylvania, the townships of Heidelberg and Jackson in Lebanon County have enacted a manure storage ordinance, requiring farmers to obtain approval from the conservation district commissioners and the township supervisors of construction plans for manure storage facilities prior to construction. Following construction, the completed project is inspected for compliance. In addition, crop management associations have been formed and have served as vehicles for the transfer of special information on the use of fertilizers.

In Virginia, town and county officials have increased efforts to upgrade sewage treatment plants and reduce urban nonpoint source runoff. The Hampton Roads

Sanitation District, like many of the utilities in the area, has implemented a number of programs which are contributing to the Chesapeake Bay cleanup: industrial pretreatment, toxics and industrial discharge monitoring, infiltration/inflow correction, and expansion and upgrading of several area facilities. Through Virginia Beach's Capital Improvements Program, two deteriorated private sewage treatment plants have been eliminated. Presently, the RADCO (Fredericksburg/Rappahannock area) Planning District is engaged in an urban nonpoint source study to determine the impact of urban nonpoint source pollutants on the water quality of the Rappahannock River.

Restoration efforts in the Nation's Capitol area have been the responsibility of the District of Columbia. They have been assisted by the Metropolitan Washington Council of Governments (COG) regional planning. With its 16 member jurisdiction, COG helps to track, coordinate and give technical and policy guidance in dealing with the problems of the Anacostia River. Projects to date include monitoring, retrofitting storm drainage pipes, modeling, and generating public awareness.

The involvement of private groups and industry in the Bay's restoration and protection is a significant addition to the Plan. Two of the most successful areas of direct citizen participation are the Citizens Monitoring Program and the Citizens SAV Groundtruthing Project. Through weekly water quality monitoring, citizens have confirmed state collected data and conducted monitoring in locations and at times which the state monitoring programs could not accommodate. More than 60 individuals are involved in monitoring under the coordination of the Citizens Program for the Chesapeake Bay, Inc. (CPCB). As part of the Citizens SAV Groundtruthing Project last year, 150 citizens sampled local Bay grasses under the sponsorship of the Maryland Department of Natural Resources (MDNR) and the U.S. Fish & Wildlife Service (USFWS).

In addition to this active individual participation, a variety of citizen organizations make notable contributions. Chesapeake Bay Foundation (CBF), with 40,000 members, plays an essential role in watchdogging the implementation of Bay programs, overseeing enforcement, preserving valuable land, and educating children. The CBF has provided public participation and information services to the Chesapeake Bay Program for nine years. CBF's coalition of member organizations include businesses, farmer groups, civic associations, conservation organizations, boating groups and fishermen. The Chesapeake Bay Yacht Club Association participating in the 1986 SAV Groundtruthing Project and regularly contributes $1,000 stipends to research institutions to support worthwhile studies. Hundreds of other organizations participate in cleanup through educational programs, research and lobbying of local government officials.

Corporations have the technical and financial resources to contribute significantly to the Bay program. Several of area corporations provided a total of $107,000 to the Chesapeake Bay Trust, a funding mechanism set up by the

Maryland legislature to channel private funds to worthy groups and projects related to the Bay. Baltimore Gas & Electric Company, Delmarva Power and Light Company, and Potomac Electric Power Company have invested nearly $2 million to date in fish hatching projects. Baltimore Gas & Electric Company transferred tapes containing $13 million worth of monitoring data from Calvert Cliffs to the Chesapeake Bay Program computer in Annapolis. Several paper companies contributed to the reprinting of "Baybook: A Guide to Reducing Water Pollution at Home."[6] Chesapeake Corporation and Westvaco contributed $2000 each, while Camp Corporation provided the paper. One of the more unique projects is Nevamar's wetlands nursery. Nevamar, a small corporation in Anne Arundel County, created a nursery where both marsh grass and submerged aquatic vegetation are grown for transplant to other sites.

THE CHESAPEAKE BAY PROGRAM

Organization

The Chesapeake Bay Program is a composite of many state, federal and regional efforts united by a common purpose and commitment: the restoration and protection of the water quality and living resources of the Bay and its tributaries. That clear goal enables states with differing institutions and political backgrounds, and federal agencies with differing mandates to work together to solve common problems while retaining the individual character of their efforts.

Since the 1984 Chesapeake Bay Agreement, the Executive Council has provided the leadership and focus that has enabled these efforts to become a coordinated, cooperative partnership that functions as a single program. The Executive Council has the following: the State of Maryland, the District of Columbia, the Commonwealths of Virginia and Pennsylvania, and the United States Environmental Protection Agency. The Executive Council has operated on a consensus basis since it was formed in January 1984. It performs three functions: coordinating, planning and assisting the jurisdiction to implement their activities.

The operational arm of the Council is its Implementation Committee which also functions through consensus. Committee membership is composed of six federal representatives, the Chesapeake Bay Commission and delegates from the jursidictions. The Committee's Planning, Nonpoint Source, Monitoring, Modeling and Research, and Data Management Subcommittee and its Living Resources Task Force work to coordinate aspects of the Bay Program crossing agency and state lines. A Scientific and Technical Advisory Committee (STAC) assists the Implementation Committee through scientific and technical appraisal of the Programs. Its members are directors of major research institutions throughout the region. Each jurisdiction has up to four members on STAC;

there are federal representatives as well. Through an EPA grant, the Chesapeake Research Consortium (CRC) at the Virginia Institute of Marine Sciences gives technical assistance to STAC.

The Executive Council has a Citizens Advisory Committee (CAC) of 25 members; four nominated by the chief executive of each jurisdiction and nine at-large members nominated by the Citizens Program for Chesapeake Bay (CPCB). The CAC advises the Council on the public policy and citizen participation aspects of its work. It functions through task forces organized to study, report, and make recommendations on nutrients, toxic substances, alternative finance, program tracking and integration, and land use. The CPCB provides support to the CAC and its task forces, and performs public information and public involvement services for the Chesapeake Bay Program under a grant from the EPA.

The EPA Chesapeake Bay Liaison Office in Annapolis provides administrative, technical, and public information support to all the organizations formed under the Chesapeake Bay Agreement. The Liaison Office administers the $7.25 million in state nonpoint source control grants, grants to the states to perform mainstem monitoring of the Bay, grants to universities to perform small research projects, and the grants to CPCB and CRC. It also manages contracts which provide data quality management and computer services as well as other contracts for watershed and hydrodynamic modeling. Staff provide technical support and coordination for all special projects such as the Point Sources Atlas and tributyltin monitoring in 1986 in addition to other activities of the Executive Council's Program Development Plan.[7]

The Executive Council's Program Development Plan begins a new phase in the Chesapeake Bay Program. Following completion of the Study in 1983, the first phase of the implementation of nonpoint source and other control programs began. These efforts were detailed in the 1985 Chesapeake Bay Restoration and Protection Plan. The Plan Supplement expands the initial efforts and represents the initiation of Phase II, the new Program Development Plan. Through the Program Development Plan, the jursidictions will participate in the development of living resources and habitat criteria. These criteria, when combined with monitoring data, mathematical models, and cost information, will help managers set tributary and Bay nutrient loadings. Once target loads have been set, Executive Council members (both jointly as a Council and separately as jurisdictions) and EPA will plan how best to meet these loadings with future pollution control strategies. These strategies will form the basis of step three; the Chesapeake Bay Program for the 1990s.

State Programs

Even before the Chesapeake Bay Agreement was signed, the four Bay jurisdictions had reviewed the Bay Study, begun to develop plans to remedy the problems and were prepared to launch programs early in 1985. They supported these initiatives with more than $250 million over the last several years. EPA's grants to the states supplement the efforts and have enabled the jurisdictions to expand their programs especially in nonpoint control activities. Through their participation in Agreement groups, the jurisdictions shared information on their successes in improving technologies, experiments/demonstrations, implemention of nonpoint best management practices, and innovative citizen involvement and information efforts.

The states' initiatives include work to restore the living resource through oyster seeding, hatcheries for striped bass and other species, the replanting of submerged aquatic vegetation, and research to better understand the habitat requirements of desired species of plants and animals. Other approaches to restore the living resources have been the improvement of septic systems so that there is a decrease in the bacterial contamination of oyster beds and establishment of a vegetative cover to reduce shoreline erosion and sedimentation in aquatic habitat areas. Further, both Maryland and the District of Columbia adopted phosphate detergent bans to minimize nutrient input to sewage treatment plants and waterways. Maryland's ban is contributing to significant cost savings in treatment plant operations in addition to a reduction of nutrient loadings.

The states have been active in developing and expediting adoption of best management practices to reduce both agricultural and urban nonpoint pollution through erosion control. Pennsylvania's Mobile Nutrient Laboratory and Virginia's Rainfall Simulator have effectively demonstrated to the agricultural community the nutrient requirements of their crops and how to reduce losses of these nutrients through specific farming practices. In addition, there have been experimemts with porous pavement, dechlorination, and biological treatment of wastewater. Through advisory committees, publications, television, radio and print media, and large public conferences the states have worked to increase public awareness of the Bay's problems, solutions to those problems, and the role of the individual in Bay improvement.

Maryland's Critical Areas Commission is a state imposed land use measure which restricts land development within a 1,000 foot zone surrounding the Bay, and extending up the tributaries to the head of tide.

Federal Programs

Six federal agencies have signed special agreements with the EPA, pledging to coordinate their activities on the Bay with the Bay Program. The agencies are the Soil Conservation Service (SCS) in the Department of Agriculture, the National Oceanic and Atmospheric Administration (NOAA) in the Department of Commerce, U.S. Fish & Wildlife Service (U.S.F.W.S.) and the U.S. Geological Survey (USGS) in the Department of Interior, the U.S. Army Corps of Engineers (U.S.A.C.E.) and the Department of Defense (DOD). Each agency serves on Agreement committees, contributes its experience and expertise to the Program and the states, and helps to foster public awareness of the Bay efforts. Some have specific programs for Chesapeake Bay; all have focused the regional portion of their national efforts on the Bay.

Early in 1985, the SCS placed a Bay coordinator in the EPA Bay office to ensure that the many field personnel of the Service in the region were closely aligned with the Bay Program. Through them, the SCS has assisted the states to reaching many farmers with information and technical assistance about best management practices. NOAA has been an active Agreement participant, and has contributed further through its Sea Grant Program research on the Bay and Stock Assessment work of key Bay finfish species.

The USGS manages all fall line monitoring on Bay tributaries and works with the states to perform intensive special monitoring of pilot watersheds and plots of land to demonstrate the impact of best management practices. The U.S.F.W.S. has been particularly effective in developing and disseminating information about the Bay's living resources. It has been instrumental in implementing the citizens effort to "groundtruth" the submerged aquatic vegetation beds of the Bay and its tributaries.

The Corps of Engineers, which has been working with the U.S.F.W.S., EPA, Virginia, and Maryland to map the distribution and diversity of SAV, also contributed its modeling expertise to assist in both the development of the steady-state model and the initiation of a time variable model. The Corps performed a major study of the impacts of low flows on living resources, and also contributed to water quality and habitat enhancement with demonstration activities under its shore erosion program.

The Department of Defense was the first federal agency to join with EPA in the Bay Program in September 1984. The DOD pledged to appraise its installations around the Bay and institute land management and point source controls as necessary to remedy pollution. In the process, EPA was to strengthen permits granted to the facilities. DOD completed the study of many of its facilities, reduced pollution and began to implement new procedures to further reduce pollution from point and nonpoint sources. DOD will continue the appraisal of additional operations until all Bay region facilities have been reviewed and EPA has further strengthened DOD permits.

EPA has worked to give a Bay focus to its other mandated programs in Region III. The goals of the Bay Program are pursued through Superfund, RCRA, FIFRA, and the NPDES programs. Wetland protection through Clean Water Act section 404 permits, construction grants to sewage treatment plants, and water quality management grants to the states through sections 106 and 205 provide the base for strong state pollution control programs.

CONCLUSION

The Chesapeake Bay Program builds its strong commitment to coordinated governmental action on an extensive research program and a broad base of public support. We can expect this interest and effort to continue. Programs in the Bay will be aided by provisions of the Clean Water Act of 1987 which, for the first time, explicitly provides for continued Environmental Protection Agency participation in the Bay Program and continues support for the Bay's efforts in nonpoint source control.

REFERENCES

1. U.S. Environmental Protection Agency. 1983. Chesapeake Bay: A Framework for Action. U.S. Environmental Protection Agency. Chesapeake Bay Program. Annapolis, MD.
2. Chesapeake Executive Council. 1985. Chesapeake Bay Restoration and Protection Plan. U.S. Environmental Protection Agency. Chesapeake Bay Program. Annapolis, MD.
3. Scientific and Technical Advisory Council. 1986. Nutrient Controls in the Chesapeake Bay. U.S. Environmental protection Agency. Chesapeake Bay Program. Annapolis, MD.
4. Anonymous. 1980. Federal Insecticide, Fungicide and Rodenticide Act, 7U.S.C.A., Section 136 *et SEQ*, Washington, D.C.
5. U.S. Environmental Protection Agency. 1985. Technical Support Document For Water Quality-Based Toxics Control. Washington, D.C.
6. Citizens Programs for the Chesapeake Bay. 1986. Baybook: A Guide to Reducing Water Pollution at Home. Baltimore, MD.
7. Chesapeake Executive Council. 1987. Program Development Plan—A Description of the Chesapeake Bay Program's Phase II Implementation Program. Annapolis, MD.

Contaminant Problems and Management of Living Chesapeake Bay Resources. Edited by S. K. Majumdar, L. W. Hall, Jr. and H. M. Austin. © 1987, The Pennsylvania Academy of Science.

Chapter Twenty-Six

ACTIONS NEEDED TO REDUCE CONTAMINATION PROBLEMS IMPAIRING CHESAPEAKE BAY FISHERIES

L. EUGENE CRONIN
12 Mayo Avenue, Bay Ridge
Annapolis, Maryland 21403

ABSTRACT

There is strong evidence that there are serious contamination problems deleteriously affecting the fish and fisheries of the Chesapeake Bay. Despite the existence of laboratory data on acute toxicity, there is little information on chronic effects, synergistic actions or reports on communities and ecosystems. Transfer of predictions to field conditions is seriously inadequate, so that we cannot connect contaminants to the kind, degree or importance of injuries to fish or other biota. Achievement of reliable prediction is of paramount importance. The difficulties, which are stated, are complex and daunting, but two approaches are recommended for simultaneous application for Chesapeake Bay and other estuarine systems. The first is to develop and implement a substantial program of *structured research* to produce and apply quantitative information on the effects of the probable pollutants on the significant species, communities and processes and designed to protect the ecosystem. This is considered to be feasible for the Chesapeake, but requires time and substantial funds. The second method suggested is the conservative construction of *new standards* for probable contaminants. These new standards would be based on expert judgment of thresholds of damaging effects, with margins of safety directly proportional to ignorance and combined with rigorous enforcement and continuing research to improve the standards. This could be initiated at once. These two approaches can greatly improve the protection of Chesapeake Bay fish and other biota from contaminants.

INTRODUCTION

It is appropriate to review the evidence and the indicators related to the hypothesis that there are indeed serious contamination problems affecting the fish and fisheries of the Chesapeake Bay. Much of the basis has already been presented in this book.[1, 2, 3]

- Contaminants are numerous and sometimes abundant in the waters, sediments and biota of the bay system.[2]
- Captured fish sometimes bear significant burdens of contaminants.[2] The degree of injury is sometimes, but not always, visible and it is pertinent to note that we usually see only the survivors of any mortality that may have occurred.
- The bay system, especially the fresh and oligohaline regions, has historically had enormous value as a hatchery and nursery area!
- The early life history stages of fish are the most susceptible to contaminant effects.
- Anadromous fish, dependent on the areas which receive most of the contaminants, have generally and severely declined in abundance and availability.[4, 5]
- Net reproductive success is poor for these species and contaminants may play one or more roles in such failures. Obviously, other factors may also be important.[4]

The hypothesis appears to be strongly supported, and corrective actions are merited.

How adequate for rational management is our present knowledge linking the contaminants to wild stocks of fish and other biota? Briefly summarized:

- There is a considerable array of laboratory data on acute toxicity of specific elements and compounds.[3, 6, 7]
- There is very limited data on chronic effects, on the effects of mixed contaminants, on synergistic actions, or on impacts on communities, assemblage and the relevant ecosystems.[8]
- Transfer of predictions and projections to field conditions has been notoriously inadequate.[8]
- Therefore, we cannot now connect contaminants to the kind, degree or importance of injuries to fish which we suspect in Chesapeake Bay and other estuaries.

Several authors have reached the same conclusion, including:

- MacKiernan, writing on nutrients, "There is considerable difficulty in demonstrating unequivocal 'cause and effect' between changes in environmental quality and trends in living resources."[9]
- Helz, "It has not been possible to prove a causative connection between the existence of regionally distributed contaminants and the existence of regional deterioration in fisheries."[2]
  and "Although we know quite a bit about their (the contaminants')

geochemical behavior, we still know rather little about their biological impact. Measurements of sediment concentrations do not provide a direct indication of the availability of toxic materials to organisms. The techniques needed to assess availability are only now being developed."[2]

and "A link between (regional) contaminants and changes in biota has not been established."[2]

- Report by the National Academy of Science and National Academy of Engineering on *Wastes Management Concepts for the Coastal Zone,* "It is now impossible to make quantitative predictions on the effect of waste disposal in the oceans, or even to measure the full effects of such introductions as have occurred."[10]
- Goldberg, Editor, in workshop report on *Scientific Problems Relating to Ocean Pollution,* from the Panel on Biological Effects, "Our information on extent of pollution effects is based on laboratory and field studies, however, it is inadequate to support even the simplest comment about the impact of marine pollution, particularly low-level contamination."[11]

We cannot, except in a small number of cases, measure, predict or evaluate the effects of contaminants on fish or other biota in the Chesapeake Bay or other estuarine systems.

I am convinced that the most urgent and important task facing our efforts to restore the quality and uses of the Chesapeake Bay system is that of linking correctly pollutants and their biological effects. Only then can we determine the present degree of injury, define habitat requirements, establish meaningful goals, monitor the Bay so as to obtain the most useful data, select the most effective corrective measures, quantify improvements and justify the costs. That essential task has hardly begun.

THE DIFFICULTIES

This necessary achievement is dauntingly complex and difficult to complete. It is perhaps the most challenging problem in coastal science, for the following reasons:
- The sources of contaminants are many, with over 5,000 industrial and municipal point sources in the watershed plus all of the non-point run-off. These inputs vary over time, and sampling and measurement are expensive and laborious.
- This is a dynamic estuarine system, with complicated physical circulation, variability on many scales of time and place, complex chemical processes and hundreds of biological and biochemical interactions.
- Many of the species, especially the fish, are highly mobile or migratory.
- Our knowledge base on many of the important linkages and processes is poor,

and many kinds of disciplinary and multi-disciplinary research are required. There are exceptionally serious problems in establishing, funding and sustaining multi-disciplinary cooperative programs.[12]
- It will be necessary to develop adequate knowledge of the sources, routes, rates, sinks, chemical forms, bio-availability and, most important, biological effects on species, populations and the ecosystem. Research on each of these is important, but only the completion of the entire set promises to be sufficient.
- We are now considering spending more than $3 million on the development of hydrologic and transport models. Good models of the significant effects of contaminants may be at least an order of magnitude more difficult to develop. Funding is not now visible.

There is a serious temptation to terminate this paper at this point and expect that the task will never be completed to a useful degree, or not before we have destroyed most of the value of the Chesapeake Bay. I am convinced, however, that a rational and tractable approach *can* and *must* be made if we are to prevent serious loss through ignorance and risky public policies.

A PROPOSAL IN TWO PARTS

There are two courses of action which can result in substantial reduction in the damage that contaminants cause in Chesapeake Bay fish and fisheries. One is a major program of *structured research* to provide the quantitative information necessary for knowledgeable control of such pollution. The second involves the establishment and effective enforcement of *new standards,* based on the acceptance and use of new policies in the control of release of contaminants into the Chesapeake Bay system. The potential for value of these actions is not limited to the Chesapeake, since they are of national and international significance. Neither are these actions limited to fish, which are exceptionally important, since they are based on reduction of serious impacts on the complete ecosystem and its biological health. In my opinion, that holistic approach is critical to the protection of fish, which live, feed, move and reproduce in the ecosystem of the estuary. They cannot be singled out for treatment alone with full success, although some studies on fish are urgently needed and important.

Many agree that the ecosystem approach is required, including Joel Hedgpeth,[13] the Panel on *Estuarine Research Perspectives* set up by the National Research Council,[12] Oviatt, White and Robertson in their summary and synthesis of the volume *Concepts in Marine Pollution Measurements,*[7] the members of the Fisheries Management Workshop for *Choices for the Chesapeake: An Action Agenda,* [14] and Cronin and Roberts, writing on testing of the biological effects of pollutants in *Ten Critical Questions for Chesapeake Bay in Research and Related Matters.*[15]

A RESEARCH PROGRAM

Many research studies have been made on the toxicity of chemicals in the estuarine environment.[7,16] Much of this work has been of high quality. There have also been several pertinent summaries of research needs, and these also have high value.[7,11,12,14,15,17,18,19]

There has not, however, been a detailed design of or commitment to a comprehensive sequential research program which would provide definitive answers to the question of the effects of contaminants on the fish and other biota of the Chesapeake Bay. There is no statement of the probable step-by-step posing of hypotheses, experimental research, replication, verification, interpretation and progress in a program to (1) adequately measure contaminant introductions and reservoirs, (2) trace them in the system, (3) determine the biological effects with prediction value, and (4) assist effectively and efficiently in preventing deleterious effects.

For the Chesapeake Bay, there are no research designs of this level of complexity—or adequacy for knowledgeable management. A detailed program was proposed for determining the full effects of releases from sewage treatment plants, but it was lost in the bureaucratic maze in Washington. Statements of nine hypotheses relating to the decline of striped bass,[20] later expanded to 14,[21] promised to lead to definitive research. Significant attention has been given to research related to some of these hypotheses, so that some have been discarded as major contributors to decline and the complex roles of other factors have been partially illuminated.[22] Much remains to be done.[23] The hypotheses included contaminants, spawning stocks, predation, climatic events, power plants, disease, and other possible factors. It is most regrettable that strong and coherent programs were not completed and interpreted. Perhaps we would now know what we must do to restore the stocks.

The most relevant successful program that I have seen is the *Coordinated Mediterranean Pollution Monitoring and Research Program* (MED POL),[24] which has established a set of 11 steps from baseline surveys through monitoring; research on the effects on species, populations, communities and ecosystems; and to establishment of effective controls (UNEP 1984). This is a central and essential part of the "Mediterranean Action Plan" described in an excellent issue of the journal *Ambio,* from the Royal Swedish Academy of Sciences in 1977.[25]

In such a program, several steps are essential:

- Goals and objectives must be clearly defined but open to modification.
- The extensive relevant existing knowledge must be brought into focus.
- A rational but flexible decision tree for the research must be designed, ranging from pilot studies through necessary projects to decisions.
- Progress must be evaluated in the continuous evolution of the program as

hypotheses are accepted, modified, or rejected.
- Funding must be sufficient, continuing and responsive to new knowledge.
- The products must be effectively communicated to those responsible for management and others who need to know.

Such a structured but flexible program is feasible. The most common analogue is that such a program took us to the moon and beyond, but there are other precedents. The design, oversight and management of such a program could be accomplished by an excellent small interdisciplinary team from academic and management institutions. They must act above parochial interests and stimulate both funding and the participation of scientists of very high quality. The research must be so well done that each step is definitive and provides the basis for next steps. This must combine the freedom such scientists require for best work with timely completion of essential projects, which is not an easy achievement. For the Chesapeake Bay, the Chesapeake Research Consortium can effectively lead this program, in close cooperation with top quality persons from the National Oceanic and Atmospheric Administration, Fish and Wildlife Service, Geological Survey, and Environmental Protection Agency, and with valuable guidance from pertinent state agencies and public interest groups.

The Chesapeake would be a superb site for such a program because of its strong scientific community and the exceptional existing knowledge base. Obviously, successful progress and achievements would have immediate transfer value around all coasts in this and other nations.

I will not attempt to detail this structured research program at this time. The keys for initiation lie in selecting the right persons for program design and in the development of commitment of adequate funding over the decade or more which will be required.

Many tools are available for contamination hazard assessment. There are field observations of concentrations and effects from pollution and from accidents. There is a significant literature from laboratory studies of acute and chronic toxicity on individuals and species—but very little on communities and virtually none on the ecosystem. Some valuable process studies are underway. Recently, the Maryland Department of Health and Mental Hygiene, responsible for environmental regulation, has sponsored an organism-based testing program of the Applied Physics Laboratory of the Johns Hopkins University. Toxicity data (LC50s) for freshwater and estuarine biota will be used in establishing industrial discharge permits. This should produce substantial achievements. There is experience in the use of microcosms and mesocosms, which expand knowledge toward community effects. There are potentials for manipulations in the field to provide the critically important extrapolation of more limited experiments. Analytical chemistry now permits both high accuracy and the processing of a large number of samples. Primitive models of the Chesapeake ecosystem exist[26] and improved ones are under development.[27] A listing of 126 "important" species has been achieved based on commercial or recreational value, known ecological significance or abundance;[28] and the program can be designed around a refinement of that list. Of these, 21 are fish, including 11 commercial species, 4 predators and 6 which are important in the food web. Other relevant lists and models have been developed. A list of 126 "priority industrial pollutants" has been prepared[29] and it can be improved to be more pertinent to present and potential contaminants in the Chesapeake. The broad structure of the Chesapeake Research Consortium exists, with considerable successful experience in multi-institutional programs so that we do not need to create it.

Why has such a program not been done? Why have we never had such a structured program on contaminant problems? How could we have spent tens of millions, perhaps a hundred million dollars, in 15 years of programs related to the Chesapeake Bay and missed this key question of predicting the effects of contaminants? There are at least 5 reasons:

- The task is complex and very difficult.
- "Structured research" conflicts seriously with our habits and prejudices about excellent research.
- The academic reward system has not highly rewarded team research but favored individual research.
- It is extraordinarily difficult to create and sustain any well-coordinated research program which involves state and federal institutions as well as diverse academic institutions because of conflicting policies and practices, budgetary and administrative complexity, and self interests.
- Budgets. Such a program requires flexible use of large sums over an extended period, available for interdisciplinary use. These are formidable obstacles for funding agencies.

Of course it can be done. Dr. Robert Livingston in Florida has achieved a program of comparable complexity, drawing on many individuals and opportunistically from many sources of support. Even designing such a sequential program provides a road map for those who can undertake the hundreds of necessary sub-projects.

Surely, if 17 Mediterranean nations, including Israel, Egypt, Lybia, Syria, France, Turkey, Italy, Spain and Yugoslavia can agree on and implement the *Mediterranean Action Plan* and the *Pollution Monitoring and Research Program,* the Commonwealth of Virginia, State of Maryland and a small number of federal agencies can broaden the present Chesapeake Bay Program to conduct and apply a *Program for a Living Chesapeake.*

NEW STANDARDS FOR CONTAMINANTS

Water criteria and standards exists, and they are under constant review by the states and federal government. They appear to be seriously deficient for

the Chesapeake Bay and other estuaries for the following reasons:[15]
- Toxic levels have frequently been established for freshwater species and transferred to estuaries without recognition that the environment, biota and processes are fundamentally different.
- The toxic effects which have been determined for estuarine species have frequently ignored the substantial effects of salinity on the organism's response, either because salinity affects bio-availability or can stress the test organism.
- EPA has noted that toxicity data on the 126 "priority pollutants" are completely lacking for at least half of the most important fish species of Chesapeake Bay.
- The usual species response tests fail to take into account community interactions such as food supply and predation, sensory disruption, etc., and therefore community and ecosystem effects cannot be predicted in the open bay.
- Of the array of chemicals which have already reached the bay, including several hundred organic compounds and scores of inorganic materials, only a small number, perhaps 10%, have been tested against any bay biota.
- Probably fewer than 2% of the species of the bay have been used in tests of pollutant effects.

We have relied excessively on arbitrary choices of numbers and on available technology rather than the sound scientific basis urged in the previous section of this paper. That basis cannot be achieved quickly, so I propose a complementary approach, designed to produce rather quickly a new set of standards for contaminants. The purpose is to assure that we will not knowingly cause damage to fish and other biota from the contaminants.

The first step is the full recognition that the Chesapeake Bay is a vastly valuable living resource system, providing food, recreation and aesthetic enjoyment. The biological quality which sustains these uses must be preserved above all other characteristics. We must determine that we will not permit biological damage.

Toward that target, standards must be freshly defined. In some cases, old standards will be confirmed and retained. In many cases, it is probable that entirely new standards must be justified and adopted. Even in the absence of the strong scientific evidence we wish, the following steps can protect the bay. The basis can be established within a year.

- Accept the 126 important species and their interactions as targets for protection.
- Develop a listing of the present and probable contaminants of the bay's waters, sediments and biota.
- Establish an expert panel of the aquatic toxicologists, chemists, ecosystem ecologists and other scientists who have the very best knowledge of the effects of these contaminants on the important species, even though it is severely limited. Draw from academic, industrial and governmental personnel, and provide them with sufficient time and support.

- Further develop and standardize accurate analytical techniques for organic and inorganic contaminants.
- For each contaminant, develop the best possible estimate of the level of concentration which will be injurious to the most susceptible stage of each species (and assemblage if feasible). When possible, additional estimates should be added on the effects of seasons, salinity and other factors, always seeking the lower edge of injurious concentrations.
- For each contaminant, have the panel establish a numerical standard lower than the estimated threshold of injury. They should provide a margin of safety as wide as is proportionate to our ignorance about the full effects of the contaminant.
- Vigorously apply and enforce these standards, far above the present pattern of NPDES permits. Dependence on data from the potential polluters, thin enforcement efforts and slow procedures for assuring compliance and punishing violators must be minimized.

To the degree possible and feasible, we must also clean up old messes like Baltimore Harbor, Elizabeth River and other sites, since the principal contaminant load is the accumulated one, not present input. This will require significant funding and careful approach, as was demonstrated in an assessment of the alternatives for corrective action in Baltimore Harbor, where the best present action for protecting the environment was found to be leaving the pollutants in place in the sediments without disturbance.[30]

We must also pursue an expanded research program to improve the standards. New learning on biotoxicity involving communities, species and the ecosystem will permit the standards to be increasingly accurate. For instance, we need *in situ* field studies of the effects of contaminants over the wide range of environmental conditions experienced by anadromous fish and resident species. This will require simultaneous toxicity observations at multiple spawning sties probably by use of mobile laboratories. We also lack sufficient knowledge of the life stages and their water quality requirements in our nursery areas. We must learn much more about the bio-availability of various forms of the chemical contaminants. There are many other important and researchable questions.

CONCLUSION

It is imperative that we act effectively to reduce damages done by contaminants to Chesapeake Bay fish and other biota. We cannot now predict with useful accuracy the effects of contaminants and therefore cannot manage either pollution or the fisheries rationally. Two complementary approaches are recommended for simultaneous use in the Chesapeake Bay as a model for estuarine protection in the nation.

- Structured research on a scale large enough to be definitive, with statement

of goals, development of new knowledge in a rational sequence, and full adequacy to permit prediction of potential biological effects and therefore effective action.
- Construction of new standards for all present and probable contaminants, based on expert judgement of their thresholds of effects and providing conservative margins for error proportional to our ignorance, and a new level of enforcement. Excellent research on relevant topics can continuously improve these standards.

Together, these approaches can contribute substantially to achievement of a characterization which I have modified from Joel Hedgpeth[13]—*The Chesapeake Bay is a place for life, not for death from contamination from human activities.*

REFERENCES

1. Setzler-Hamilton, E. 1987. Utilization of Chesapeake Bay by early life stages of fish. pp. 63-93. *In:* S.K. Majumdar; L.W. Hall, Jr.; and H.M. Austin (Eds.). *Contaminant Problems and Management of Living Chesapeake Bay Resources.* The Pennsylvania Academy of Science, Publication, Easton, PA. pp. 573.
2. Helz, G.R. and R.J. Huggett. 1987. Contaminants in Chesapeake Bay: The Regional Perspective. pp. 270-297. *In:* S.K. Majumdar; L.W. Hall, Jr.; and H.M. Austin (Eds.). *Contaminant Problems and Management of Living Chesapeake Bay Resources.* The Pennsylvania Academy of Science, Publication, Easton, PA. pp. 573.
3. Klauda, R. and M.E. Bender. 1987. Contaminant effects on Chesapeake Bay finfishes. pp. 321-372. *In:* S.K. Majumdar; L.W. Hall, Jr.; and H.M. Austin (Eds.). *Contaminant Problems and Management of Living Chesapeake Bay Resources.* The Pennsylvania Academy of Science, Publication, Easton, PA. pp. 573.
4. Austin, H.M. 1987. Chesapeake Bay fisheries—An Overview. pp. 33-53. *In:* S.K. Majumdar; L.W. Hall, Jr.; and H.M. Austin (Eds.). *Contaminant Problems and Management of Living Chesapeake Bay Resources.* The Pennsylvania Academy of Science, Publication, Easton, PA. pp. 573.
5. Rothschild, B.J., P.W. Jones and J.S. Wilson. 1981. Trends in Chesapeake Bay fisheries. *Trans. 46th N.A. Wildlife and Nat. Res. Conf.* pp. 284-298.
6. Costlow, J.D. 1982. Impact of toxic organics on the coastal environment. pp. 86-95. *In:* T.W. Duke (Ed.) *Impact of Man on the Coastal Environment.* U.S. Environmental Protection Agency, Washington, D.C. EPA-600/8-22-021, 114 pp.
7. White, H.H. (Ed.). 1984. *Concepts in Marine Pollution Measurements.* Maryland Sea Grant College, College Park, MD. 743 pp.

8. National Research Council, 1981. *Testing for Effects of Chemicals on Ecosystems*. Report by the Committee to Review Methods for Ecotoxicology, Comm. on Nat. Res. National Academy Press, Washington, D.C. 103 pp.
9. MacKiernan, G.B. 1985. Sources and impacts of nutrients in the Chesapeake Bay, pp. 1-20. *In:* C.W. Coale, J.P. Marshall and W.R. Kerns (Eds.). *Land Use and the Chesapeake Bay*. Proceedings of a conference. Va. Coop. Ext. Service, Blacksburg, VA. 139 pp.
10. National Academy of Sciences and National Academy of Engineering. 1976. *Wastes Management Concepts for the Coastal Zone — Requirements for Research and Investigation*. Chapter 5, Biological Effects, pp. 61-83. NAS/NAE, Washington, D.C. 126 pp.
11. Goldberg, E.D. (Ed.) 1979. Proceedings of a Workshop on Scientific Problems Relating to Ocean Pollution, Estes Park, Colo: Environmental Research Laboratories, National Oceanic and Atmospheric Administration, Boulder, Colo. 225 pp.
12. National Research Council Panel on Estuaries Research Perspectives, 1983. *Fundamental Research on Estuaries: The Importance of an Interdisciplinary Approach*. Chapter 2, Environmental Effects, pp. 17-38. National Academy Press, Washington, D.C. 79 pp.
13. Hedgpeth, J.W. 1977. Seven ways to obliteration: Factors in estuarine degradation, pp. 723-737. *In: Estuarine Pollution Control and Assessment: Proceedings of a Conference*. U.S. Environmental Protection Agency, Washington, D.C. 756 pp.
14. Citizens Program for the Chesapeake Bay. 1983. *Choices for the Chesapeake: An Action Agenda*. Workshop Recommendations on Fisheries Management. Citizens Program for Chesapeake Bay. 85 pp.
15. Cronin, L.E. and M.H. Roberts, Jr. 1983. How can we best test the effects of pollutants and changes on the animals and plants of Chesapeake Bay?, pp. 31-49. *In:* L.E. Cronin (Ed.) *Ten Critical Questions for Chesapeake Bay in Research and Related Matters*. Contr. 113, Chesapeake Research Consortium, Gloucester Pt., VA. 156 pp.
16. Perkins, E.J. 1974. *The Biology of Estuaries and Coastal Waters*. Chapter 16, The biological effects of waste disposal, pp. 513-602 (with 402 references). Academic Press, London and New York. 678 pp.
17. Kohlenstein, L.C. 1983. What are the causes of serious declines in striped bass, shad, white perch, herrings—which spawn near the heads of the Bay and tributaries?, pp. 8-30. *In:* L.E. Cronin (Ed.) *Ten Critical Questions for Chesapeake Bay in Research and Related Matters*. Contr. 113, Chesapeake Research Consortium, Gloucester Pt., VA. 156 pp.
18. Oviatt, C.A., H.H. White and A. Robertson. 1984. Summary and synthesis, pp. 725-735. *In:* H.H. White (Ed.)*Concepts of Marine Pollution Abatement Measurement*. MD Sea Grant College, College Park, MD. 743 pp.

19. McClellan, W.D. and M.J. Jaber. 1985. Assessing the potential hazard of new chemicals in coastal ecosystems, pp. 183-196. *In:* N.L. Chao and W. Kirby-Smith (Eds.). *Proceedings of the International Symposium on Utilization of Coastal Ecosystems: Planning, Pollution and Productivity,* Vol. 1. Fundação Universidade de Rio Grande, Rio Grande, RS, Brazil and Duke University Marine Laboratory, Beaufort, NC. 496 pp.
20. USDOI and USDOC (U.S. Department of the Interior and U.S. Department of Commerce). 1982. Emergency striped bass research study, 1981 Annual Report. Washington, D.C.
21. Cronin, L.E. 1981. Preliminary proposal for a project under P.L. 96-118 amendment for striped bass research: Proposal to evaluate a series of hypotheses related to the decline of striped bass along the Atlantic Coast. Letter and attachments to Norris B. Jeffrey. Chesapeake Research Consortium, Gloucester Pt., VA. 6 pp.
22. USDOI and USDOC (U.S. Department of the Interior and U.S. Department of Commerce). 1986. Emergency striped bass research study, Report for 1985. Washington, D.C. 67 pp.
23. USDOI and USDOC (U.S. Department of the Interior and U.S. Department of Commerce). 1985. Emergency striped bass research study, 1984 Annual Report. Washington, D.C. 58 pp.
24. United Nations Environmental Programme. 1984. Coordinated Mediterranean Pollution Monitoring and Research Program (MED POL)—Phase I: Programme Description. UNEP Regional Seas Reports and Studies No. 23. Geneva. 223 pp.
25. Royal Swedish Academy of Sciences/Universitetsforlanget. 1977. The Mediterranean—A special issue. *Ambio IV* (6): 299-378.
26. Green, K.A. 1978. A conceptual ecological model for Chesapeake Bay. Coastal Ecological Project, Office of Biological Services, U.S. Fish and Wildlife Service. FWS/OBS 78/69. 22 pp.
27. Ulanowicz, R.E. and D. Baird. 1986. A network analysis of the Chesapeake Bay ecosystem. Interim Report to the Tidewater Administration, Department of Natural Resources, State of Maryland. Univ. Md. Ches. Biol. Lab. Ref. No. (UMCEES) CBL 86-79. 14 pp. plus appendices.
28. Pfitzenmeyer, H.T. 1977. Important Chesapeake Bay species identification. *Chesapeake Bay Future Conditions Report,* Baltimore District. Chapter IV in Volume II, *Biota.* U.S. Army Corps of Engineers. pp. 61-69.
29. Keith, L.H. and W.A. Telliard. 1979. Priority pollutants: I—A perspective view. *Environmental Science and Technology* 13 (4): 416-423.
30. Trident Engineering Associates, Inc. 1977. Evaluation of the Problem Posed by In-Place Pollutants in Baltimore Harbor and Recommendation of Corrective Action. U.S. Environmental Protection Agency Office of Water Planning and Standards. EPA—440/5-77-015B, Principal Report, 78 pp. and EPA—440/5-77-015A, Appendices.

Subject Index

Acidic Deposition, 346
Acute Toxicity of Kepone, 380
Agriculture, Runoff, 506
Algae Bloom, 507
Algae Respond to Nutrients, 300
Ambio, 559
American Black Duck, 101
American Oyster, 165
American Wigeon, 105
Anadromous Finfish Species in Chesapeake Bay, 321
Anadromous Fish, 556
Anatomy of the American Oyster, 167
Apalachiola Bay, 193
Applied Physics Laboratory of Johns Hopkins, 560
Aquaculture, 211
Aquaculture of Clams, 214
Aquatic Toxicologists, 562
Aquatic Vegetation, 117
Aquatic Vegetation in the Chesapeake Bay, 116
Arsenic, 341
Artificial Reefs, 495
Atlantic Menhaden, 48
Atmospheric Fallout, 194
Atrazine, 196
Autumn Season, 13

Bacterial Abundance and Oxygen Consumption, 463
Bacterial Production, 457
Bacterioplankton Densities, 134
Baltimore Harbor, 24, 292, 504
Baltimore River, 563
Bay Floor, 30
Benthic, 158
Benthic Communities, 159
Benthic Environment, 158
Benthic Environments of the Chesapeake Bay, 160
Benthic Herbivory, 150
Benthic Invertebrates, 160
Benthic Resources of the Chesapeake Bay, 158
Benthic Suspension Feeders, 150
Benthos, 145
Biocides, 325
Bioenergetics Models, 354
Biological Water Quality Test, 429
Biologically-Based Whole Effluent Toxicity Limits, 523
Biology, Life History and Distribution of the Blue Crab, 180

Biomass of Microzooplankton, 254
Biota, 556
Blue Crab Fisheries, 179
Blue Crab in Chesapeake Bay, 177
Blue Crab Megalopa, 182
Blue Crab Zoeae, 181
Bottom Fishing, 39
Bufflehead, 108

Cadmium, 341
Calvert Cliffs Nuclear Power Plant, 349
Canada Goose, 99
Canvasback, 105
Carbon Dynamics and Flux, 134
Carbon Flow, 134
Carbon, Oxygen and Nutrient Dynamics of Chesapeake Bay, 135, 149
Categories of Primary Producers, 395
Characteristics of the Flow of the Susquehanna River, 5
Chemical and Physical Pollutants in the Chesapeake Bay, 192
Chesapeake Bay, 1, 4, 94, 165, 218, 555, 562
Chesapeake Bay Advisory Committee, 483
Chesapeake Bay Anoxia, 442
Chesapeake Bay Blue Crab, 180
Chesapeake Bay Critical Area Commission, 498
Chesapeake Bay Deadrise, 46
Chesapeake Bay Ecosystem, 502
Chesapeake Bay Finfishes, 321
Chesapeake Bay Finfish Kills, 315
Chesapeake Bay Fisheries, 57
Chesapeake Bay Foundation, 508
Chesapeake Bay Management of the Oyster Fishery, 171
Chesapeake Bay as Nursery Grounds, 63
Chesapeake Bay Nutrients, 306
Chesapeake Bay Oyster Bars, 376
Chesapeake Bay Plankton, 217
Chesapeake Bay Program, the, 550
Chesapeake Bay Restoration and Protection Plan Goals, 543
Chesapeake Bay Tributaries, 503
Chesapeake Bay Trust, 498, 504
Chesapeake Bay Waterfowl, 94
Chesapeake Bay Water Quality Monitoring Program, 248
Chesapeake Bay Research Consortium, 560
Chester River, 499
Chlorinated Hydrocarbons, 277
Chlorinated Phenols, 196
Chlorination Effects on Bivalve Larval Settlement, 428

Chlorination of Wastewater, 195
Chlorine, 330, 505
Chlorine Discharge Sources Around Chesapeake Bay, 283
Choptank River, 3, 248, 501
Chronic Effects of Kepone in Some Marine Animals, 381
Circulation Patterns According to Direction of Flow, 26
Commercial and Recreational Fisheries, 63
Commercial Chesapeake Bay Fisheries, 54
Commercial Fin Fishery, 34
Commercial Fisheries, 39
Commercial Species of Chesapeake Bay, 57
Common Benthic Invertebrates from the Chesapeake Bay, 162
Common Goldeneye, 108
Concentration of Kepone, 196
Condemned Shellfish Areas in MD and VA, 377
Consumers of Blue Crabs, 189
Contaminant Effects on Chesapeake Bay Shellfish, 373
Contaminant Effects on Primary Producers in Chesapeake Bay, 394
Contaminants, 29, 556, 559
Contaminants Dispersal and Accumulation, 29
Contribution of the Bay Fisheries, 56
Copper, 341
Copper Residues in Oysters, 389
Crab Trot line, 45
Critical pH Values, 349
Current Status of Submerged Aquatic Vegetation in Chesapeake Bay, 123
Cycling Nutrients, 161

2, 4-D, 196
DDE, 111
DDE Residues, 112
DDT, 196
DDT and Its Derivatives, 196
Dechlorination Facilities, 505
Decline of Submerged Aquatic Vegetation, 487
Declining Resources and Environmental Degradation, 59
Deficiency of Dissolved Oxygen, 194
Delaware Bay, 170
Delaware Canal, 501
Dendritic River Valley System, 3
Dibenzofurans, 341
Dicamba, 196
Dieldrin, 196
Diffuse Sources of Nutrients, 316
Dioxins, 341

Distribution of Microzooplankton in the Chesapeake Bay, 251
Distribution of Polychlorinated Biphenyls, 290
Distribution of Temperature of the Chesapeake Bay, 22
Distributions of Phytoplankton, 135
Dominant Zooplankton in the Virginia Portion of the Chesapeake Bay, 235
Drainage Basin of the James River, 7
Drainage Basin of the Potomac River, 7
Drainage Basin of the Susquehanna River, 7

Economic Importance and Characteristics of the Chesapeake Bay Marine Resources, 54
Ecosystems, 555
Ecotoxicological Response, 418
Effects of Contaminants on Estuarine Zooplankton, 417
Effects of Heavy Metals, 419
Effects of Pollutants on Chesapeake Bay Marshes, 396
Elemental Composition of Chesapeake Bay Sediments, 273
Elizabeth River, 275, 292, 563
Elk River, 499
Emergent Aquatic Vegetation, 124
Emergent Wetlands, 117
Enrichment of Zinc in Sediments of Baltimore Harbor, 274
Evanescent Contaminants in the Chesapeake Bay, 285
Evanescent Contamination, 282
Environmental Degradation, 61
Environmental Pollution, 61
EPA Chesapeake Bay Study Report, 34
Erosion Buffer, 130
Estuarine Spawners, 74
Excess Nutrients, 487
Excess Toxics, 488
Exposure to Kepone, 329
Eyed Larvae, 213

Federal and State Effluent Standards, 522
Finfish, 43, 494
Finfish Mortality, 315
Finfish Spawning, 322
Flood Buffer, 130
Freshwater Inflow, 3
Freshwater/Oligohaline Spawners, 85
Freshwater Species, 87
Functions of Emergent Wetlands, 127

Gadwall, 104
Geographic Initiation of Anoxia, 457
Geological Survey, 560

Gill Net, 52
Grass Bed Spawners, 85
Gunpowder River Estuary, 147

Habitat Improvement, 493
Halogenated Solvents, 285
Halogenated Substances, 195
Headboat Fishery, 34
Heavy Metal, 508
Heptachlor, 196
Herbicides, 292, 508
Hickory Shad, 501
Highly Zinc-Enriched Sediments, 277
Holding Pen for Peeler Crabs, 47
Holistic Approach, 558
Hydraulic Escalator Dredge, 45
Hydrologic and Transport Models, 558

Impact of Contaminants on Zooplankton, 418
Inorganic and Organic Contaminant Mixtures, 339
Intertidal Marshes, 396

James River, 3, 280, 373

Kepone, 195, 270, 325, 379
Kepone Concentrations in James River Sediments, 327
Kepone Concentrations in Sediments, 326
Kepone Concentrations in Sediment Cores, 287
Kepone Contamination of the James River, 280, 373
Kepone Residues in Atlantic Croaker and Spot, 328
Kepone Residues (Wet Wt) in Blue Crabs, vs. Effects Levels, 374
Kepone Residues (Wet Wt) in Male and Female Blue Crabs, vs. Time, 378

Lake Pontchartrain, 190
LC50 (Toxicity Data), 560
Lead, 341
Lesson Learned from the Chesapeake Bay Program, 516
Levels of Chlorine Produced Oxidants, 332
Life Cycle of the American Oyster, 169
Living Resources, 493
Lower Chesapeake Bay, 223
Lower York River, 122

Macrozooplankton, 150
Macrozooplankton Hervibory, 147
Main Stem of the Bay, 18
Major Commercial Species of Chesapeake Bay, 57

Major Taxonomic Groups of the Benthic Invertebrates, 160
Mallard, 102
Management of Chesapeake Bay, 300
Map of the Chesapeake Bay, 2, 272
Marine Resources of the Chesapeake Bay, 33
Marine Spawners, 67
Marshes, 126
Maryland, 498
Maryland Bay Restoration Program, 498
Maryland Department of Agriculture, 507
Maryland Department of Health, 290
Maryland Environmental Trust, 503
Mean Acute Toxicity of Chlorine, 333
Mean Density of Major Zooplankton Taxa Averaged Over the Virginia Study Area, 241
Mean Monthly Salinity in the Virginia Chesapeake Bay, 233
Mean Salinity in the Virginia Chesapeake Bay, 232
Mediterranean Action Plan, 559
Menhaden, 44, 57
Menhaden Fishery, 57
Menhaden Vessels, 40
Mesocosms, 560
Mesohaline Zone, 258
Mesozooplankton, 219, 263
Mesozooplankton in the Chesapeake Bay, 217
Mesozooplankton of the Chesapeake Bay, 219
Methoxychlor, 196
Microbial Metabolism, 314
Microcosms, 560
Microheterotrophic Assemblages, 134
Microzooplankton Distributions, 258
Microzooplankton in the Chesapeake Bay, 245, 246
Microzooplankton Taxa, 260
Migratory Waterfowl, 94
Mirex, 196
Mobjack Bay Area, 122
Model Prediction of the Separate and Combined Effects of Stresses, 398
Mollusk Culture for the Chesapeake Bay, 210
Mollusk Fisheries of the Bay, 211
Mollusk Harvest, 211
Mortality from Pollution, 61
Mortality of the Blue Crab, 183
Municipal Wastewater Discharge, 194
Mute Swan, 98

Nanticoke River, 3
National Pollution Discharge Elimination System (NPDES) Permit, 518, 563

1979-1983 EPA Bay Program, 33
Nitrate Concentrations, 310
Nitrogen and Phosphorus Regeneration in Bay Sediments, 149
Nitrogen Recycling Pathways, 314
NOAA's Program, 356
Non-Nitrate Nitrogen, 316
Nonpoint Source Initiative, 474
Non-Point Source Pollution Control, 500, 506
Northern Pintail, 105
Nuclear Regulatory Commission (NRC), 349
Nutrient Discharge Reductions, 544
Nutrient Enrichment of the Bay System, 398
Nutrient-Limitation, 142
Nutrient-Limited Phytoplankton Production, 142
Nutrient Recycling in Chesapeake Bay, 311, 313
Nutrients, 507
Nutrients from Point Sources, 310
Nutrients in Chesapeake Bay, 298
Nutrient Sources, Reservoirs, and Sinks for the Chesapeake Bay, 312

Occurrence of Cataracts in Atlantic Croaker and Weakfish, 338
Oligohaline Zone, 258
Organic and Industrial Effluents, 404
Organism based testing program, 560
Organochlorine Pesticides, 341
Organotins, 334, 383
Overfishing, 60
Oxidants, 423
Oxygen Demand in the Benthos, 146
Oxygen Depletion in the Bay, 445
Oyster Growing Areas in Chesapeake Bay, 166
Oyster Hatchery, 496
Oyster Landings in Maryland from Public Beds, 172
Oyster Larvae, 212
Oyster Production in the Chesapeake Bay, 175
Oyster Reef Spawners, 80
Oyster Reproduction, 500
Oyster Rock Repletion, 496

PAH Concentrations in Baltimore Harbor, 279
Patent Tongs, 43
Patterns of Nutrient Concentrations in Chesapeake Bay, 303
Patuxent River, 3, 271, 258
Patuxent River Estuarine Model, 499
Patuxent River Policy Plan, 499
Patuxent Type Tributary Estuary, 24
Patuxent Watershed Act, 499
Paraquat, 196
PCB Concentrations in Oysters from the Elizabeth River, 277
PCB Congeners, 278
PCBs, 341
PCB's in Oysters, 278
PCB's in Surface Sediments of the Upper Chesapeake Bay, 289
Peach Bottom Atomic Power Station, 352
Peeler Crab Pound Net, 47
Pennsylvania Association of Conservation District Directors, Inc, 477
Pennsylvania Department of Environmental Resources, 472
Pennsylvania Farmers Association, 477
Pennsylvania Farmers Union, 477
Pennsylvania State Conservation Commission, the, 483
Pesticide, 508
Pesticides, 421
Petroleum Hydrocarbons, 341
Phosphorus, 504
Phosphorus Discharge, 505
Phosphorus Recycling Pathways, 313
Phosphorus Retention, 316
Phthalate Esters, 291
Physical Description of the Chesapeake Bay, 1
Phytoplankton, 395, 399
Phytoplankton and Bacteria, 145
Phytoplankton Bloom, 263
Phytoplankton and the Macrobenthic Community, 147
Phytoplankton and Zooplankton, 146
Phytoplankton Biomass, 136, 455
Phytoplankton in Chesapeake Bay, 134, 135
Phytoplankton Responses to Bay Circulation, 142
Phytoplankton Response to Contaminants, 399
Phytoplankton Sensitivity to Stress, 406
Point Source and Other Initiatives, 479
Point Sources of Nutrients, 316
Point Source Pollution Control, 503
Pollutant Effects, 425
Pollutants of Concern to the Bay from Pennsylvania, 473
Pollutants-Non-Point Source Run Off, 33
Pollutants-Point Source Discharges, 33
Pollutants and Waterfowl, 111
Pollutants in Chesapeake Bay, 112
Pollution Abatement, 490
Polychlorinated Biphenyls (PCB's), 111, 196, 270
Polynuclear Aromatic Hydrocarbons, 195, 337

Population Dynamics of Finfishes, 348
Population Trends (1948-86) of Black Ducks and Mallards in Chesapeake Bay, 103
Population Trends (1948-86) of Canada Geese in Chesapeake Bay, 99
Population Trends (1948-86) of Canvasback and Redhead in Chesapeake Bay, 107
Population Trends (1948-86) of Gadwall in Chesapeake Bay, 103
Population Trends (1948-86) of Goldeneye and Bufflehead in Chesapeake Bay, 109
Population Trends (1948-86) of Ruddy Duck and Scoter in Chesapeake Bay, 110
Population Trends (1948-86) of Scaup in Chesapeake Bay, 108
Population Trends (1948-86) of Tundra Swans in Chesapeake Bay, 97
Population Trends (1948-86) of Wigeon and Pintail in Chesapeake Bay, 104
Possible Interactions of Dissolved Metal Ions with Phytoplankton, 405
Potomac River, 3, 271, 258, 499, 501
Predation Intensity on the Blue Crab, 187
Predators and Diseases of the American Oyster, 170
Primary Producers in Chesapeake Bay, 395
Priority Industrial Pollutants, 561
Priority Pollutants, 562
Propanil, 196
Public and Leased Oyster Grounds, 375
Public Involvement in the Chesapeake Bay Program, 512
Public Involvement on the Chesapeake, 514
Pycnocline and Sub-Pycnocline Waters, 144

Radionuclides, 349
Rappahannock River, 3, 158, 271
Recreational Fisheries, 36, 59
Recreational Shellfishery, 34
Redfield Ratio, 302
Redhead, 107
Redox Reactions, 292
Relationships Between Phytoplankton and Nutrient Cycling, 148
Research on Emergent Wetlands, 126
Resin Acids (Dry Wt) in Oysters and Rangia, 387
Resources, 34
Restoration of Living Chesapeake Bay Resources, 541
Riparian Forest Buffer Strips, 316
Ruddy Duck, 109

Salinity, 163
Salinity Ranges of Abundant Killfishes in Chesapeake Bay, 80
Scaup, 107

Scientific Limits, 525
Scoter, 110
Seasonal and Habitat Overview, 65
Seasonal Distribution of Microzooplankton, 253
Seasonal Distributions of Phytoplankton in Chesapeake Bay, 137
Seasonal Oxygen Dynamics, 446
Seasonal Patterns in Diversity of Fish Eggs/Larvae by Habitat in Chesapeake Bay, 66
Seasonal Patterns in Diversity of Fish Eggs/Larvae in Chesapeake Bay, 64
Sediment Oxygen Demand, 146
Sediment Type, 163
Selected Characteristics of the Chesapeake Bay Commercial Fishing Fleet, 55
Selection of Appropriate Test Species, the, 533
Selenium, 341
Sensitivity of Phytoplankton to Arsenic or Cadmium, 400
Shellfish, 41, 57
Shellfish as Indicators of Pollution, 384
Shellfish Grounds, 495
Short-term Oxygen Dynamics, 449
Significance of Benthic Invertebrates, 161
Sink Gill Net, 51
Snow Goose, 100
Socio-Economic Overview of the Chesapeake Bay Fisheries, 54
Soil Conservation Service, 477
Spawning Habitat, 63
Special Problems with Estuaries, 513
Staked Gill Net, 51
Statutory Limitations, 524
Striped Bass, 501
Submerged Aquatic Species in Chesapeake Bay, 118
Submerged Aquatic Vegetation, 95, 117, 394, 397, 494
Sulfur Cycling, 465
Surface Runoff, 194
Surface Salinity Distribution in the Chesapeake Bay, 11, 13
Surface Sediment Concentrations of Benzo(a)pyrene, 336
Susquehanna Flats, 271
Susquehanna River, 1, 4, 499
Susquehanna River Flow, 142
Susquehanna River in Pennsylvania, 352
Symbionts of the Blue Crab, 184, 185

Terminology of Graded Oysters, 42
Tides and Tidal Currents, 9
Tides in the Chesapeake Bay, 9

Total Pyrogenic PAH's in Surface Sediments from Chesapeake Bay, 288
Toxicity Data With Respect to Heavy Metals, 420
Toxicity Reduction Efforts, 546
Trace Metals, 390
Trace Metals and Other Inorganics, 402
Trends in Emergent Wetland Distribution, 131
Tributary Estuaries to the Chesapeake Bay, 22
Tributyltin (TBT) on Chesapeake Bay Biota, 334
Tri-States' Governors' Conferences, 34
Tropical Storm Agnes, 30
Tundra Swans, 98
Types and Extent of Emergency Wetlands in Chesapeake Bay, 124
Types of Emergent Tidal Wetlands, 125

U.S. Environmental Protection Agency, 119, 498
Uses and Purposes of Aquatic Bioassays, 528

Vertical Distributions of Chlorophyll, 139
Vertical Profiles of Cu, Zn, and Pb, 286
Virginia and Maryland Hard Crab Landings and Value, 178
Virginia Fisheries Management Plan, 175
Virginia Resources Commission, 175
Virginia's Chesapeake Bay Initiatives, 490

Waterfowl of Chesapeake Bay, 94
Waterfowl Species in Chesapeake Bay, 96
Water Quality Loan Fund, 504
Water Quality of the Chesapeake Bay, 210
Water Quality Standards, 521
Watershed, 307
Watershed Nutrient Discharge, 308
Wetlands, 129
Wetlands in the Chesapeake Bay System, 128
Wetlands, Nontidal, 503
Wetlands, Tidal, 503
Wildlife, Corridors, 507

Yearly Mean Kepone Residues (Wet Wt) in Oysters, 379
Yellow Perch, 501
York River, 73, 271
Youghiogheny River, 498

Zooplankton Species Composition for Maryland Mesohaline Regions, 230
Zooplankton Species Composition for Maryland Oligohaline Regions, 229
Zooplankton Species Composition for Maryland Polyhaline Regions, 231
Zooplankton Species Composition in Maryland, 226, 227

OFFICERS OF THE PENNSYLVANIA ACADEMY OF SCIENCE

SHYAMAL K. MAJUMDAR, President, Professor of Biology, Lafayette College, Easton, Pennsylvania 18042

KURT C. SCHREIBER, President Elect & Director, Science Talent Search, 1812 Wightman Street, Pittsburgh, Pennsylvania 15217

GEORGE C. SHOFFSTALL, Immediate Past-President/Executive Secretary, 502 Misty Drive, Suite 1, Lancaster, Pennsylvania 17603

SHERMAN S. HENDRIX, Treasurer, Department of Biology, Gettysburg College, Gettysburg, Pennsylvania 17325

RALPH A. CAVALIERE, Assistant Treasurer, Department of Biology, Gettysburg College, Gettysburg, Pennsylvania 17325

LEONARD M. ROSENFELD, Recording Secretary, Department of Physiology, Jefferson Medical College of Thomas Jefferson University, Philadelphia, PA 19127

HOWARD S. PITKOW, Corresponding Secretary, Professor of Physiology, Pennsylvania College of Podiatric Medicine, Eighth at Race Street, Philadelphia, Pennsylvania 19107

DANIEL KLEM, JR., Editor of the Proceedings, Department of Biology, Muhlenberg College, Allentown, Pennsylvania 18104

FRED J. BRENNER, Newsletter Editor, Biology Department, Grove City College, Grove City, Pennsylvania 16127

J. ROBERT HALMA, Historian, Department of Biology, Cedar Crest College, Allentown, Pennsylvania 18104

EDWARD TESTA, Director of Junior Academy, Valley View High School, Archbald, Pennsylvania 18403

JUSTICE JOHN P. FLAHERTY, Advisory Council Chairman

SISTER M. GABRIELLE MAZE, Fund Raising, Past-President, Grove and McRobert Road, Pittsburgh, Pennsylvania 15234